Studies in Systems, Decision and Control

Volume 27

Series editor

Janusz Kacprzyk, Polish Academy of Sciences, Warsaw, Poland
e-mail: kacprzyk@ibspan.waw.pl

About this Series

The series "Studies in Systems, Decision and Control" (SSDC) covers both new developments and advances, as well as the state of the art, in the various areas of broadly perceived systems, decision making and control- quickly, up to date and with a high quality. The intent is to cover the theory, applications, and perspectives on the state of the art and future developments relevant to systems, decision making, control, complex processes and related areas, as embedded in the fields of engineering, computer science, physics, economics, social and life sciences, as well as the paradigms and methodologies behind them. The series contains monographs, textbooks, lecture notes and edited volumes in systems, decision making and control spanning the areas of Cyber-Physical Systems, Autonomous Systems, Sensor Networks, Control Systems, Energy Systems, Automotive Systems, Biological Systems, Vehicular Networking and Connected Vehicles, Aerospace Systems, Automation, Manufacturing, Smart Grids, Nonlinear Systems, Power Systems, Robotics, Social Systems, Economic Systems and other. Of particular value to both the contributors and the readership are the short publication timeframe and the world-wide distribution and exposure which enable both a wide and rapid dissemination of research output.

More information about this series at http://www.springer.com/series/13304

Aly El-Osery · Jeff Prevost
Editors

Control and Systems Engineering

A Report on Four Decades of Contributions

Springer

Editors
Aly El-Osery
Electrical Engineering Department
New Mexico Institute of Mining and Technology
Socorro New Mexico
USA

Jeff Prevost
Electrical and Computer Engineering Department
The University of Texas at San Antonio
San Antonio Texas
USA

ISSN 2198-4182 ISSN 2198-4190 (electronic)
Studies in Systems, Decision and Control
ISBN 978-3-319-14635-5 ISBN 978-3-319-14636-2 (eBook)
DOI 10.1007/978-3-319-14636-2

Library of Congress Control Number: 2014958981

Springer Cham Heidelberg New York Dordrecht London

Printed on acid-free paper

Springer International Publishing AG Switzerland is part of Springer Science+Business Media (www.springer.com)

Forewords

Foreword by Janusz Kacprzyk

The very purpose of this short note is to say some words about Professor Mo Jamshidi to whom this volume is dedicated. While doing so, I will mention both his great research and scholarly achievements, his service to the scientific and professional communities, his deep understanding and insight into processes that have shaped the landscapes of not only science and technology but various social, environmental, etc. processes, and many other aspects of his long time activities and career we all should be grateful for.

To start, it may be natural to ask a simple question on how long I know Mo Jamshidi. The answer is not so simple. First, many years ago, as a student in automatic control and large scale systems, that were one of hot topics at that time, I had a chance to use his seminal works, notably in my M.Sc. dissertation. Of course, at that time we were not involved in any kind of fuzzy control and related topics.

I could not even imagine at that time that I would some time meet him personally. That time came probably in the 1980s when, while visiting my mentor, Professor Lotfi A. Zadeh at the University of California, Berkeley, I was told about a great admiration he had for the works of Mo Jamshidi. Thereafter, I had had a chance to regularly meet Mo and follow his great works, both related to the foundations and applications of fuzzy logic and more general.

In a short note like this it is not possible to list all those contributions of Mo Jamshidi that have made him one of the most respected and influential researcher and scholar not only in the USA but worldwide. Therefore, I will try to just briefly summarize his achievements that, in my honest opinion and following my own interests, are the most relevant.

As I have already mentioned, Mo has been since the beginning of his illustrious career a "systems man". In particular, he has since the very beginning recognized that the world is a very complicated system and its analysis should proceed using proper tools and techniques. In the very beginning, these were perspectives, and tools and techniques of large scale systems. They were quite successful and provided means to model and analyze z multitude of quite complex situations and systems. However, as it is always the case in science, at some time it turned out that the complexity of real world systems is just to high for the tools and techniques employed so far, and something new should be conceptualized and developed.

In this concept, Mo Jamshidi had shown an extraordinary vision and started promoting novel points of view, first those related to the use of elements of complex systems theory, and then the new concept, or paradigm, of the so called "systems of systems", or – to me more specific – "SOSE" (systems of systems engineering). To many younger people, who just do not remember other great achievements of Mo, he is just the greatest researcher and scholar behind the SOSE. I am deliberately writing both researcher and scholar because he has not only contributed to the new field with many original works, but have although devoted much of his time and effort to the education of many research communities in that new area. This is extremely relevant because of the human nature, that is, to promote anything new one should not only show good or better results but only somehow overcome a natural human reluctance to adopt something new when the old somehow works. In recent years, he has been also active in new directions related to sustainable development, smart grids, etc.

Needless to say that Mo's publication record is immense and contains some 700 publication, including many books part of which have become standard texts used all over the world.

Those great works of Mo that resulted, first, from his extraordinary vision, and then from his brilliant works, complemented with real world applications and implementation, has been widely recognized. Distinctions, awards, etc. he was presented are too numerous to list but let me just mention those which have been given to him by the Institute of Electrical and Electronics Engineers (IEEE), the largest engineering professional organization in the world. Just to mention a few, he is an IEEE Fellow, and the recipient of some of the most prestigious IEEE medals and awards, exemplified by the IEEE-USA Career Award on Systems Engineering, IEEE SMC Norbert Weiner Research Excellence Award, IEEE Centennial Medal, IEEE CSS Distinguished Member Award, etc. These are great distinction given only to the best of the best in the community. Many top organizations ad companies, notably NASA, IBM, etc., have taken advantage over the years of his knowledge and vision. At his university, he has received the prestigious Lutcher Brown Endowed Distinguished Chair.

Mo Jamshidi's service to the research community has been extraordinary and immense too. He has founded many journals, edited volumes, organized a lot of conferences, notably launching an extremely influential and successful series of the World Automation Congresses, just to list a few.

The record of Mo Jamshidi's successes in research, engineering, teaching, knowledge dissemination, etc. which I have just briefly mentioned above, does not give a full picture of his great performance and what he has really done to the society. In this respect, the work of Mo is great too. Basically, he has been all his life aware of something which has been known since the beginning of mankind from a multitude of statements by prophets, great thinkers, philosophers, statesmen, etc. that the more gifted and prominent a person is, the more obligations to the society he or she has.

In this context, one should clearly mention Mo's extremely socially conscious long time efforts to involve to a larger extend in the higher education and then in research students from both ethnic minorities, less privileged social classes and from less developed countries. His vision in this respect, that has been pursued and implemented for some decades, has proven to be just the proper solution that has provided the American science (but this concerns virtually all other countries as well) with a large number of young and gifted people who, if given opportunities, can contribute to a large extent to the development and success of their home or new country. Mo's supervising of so many M.Sc and Ph.D. theses, as well as mentoring many people from ethnic minorities are worth mentioning. Just as a remark, when he started those activities, nobody could predict that pretty soon they could be about a half of the American population with an obvious implication of an increased role in all areas of social involvement, including the top level education and research.

To summarize this brief note, I wish to recapitulate what I have said. First of all, Professor Mo Jamshidi is a very special person who somehow escapes a clear cut assignment in the following sense. He is, first, a great researcher who has not only contributed to so many areas, but has always shown vision by recognizing what will be promising to both advance science and solve real world problems. Second, he has been a great "professor" in the sense of the traditional university system that can be traced to at least the Middle Ages when the oldest European universities were founded. Namely, being aware of his deep knowledge and expertise, he has been always willing to share it with the young generation. Third, he has shown a deep social responsibility that is so characteristic for great individuals, that is, has always tried to support and help gifted young people from underprivileged social groups, less developed countries, etc. to avoid the loss of talent that could be detrimental to the society.

I think that I can express a deep conviction of the entire research and scholarly society that Professor Mo Jamshidi has deserved all kinds of appreciations he has been given. This volume, dedicated to him, is just another token of appreciation we do owe to him.

Janusz Kacprzyk Professor Janusz Kacprzyk, Ph.D., D.Sc.
Fellow of IEEE, IFSA
Full Member, Polish Academy of Sciences
Foreign Member, Bulgarian Academy of Sciences
Foreign Member, Spanish Royal Academy of Economic and Financial Sciences
President, Polish Operational and Systems Research Society
Past President of IFSA (International Fuzzy Systems Association)
Systems Research Institute

Foreword by Asad Madni

A Tribute To Mo Jamshidi

It is a true pleasure for me to wish my dear friend Mo Jamshidi a happy 70th birthday. I have known Mo for more than two decades and cherish his friendship as a family treasure. As a 70 year young man, Mo has demonstrated incredible energy and passion in everything he does. I have tried, untiringly albeit unsuccessfully, to get him to divulge this secret source of energy but, as usual, he has provided masterful answers without any of them providing the needed insight.

Remarkable and valuable as Mo's contributions have been to the field of control systems, large scale systems and system of systems (SoSE) engineering from both a teaching and research perspective, his noteworthy contributions in the area of minority education cannot be overemphasized. He has developed innovative educational models that have proven to be most effective as is demonstrated by his numerous minority students who upon graduation have gone on to hold senior positions in academia and in industry.

Every student that comes in contact with Mo soon realizes that he or she has become an integral part of his family. His caring attitude, compassion, thoughtfulness, wisdom, and experience have benefited hundreds of students that have studied directly under him or have had the privilege of being in close contact with him. I remember the time when Mo and I jointly guided Prasanna Sridhar's (now Dr. Prasanna Sridhar of Microsoft Corporation) doctoral research in the area of wireless sensor networks. The resulting research provided seminal contributions in some of the most important and difficult challenges in this field. The innovations were recognized in numerous conference papers and archival journals and till today stand as landmark contributions in the area of wireless sensor networks. Exciting and productive as our collaboration was, even more remarkable was the close personal relationship that the three of us developed with each other and which eventually permeated within our families. Unique as this sounds, it is Mo's normal modus operandi.

Another great legacy and a brain child of Mo is the World Automation Congress (WAC) which he founded together with his wife Jila. After more than two decades this biennial conference continues to attract the brightest and the best scholars and researchers from around the world and provides a unique venue for the dissemination of new and innovative ideas and research results. Besides being a gathering of technical minds, WAC has become an occasion of re-union for friends. My wife Taj, my son Jamal and I look forward to every WAC and especially to meeting the loyal friends who regularly attend.

There is much to be said about Mo, but I am afraid that 70 years of dedication, devotion and hard work cannot be done justice to in a foreword such as this. So I will end with my best wishes to Mo, Jila, Ava, Nima, and the rest of the members of this wonderful family. May Mo and his family be blessed with the best that life has to offer as we look forward to another 70 years of contributions from Mo.

Asad M. Madni, Ph.D., NAE
President, BEI Technologies, Inc. Emeritus
Distinguished Adjunct Professor/Distinguished Scientist at UCLA

Foreword by Clarence de Silva

I have known Professor Mo Jamshidi personally for over twenty five years, having first met him at a technical conference organized by him, even though I had known about his significant contributions to the profession and field of engineering for a longer period. With intimate knowledge of his accomplishments and contributions, it is indeed a great pleasure, privilege and honor to write a foreword for a book that is published as a tribute to Professor Jamshidi on his 70th birthday.

It is also fitting to state that the book is a timely and welcome addition to the knowledge base in the field of control systems and applications. The book has brought together a group of distinguished scholars and researchers from several disciplines and institutions, with the intent of advancing the technical knowledge in the subject of control system, including fundamentals, analysis, and application particularly in robotics, networked sensing, and communication. As needed, many chapters take an application emphasis. The coverage is sufficiently broad while the treatment is in depth where needed. I am further delighted to see that the present book sufficiently addresses a broad set of analytical and practical issues that are close to the heart of Professor Jamshidi. Indeed, he has contributed vastly to these very same areas during his long career.

No matter in what perspective I view his accomplishments, Professor Jamshidi comes up as a leading figure. Concerning Professor Jamshidi's research excellence, I have read, used, and quoted his papers and books in my own research activity, and have highly recommended them to my colleagues and students. Of particular interest to me has been his work in Intelligent Control and Automation, Large-scale Systems, Manufacturing Systems, and Control System Design, and his pioneering work in the new field of Systems of Systems. In these fields, his work has been highly authoritative, original, precise, and above all, quite applied. I have not seen anywhere in his publications, he had not related his theories and analytical contributions to real-life applications. He does not stop there. He himself implements his techniques in laboratory systems, demonstrates them, and then transfers the technology for general use. Of course, I have not had the opportunity to read all his books (12 textbooks and 56 edited volumes), journal papers (hundreds) and conference papers (several hundred). But, what I have read and understood have been of a highly superior quality and enormous practical value. It is not the quantity but the quality of his work that amazed me. This has been justified through patents and numerous foreign- language translations of his work. He has provided a fresh outlook to the applications of science and technology in fields like energy, environment, security and economy. His leadership in the field of "system of systems engineering - SoSE" is bringing about direct ramifications on the applications of systems engineering for an efficient and robust outlook of energy, environment, security, and economic management of a nation. He has globalized this effort enabling a number of countries to benefit from SoSE .

His leadership skills are second to none. He has led a world renowned and highly distinguished Research Center at the University of New Mexico with the sponsorship of NASA. Also, for his distinguished research achievements and stature he had been appointed to the AT&T Endowed Chair and Regents Professor. At present he holds the Lutcher Brown Endowed chaired Professorship of the University of Texas at San Antonio. The funding level for his research has been about $10 million during the past 15 years. Management of such facilities and funds, with a very large group of research personnel, students, and support staff, is not a trivial task. His leadership as the founding organizer of tremendously successful and highly respected international conferences, most notably, the World Automation Congress; and the International Symposium on Robotics and Manufacturing, since 1986, also should bear excellent testimony to his organizational skills. Furthermore, he has served as founding editor for highly respected international journals in applied sciences and engineering; more recently as the founding Editor- in-Chief if the IEEE Transactions on Systems of Systems. He has been on the boards of several organizations including companies that he himself has established, that are dedicated to engineering education, and research and development.

I have heard Mo giving technical presentations to very large audiences. Also, I have had the privilege of having him Vancouver as a keynote speaker and an expert instructor for workshops and short courses. In all these occasions, he displayed his fine capability to capture an audience, regardless of the level of education or experience, and get the technical message across in a very clear and precise language, without making the presentation too complex or uninteresting. In all such occasions, the feedback that I received from the participants had been excellent. Above all, he always came well prepared, with courseware including a complete set of notes, software, hardware, demonstrations, videos, etc. Then, he would personally setup equipment, check photocopies of handouts for accuracy, and would assist the course facilitators for conducting exercises and training sessions. I have not seen a more skillful and dedicated teacher anywhere during my academic career of over 35 years.

I have no doubt that the present book will lead to further insights, new research and developments, and increased practical applications in the subject area. It will provide a valuable source of knowledge for researchers, students, and practicing professionals alike. Above all, it is a fitting recognition of the immense professional which professor Jamshidi has made through his long career.

Happy Birthday My Friend!

Clarence W. de Silva, Ph.D., D.Eng. (hon), FRSC
Professor, Senior Canada Research Chair in
Mechatronics and Industrial Automation,
and Peter Wall Scholar
Department of Mechanical Engineering
The University of British Columbia
Vancouver, Canada
www.mech.ubc.ca/~ial

Foreword by Soroosh Sorooshian

Foreword for Tribute Book honoring Professor Mohammad "Mo" Jamshidi

I was truly honored recently to receive a call and be asked if I was willing to provide a foreword for this collection honoring my distinguished colleague and friend Professor Mohammad "Mo" Jamshidi.

I first got an opportunity to meet Mo back in the early 1980's in an annual meeting of the International Federation of Automatic Control (IFAC). I had heard of Mo Jamshidi when I was a student at UCLA through other colleagues who referred to Mo as one of the "up and coming" Iranian scholars and engineers who has gone back to Iran to educate a new generation of academics and engineers.

The wheel of history and the circumstances after the Iranian revolution resulted in Mo moving back to the U.S. which was a great gain for the American engineering community. I need not say much about his career path, which is well documented, except that he was at the University of New Mexico for nearly 25 years until he moved to the University of Texas, San Antonio, where he is currently the Lutcher Brown Chair of Engineering. The reason I point out these two institutions in his career is to speak of his loyalty and commitment everyplace he has been to ensure sufficient time to have an impact on his workplace. In this respect, let me provide the following observations organized within a few categories as listed below:

Contributions to research in Engineering and Science

Professor Jamshidi's scientific contributions to the general area of Systems Engineering and Control Theory have been seminal and powerful. His developments in modeling, optimization, CAD and control of large-scale systems (LSS) are well documented in the literature. His work on "multi-scale property" of large scale systems led to the development of "multi-scale system design" based on sensitivity of system variables with respect to parameters variations.

His impact on system engineering is best represented thorough his pioneering contributions to the theory of System of System Engineering. This approach has found its way into applications like robotics, mechatronics, applied optics, steel mills, space structures, auto engines, etc. He has played a key role in advancing the concept of "System of System Engineering (SoSE)" globally. His 1983 book on complex systems is the first text book on this subject.

Application of his "multi-scale system design" concept in Labs and industries like General Motors Technical Center (Saturn engine) and Cold Rolling mills (Alleghany Ludlum Steel) are noteworthy. His co-invention related to the application of fuzzy logic for creating quality prints from video, resulted in a patent.

His nearly 200 journal papers, 11 books, over 150 other significant writings as book chapters and conference papers with his students support my comments in terms of his long-lasting impact in the field of Engineering. It is also worth mentioning his ability to obtain millions of dollars of funding in pursuing his scientific vision.

Contributions to Educations and Capacity-building

His contribution to minority graduate education has been remarkable. He has mentored and successfully graduated over 80 MS and 30 PhDs (Over 50 Hispanics; 30 African Americans; 15 Native Americans). His minority mentorship success has been featured in national magazines as "Outlook on Hispanic Education" (2002 and 2006). His former students are holding prestigious positions in academia (e.g., Air Force Academy, John Hopkins, Georgia Tech., University of Texas at El Paso, University of New Mexico, University of Colorado Colorado Springs, Southwestern Indian Polytechnic Institute); government laboratories (e.g., NASA Jet Propulsion Laboratory, Sandia and Los Alamos National Laboratories, US Air Force Research Laboratory); and private industry (e.g., Boeing Aerospace Corporation and Honeywell among others).

Most importantly, his contributions to minorities graduate engineering education were initiated in the early 80's with his tireless work at the University of New Mexico (UNM, a Hispanic Serving and Tier-I Institution). He developed and has successfully implemented a minority education model called VI-P® Pyramid (K-12 at the bottom of the pyramid and doctoral at the top of the pyramid).

Contributions to the Profession I know of no other colleague who has given so much to the promotion and advancement of Systems Engineering and on the profession. The best evidence of that is the number of Journals, for which Mo Jamshidi served as the founding editor-in-chief, notably the IEEE Control System Magazine and the IEEE Systems Journal. Recognizing the commitment and the vision of Mo Jamshidi, IEEE entrusted him to start up these two publications and the rest is history.

In addition to his role on editorial boards, Mo has also been a leader in various capacities in professional societies, advisory boards of numerous governmental agencies including NASA. His tireless efforts of organizing short courses, conferences, workshops, etc., has been remarkable with a long-lasting impact of bringing together from all corners of the world, well-known experts in the field and the younger scientists to learn from the giants and mingle with them.

Contributions to the Global Community

Mo has received numerous honorary degrees and recognitions from many corners of the world. He has rarely turned down the invitations to deliver keynote talks at various conferences and universities to share his knowledge of Engineering Systems and spend time with students and junior faculty. He has been a remarkable ambassador of "scientific communication" devoid of politics and other issues which otherwise could hamper dialogue and promote knowledge transfer. As an Iranian I am so honored to have known Mo and learn so much from him with respect to acknowledging the contributions of others and making sure their work gets the proper recognition.

It has been a privilege to know you and to learn from you as you are one of the best role models for us all. Congratulations on your accomplishments as an engineer, scholar, a committed ambassador for the promotion of science and knowledge, and for being a true friend to all of us.

Needless to say that you, more than anyone, acknowledge that your success has come with the support of Jila, your wonderful wife and partner in life.

Soroosh Sorooshian, Ph.D., NAE
UCI Distinguished Professor and
Director Center for Hydrometeorology and Remote Sensing (CHRS)
University of California, Irvine

Foreword by James Tien

In Tribute To My Friend and Colleague, Professor Mo Jamshidi

I am honored to be given this privilege to write a Foreword for this compiled book, in recognition and celebration of Professor Mo Jamshidi on his seventieth birthday. As detailed in the contents herein, Professor Jamshidi has dedicated his technical prowess to formalizing and extending the complex field of large-scale systems (LSS), ranging from modeling and design, to control and optimization, to establishing a new field in system of systems engineering (SoSE). His publications, including books, on complex LSS and SoSE have greatly expanded our understanding of systems and systems engineering. He has over 700 publications to his credit, including 12 textbooks, 56 edited volumes and research monographs, 190 journal papers, 352 conference papers, etc. As the Lutcher Brown Endowed Distinguished Chaired Professor at the University of Texas, San Antonio, he remains highly productive in research and education; since 2006, he has been the Director of the Cyber-Physical Systems - ACE Laboratory and since 2008, he has led a Sustainable Energy Research Group on Smart Grids. Professor Jamshidi is most proud of his contributions to the engineering education of ethnic minorities; this commenced in 1980 with a NASA JPL grant at the University of New Mexico, a Tier-I Hispanic Serving Institution. Supported by several additional major federal grants, he formalized a model for the productive education of minorities; called VI-P Pyramid, the model starts with K-12 students (as the first layer of the pyramid) to doctoral students (as the final layer of the pyramid), all progressing in a seamless manner, as upper level students mentor lower level ones. Since 1980, he has graduated over 114 minority students, consisting of 62 Hispanics, 34 African-Americans, 15 Native Americans, and 3 Pacific Islanders. Professor Jamshidi's dedication to and mentorship of a large number of minority students in engineering fields at all levels have been featured in several national magazines (e.g., 2002 and 2006 issues of Outlook on Hispanic Education). His graduates are holding prestigious positions in the U.S. and abroad; those in academia are following his example by producing their own graduates drawn from the pool of minority students. He has provided expert opinion to organizations like NASA, NRC, DOE, USAF, LMC, NATO, and EU. As an example, he advised NASA JPL on the intelligent behavior of the autonomous Mars rover "Sojourner", a program in which an African-American doctoral student of his was a part of – specifically, the 2004 Mars missions.

Professor Jamshidi should also be recognized for his many honors and awards, including being an elected Fellow of IEEE, ASME, AAAS, NYAS, TWAS, and Hungarian Academy of Engineering. He has served on the Review Boards of the MacArthur Foundation, NASA, JPL, UNESCO, NSF, DOE, EBSCOR Research Centers, INDO-US Science Technology Forum, and INDO-UK Education and Research Initiative. He was the Founding Editor-in-Chief of the IEEE Control Systems Magazine (1980), the Elsevier Computers and Electrical Engineering Journal (1989), Autosoft Journal

(2004) and the IEEE Systems Journal (2007). He holds honorary doctorates from the University of Waterloo, Canada (2004) and the Technical University of Crete, Greece (2004). Other recognitions include IEEE-USA Career Award on Systems Engineering, NASA Headquarter National Service Award for Minority and Women-Owned Business Utilization, and the IEEE SMC Norbert Weiner Research Excellence Award.

James M. Tien, PhD, NAE
Distinguished Professor and Dean
University of Miami

Foreword by Lotfi Zadeh

It is a real pleasure to write a brief foreword to **Control And Systems Engineering – A Report on Four Decades of Contributions**, or CSE for short, honoring my close friend, Professor Mo Jamshidi, on his 70th birthday. The contents of CSE reflect Mo's far-ranging contributions to a broad spectrum of fields ranging from systems analysis and control to information processing and decision analysis. A contribution which stands out is the major role which Mo has played in the development of the concept of system of systems and its applications. The concept of system of systems serves as a basis for a better understanding of the complex world we live in –a world of interdependence and globalization. A striking side-effect of globalization is the emergence of economic sanctions as an instrument of economic warfare– an instrument which can bring a country to its knees, with no shots fired. My acquaintance with Mo goes back to his student days at the University of Illinois. I have been in close professional contact with him for the past four decades.

CSE is a tribute to Mo's remarkable achievements as a researcher, an educator and administrator. In the course of his illustrious career, Mo has authored or co-authored over 700 publications: 12 textbooks, 56 edited volumes and research monographs, 190 journal papers, 352 conference papers, etc. These numbers speak for themselves. Scientifically, his main contributions center on the analysis of large-scale systems and, more recently, on the development and applications of systems of systems. Mo has an unusual ability to apply basic theories to the solution of practical problems, an example being his US patent for a fuzzy logic video printer. Another significant example is his design of an adaptive control system for multi-aperture telescopes.

Mo is a man of vision and a true leader. He is always of help to others. At the University of New Mexico, he founded the NASA Center for Autonomous Control which provided minority students with opportunities to pursue careers in science and engineering. This Center has proved to be an unqualified success in all respects.

Through his numerous textbooks and inspired teaching, Mo has made a major contribution to engineering education. He works closely with his graduate students, many of whom have gone on to achieve prominence in their fields. Professionally, Mo has been extremely active as a consultant and organizer. In particular, he achieved major success with his regularly held biennial World Automation Congress (WAC). WAC draws over one thousand participants from all over the world. Mo founded the IEEE Control Systems magazine and serves on the editorial boards of over twenty journals. After joining the University of Texas, San Antonio in 2006, he initiated a number of research projects focused on the analysis of large-scale systems and system of systems.

CSE is more than a tribute to Mo; it is also a significant contribution to the literature of systems and control. Mo, the contributors to CSE and the publisher deserve our thanks and congratulations on producing a volume which treats with authority and skill some of the principal applications of modern methods of systems analysis and control.

Lotfi A. Zadeh, Ph.D., NAE
Professor Emeritus, UC Berkeley
Professor in Graduate School
Director, Berkeley Initiative on Soft Computing (BISC)
UC Berkeley
Berkeley, CA, USA

Preface

This book is a tribute to 40 years of contributions by Professor Mo Jamshidi who is a well known and respected scholar, researcher, and educator.

Dr. Mo Jamshidi (Fellow IEEE, Fellow ASME, A. Fellow-AIAA, Fellow AAAS, Fellow TWAS, Fellow NYAS, Member HAE, Member, Russian Academy of Nonlinear Systems) received BS in EE, Oregon State University, Corvallis, OR, USA in June 1967, the MS and Ph.D. degrees in EE from the University of Illinois at Urbana-Champaign, IL, USA in June 1969, and February 1971, respectively. He holds honorary doctorate degree from University of Waterloo, Canada, 2004, Technical University of Crete, Greece, 2004 and Odlar Yourdu University, Baku, Azerbaijan, 1999. Currently, he is the Lutcher Brown Endowed Distinguished Chaired Professor at the University of Texas, San Antonio, TX, USA. He has been an advisor to NASA (including 1st MARS Mission), USAF Research Laboratory, USDOE and EC/EU (Brussels). He has advised over 60 MS and 50 PhD students. He has also directly or indirectly through funded research helped 120 America's minority graduate students. He has over 710 technical publications including 68 books (11 text books, see www.wacong.org/freepublicationsbymojamshidid), research volumes, and edited volumes in English and 7 foreign languages. He is the Founding Editor or co-founding editor or Editor-in-Chief of 5 journals including IEEE Control Systems Magazine and the IEEE Systems Journal. He is an Honorary or Visiting Professor at Deakin University (Australia), Birmingham University (UK), Obuda University (Hungary), Loughbrough University (UK) and at three Chinese Universities (East Normal, Nanjing and Xi'an, China). He has received numerous honors and awards, including IEEE Centennial Medal, IEEE Millennium Awards, and the IEEE's Norbert Weiner Research Achievement Award, the 2014 IEEE-USA Systems Engineering Career Award, and 2014 WAC Medal of Honor, among others. He is a member of the University of the Texas System Chancellor's Council since 2011. He is currently involved in research on system of systems control engineering with emphasis on cloud computing, robotics, UAVs, biological and sustainable energy systems, including smart grids and big data analytic.

This book has contributed chapters by colleagues and former, and current students of Mo Jamshidi. In addition, to the contributed chapters, the content includes personal notes from good friends of Mo Jamshidi. Furthermore, forewords to the book include:

- Janusz Kacprzyk
- Asad Madni
- Soroosh Sorooshian
- Clarence W. de Silva
- James Tien
- Lotfi Zadeh

The editors would like to thank everyone who made this work possible.

Aly El-Osery
Albuquerque, NM

Jeff Prevost
San Antonio, TX

Dr. Aly El-Osery received his B.S. in 1997, M.S. in 1998 and Ph.D. in 2002 in electrical engineering from the University of New Mexico. From 1997 to 2002 he was a research assistant at the Autonomous Control Engineering Center at the University of New Mexico. Dr. El-Osery has received many awards among them Outstanding Junior (1996) and Outstanding Graduate Student (1998) from the Department of Electrical Engineering and the School of Engineering award for Outstanding Graduate student in Electrical and Computer Engineering (1998–1999).

In 2002 he joined the Electrical Engineering Department at New Mexico Institute of Mining and Technology, Socorro, New Mexico as an Assistant Professor. In 2007, Dr. El-Osery received his tenure and promotion to Associate Professor. His research interests are in the areas of multi-agents robotics, wireless communications, control systems, sensor networks and soft computing. He has over 50 journal, book chapters, and conference publications in these areas. He is the co founder of the Intelligent Systems and Robotics Group (ISRG) at New Mexico Tech. He is a senior member of IEEE and several of its societies including Systems Man and Cybernetics Society, and Communication Society.

Dr. John Jeffery Prevost received his first B.S. degree from Texas A&M in Economics in 1990. He received his second B.S. degree in Electrical Engineering from the University of Texas at San Antonio, where he graduated Magna Cum Laude in December, 2009. In 2012 he received his M.S. degree in Electrical Engineering, also from the University of Texas at San Antonio along the way to earning his Ph.D. in Electrical Engineering in December, 2013. During his academic career he received two awards for Best Conference Paper, first at the World Automation Congress 2012 conference (Puerto Vallarta, Mexico) and again at the System of Systems Engineering 2013 conference

(Maui, HI). Since 2011, he has taken the role of Executive Coordinator for the Cloud Technology Initiative where he is responsible for assisting UTSA's efforts at promoting and adopting cloud related activities in education and research. His formal position with the university is Assistant Research Professor in the department of Electrical and Computer Engineering. He is also a proven technical leader in the computer hardware and software industry, with a 20 year career. He has served in positions of Director of Product Development, Director of Information Systems and Chief Technical Officer. He is an active consultant in the areas of complex systems and maintains strong ties with industry leaders. He has served in the various capacities at professional conferences such as General Chair, Publications Chair, and the Tutorial and Organized Session Chair. His is a member of Tau Beta Pi, Phi Kappa Phi and Eta Kappa Nu Honor Societies, and has been a member of IEEE since 2006. He has published 9 conference and journal papers. Since July, 2012 he has been representing the 80/20 Foundation's efforts to help transform the University of Texas at San Antonio into one of the premier research and education institutions in the area of Open Cloud. In this capacity he has been a major factor in bringing $4.5M in gifts to UTSA, which will, or has, resulted in total matching of $5.5M. His efforts have resulted in approximately $10M of additional funding to UTSA so far. He has also secured an additional $5M in funding for a cloud datacenter to be constructed on the UTSA main campus. His current research interests include energy aware cloud optimization, cloud controlled robotics, cloud based communications, and quantum cloud computing.

Contents

Reflection on Four Decades of Contributions of My Graduate Students

Mo Jamshidi

Lutcher Brown Endowed Distinguished Chaired Professor,
The University of Texas,
San Antonio, TX 78249, USA

1 Introduction

I was born in Shiraz, Iran on May 10, 1944 and after my K-12 education in 1962 I received a tuition scholarship from Oregon State University (Corvallis, OR) and came to US for my higher education. At Oregon State in 1963 I majored in electrical engineering and graduated in June 1967 with honors. In fall 1967 I entered my graduate education at the University of Illinois at Urbana-Champaign and received my master and doctorate degrees, both with theses, in June 1969 and February 1971, respectively.

Early on during my undergraduate degree at Oregon State received my first control theory education from late Prof. Solon Stone, enjoyed the topic immensely and decided to major in control theory. At Illinois I joined the control systems group at the Coordinated Science Laboratory (CSL), headed by Joe Cruz, Jr., an imminent professor of control and systems engineering whose own work was done under the supervision of late Professor Mac Van Valkenburg, another eminent professor of network theory.

Other eminent members of the group at CSL were Bill Perkins (a Stanford graduate from Gene Franklin's group) and a visiting professor Petar Kokotovic (a Moscow State University graduate, USSR from late Yakov Tsypkin, a legend in control theory) from Yugoslavia. I received my MS and Ph.D. from Joe Cruz, Jr. and Petar Kokotovic, respectively at Illinois.

Having lost my father before I was born, I have always had a deep desire to look at all my younger people, be it my own children or my students as my own children and tried my best to treat them with respect and care for the past 43 years.

I stayed 8 more months at Illinois as a post-doctoral fellow and by August 4, 1971 I left USA to go to Iran and join Pahlavi (now Shiraz) University in my own hometown Shiraz, Iran. Yes, the same place famous for its grapes and production of "Shiraz Red Wine".

Three years later I found the love of my life in Jila Salari, a student at Pahlavi University and we got married on June 21, 1974. During the summer of 1974, after a short stay at the University of Stuttgart I joined IBM T J Watson Research Center in Yorktown Heights, NY until April, 1977. As Jila and I were returning home, we stayed in Lyngby, Denmark, where I was doing research on energy systems forecasting at the Technical University of Denmark.

A. El-Osery and J. Prevost (eds.), *Control and Systems Engineering*,
Studies in Systems, Decision and Control 27, DOI: 10.1007/978-3-319-14636-2_1

Returning home in July, 1977, the Iranian Nation was undergoing a revolution that no one has seen before. Like almost all revolutions, Iran was undergoing turmoil and chaos everywhere. By July 1979, I could no longer had a peace of mind to work on my research and our lives were dominated each day by political fever and unexpected outcomes. We (Jila, I and our 14-month old daughter Ava) decided to head back to US again.

In August 1979 I joined the University of New Mexico in Albuquerque – a City that reminded me of my home town Shiraz at the invitation of my late friend and colleague Peter Dorato at the University of New Mexico (UNM). In 1980 I received a tenured full professor position in ECE Department there and stayed there for 27 years. During my tenure at UNM Our son Nima was born in Albuquerque in July 1982. I was happy to spend sabbatical visits at George Washington University, University of Virginia, French NSF (CNRS) LAAS Laboratory (Toulouse), and Hong Kong Polytechnic University and National University of Singapore.

Around mid-2005 I was approached by my friend Professor Mehdi Shadaram from the University of Texas at San Antonio and Jila and I have moved to Texas since January 2006 as an endowed distinguished chaired professor.

However, this book is not so much about me, but about my approximately 120 current and former graduate students who have played a very important role in forming my career right equal to my own immediate family (Jila, Ava and Nima).

2 Graduate Students

In my 45-year career, so far, I have spent time at several institutions: Shiraz University (Iran, 1970-79), University of New Mexico (USA, 1979-2005), University of Texas (San Antonio, USA, 2006-present). In addition, I have had the pleasure of visiting many other institutions like University of Stuttgart (Summer 1975), George Washington University (USA 1987-1988), LAAS and University of Toulouse (France, 1994-95), Polytechnic University of Hong Kong (Spring 2003), Deakin University (Australia, Summer, 2006), and Universidad Polytechnic du Madrid (Spain, 2007). During these assignments and work opportunities, I was very fortunate to cross paths with over 120 graduate and thousands of undergraduate students around the globe. Below are a near-complete list of my graduate students from 1984-2014 by country and name in chronological order:

2.1 MS Students

Algeria

1. Rafik Benmansour, 2014 (Last position: unknown)
   MS thesis: "TSP problem via mobile robot navigation"

Bolivia

1. Aldo Jaime, 2009 (Last position unknown)
   MS thesis: "Analysis, design and implementation of UAV's and ground stations"

China

1. Huimin Xue, 1994 (Last position unknown)
   MS thesis: "Adaptive fuzzy control of electric power systems"
2. X. Zhu (Mechanical Engineering), 2000 (Last position: Unknown)
   MS thesis: "Numerical Solutions of Dynamic Mode Inverse Kinematics Problem for Robotic Manipulators"
3. Y. Lu, 2002 (Last position unknown)
   MS thesis: "Fuzzy control of electric power systems"
4. Shan Xia, 2004 (Last position unknown)
   MS thesis: "A genetic algorithm optimized fuzzy control of intelligent agents"

Croatia

1. Denis Barak, 1993 (Last position: Unknown)
   MS thesis: "Modeling and control of fuzzy control systems with applications to industrial systems"

Egypt

1. Aly El-Osery, 1998 (Last position: NM Institute of Technology, USA)
   MS thesis: "Design and implementation of expert systems for digital and analog image enhancement"

France

1. Theiry Portas, 1985. (Last known position: French Diplomatic Corp)
   thesis: "Hierarchical model and structural properties of large-scale energy systems"
2. Olivier Pages (INSA France), 1997 (Last position: Unknown)
   MS thesis: "Intelligent simulation for cooperative robots"
3. Daniel Aznar (INSA, France), 1998 (Last position: Unknown)
   MS thesis: "SoftLab©- Neural network applications"
4. Remi Lecointe(INSA, France), 1999 (Last position: Unknown)
   MS Thesis: "Intelligent control of electric power systems"
5. Francois Lhomme (INSA, France), 1999 (Last position: Unknown)
   MS Thesis: "SoftLab©- Neuro-computing and Adaptive Fuzzy Systems"

Germany

1. A. Rommel (University of Seigen, Germany) 2001 (Last position: Unknown)
   MS thesis: "Dynamic modeling and optimal control of satellite arrays"

Iceland

1. Elli Kristjansson, 1993 (Last position: Unknown)
   MS thesis: "Nonlinear and fuzzy control of electric power systems"

India

1. Jay Bhata, 1996 (Last position: Unknown)
   MS thesis: "A software environment for risk assessment"
2. V. Vakipuram, 2002 (Last position: Unknown)
   MS thesis: "Fuzzy control of water Desalination systems"
3. S. Murali, 2002 (Last position unknown)
   MS thesis: "Stability of fuzzy control systems with application to power systems"
4. Prasanna Sridhar, 2003. (Last position Microsoft Corp., USA)
   MS thesis: "A virtual discrete-event simulation and modeling of intelligent agents"
5. Umesh Dole, 2004. (Last position unknown)
   MS thesis: "Intelligent Navigation of All-Train Rovers"
6. Vikraman Raghavan, 2007. (Last position unknown, USA)
   MS thesis: "Autonomous control and sensor fusions of robotic agents"
7. Kranthimanoj Nagothu, 2009 (Last position Theatro Corp., USA)
   MS thesis: "Underwater communication among Rovers"
8. Srujana Eega, 2009 (Last position unknown, USA)
   MS thesis: "Design and Simulation of a DC Thruster Motor"
9. Satish Vaishnav, In progress 2014
   MS thesis: "Quadcopter Swarm Localization and Control"
10. Mohan Kumar Muppidi, In progress 2014
    MS thesis: "Robotic Navigation using visual SLAM and Image Processing"
11. Chetan Manikanta Puppala, In progress, 2014
    MS thesis: "SLAM based navigation of quadrotors"

Iran

1. Ali Jadbabaie, 1997. (Currently, Endowed Professor, University of Pennsylvania, USA)
   MS thesis: "On the stability of fuzzy logic control systems"
2. Shahab Sheikh-Bahaei, 2003 (Last position unknown)
   MS thesis: "Fuzzy logic applications of mobile rovers"
3. Alireza Naddaf, 2004. (Last position unknown)
   MS thesis: "Predictive control of intelligent agent systems"
4. Peymon Gazi, 2010 (Last position unknown)
   MS thesis: "Swarm robotics via Network Control"
5. Amir Rajaee, 2012 (Last position: Qualcom Corp., USA)
   MS thesis: "Optimal management of Smart Grids"
6. Kaveh Kheradmand, Current 2014 (Last position unknown)
   MS Thesis: "Wind Energy Control"

7. Maryam Ezell, In Progress, 2014
   MS thesis: "Robotic Control and Sensing"
8. T. John Parsi, DDS, Current
   MS Thesis: "Robotics application in dentistry"

Mexico

1. Gerardo Trevino, 2012 (Last position: Southwest Research Institute, USA)
   MS thesis: "Cyber-Physical Sustainable Energy and Electric Cars"
2. Jose Gomez, 2009 (Last position US Navy)
   MS thesis: "Design and implementation of hybrid fuzzy and adaptive autopilot for UAVS"
3. Jacqueline Diaz, 2009 (Last position US Navy)
   MS thesis: "Design and implementation of hybrid fuzzy and adaptive autopilot for UAVS"

Morocco

1. Meryem Fennich, 2013(Last position unknown)
   MS thesis: "Management of PV - Wind Energy"

Netherlands — Advised at CNRS LAAS, France

1. Jasper Bruinzeel, 1995 (Last position: Unknown)
   MS thesis: "Fuzzy control of complex systems using rule hierarchy and sensory fusion"

Norway — Advised at CNRS LAAS, France

1. Finnur Olaffsen, 1995 (Last position: Unknown)
   MS thesis: "Fuzzy control of multi-stage flash desalination systems"

Taiwan

1. C.-M. Wang, 1988. (Current position: Unknown)
   MS thesis: "Optimization of large-scale non-linear systems with time-delay"
2. Tom C. Yenn, 1990 (Last Known position: Taiwan Atomic Energy Commission)
   MS thesis: "On the computational aspects of Kalman filtering"
3. Chung-Shi Tseng, 1991 (Last known position: Professor, Ming Hsin Universiy of Science and Technology,Taiwan)
   MS thesis: "On the computer aided robust decentralized control design of a five-axis robot"

Turkey

1. Levent Sari, 2014 (Last position: Ministry of Energy, Turkey)
   MS thesis: "Energy Forecasting of Turkey up to 2023"
2. Halid Kaplan, Graduated, May 2014 (PhD Program)
   MS thesis: "Data Analytic Studies for Turkeys Energy Forecast"

USA

1. Daniel A. Pritchard, 1986. (Last known position: Sandia National Laboratory)
   MS thesis: "Sun tracking by peak power positioning for photovoltaic concentrator arrays"
2. Jane M. Brideau, 1987. (Last known position: Sandia National Lab)
   MS thesis: "Hierarchical control of time-delay discrete-time systems"
3. Robin S. Morel, 1989. (Last known position: Los Alamos National Lab)
   MS thesis: "Software engineering design of linear control systems"
4. Gerald L. Schotik, 1990. (Last known position: Los Alamos National Lab)
   MS thesis: "Software engineering design and analysis of multivariable control systems"
5. John T. McGuffin, 1990 (Last Known position: Boeing Company)
   MS thesis: "The design of device independent and system independent computer application packages"
6. Steven R. O'Neill, 1991 (Current position: Unknown)
   MS thesis: "Robot-S: An interactive design and simulation language for robot manipulators"
7. William Horne, 1992 (Last position: Unknown)
   MS thesis: "On the connection-based control architecture for robot manipulators"
8. Joseph A. Meinhardt, 1993 (Last position: Unknown)
   MS thesis: "Modeling and control of an optical phase array imaging telescope with a wide field of view"
9. William Honey, 1992 (Last position: Own his company)
   MS thesis: "A MATLAB-based ToolBox for robot manipulators"
10. Scott Peterson, 1993(Last position: Unknown)
    MS thesis: "Adaptive Optics System Control Using Linear Quadratic Methods"
11. Douglas Miller, 1993 (Last position: Unknown)
    MS thesis: "Applications of fuzzy logic to stock markets and financial planning"
12. Alex Martinez, 1993 (Last position: Unknown)
    MS thesis: "Fuzzy Control of Automotive Engines Idle Speed"
13. Steve Baugh, 1995 (Last position: Unknown)
    MS thesis: "A control approach for laser guidance systems"
14. Tanya Lippincott, 1999 (Last position: Honeywell Corp.)
    MS thesis: "Intelligent navigation of mobile robots"

15. Scott Beatty, 2005. (Last position: Lockheed Martin)
    MS thesis: "Simulation of a spacecraft electrical power distribution system using the simulink power system block set and soft computing techniques"
16. Patrick Benavidez, 2010 (Last position To come)
    MS thesis: "Design of a swarm of autonomous ground vehicles for use in remote sensing applications"
17. Joaquin Labrado, 2013 (Last position: SRC, Inc. ,USA)
    MS thesis: "Control of Space Debris"
18. Barney Tannahill, 2013 (Last position: Southwest Research Institute, USA)
    MS thesis: "Big Data analytic techniques: Predicting renewable energy capacity to facilitate the optimization of power plant trading algorithms"

2.2 PhD Students

Australia

1. Ben Horan, 2009, (Co-advisor, S. Nahavandi, Deakin University, Australia), (Last known employer: Deakin University, Australia)
   Ph.D. Dissertation: "Haptic control of a rover in a system of robots framework"
2. Matthew Joordens (Co-advisor, S. Nahavandi, Deakin University, Australia), Completed, 2009 (Last known employer: Deakin University, Australia)
   Ph.D. Dissertation: "Design and implementation of a system of underwater rovers"

Bangladesh

1. Purnendu Sarkar (Last known employer: Tyco Healthcare Valleylab Corp., 2002 Ph.D. Dissertation: "Modeling, simulation, design and control of a two-stage desalination pilot plant"

Brazil

1. Marco De Oliveria, 2001, (Last known employer: Brazilian National University, Brazil)
   Ph.D. Dissertation: "Hierarchical intelligent control of multi-agent systems"

China

1. Z. Jason Geng, 1988, (Last known employer: Chinese Academy of Sciences)
   Ph.D. Dissertation: "Control and stability of two-dimensional systems"
2. Tao Song, 2004, (Last known employer: University of California, San Diego)
   Ph.D. Dissertation: "Intelligent enhancement and recognition in magnetic resonance imaging"
3. Yan Wang, 2004, (Last known employer: Searching employment), Ph.D. Dissertation: "Intelligent pattern recognition and remote sensing"

4. Jingyu Liu, 2004, (Last known employer: Mine Institute, University of New Mexico, Albuquerque, NM, USA)
   Ph.D. Dissertation: "Cardiac output modeling and simulation using soft computing"

Bolivia

1. Aldo Jaime, current, UTSA, 2014 Ph.D. Dissertation topic: "Machine vision and intelligence with humanoid robots"

Egypt

1. Aly El-Osery, 2002,(Last known employer: New Mexico Institute of Technology, USA)
   Ph.D. Dissertation: "Optimal Power Control of CDMA Based Cellular Systems"

Ghana

1. Francis Asamoah, 1985, (Last known employer: West Indies University, Caribbean Islands)
   Ph.D. Dissertation: "On Lyapunov stability of bilinear large-scale systems"

India

1. Kishan Kumbla, 1997, (Last known employer: Hitachi Corporation, USA)
   Ph.D. Dissertation: "A neural-network fuzzy logic controller for fault-tolerant robot manipulators"
2. Prasanna Sridhar, 2007, (Last known employer: Microsoft Corporation, Seattle, WA, USA)
   Ph.D. Dissertation: "Hierarchical Aggregation and Intelligent Monitoring and Control in Fault-Tolerant Wireless Sensor Networks"
3. Anjan Kumar Ray, 2009, Co-advisor: Laxmidhar Behera, Indian Institute of Technology, Kanpur, India), (Current Employer: PDPM- IIITDM Jabalpur, India)
   Ph.D. Dissertation: "Navigation and control of robotic swarms in unstructured environments"
4. Kranthimanoj Nagothu, 2013, (Last Employer: Theatro Labs, Dallas, TX, USA)
   Ph.D. Dissertation: "Cloud Centers in Smart Grid Data Analysis"

Iran

1. Mahmoud Tarokh, 1987, (Last known employer: San Diego State University, USA)
   Ph.D. Dissertation: "On decentralized pole placement problem with application to robotics"

2. Nader H. Vadiee, 1992,(Last known employers: University of New Mexico and Southwest Indian Polytechnic Institute, USA)
   Ph.D. Dissertation: "On a Programmable fuzzy logic array based on a new soft fuzzy reasoning paradigm"
3. Mohammad-R Akbarzadeh-T, 1998,(Last known employer: Ferdowsi University at Mashad, Iran)
   Ph.D. Dissertation: "Fuzzy control and Evolutionary optimization of Complex Systems"
4. Ali Asgharzadeh, 1999. (Last known employer: UAE Telecom, Dubai, UAE)
   Ph.D. Dissertation: "An Intelligent Approach to Image Enhancement"
5. Somayeh Bakhtiari, graduated, (Co-advisor: SoS Agaian), 2012. Last known Employer: Sony, Inc.
   Ph.D. Dissertation "Image processing Advances for Underwater"
6. Dariush Shahgoshtasbi, 2012, (Last known employer: ServiceNow Co, Seattle, WA, USA)
   Ph.D. Dissertation: "Energy efficiency in a smart home with an intelligent neuro-fuzzy paradigm"
7. Yashar Manjili, 2014, in Progress
   Ph.D. Dissertation: "Adaptive intelligent energy control framework for electrical micro-grids based on energy market and solar energy forecast"
8. Elmira M. Bonab (Current student)
   Ph.D. Dissertation: "System identification of Complex Systems"
9. Arezou Mousavi Khalkhali, Current, 2015,
   Ph.D. Dissertation: "Deep architecture and learning of constructing a deep Regression model utilizing sparse auto-encoders and stochastic gradient descent"
10. Morteza Dabaghmanesh, Current, 2015 Ph.D. Dissertation: "Cyber-physical control systems"
11. Peyman Najafirad, Current, 2015 Ph.D. Dissertation: "Open Source Cloud Architecture"

Israel

1. Sigal Berman, 2003, (Last known employer: Ben-Gurion University, Israel)
   Ph.D. Dissertation: "Intelligent Navigation of Automatic Guided Vehicles for Flexible Manufacturing"

Korea

1. Young-Tae Kim, 1986, (Last known employer: Donggu University)
   Ph.D. Dissertation: "Modeling and multivariable control of multi-link robot manipulators"
2. Byong J. Oh, 1988, (Last known employer: Hannam University)
   Ph.D. Dissertation: "On adaptive control of robot manipulators"
3. Soy R. Lee, 1990, (Last known employer: Korean Army Academy)
   Ph.D. Dissertation: "On the Lyapunov stability of large-scale nonlinear systems with time delay"

Malaysia

1. N. Alang-Rashid, 1993, (Last known employer: Malaysian Atomic Energy Agency, Malaysia)
   Ph.D. Dissertation: "Monitoring and control of nuclear reactors via a parameter estimation tuning fuzzy controller"

Mexico

1. Jorge Benitez-Read, 1992, (Last known employer: Mexican Nuclear Energy Agency)
   Ph.D. Dissertation: "Advance control architectures for nuclear reactors"
2. Miguel A. P. Garza, 2011, (Co-advisor: Jose A. de la O Serna, UANL, Mexico)
   Ph.D. Dissertation: "Implementation of harmonic estimators based on maximally flat FPGA target"

Russia

1. Aleksander Panchul, 2010, (Co-advisor, D. Akopian, UTSA, Last Employer: Unknown) Ph.D. Dissertation: "Design of Virtual environments for Simulation of Wireless Networks"

Taiwan

1. John N. Lieu, 1991. (Deceased)
   Ph.D. Dissertation: "On adaptive robust force control of robot manipulators"

Turkey

1. Yunus Yetis, Current student at UTSA, 2014 Ph.D. Dissertation: "Complex energy systems modeling and control with Application to Turkey"
2. Berat Erol, Current student at UTSA, 2014 Ph.D. Dissertation: "Machine intelligent Robots"
3. Halid Kaplan, Current student at UTSA, 2014 Ph.D. Dissertation: "Modeling and Management of Renewable Energy Systems"
4. Abdurrahman Akuzum, Current student at UTSA, 2014 Ph.D. Dissertation: "Solar-Based Energy Efficient Electric Vehicles"

USA

1. Roger E. Salters, 1984. (Last known employer: University of Denver, USA, Deceased)
   Ph.D. Dissertation: "On the extensions of Aoki's aggregation conditions to large-scale stochastic systems"

2. John M. Santiago, 1985, (Last known employer: Colorado College, Colorado Springs, CO, USA)
   Ph.D. Dissertation: "On the extensions of the balanced approach of model reduction with applications to large flexible space structures"
3. Mark Jacobus, 1991, (Last known employer: Sandia National Laboratories, USA)
   Ph.D. Dissertation: "Strong stabilization using fixed-order dynamic compensators"
4. Richard A. Carreras, 1994, (Last known employer: US Air Force Research Laboratory, USA)
   Ph.D. Dissertation: "Image restoration using nonlinear optimization techniques on an imaging system"
5. Edward Tunstel, 1996, IEEE Fellow, (Last known employer: Applied Physics Laboratory, John Hopkins University, USA)
   Ph.D. Dissertation: "Adaptive hierarchy of distributed fuzzy control: Application to behavior control of Rovers"
6. Nancy Miller, 1992, (Last known employer: National University of Malaysia, Malaysia)
   Ph.D. Dissertation: "A neural network for Phase Diversity: Simulation and Experiment"
7. Mark Johnson, 2002, (Last known employer: Aerospace Corporation, USA)
   Ph.D. Dissertation: "Intelligent modeling and control of Flexor-Tendon Repairs via Soft Computing"
8. Paul De Rego, 2003, (Last known employer: Honeywell Corporation, USA)
   Ph.D. Dissertation: "Optimal control for the autonomous deployment of a remote sensing spacecraft array"
9. Ana Martinez, 2003, (Last known employer: Sandia National Laboratories)
   Ph.D.: Dissertation: "Antenna baseline estimation coherent interferometric synthetic aperture radar image registration"
10. W. Gerry Parkinson, 2004, (Last known employer: Los Alamos national Laboratory, USA, Retired in 2004.)
    Ph.D. Dissertation: "Intelligent control of industrial autonomous systems"
11. Jonathan Lucero, 2004 (Last known employer: Arizona Public Service, USA)
    Ph.D. Dissertation: "Fuzzy logic solutions of structural engineering systems"
12. Ted Shaynefelt, 2012, (Last known employer: University of Hawaii, Hilo)
    Ph.D. Dissertation: "Hypercomplex number based automated robotic vanilla pollination system with vision sensing"
13. J. Jeff Prevost, 2013, (Last known employer: University of Texas, San Antonio)
    Ph.D. Dissertation: "Optimization Model for Low Power Computing in Cloud Data Centers"
14. Amy Daali, 2014, (Current employer: Trinity University, USA)
    Ph.D. Dissertation: "Modeling and Simulation of Brian Tumors"
15. Patrick Benavidez, (Current at UTSA), 2015 Ph.D. Dissertation: "Mobile robot localization and navigation via visual SLAM"

2.3 Visiting Overseas Ph.D. Students

1. Ben Horan (PhD at Deakin University, Co-advisor
   Ph.D. completed in 2008, 6 months visit at UTSA, 2006)
2. Matthew Joordens (PhD at Deakin University, Co-advisor
   Ph.D., completed in 2010, 14 months visit at UTSA, 2008-2009)
3. Anjan Kumar Ray (PhD at Indian Institute of Technology- Kanpur, Co-advisor
   Ph.D. completed in 2009, 6 months visit at UTSA, 2008)
4. Sami Al-Abrabbuh (BS at King Fahd University of Petroleum and Minerals, Saudi Arabia, July – October 2008)
5. Luis Vega (MS at CIVESTA, Mexico, visit in Spring 2009, Co-advisor
   MS to be completed in 2010)
6. Aleksander Jevic (PhD at Universidad Polytechnic de Madrid) Co-advisor
   Ph.D. completed in 2011, 3 months visit at UTSA, Fall 2009
7. Miguel A P Garza, PhD at UANL, Mexico, Co-advisor to be completed in 2011, 3-months visit
8. C-H. Huang, PhD at National Central University, Taiwan, Co-advisor, Ph.D., current (April 2011 – February 2012)
9. Lydie Roine, B.S., ENSIG, France (Summer 2012)
10. Marjorie Tixier, B.S, Lille Polytechnic, France (Summer 2014)

3 Conclusions

My professional career, consisting of 45 years in academia, consulting, advisorship and industries, has been full of challenges and opportunities that I have met and used with a lot of enthusiasm and energy. However, nothing has been more enjoyable than an opportunity to impact the lives and careers of hundreds young graduate and undergraduate students from 24 nations of the world. To all those students this chapter and book is dedicated.

Proportional-Integral Observer in Robust Control, Fault Detection, and Decentralized Control of Dynamic Systems

Bahram Shafai[1] and Mehrdad Saif[2]

[1] Northeastern University, Boston MA 02115, USA
shafai@ece.neu.edu
[2] University of Windsor, Windsor ON N9B 3P4, Canada
msaif@uwindsor.ca

Abstract. This chapter initially reviews observer theory as it was developed over the past few decades. The state observer and its order reduction including functional observer in connection to state feedback control design are briefly discussed. The robustness of observer-based controller design is also explored. The loss of robustness due to the inclusion of observer in optimal linear quadratic regulator (LQR) and its recovery procedure (LTR) are summarized. The subsequent development of new observer structures such as disturbance observer (DO), unknown input observer (UIO), and proportional-integral observer (PIO) for disturbance estimation and fault detection is highlighted. Throughout the chapter we concentrate mainly on important advantages of PI-observer. Finally, we consider the problem of designing a decentralized PI observer with prescribed degree of convergence for a set of interconnected systems. Under the assumption of linear interactions, we provide a direct design procedure for the PI observer which can effectively be used in disturbance estimation and observer-based control design enhancing the robustness properties. In this connection we also extend the results to the case of designing controllers that attenuate the disturbance while preserving the stability. It is shown that the design can be formulated in terms of LMI which efficiently solve the problem.

1 Introduction

It is well-known that control of a dynamical systems represented by state-space description has resulted in several distinct design techniques. Among them the state feedback control design has been used extensively to stabilize an unstable system and improve its performance.

Many of the approaches, such as pole-shifting or eigenvalue assignment and optimal control using a quadratic performance criterion known as linear quadratic regulator (LQR), assume that the system states are all available for measurement and feedback purposes. This assumption often does not hold in practice either because the state variables are not accessible for direct measurement or because the number of measurable devices is limited.

A. El-Osery and J. Prevost (eds.), *Control and Systems Engineering*,
Studies in Systems, Decision and Control 27, DOI: 10.1007/978-3-319-14636-2_2

Due to these facts, a dynamical system known as state estimator or state observer can be designed driven by the available input and output of the system in order to produce the estimates of the states.

The observer was first introduced by Luenberger in 1964 [1], who also reported its extension in [2], and since then it became a major component of not only feedback control design but also a vehicle for disturbance estimation and fault detection. The development of observer theory was rapidly evolved according to system classification (linear, nonlinear, time invariant or time-varying, etc.) observer orders, different types of observer structures (proportional, proportional-integral, sliding mode, adaptive, etc.) or considered for special class of systems (delay, positive, singular or descriptor, large-scale, distributed, etc.).

Observers were also classified based on the goals of the designer or the requirements imposed by the design objectives. The first and the most important use of observer was to realize observer-based controller design, in which a state feedback control law had to be implemented when the states of the system were not accessible. A subsequent development of this was the requirement of making the observer optimal and robust due to inevitable presence of uncertainty in the system. The application of observers was also evident in the disturbance estimation, and fault detection and isolation (FDI). Consequently, disturbance observers (DO), unknown input observers (UIO), and proportional-integral Observers (PIO) were extensively used in different scenarios with proven advantages. The reference list at the end of the chapter provides a subset of the contributions made in the development of observer theory and its applications. We will refer to the relevant ones pertinent to the subjects covered in the following sections.

This chapter starts with fundamental results on observer theory. The most important observer structures are defined and design procedures are provided in Section 2. The robustness of observer-based control design using proportional observer (PO) and proportional integral observer (PIO) are discussed in Section 3. In Section 4, the connection of PIO with other types of observers such as DO and UIO are given with application to disturbance estimation and fault detection. The design of PI observer in decentralized control of interconnected large-scale systems is provided in Sections 5 and 6. Finally, concluding remarks are given in Section 7.

2 Observer Fundamentals

In this section, we provide the most important results in observer theory. The state and functional observers and the associated design procedures are discussed based on [3–6].

2.1 Full-Order Observer and State Feedback

Consider the n-dimensional linear time-invariant system

$$\dot{x} = Ax + Bu \tag{1}$$

$$y = Cx \tag{2}$$

where the state $x \in \mathbb{R}^n$, the input $u \in \mathbb{R}^m$, and the output $y \in \mathbb{R}^p$ with the usual assumption of (A, B) controllable and (A, C) observable.

A Luenberger state observer to estimate the states of the above system is defined by

$$\dot{\hat{x}} = A\hat{x} + Bu + L(y - C\hat{x}) \tag{3}$$

or equivalently

$$\dot{\hat{x}} = (A - LC)\hat{x} + Ly + Bu \tag{4}$$

where L is the Luenberger gain matrix and it needs to be designed such that the observer becomes stable.

If one defines the error vector $e = x - \hat{x}$, then the error dynamics is governed by

$$\dot{e} = (A - LC)e \tag{5}$$

and $\lim_{t\to\infty} e(t) = 0$ provided that the matrix $A - LC$ is stable.

Since the pair (A, C) is observable, it is always possible to find L such that all eigenvalues of $A - LC$ are arbitrarily assigned in the left half of complex plane. There are various methods to achieve this goal by two distinct approaches. One class of methods requires the transformation of the pair (A, C) to a particular design coordinate to facilitate the design procedure and the other class is based on certain direct design procedure without requiring any coordinate transformation on the original system description. These two approaches have also been applied for the design of state feedback control law.

$$u = r + Kx \tag{6}$$

to make the closed-loop system

$$\dot{x} = (A + BK)x + Bv \tag{7}$$

stable with desired performance. In this case, one can find the gain matrix K, since the pair (A, B) is assumed controllable. Note that, due to duality principle, one can apply any state feedback design method for the design of observer as well. Indeed, if $\{A, C\}$ is observable, the pair $\{A^T, C^T\}$ becomes controllable and one can write $\lambda(A - LC) = \lambda[A^T + C^T(-L)^T]$. Now, by defining $\tilde{A} = A^T$, $\tilde{B} = C^T$, and $\tilde{K} = -L^T$, the eigenvalue assignment with respect to $A - LC$ is reduced to the one for $\tilde{A} + \tilde{B}\tilde{K}$.

It is clear that the dimension of this observer is n, which is the same as the dimension of the system and it is called full-order proportional observer (P-observer). The structure of this observer along with the stale feedback implementation is shown in Figure 1.

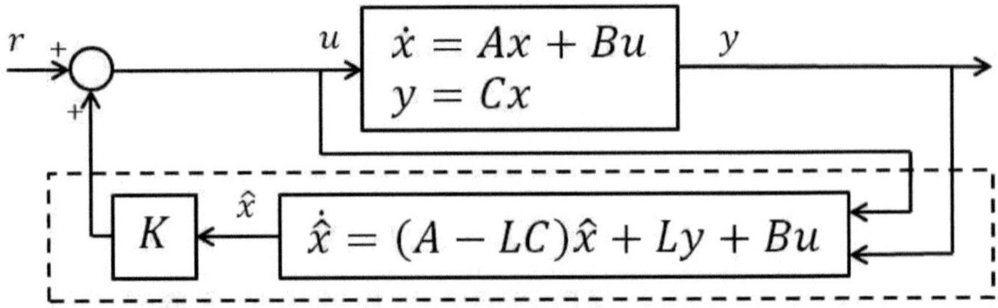

Fig. 1. P-Observer-based state feedback control system

To guarantee that the overall observer-based controller is stable, one combines (1)-(4) with $u = v + K\hat{x}$

$$\begin{bmatrix} \dot{x} \\ \dot{\hat{x}} \end{bmatrix} = \begin{bmatrix} A & BK \\ LC & A - LC + BK \end{bmatrix} \begin{bmatrix} x \\ \hat{x} \end{bmatrix} + \begin{bmatrix} B \\ B \end{bmatrix} r \tag{8}$$

$$y = \begin{bmatrix} C \, 0 \end{bmatrix} \begin{bmatrix} x \\ \hat{x} \end{bmatrix}$$

and by using $e = x - \hat{x}$, the stability status of (8) can equally be checked by the stability of

$$\begin{bmatrix} \dot{x} \\ \dot{e} \end{bmatrix} = \begin{bmatrix} A + BK & -BK \\ 0 & A - LC \end{bmatrix} \begin{bmatrix} x \\ e \end{bmatrix} + \begin{bmatrix} B \\ 0 \end{bmatrix} r \tag{9}$$

which is clear since $A + BK$ and $A - LC$ are stability matrices.

Thus, the union of state feedback and observer eigenvalues are the eigenvalues of the closed-loop system leading to the so-called separation property [2].

2.2 Reduced-Order Observer

In most cases the matrix C in the output equation is of full row rank p and by a similarity transformation $\bar{x} = Px$ one can transform the system (1), (2) to an output identifiable form written as

$$\begin{bmatrix} \dot{\bar{x}}_1 \\ \dot{\bar{x}}_2 \end{bmatrix} = \begin{bmatrix} \bar{A}_{11} & \bar{A}_{12} \\ \bar{A}_{21} & \bar{A}_{22} \end{bmatrix} \begin{bmatrix} \bar{x}_1 \\ \bar{x}_2 \end{bmatrix} + \begin{bmatrix} \bar{B}_1 \\ \bar{B}_2 \end{bmatrix} u \tag{10}$$

$$y = \begin{bmatrix} I_p \, 0 \end{bmatrix} \begin{bmatrix} \bar{x}_1 \\ \bar{x}_2 \end{bmatrix} \tag{11}$$

The transformation matrix P is easily constructed by $P = \begin{bmatrix} C^T \; R^T \end{bmatrix}^T$, where R is an $(n-p) \times n$ arbitrary matrix making P nonsingular. It can also be shown that $\{A, C\}$ is observable if and only if $\{\bar{A}_{22}, \bar{A}_{12}\}$ is observable. Now, with the aid of $y = \bar{x}_1$, one can write (10), (11) as a reduced order system

$$\dot{\bar{x}}_2 = \bar{A}_{22}\bar{x}_2 + \tilde{B}\bar{u} \tag{12}$$

$$\bar{y} = \bar{A}_{12}\bar{x}_2 \tag{13}$$

where $\tilde{B} = \begin{bmatrix} \bar{A}_{21} & \bar{B}_2 \end{bmatrix}$, $\bar{u} = \begin{bmatrix} y \\ u \end{bmatrix}$ and $\bar{y} = \dot{y} - \bar{A}_{11}y - \bar{B}_1 u$. Since $\{\bar{A}_{22}, \bar{A}_{12}\}$ is observable, one can construct an observer of order $n - p$ for (12), (13)

$$\dot{\hat{\bar{x}}}_2 = (\bar{A}_{22} - \bar{L}\bar{A}_{12})\hat{\bar{x}}_2 + \bar{L}\bar{y} + \tilde{B}\bar{u} \tag{14}$$

such that the eigenvalues of $\bar{A}_{22} - \bar{L}\bar{A}_{12}$ are arbitrarily assigned by a proper choice of $\bar{L}$.

To avoid the derivative term of the output in $\bar{y}$, we introduce a new variable $\bar{z} = \hat{\bar{x}}_2 - \bar{L}y$, which reduces (14) to

$$\dot{\bar{z}} = (\bar{A}_{22} - \bar{L}\bar{A}_{12})\bar{z} + [(\bar{A}_{22} - \bar{L}\bar{A}_{12})\bar{L} + (\bar{A}_{21} - \bar{L}\bar{A}_{11})]y + (\bar{B}_2 - \bar{L}\bar{B}_1)u \tag{15}$$

and the state estimate $\hat{x}$ is obtained from

$$\hat{x} = P^{-1}\begin{bmatrix} I_p & 0 \\ \bar{L} & I_{n-p} \end{bmatrix}\begin{bmatrix} y \\ \bar{z} \end{bmatrix} \tag{16}$$

The above derivation of (15) and (16) suggest that the general reduced-order observer may also be defined by

$$\dot{z} = Fz + Gy + Hu \tag{17}$$

$$\hat{x} = Mz + Ny \tag{18}$$

where the state $z \in \mathbb{R}^{(n-p)\times(n-p)}$ of the reduced-order observer is related to the state x of the system by the observer reconstruction error vector

$$\epsilon(t) = z(t) - Tx(t) \tag{19}$$

where $T \in \mathbb{R}^{(n-p)\times n}$.

Taking the derivative of (19) and using (1), (2), (17), (18) yields

$$\dot{\epsilon}(t) = F\epsilon(t) + (FT - AT + GC)x + (H - TB)u \tag{20}$$

Thus, the state of an observable system (1), (2) can be estimated with an $n - p$ dimentional observer described by (17), (18) in the sense that $\lim_{t\to\infty} \epsilon(t) = 0$ for any $x(0)$, $z(0)$, and $u(t)$ if and only if

$$F \in \mathbb{R}^{(n-p)\times(n-p)} \quad \text{Hurwitz stable matrix} \tag{21}$$

$$TA - FT = GC \tag{22}$$

$$MT + NC = I \tag{23}$$

$$H - TB = 0 \tag{24}$$

It is not difficult to show that $\epsilon(t)$ is directly related to $e(t)$ and $\lim_{t\to\infty} \epsilon(t) = 0$ implies $\lim_{t\to\infty} e(t) = 0$.

The design of the observer requires to solve (22) for T under the stability condition of F. There are many methods to solve (22). One simple approach is to consider the system to be in output identifiable form with the partitioned pair $\{A, C\}$ as

$$A = \begin{bmatrix} A_{11} & A_{12} \\ A_{21} & A_{22} \end{bmatrix}, \quad C = \begin{bmatrix} I & 0 \end{bmatrix} \tag{25}$$

Using (23), one possible structure for matrix T is given by

$$T = \begin{bmatrix} L & I \end{bmatrix} \tag{26}$$

and equation (22) reduces to

$$A_{22} + LA_{12} = F \tag{27}$$

$$A_{21} + LA_{11} - FL = G \tag{28}$$

Since the pair $\{A_{22}, A_{12}\}$ is observable, one can use any technique to obtain L such that $A_{22} + LA_{12}$ becomes Hurwitz stable. Consequently, T is determined and the observer parameter matrices F, G, H are easily computed. The matrices M and N are also obtained by

$$\begin{bmatrix} M & N \end{bmatrix} = \begin{bmatrix} T \\ C \end{bmatrix}^{-1} \tag{29}$$

Similarly, it is possible to show that the reduced-order observer integrated with the state feedback control admit the separation property.

2.3 Functional Observer

In certain cases, it is not necessary to estimate all state variables, rather to estimate some functions of the states. A generic situation is to estimate directly the control law $u = Kx$. This avoids two separate steps of estimating x by $\hat{x}$, and then using state feedback control $u = K\hat{x}$ for implementation. It has been recognized that the order r of this so-called functional observer may be reduced considerably ($r < n - p$).

Depending on estimating one scalar function of the states or a vector function of the states, we refer them as a single-functional observer or a multi-functional observer, respectively. The functional observer can be designed with desired eigenvalues (fixed) or stable eigenvalues (free) [7],[8].

A. Fixed Design of Functional Observer:

Let us consider a system represented in output identifiable form with the pair $\{A, C\}$ defined by (25). We provide a method of designing a single functional observer with desired eigenvalues to estimate $w = kx$, where k is a given $1 \times n$

constant vector. The vector k may have resulted from a state feedback design or an arbitrary linear function of the states.

The functional observer is chosen with the following structure

$$\dot{z} = Fz + Gy + Hu \tag{30}$$

$$\hat{w} = Mz + Ny \tag{31}$$

where $F \in \mathbb{R}^{r\times r}$, $G \in \mathbb{R}^{r\times p}$, $H \in \mathbb{R}^{r\times m}$, $M \in \mathbb{R}^{1\times r}$, and $N \in \mathbb{R}^{1\times p}$.

The problem is to design such an observer with lowest possible order such that $\hat{w}$ approaches $w = kx$ in the sense of $\lim_{t\to\infty} \hat{w}(t) = w$. Following the same steps as the design of a reduced-order observer, it is straightforward to arrive at similar conditions for the existence of a functional observer, i.e.

$$F \in \mathbb{R}^{r\times r} \quad \text{Hurwitz stable matrix} \tag{32}$$

$$TA - FT = GC \tag{33}$$

$$MT + NC = k \tag{34}$$

$$H = TB \tag{35}$$

Now partitioning $T \in \mathbb{R}^{r\times n}$ as $\begin{bmatrix} T_1 \; T_2 \end{bmatrix}$, where $T_1 \in \mathbb{R}^{r\times p}$ and $T_2 \in \mathbb{R}^{r\times(n-p)}$, and substituting it along with partitioned pair $\{A, C\}$ defined by (25) in (33) leads to

$$T_1A_{11} + T_2A_{21} - FT_1 = G \tag{36}$$

$$T_1A_{12} + T_2A_{22} - FT_2 = 0 \tag{37}$$

If we also partition k as $k = \begin{bmatrix} k_1 \; k_2 \end{bmatrix}$, where $k_1 \in \mathbb{R}^{1\times p}$ and $k_2 \in \mathbb{R}^{1\times(n-p)}$, then (34) becomes

$$MT_1 + N = k_1 \tag{38}$$

$$MT_2 = k_2 \tag{39}$$

Next we choose an arbitrary observable pair $\{F, M\}$ to be defined by

$$F = \begin{bmatrix} 0 & 1 & 0 & \dots & 0 \\ 0 & 0 & 1 & \dots & 0 \\ \vdots & \vdots & \vdots & & \vdots \\ 0 & 0 & 0 & \dots & 1 \\ -\alpha_r & -\alpha_{r-1} & -\alpha_{r-2} & \dots & -\alpha_1 \end{bmatrix}, \quad M = \begin{bmatrix} 1\,0\,0 \dots 0 \end{bmatrix} \tag{40}$$

where F is a stable matrix with desired eigenvalues.

We show that by choosing $r = \nu - 1$, where ν is the observability index of the pair $\{A, C\}$, a functional observer of order $r = \nu - 1$ exists. Let $t_{ij};\ j = 1, 2, \dots, r$ be the $j-$th row of $T_i;\ i = 1, 2$. Then using (40), one can expand (37) and (39) as follows

$$\begin{cases} t_{21} = k_2 \\ t_{11}A_{12} + t_{21}A_{22} = t_{22} \\ \vdots \\ t_{1r}A_{12} + t_{2r}A_{22} = -\alpha_r t_{21} - \alpha_{r-1}t_{22} - \dots - \alpha_1 t_{2r} \end{cases} \tag{41}$$

These equations, except the last one, can express t_{2j}; $j = 1, 2, \ldots, r$ in terms of t_{1j} and substituting them in the last equation yields

$$\begin{bmatrix} t_{1r} \; t_{1,r-1} \; \ldots \; t_{12} \; t_{11} \end{bmatrix} \begin{bmatrix} I & 0 & \ldots & 0 & 0 \\ \alpha_1 I & I & \ldots & 0 & 0 \\ \alpha_2 I & \alpha_1 I & \ldots & 0 & 0 \\ \vdots & \vdots & & \vdots & \vdots \\ \alpha_{r-1} I & \alpha_{r-2} I & \ldots & \alpha_1 I & I \end{bmatrix} \begin{bmatrix} A_{12} \\ A_{12}A_{22} \\ \vdots \\ A_{12}A_{22}^{r-1} \end{bmatrix} = -\tilde{k} \tag{42}$$

where $\tilde{k} = [k_2 A_{22}^r + \alpha_1 k_2 A_{22}^{r-1} + \cdots + \alpha_{r-1} k_2 A_{22} + \alpha_r k_2]$. Since $\{A_{22}, A_{12}\}$ is an observable pair with the observability index $\nu - 1$, $\left[A_{12}^T \; (A_{12}A_{22})^T \; \ldots \; (A_{12}A_{22}^{r-1})^T \right]^T$ has rank $n - p$, and for any k and α_i; $i = 1, \ldots, r-1$ with $r = \nu - 1$, a solution for T_1 exists in (42).

Once T_1 is known with t_{1j}; $j = 1, \ldots, r$, one can compute t_{2j}; for all j from (41), which specifies T_2 and the remaining observer parameter matrices G, H, and N can be obtained from (36), (35), and (38). It should be pointed out that for a properly chosen set of eigenvalues and depending on the specified functional, it is possible to have other solutions for (42) leading to functional observers of lower order $r < \nu - 1$. It is also clear that for estimating a vector function of the states, where $W \in \mathbb{R}^{l \times 1}$, one can design multi-functional observer of order $l(\nu - 1)$ using the above technique. There exists other design techniques with additional assumption to further reduce the order, which have been reported in the literature (see [6] ,[7–12] and the references therein.).

B. Free Design of Functional Observer:

For the free design problem it is required to determine the lowest possible order functional observer (possibly minimal order) such that the constraint equations (32) - (35) are satisfied with the sole restriction that the observer eigenvalues lie in the open left half of complex plane without prespecifying them. Note that the constraint equation (32) - (35) can be written in exactly the same way for general multi-functional observer problem. In this case $W = Kx$, where $K \in \mathbb{R}^{l \times n}$ and the remaining matrices have appropriate dimensions. An equivalent problem is to find the coefficient matrices of the observer and the structure matrix T such that $z(t)$ approximates $Tx(t)$ and from (34)

$$K^T \in R(\Gamma^T) \tag{43}$$

where $\Gamma = \left[T^T \; C^T \right]^T$. An elegant approach, which solves even the multi-functional observer problem, is given here for single-output system. Later on suggestions are made to generalize the approach for multi-output systems as well.

Let us assume that the system is in observable canonical form with the pair $\{A, C\}$ defined by

$$A = \begin{bmatrix} 0\,0 \cdots & -\alpha_0 \\ 1\,0 \cdots & -\alpha_1 \\ 0\,1 \cdots & -\alpha_2 \\ \vdots\,\vdots & \vdots \\ 0\,0 \cdots & -\alpha_{n-1} \end{bmatrix}, C = \begin{bmatrix} 0\,0 \cdots\, 1 \end{bmatrix} \tag{44}$$

If the original system is not in this form, then one needs to transform it to this form by the well-known similarity transformation. The characteristic polynomial of A is given by

$$\Delta(\lambda) = \lambda^n + \alpha_{n-1}\lambda^{n-1} + \ldots + \alpha_1\lambda + \alpha_0 \tag{45}$$

In this situation the condition (33) becomes

$$TA - FT = gC \tag{46}$$

where $T \in \mathbb{R}^{r \times n}$, $g \in \mathbb{R}^{r \times 1}$, and r is the minimal order yet to be determined. Writing T in terms of its column

$$T = \begin{bmatrix} t_1\; t_2 \cdots\, t_n \end{bmatrix} \tag{47}$$

we get the following equations by substituting (44), and (47) into (46)

$$t_i = F^{i-1}t_1 \qquad i = 1, 2, \ldots, n \tag{48}$$

$$g = -\Delta(F)t_1 \tag{49}$$

Thus, T has the general form

$$T = \begin{bmatrix} t_1\; Ft_1 \cdots\, F^{n-1}t_1 \end{bmatrix} \tag{50}$$

and

$$\Gamma = \begin{bmatrix} t_1 & Ft_1 & \cdots & F^{n-1}t_1 \\ 0 & 0 & \cdots & 1 \end{bmatrix} \tag{51}$$

It is clear that the matrix $\Gamma \in \mathbb{R}^{(r+1) \times n}$ has full rank $r+1$ if and only if the pair $\{F, t_1\}$ is controllable. Therefore the functional observer problem is reduced in finding t_1 and F such that $K^T \in R(\Gamma^T)$. Now, suppose for some r and t_1 the matrix Γ in (51) has full rank $r+1$. Then it can be shown that $K^T \in R(\Gamma^T)$ if and only if

$$S_\beta^r K^T = 0 \tag{52}$$

where the $(n+r-1) \times n$ matrix S_β^r is defined by

$$S_\beta^r = \begin{bmatrix} \beta_0 & \beta_1 & \beta_2 & \cdots & \beta_{r-1} & 1 & 0 & 0 & \cdots & 0 \\ 0 & \beta_0 & \beta_1 & \cdots & \beta_{r-2} & \beta_{r-1} & 1 & 0 & \cdots & 0 \\ \vdots & \vdots & \vdots & & \vdots & \vdots & & & & \vdots \\ 0 & 0 & 0 & \cdots & \beta_0 & \beta_1 & \cdots & \beta_{r-1} & 1 & 0 \end{bmatrix} \tag{53}$$

for $r = 1, 2, \ldots, n-2$ and β_i's are the coefficients of the characteristic polynomial of F i.e.

$$\beta(\lambda) = \lambda^r + \beta_{r-1}\lambda^{r-1} + \ldots + \beta_1\lambda + \beta_0 \tag{54}$$

Note that $S_\beta^0 = \begin{bmatrix} I_{n-1} \, 0 \end{bmatrix}$ and $S_\beta^{n-1} = 0$. Thus, if $\beta(\lambda)$ is stable then there exists a functional observer of dimension r for the single-output system (44) with $\hat{w}(t)$ estimating $w(t) = Kx(t)$ if and only if (52) is satisfied. To obtain the minimal order functional observer, one should starts with $r = 1$ and increase r until the condition (52) is satisfied.

For multi-output case, one has several options. One option is to reduce the system to p coupled single-output systems and apply the same procedure. Another option is to create a linear combination of the output $y_n = [\sum \eta_i c^i]x = c_n x$ so that $\{A, c_n\}$ is an observable pair and then apply the same procedure. Other direct methods have also been reported in the literature (see [6] ,[7–12] and the references therein).

Before closing this section, we should state that the introduction of balanced realization and model reduction techniques caused less interest in trying to find minimal order functional observer design. A major development of observer theory was devoted to extend the above techniques [3–6] and apply them to various systems based on their classification (Time-varying, Nonlinear, etc. [13–20]) or other special class of systems (Delay, Singular, Positive, etc. [21–30]). Also, numerical consideration for both state feedback and observer design was an important factor specially when dealing with large size systems (see [31–34]).

3 Robustness with Observers

3.1 P-Observer and Loop Transfer Recovery

The term robustness refers to the preservation of desirable system properties, mainly stability and performance under structured and unstructured uncertainties. The design of state feedback through optimal control was found to have desirable robustness properties. To be more specific, seeking control law for $u(t)$ in (1) such that the closed-loop system is stable and minimizes the quadratic performance criterion

$$J = \int_0^\infty \left(x^T Q x + u^T R u\right) dt \tag{55}$$

leads to the following result

$$u(t) = Kx(t) \quad , \quad K = -R^{-1}B^T P \tag{56}$$

where P is the solution of the algebraic Riccati equation (ARE)

$$A^T P + PA - PBR^{-1}B^T P + Q = 0 \tag{57}$$

The above state feedback control law provides impressive robustness in terms of gain and phase margins and cope with parametric and other types of uncertainties. As it has been recognized for many years, these robustness properties are lost when an observer is incorporated in the feedback loop. The classical papers [35], [36] showed that, for robust recovery, the observer dynamics (4) should satisfy the following condition

$$L\left[I + C(sI - A)^{-1}L\right]^{-1} = B\left[C(sI - A)^{-1}B\right]^{-1} \tag{58}$$

Observers that satisfy the above condition are called "Robust Observers" and when combined with state feedback control law, we refer the overall design as "Robust Observer-Based Control Design".

In the paper [35] a robust recovery procedure was suggested by selecting the following Kalman-type filter gain

$$L(q) = P(q)C^T R^{-1} \tag{59}$$

where $P(q)$ satisfies the ARE

$$AP + PA^T - PC^T R^{-1}CP + Q(q) = 0 \tag{60}$$

for which

$$\begin{aligned} Q(q) &= Q_0 + q^2 BVB^T, V > 0 \\ R &= R_0 \end{aligned}$$

where Q_0 and R_0 are noise covariance matrices.

This parametrized gain selection in terms of q guarantees that (58) to be satisfied asymptotically as $q \to \infty$. This approach was further investigated and led to a design procedure known as LQG/LTR [37]. Due to limitation of increasing gain and solving parametrized ARE as well as numerical issues, other approaches based on reduced-order observer and proportional integral observer have been proposed (see [38], [41] and the references therein). Simply stated, the loss of robustness in observer-based controller design is due to the signal transfer from the control u to the observer through the control distribution matrix. To eliminate this effect in the reduced-order observer case, the matrix $H = TB$ in (17) must be zero or equivalently with the aid of (26) L must satisfy $B_2 - LB_1 = 0$. Thus, for square systems and under the non-singularity assumption of B_1, we have $L = B_2 B_1^{-1}$. Since $C = \left[I\ 0\right]$, the non-singularity assumption is equivalent to $\det CB \neq 0$. It can be shown that with this selection of L, the poles of the robust observer coincide with transmission zeros of the plant. Thus, for a minimum phase system, the exact recovery can be realized. When the system is not minimum phase, then exact recovery is not possible and one can only achieve recovery asymptotically. A comprehensive treatment of LTR can be found in [38–42].

3.2 PI-Observer and Robustness

Consider again the system (1),(2) with usual assumption of $\{A, B\}$ controllable and $\{A, C\}$ observable. A Proportional-Integral observer (PI-observer) is defined by

$$\dot{\hat{x}} = (A - L_P C)\hat{x} + L_P y + Bu + Bv \tag{61}$$
$$\dot{v} = L_I(y - C\hat{x}) \tag{62}$$

Figure 2 shows this observer implemented with the state feedback control law $u = r + K\hat{x}$.

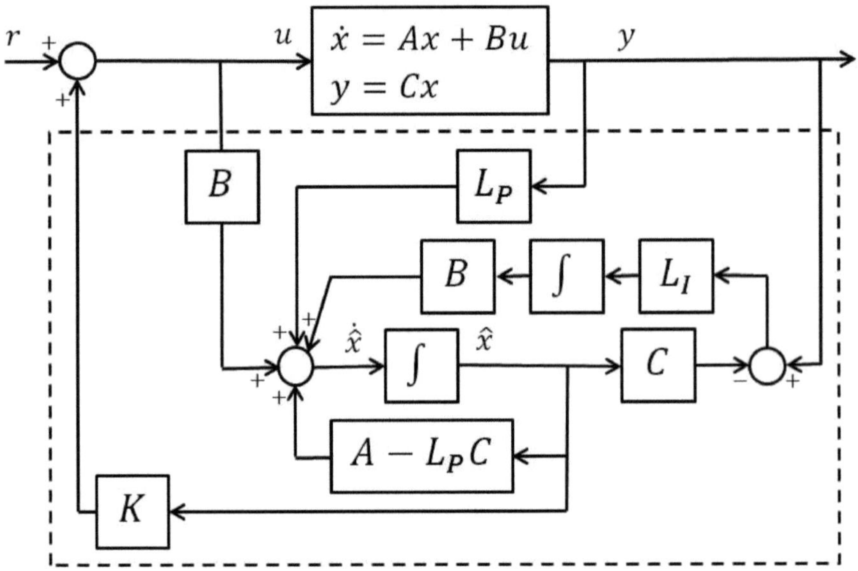

Fig. 2. PI-Observer-based state feedback control system

The PI-observer can be written in augmented form by defining $z = \begin{bmatrix} \hat{x}^T & v^T \end{bmatrix}^T$ as

$$\dot{z} = (A_X - L_X C_X)z + L_X y + B_X u \tag{63}$$

where

$$A_X = \begin{bmatrix} A & B \\ 0 & 0 \end{bmatrix}, B_X = \begin{bmatrix} B \\ 0 \end{bmatrix}, C_X = \begin{bmatrix} C & 0 \end{bmatrix}, L_X = \begin{bmatrix} L_P \\ L_I \end{bmatrix}$$

and the adjusted control law $u = r + K_X\hat{x}$ with $K_X = \begin{bmatrix} K & 0 \end{bmatrix}$. Since $\{A_X, C_X\}$ is an observable pair, any eigenvalue assignment techniques can be employed to find L_X such that $A_X - L_X C_X$ is stable.

Combining observer and state feedback, we get

$$\begin{bmatrix} \dot{x} \\ \dot{\hat{x}} \\ \dot{v} \end{bmatrix} = \begin{bmatrix} A & BK & 0 \\ L_P C & A - L_P C + BK & B \\ L_I C & -L_I C & 0 \end{bmatrix} \begin{bmatrix} x \\ \hat{x} \\ v \end{bmatrix} + \begin{bmatrix} B \\ B \\ 0 \end{bmatrix} r \tag{64}$$

Defining the state estimation error by $e(t) = \hat{x}(t) - x(t)$, one can define the transformation matrix

$$P = \begin{bmatrix} I & 0 & 0 \\ -I & I & 0 \\ 0 & 0 & I \end{bmatrix}$$

and transform (64) to

$$\begin{bmatrix} \dot{x} \\ \dot{e} \\ \dot{v} \end{bmatrix} = \begin{bmatrix} A + BK & BK & 0 \\ 0 & A - L_P C & B \\ 0 & -L_I C & 0 \end{bmatrix} \begin{bmatrix} x \\ e \\ v \end{bmatrix} + \begin{bmatrix} B \\ 0 \\ 0 \end{bmatrix} r \tag{65}$$

which establishes the separation property of PI-observer-based state feedback control system.

One advantage of PI-observer in robust control is evident due to additional integral path. This makes it possible to have extra degrees of freedom to choose observer gains, which can improve robustness. Another advantage of LTR design as compared to P-observer case is that both exact and approximate LTR can be achieved with relatively low PI-observer gains. A slight generalization of PI-observer structure is also possible, which allows additional degrees of freedom. It is called Proportional-Integral Fading observer (PIF), which discounts the integral term over time and improves transient response. The structure of PIF is defined as follows

$$\begin{aligned} \dot{\hat{x}} &= (A - L_P C)\hat{x} + L_P y + Bu + Ev \\ \dot{v} &= Fv + L_I(y - C\hat{x}) \end{aligned} \tag{66}$$

where PIF consists of four design parameters L_P, L_I, E, and F. Note that in general $v \in \mathbb{R}^{l \times 1}$, however we assume for simplicity $l = m$ as in the case of (61) or $l = p$ depending on different scenarios. In any event for the PI observer to exist, it is required that the number of integrators to be less than or equal to the number of outputs, i.e. $l \leq p$. In this case the augmented form of the PIF becomes

$$\dot{z} = (A_Z - L_Z C_Z)z + L_Z y + B_Z u \tag{67}$$

where

$$A_Z = \begin{bmatrix} A & E \\ 0 & F \end{bmatrix}, B_Z = \begin{bmatrix} B \\ 0 \end{bmatrix}, C_Z = \begin{bmatrix} C & 0 \end{bmatrix}, L_Z = \begin{bmatrix} L_P \\ L_I \end{bmatrix}$$

Again one can use any eigenvalue assignment technique to obtain L_Z, provided that the design parameters E, F are chosen such that

$$rank \begin{bmatrix} A & E \\ 0 & F \\ C & 0 \end{bmatrix} = n + l$$

It is also possible to derive a reduced-order PI-observer, which has an interesting connection to exact LTR condition of P-observer case (see [41] for more details). It is shown that the reduced-order PI-observer achieves exact LTR if and only if its corresponding reduced-order P-observer achieves exact LTR. The condition of exact LTR for reduced-order PI-observer leads to

$$B_2 - L_P B_1 = 0 \tag{68}$$

$$L_I B_1 = 0 \tag{69}$$

which enforces $L_I = 0$ and (68) is exactly the same as reduced-order P-observer case. In the same paper, recovery matrices are defined for full- and reduced-order P- and PI-observers. For example, recovery matrices for full-order cases are defined as follows

$$M_P(s) = K(sI - A + L_P C)^{-1} B \tag{70}$$

$$M_{PI}(s) = K_X(sI - A_X + L_X C_X)^{-1} B_X \tag{71}$$

The condition of exact recovery $M_P(s) = 0$ or $M_{PI}(s) = 0$ cannot always be satisfied. Consequently, for approximate or asymptotic recovery, the recovery matrices should be made small in some sense. If the controller is parameterized in terms of the observer gain by $L_P(q)$ then to obtain asymptotic LTR we seek an $L_P(q)$ such that for all ω

$$M_P(\jmath\omega) = K(\jmath\omega I - A + L_P(q)C)^{-1} B \to 0 \quad \text{as} \quad q \to \infty$$

for P-observer and similarly for PI-observer case $M_{PI}(\jmath\omega) \to 0$ as $q \to \infty$. A more depth analysis of LTR can be found in [41].

4 Disturbance Estimation and Fault Detection with Observers

As pointed out in the introduction, Disturbance Observer (DO), Unknown Input Observer (UIO) and Proportional-Integral Observer (PIO) have found proven advantages in disturbance estimation and fault detection [43–53]. In this section, we give a brief overview of DO, UIO, and elaborate more on PIO for the purpose of its applicability when disturbance and fault are considered.

4.1 Disturbance Observer (DO)

First consider a system with constant disturbance

$$\begin{cases} \dot{x} = Ax + Bu + Ed \\ \dot{d} = 0 \\ y = Cx \end{cases} \tag{72}$$

The above system can be written in augmented form as

$$\begin{bmatrix} \dot{x} \\ \dot{d} \end{bmatrix} = \begin{bmatrix} A & E \\ 0 & 0 \end{bmatrix} \begin{bmatrix} x \\ d \end{bmatrix} + \begin{bmatrix} B \\ 0 \end{bmatrix} u$$

$$y = Cx \tag{73}$$

If we denote

$$z = \begin{bmatrix} x \\ d \end{bmatrix}, A_D = \begin{bmatrix} A & E \\ 0 & 0 \end{bmatrix}, B_D = \begin{bmatrix} B \\ 0 \end{bmatrix}, C_D = \begin{bmatrix} C & 0 \end{bmatrix} \tag{74}$$

then we have

$$\begin{aligned} \dot{z} &= A_D z + B_D u \\ y &= C_D z \end{aligned} \tag{75}$$

A full-order observer can be constructed as

$$\dot{\hat{z}} = (A_D - L_D C_D)\hat{z} + L_D y + B_D u \tag{76}$$

and the methods of Section 2 can be employed to obtain the estimates of both states and the constant disturbances. Next, in case of disturbance model $\dot{d} = Md$, one can apply the same procedure with a minor modification of replacing A_D by $A_D = \begin{bmatrix} A & E \\ 0 & M \end{bmatrix}$.

4.2 Unknown Input Observer (UIO)

The UIO is capable of estimating the states in such a way that disturbance is decoupled in the estimation process. However, it is possible to attain the disturbance afterwards. The structure of a full-order UIO is given by

$$\begin{aligned} \dot{z} &= Fz + Gy + Hu \\ \hat{x} &= Mz + Ny \end{aligned} \tag{77}$$

where $M = I$ and $H = TB$ are assumed. By defining $G = G_1 + G_2$, the estimation error e can be derived by using (77) and the state equation (72) as

$$\begin{aligned} \dot{e} &= (A - NCA - G_1C)e + [F - (A - NCA - G_1C)]z \\ &\quad + [G_2 - (A - NCA - G_1C)N]y + [T - (I - NC)]Bu + (NC - I)Ed \end{aligned} \tag{78}$$

Now, if the following conditions hold

$$(NC - I)E = 0 \tag{79}$$

$$T = I - NC \tag{80}$$

$$F = A - NCA - G_1C \tag{81}$$

$$G_2 = FN \tag{82}$$

then

$$\dot{e} = Fe \tag{83}$$

and by imposing stability on F, $\lim_{t\to\infty} e(t) = 0$. The existence condition of the UIO depends on the solution of (79). This equation is solvable if and only if

$$rank(CE) = rank(E) \tag{84}$$

Therefore a left inverse of CE exists and

$$N = E(CE)^{\dagger} \tag{85}$$

Denoting $A - NCA$ by A_1, one can find $F = A_1 - G_1 C$ by eigenvalue assignment. The remaining observer parameters can be obtained by simple substitutions. Finally, taking the derivative of the output

$$\dot{y} = C\dot{x} = CAx + CBu + CEd \tag{86}$$

allows to obtain the disturbance estimate

$$\hat{d} = (CE)^{\dagger}[\dot{\hat{y}} - CA\hat{x} - CBu] \tag{87}$$

4.3 Proportional-Integral Observer (PIO)

In Section 3.2 we explored that PI-observer has advantages in improving robustness and robust recovery problem of observer-based controller design. It was shown that LTR can be achieved by relatively low PI-observer gains and both exact and approximate recovery were possible.

In this section we discuss PI-observer for the purpose of estimating both the original states and the unknown inputs. It will be shown that PI-observer with high gains is able to achieve this goal. Consider the system

$$\dot{x}(t) = Ax(t) + Bu(t) + Ed(x,t) + Ng(x,u,t) \tag{88}$$
$$y(t) = Cx(t) + h(t) \tag{89}$$

where $d(x,t)$ may arise from any nonlinearities such as friction in electro-mechanical system or any unknown input signal $d(t)$, $g(x,u,t)$ is the uncertainties including unmodeled dynamics and $h(t)$ is the measurement noise.

Case 1: Suppose uncertainties $g(x,u,t)$ and measurement noise $h(t)$ are eliminated from the model, and the disturbance model is known with its associated distribution matrix E. For simplicity let the disturbance be constant ($\dot{d} = 0$), then the following PI-observer

$$\begin{bmatrix} \dot{\hat{x}} \\ \dot{\hat{d}} \end{bmatrix} = \begin{bmatrix} A - L_P C & E \\ -L_I C & 0 \end{bmatrix} \begin{bmatrix} \hat{x} \\ \hat{d} \end{bmatrix} + \begin{bmatrix} L_P \\ L_I \end{bmatrix} y + \begin{bmatrix} B \\ 0 \end{bmatrix} u \tag{90}$$

will be identical to the disturbance observer (76) with $L_D = \begin{bmatrix} L_P \\ L_I \end{bmatrix}$. Similarly one can use PIF for the disturbance model of the form $\dot{d} = Md$.

Case 2: Suppose the disturbance is not known(unavailable) and $g(x,u,t)=0$, $h(t)=0$. Based on the well-known methods of disturbance rejection control (see [50] and the cited references), one can assume a fictitious system representing the unknown input by

$$\dot{v} = Mv \tag{91}$$

with the approximation $d(x,t) \approx Pv(t)$. Writing the above system in the extended form by

$$\begin{bmatrix} \dot{x} \\ \dot{v} \end{bmatrix} = \begin{bmatrix} A & EP \\ 0 & M \end{bmatrix} \begin{bmatrix} x \\ v \end{bmatrix} + \begin{bmatrix} B \\ 0 \end{bmatrix} u \tag{92}$$

$$y = \begin{bmatrix} C & 0 \end{bmatrix} \begin{bmatrix} x \\ v \end{bmatrix} \tag{93}$$

It has been shown that the choice of $M=0$ or $M \to 0$ and $P=I$, the estimation of unknown input is acceptable. This means that without exact knowledge about the dynamical behaviour of the unknown input d, a very general approach is possible by assuming the disturbance as piecewise constant. Thus, for all practical purposes, we design a high gain PI observer to estimate the unknown input as follows

$$\begin{bmatrix} \dot{\hat{x}} \\ \dot{\hat{d}} \end{bmatrix} = \begin{bmatrix} A & E \\ 0 & 0 \end{bmatrix} \begin{bmatrix} \hat{x} \\ \hat{d} \end{bmatrix} + \begin{bmatrix} B \\ 0 \end{bmatrix} u + \begin{bmatrix} L_P \\ L_I \end{bmatrix} (y(t) - \hat{y}(t)) \tag{94}$$

$$y = \begin{bmatrix} C & 0 \end{bmatrix} \begin{bmatrix} \hat{x} \\ \hat{d} \end{bmatrix} \tag{95}$$

Assuming that $d(x,t)$ is bounded, then there exists a high gain PI observer (94), (95) for the system with any initial conditions if $\{A,C\}$ is observable, $rank \begin{bmatrix} A & E \\ C & 0 \end{bmatrix} = n+l$, and $CA^iE=0$, $i=0,1,\ldots,\nu-2$, where ν is the observability index of $\{A,C\}$.

Case 3: Suppose $d(x,t) \neq 0$, $g(x,u,t) \neq 0$, and $h(t) \neq 0$. With the same assumptions for unmodeled dynamics $g(x,u,t)$ as for the unknown input $d(x,t)$ we have the following argument.

Defining the errors as $e(t) = \hat{x}(t) - x(t)$ and $\epsilon(t) = \hat{d}(t) - d(t)$, the error dynamics of the extended system becomes

$$\begin{bmatrix} \dot{e} \\ \dot{\epsilon} \end{bmatrix} = \begin{bmatrix} A - L_PC & E \\ -L_IC & 0 \end{bmatrix} \begin{bmatrix} e \\ \epsilon \end{bmatrix} - \begin{bmatrix} Ng(t) \\ \dot{d}(t) \end{bmatrix} + \begin{bmatrix} L_P \\ L_I \end{bmatrix} h(t) \tag{96}$$

The error dynamics can be described by

$$e(s) = G^{-1}(s)E\epsilon(s) - G^{-1}(s)Ng(s) + G^{-1}L_Ph(s) \tag{97}$$

$$\begin{aligned} \epsilon(s) = &-[sI + L_ICG^{-1}(s)E]^{-1}sd(s) + [sI + L_ICG^{-1}(s)E]^{-1}L_ICG^{-1}(s)Ng(s) \\ &+[sI + L_ICG^{-1}(s)E]^{-1}L_I(I - CG^{-1}(s)L_P)h(s) \end{aligned} \tag{98}$$

where

$$G(s) = [sI - (A - L_PC)]$$

The gains L_P and L_I are required to stabilize the extended system (96) and also to make sure to minimize the influence of unknown disturbance on the estimation errors $e(t)$ and $\epsilon(t)$. To achieve this, we require all eigenvalues of the system matrix in (96) to be in the left half of complex plane and $\|L_I\| \gg \|L_P\|$. From (98), it can be seen that the transfer function from $sd(s)$ to $\epsilon(s)$ should satisfy $\left\|(sI + L_I CG^{-1}E)^{-1}\right\|_\infty \leq \gamma$, $\gamma \to 0$ to minimize the influence of the disturbance on the estimation error $\epsilon(t)$. So, the condition of high gain L_I relative to L_P realizes γ to become small. Assuming the unknown inputs are bounded, the estimation error ϵ can be reduced to small value if g and h are not considered. On the other hand, the presence of g, h may increase the error ϵ. With respect to (97), with the bounded assumptions, the stability of extended system can be realized by eigenvalues assignment. Methods for improving the high gain design are possible [53].

4.4 Fault Detection

The previous section clearly suggests that the DO, UIO, and PIO can equally be used for fault detections. This means that fault may be considered as disturbance and one can estimate the fault for diagnosis purposes. A linear time invariant system without disturbance, but with actuator faults f_a and sensor faults f_s is represented by

$$\dot{x} = Ax + Bu + Bf_a \tag{99}$$

$$y = Cx + f_s \tag{100}$$

where, $f_a \in \mathbb{R}^m$ denotes the presence of actuator faults and $f_s \in \mathbb{R}^p$ denotes the sensor faults. When the state estimation is available, the residual r can be generated as:

$$r = y - C\hat{x} \tag{101}$$

If a Luenberger observer based residual generator is applied to the system described in (99), (100), the residual and the state estimation error ($e = x - \hat{x}$) are described by

$$\dot{e} = (A - LC)e + Bf_a - Lf_s \tag{102}$$

$$r = Ce + f_s \tag{103}$$

Clearly, the residual is sensitive to both actuator faults and sensor faults, which means that they can not successfully be detected. A simple threshold logic can be applied as

$$\begin{cases} \|r\| < \text{Threshold} & \text{for fault-free case} \\ \|r\| \geq \text{Threshold} & \text{for faulty case} \end{cases} \tag{104}$$

for the purpose of fault detection. However, when the system is corrupted with disturbance, Luenberger observer is no longer appropriate for robust fault detection application. As we apply Luenberger observer based residual generator to the system with disturbance, the state estimation error and residual are obtained as

$$\dot{e} = (A - LC)e + Ed + Bf_a - Lf_s \tag{105}$$
$$r = Ce + f_s \tag{106}$$

The residual is still sensitive to both actuator faults and sensor faults, and it is also sensitive to disturbance term Ed as well, which means the residual is unable to differentiate between faults and disturbance.

Robustness is essential in fault detection application, since practical systems may never be free from unknown disturbances and modeling errors. The generated residual should be robust against the system uncertainty and disturbance whilst still sensitive to the faults detection using other types of observers. It has been shown in the literature that DO, UIO, and PIO are capable of robust fault detection.

For simplicity, let us assume the disturbance is constant $\dot{d} = 0$. When a DO- or PIO-based residual generator is applied to the system, the residual, the state estimation error ($e = x - \hat{x}$) and the disturbance estimation error ($\epsilon = d - \hat{d}$) will be written as

$$\begin{bmatrix} \dot{e} \\ \dot{\epsilon} \end{bmatrix} = \begin{bmatrix} A - L_1C & E \\ -L_2C & 0 \end{bmatrix} \begin{bmatrix} e \\ \epsilon \end{bmatrix} + \begin{bmatrix} B \\ 0 \end{bmatrix} f_a - \begin{bmatrix} L_1 \\ L_2 \end{bmatrix} f_s \tag{107}$$

$$r = \begin{bmatrix} C & 0 \end{bmatrix} \begin{bmatrix} e \\ \epsilon \end{bmatrix} + f_s \tag{108}$$

Let

$$\xi = \begin{bmatrix} e \\ \epsilon \end{bmatrix}, \quad A_F = \begin{bmatrix} A - L_1C & E \\ -L_2C & 0 \end{bmatrix} \tag{109}$$

Then, with $B_D = \begin{bmatrix} B^T & 0^T \end{bmatrix}^T$, $C_D = \begin{bmatrix} C & 0 \end{bmatrix}$, and $L_D = \begin{bmatrix} L_1^T & L_2^T \end{bmatrix}^T$ we have

$$\dot{\xi} = A_F\xi + B_Df_a - L_Df_s \tag{110}$$
$$r = C_D\xi + f_s \tag{111}$$

From (110) and (111), it is clear that the disturbance effects have been decoupled from the residual. The faults can be detected by comparing the robust residual with a threshold (104).

5 Decentralized PI-Observer Design

In this section, we consider the problem of designing decentralized PI observer with prescribed degree of convergence for a set of linearly interconnected systems. Under the assumption of linear interactions, we provide a direct design procedure for the PI observer which can effectively be used in simultaneous state and disturbance/fault estimation [54].

Consider a linearly interconnected system composed of N subsystems represented by

$$\begin{aligned}\dot{x}_i(t) &= A_i x_i(t) + B_i u_i(t) + h_i(t,x) + w_i(t)\\ y_i(t) &= C_i x_i(t)\end{aligned} \qquad \text{for} \quad i = 1, \ldots, N \qquad (112)$$

where $x_i \in \mathbb{R}^{n_i}$, $u_i \in \mathbb{R}^{m_i}$, and $y_i \in \mathbb{R}^{p_i}$ are the state, input, and output vector of the ith subsystem, respectively. $w_i \in \mathbb{R}^{m_{wi}}$ is denoted disturbance or fault and $h_i(t,x)$ represent linear time-invariant interconnection function described by:

$$h_i(t,x) = \sum_{j=1}^{N} H_{ij} x_j(t) \qquad (113)$$

We shall assume the pair (A_i, B_i) is controllable and the pair (A_i, C_i) is observable. The entire interconnected system (112) can equivalently be written in composite form as

$$\begin{aligned}\dot{x}(t) &= Ax(t) + Bu(t) + h(t,x) + w(t)\\ y(t) &= Cx(t)\end{aligned} \qquad (114)$$

where $A = diag(A_1, \ldots, A_N)$, $B = diag(B_1, \ldots, B_N)$, $C = diag(C_1, \ldots, C_N)$, $x = [x_1^T, \ldots, x_N^T]^T \in \mathbb{R}^n$ $(n = \sum_{i=1}^N n_i)$, $u = [u_1^T, \ldots, u_N^T]^T \in \mathbb{R}^m$ $(m = \sum_{i=1}^N m_i)$, $w = [w_1^T, \ldots, w_N^T]^T \in \mathbb{R}_w^m$ $(m_w = \sum_{i=1}^N m_{wi})$,$y = [y_1^T, \ldots, y_N^T]^T \in \mathbb{R}^p$ $(p = \sum_{i=1}^N p_i)$, and $h(t,x) = [h_1^T(t,x), \ldots, h_N^T(t,x)]^T$. When the subsystems are decoupled; i.e. $h_i(t,x) = 0 \forall i = 1, \ldots, N$, and the disturbance or fault is absent; i.e. $w_i(t) = 0$, it is easy to construct N independent Luenberger observer for the subsystems. However, such proportional type observers (P-observers) have shortcoming when disturbance or fault is present in the above system. Therefore, we take advantage of the proportional-integral observer (PI-observer), which has proven advantages in various scenarios, and evaluate its robust performance in the interconnected systems framework. Let the decoupled subsystems incorporate disturbance terms $w_i = E_i d_i$ where $d_i \in \mathbb{R}^{m_i}$ represent constant disturbances and $E_i \in \mathbb{R}^{n_i \times m_i}$ is a known distribution matrix associated with the ith disturbance. Thus, we have

$$\begin{aligned}\dot{x}_i(t) &= A_i x_i(t) + B_i u_i(t) + E_i d_i(t)\\ y_i(t) &= C_i x_i(t)\end{aligned} \qquad \text{for} \quad i = 1, \ldots, N \qquad (115)$$

and we define the PI-observer for (115) as

$$\begin{aligned}\dot{\hat{x}}_i &= (A_i - L_{Pi} C_i)\hat{x}_i + L_{Pi} y_i + B_i u_i + E_i \hat{d}_i\\ \dot{\hat{d}}_i &= L_{Ii}(y_i - C_i \hat{x}_i)\end{aligned} \qquad \text{for} \quad i = 1, \ldots, N \qquad (116)$$

which can be written in the augmented form

$$\dot{z}_i = (A_{Xi} - L_{Xi} C_{Xi}) z_i + L_{Xi} y_i + B_{Xi} u_i \qquad (117)$$

where

$$z_i = \begin{bmatrix} \hat{x}_i \\ \hat{d}_i \end{bmatrix},\ L_{Xi} = \begin{bmatrix} L_{Pi} \\ L_{Ii} \end{bmatrix},\ A_{Xi} = \begin{bmatrix} A_i & E_i \\ 0 & 0 \end{bmatrix},\ B_{Xi} = \begin{bmatrix} B_i \\ 0 \end{bmatrix},\ C_{Xi} = \begin{bmatrix} C_i & 0 \end{bmatrix} \tag{118}$$

Denoting the state equation error by $e_i = x_i - \hat{x}_i$ and the disturbance estimation error by $\varepsilon_i = d_i - \hat{d}_i$, the error dynamics of the PI-observer is given by

$$\dot{\varphi}_i = (A_{Xi} - L_{Xi}C_{Xi})\varphi_i \tag{119}$$

where $\varphi_i = [e_i^T \ \varepsilon_i^T]^T$. Since the pair (A_{Xi}, C_{Xi}) is observable one can use any eigenvalue assignment techniques to obtain L_{Xi}, which specifies both the proportional and integral gains of the observer L_{Pi} and L_{Ii}. At this point, we show that one can design the PI-observer with a desired degree of convergence $\alpha > 0$ i.e. $\varphi(t)$ decays by the rate at least $\exp(-\alpha t)$ and $\lim_{t\to\infty} \varphi(t) = 0$. To do this, L_{Xi} must be chosen by

$$L_{Xi} = P_{Xi}C_{Xi}^T \tag{120}$$

where $P_{Xi} \in \mathbb{R}^{(n_i+m_i)\times(n_i+m_i)}$ is the symmetric positive definite solution of the Riccati equation

$$P_{Xi}(A_{Xi}^T + \alpha I_{Xi}) + (A_{Xi} + \alpha I_{Xi})P_{Xi} - P_{Xi}C_{Xi}^T C_{Xi}P_{Xi} + Q_{Xi} = 0 \tag{121}$$

where $Q_{Xi} \in \mathbb{R}^{(n_i+m_i)\times(n_i+m_i)}$ is an arbitrary symmetric nonnegative definite matrix, and I_{Xi} is an identity matrix of the same size as the matrix A_{Xi}. This result follows directly from optimal control theory as applied to our augmented structure.

It is evident that the local observers under the assumption $h_i(t,x) = 0$ will not be satisfactory. This is expected since the interconnection function or the output of the other subsystems is not used in constructing the observer for subsystems. Hence, the problem of interest is to seek modification of the above observer design, which results in the same degree of convergence when $h_i(t,x) \neq 0$.

Theorem 1. *Consider the large-scale system* (112) *with linear interactions* (113) *and $w_i = 0$, i.e.*

$$\dot{x}_i(t) = A_i x_i(t) + B_i u_i(t) + \sum_{j=1}^{N} H_{ij}x_j(t) \tag{122}$$

Let the composite matrix $H = [H_{ij}]$, $i, j = 1, \ldots, N$ satisfy

$$rank \begin{bmatrix} C \\ H \end{bmatrix} = rank\,[C] = p = \sum_{i=1}^{N} p_i \tag{123}$$

Then a set of decentralized PI-observers for the system is given by

$$\dot{z}_i = (A_{Xi} - L_{Xi}C_{Xi})z_i + L_{Xi}y_i + B_{Xi}u_i + \sum_{i=1}^{N} \tilde{L}_{Xij}y_j \tag{124}$$

where L_{Xi} *can be obtained from* (120) *and* $\tilde{L}_{Xij} = \begin{bmatrix} \tilde{L}_{ij} \\ 0 \end{bmatrix} \in \Re^{(n_i+m_i)\times p_j}$ *are the elements of the composite matrix* $\tilde{L}_X$ *given by*

$$\tilde{L}_X = H_X C_X^T (C_X C_X^T)^{-1} \tag{125}$$

where $H_X = [H_{Xij}]$ *with* $H_{Xij} = \begin{bmatrix} H_{ij} & 0 \\ 0 & 0 \end{bmatrix} \in \mathbb{R}^{(n_i+m_i)\times(n_i+m_i)}$ *ensures the degree of convergence* α *for the observer.*

Proof. Let the gain L_{Xi} be selected as (120) such that the decentralized PI observer written in a composite form

$$\dot{z} = (A_X - L_X C_X)z + L_X y + B_X u \tag{126}$$

has degree of convergence α. Under the rank condition (123), H_X can be expressed as $H_X = \tilde{L}_X C_X$ with $\tilde{L}_X$ given by (125) and

$$\rho \begin{bmatrix} C \\ CA \\ \vdots \\ CA^{n-1} \end{bmatrix} = \rho \begin{bmatrix} C \\ C(A+H) \\ \vdots \\ C(A+H)^{n-1} \end{bmatrix} \tag{127}$$

or equivalently

$$\rho \begin{bmatrix} C_X \\ C_X A_X \\ \vdots \\ C_X A_X^{n-1} \end{bmatrix} = \rho \begin{bmatrix} C_X \\ C_X(A_X + H_X) \\ \vdots \\ C_X(A_X+H_X)^{n-1} \end{bmatrix} \tag{128}$$

where $\rho(.)$ denote the rank of a matrix. Since the pair (A_i, C_i) is observable, the corresponding block diagonal pair (A, C) is observable and the left side of (127) is full column rank. Therefore the pair $(A_X + H_X, C_X)$ is observable and a gain matrix G_X can be selected to guarantee the stability of the following observer

$$\dot{z} = (A_X + H_X - G_X C_X)z + G_X y + B_X u \tag{129}$$

A choice of $G_X = L_X + \tilde{L}_X$ would result in

$$\sigma(A_X + H_X - G_X C_X) = \sigma(A_X - L_X C_X)$$

where $\sigma(.)$ denotes the spectrum of a matrix. This establishes the equivalence of (129) and (124).

The decentralized PI-observer (124) has the potential of simultaneously estimating the state and disturbance/fault. In fact, $\hat{x}_i(t) \to x_i(t)$ and $\hat{d}_i(t)$ converges to d_i for $i = 1, ..., N$ as $t \to \infty$.

A direct design of PI-observer for interconnected system

$$\dot{x}_i(t) = A_i x_i(t) + B_i u_i(t) + \sum_{j=1}^{N} H_{ij} x_j(t) \tag{130}$$

can be established by the following structure

$$\dot{z}_i = (A_{Xi} - L_{Xi} C_{Xi}) z_i + L_{Xi} y_i + B_{Xi} u_i + \sum_{i=1}^{N} M_{Xij} z_j \tag{131}$$

where $z_i = \left[\hat{x}_i^T \; v_i^T\right]^T$, $M_{Xij} = \begin{bmatrix} M_{ij} & 0 \\ 0 & 0 \end{bmatrix}$ and the rest of the parameters are defined as before. When $M_{ij} = 0$, one can design N independent local PI-observers. For example, selecting $L_{Xi} = P_{Xi} C_{Xi}^T$ where P_{Xi} is the solution of the AMRE $P_{Xi} A_{Xi}^T + A_{Xi} P_{Xi} - P_{Xi} C_{Xi}^T C_{Xi} P_{Xi} + Q_{Xi} = 0$ ensures asymptotic convergence of $\hat{x}_i$ to x_i. When $M_{ij} \neq 0$, the selection $M_{ij} = H_{ij} \forall i, j$ results in error dynamics

$$\dot{\psi}_i = (A_{Xi} - L_{Xi} C_{Xi}) \psi_i + \sum_{i=1}^{N} H_{Xij} \psi_j \tag{132}$$

where $\psi_i = \left[e_i^T \; v_i^T\right]^T$ and $M_{Xij} = H_{Xij} = \begin{bmatrix} H_{ij} & 0 \\ 0 & 0 \end{bmatrix}$.

Theorem 2. *Let $\tilde{A}_{Xi} = A_{Xi} - L_{Xi} C_{Xi}$ be a stable matrix by proper selection of L_{Xi} and let $\tilde{P}_{Xi} = \tilde{P}_{Xi}^T > 0$ be the solution of $\tilde{P}_{Xi} \tilde{A}_{Xi}^T + \tilde{A}_{Xi} \tilde{P}_{Xi} + \tilde{Q}_{Xi} = 0$ for an arbitrary selected $\tilde{Q}_{Xi} = \tilde{Q}_{Xi}^T > 0$ such that*

$$\lambda_m(\tilde{Q}_{Xi}) > \lambda_m(\tilde{P}_{Xi}) \sum_{\substack{j=1 \\ j \neq i}}^{N} \|H_{Xij}\| + \sum_{\substack{j=1 \\ j \neq i}}^{N} \lambda_m(\tilde{P}_{Xj}) \|H_{Xij}\| \qquad \forall i = 1, \ldots, N \tag{133}$$

Then (132) is asymptotically stable and guarantees the stability of the set of PI observers.

Proof. The proof follows from a routine employment of Lyapunov argument by selecting the conventional Lyapunov function for the composite system and proving the stability under the condition (133).

Before closing this section, we consider (130) with disturbance terms as follows

$$\begin{aligned} \dot{x}_i(t) &= A_i x_i(t) + B_i u_i(t) + \sum_{j=1}^{N} H_{ij} x_j(t) + E_i d_i \\ y_i &= C_i x_i \end{aligned} \tag{134}$$

We define the PI observer for the above system as

$$\begin{aligned}\dot{\hat{x}}_i &= (A_i - L_{Pi}C_i)\hat{x}_i(t) + L_{Pi}y_i + B_iu_i + \sum_{j=1}^{N} H_{ij}\hat{x}_j + E_i\hat{d}_i \\ \dot{\hat{d}}_i &= L_{Ii}(y_i - C_i\hat{x}_i) \qquad \text{for} \qquad i = 1, \ldots, N\end{aligned} \tag{135}$$

which can be written in the composite form

$$\dot{z}_i = (A_{Xi} - L_{Xi}C_{Xi})z_i + L_{Xi}y_i + B_{Xi}u_i + \sum_{j=1}^{N} H_{Xij}z_j \tag{136}$$

where $z_i = \left[\hat{x}_i^T \; v_i^T\right]^T$. Then we have the following result.

Corollary 1. *Let $\hat{A}_{Xi} = A_{Xi} - C_{Xi}L_{Xi}$ be a stable matrix by proper selection of L_{Xi} and let $\hat{P}_{Xi} = \hat{P}_{Xi}^T > 0$ be the solution of $\hat{P}_{Xi}\hat{A}_{Xi}^T + \hat{A}_{Xi}\hat{P}_{Xi} + \hat{Q}_{Xi} = 0$ for an arbitrary selected $\hat{Q}_{Xi} = \hat{Q}_{Xi}^T > 0$ such that* (133) *holds. Then the set of PI observers* (136) *is asymptotically stable and simultaneously estimates the states and disturbance.*

6 Decentralized PI-Observer-Based Control Design

In a recent paper [55] we have demonstrated that when the interconnection function $h_i(t, x)$ assumes the quadratic constraint

$$h_i^T(x, t)h_i(x, t) \leq \alpha^2 x^T H_i^T H_i x \tag{137}$$

where H_i are constant bounding matrices, one can employ PI-observer based controller that robustly regulate the overall system (112). We also provided a procedure to attenuate disturbance. The solutions of both problems were formulated in terms of LMIs.

In this Chapter we provide a robust PI-observer based feedback control design under the assumption that that the interconnection function is linear ($h_i(t, x) = \sum_{j=1}^{N} H_{ij}x_j(t)$). Thus, the class of interconnected systems (112) considered in this work can be classified by the interconnection of N linear systems as follows

$$\begin{aligned}\dot{x}_i(t) &= A_ix_i(t) + B_iu_i(t) + \sum_{j=1}^{N} H_{ij}x_j(t) + w_i(t) \\ y_i(t) &= C_ix_i(t)\end{aligned} \tag{138}$$

The composite system can then be written by the following state equation

$$\begin{aligned}\dot{x}(t) &= Ax(t) + Bu(t) + Hx(t) + w(t) \\ y(t) &= Cx(t)\end{aligned} \tag{139}$$

where all vectors and matrices defined in previous section and the interconnection matrix $H = [H_{ij}]$ with $H_{ii} = 0$ which has also been introduced in theorem 1.

6.1 P-Observer Case

Let the P-observer for each subsystem be described by

$$\begin{aligned} \dot{\hat{x}}_i(t) &= (A_i - L_iC_i)\hat{x}_i(t) + L_iy_i(t) + B_iu_i(t) \\ \hat{y}_i(t) &= C_i\hat{x}_i(t) \qquad i = 1,\dots,N \end{aligned} \tag{140}$$

and its composite form by

$$\dot{\hat{x}}(t) = (A - LC)\hat{x}(t) + Ly(t) + Bu(t) \tag{141}$$

where $\hat{x} = \left[\hat{x}_1^T \cdots \hat{x}_N^T\right]^T$ and $L = diag(L_1,\dots,L_N)$. The control law of each subsystem is given by

$$u_i(t) = -K_i\hat{x}_i(t) \qquad i = 1,\dots,N \tag{142}$$

which can also be written with respect to the composite system (139) as

$$u(t) = -K\hat{x}(t) \tag{143}$$

with $K = diag(K_1,\dots,K_N)$.

Introducing the estimation error $e_i = x_i - \hat{x}_i$, $\forall i = 1,\dots,N$ or equivalently $e = x - \hat{x}$, then the overall closed-loop system incorporating (141) in (139) and its associated error dynamics are obtained as follows

$$\begin{aligned} \dot{x}(t) &= (A + H - BK)x(t) + BKe(t) + w(t) \\ \dot{e}(t) &= (A - LC)x(t) + Hx(t) + w(t) \end{aligned} \tag{144}$$

Thus, the overall system (144) can be written in the augmented form

$$\dot{x}_a(t) = A_ax_a(t) + B_aw(t) \tag{145}$$

where $x_a = \left[x^T \; e^T\right]^T$ and

$$A_a = \begin{bmatrix} A + H - BK & BK \\ H & A - LC \end{bmatrix}, B_a = \begin{bmatrix} I \\ I \end{bmatrix} \tag{146}$$

Defining the Lyapunov function for (145) as

$$V = x_a^TPx_a \tag{147}$$

where $P = diag(P_c, P_o)$ with $P_c = diag(P_{ci})$ and $P_o = diag(P_{oi})$, then a sufficient condition for asymptotic stability of (145) is given by

$$\dot{V} = x_a^T\left[A_a^TP + PA_a\right]x_a \leq 0 \tag{148}$$

which leads to

$$\begin{bmatrix} X_{11} & X_2 \\ X_2^T & X_{22} \end{bmatrix} \leq 0 \tag{149}$$

where $X_{11} = (A + H - BK)^T P_c + P_c(A + H - BK)$, $X_{22} = (A - LC)^T P_o + P_o(A - LC)$, and $X_2 = P_c BK + H^T P_o$.

The above inequality is described in terms of variables P_c, P_o, K and L, which contains coupled terms of matrix variables (P_c, K) and (P_o, L). Thus, we have a bilinear matrix inequality and certain adjustment needs to be done for its solution. To transform the inequality (149) to a form which is affine in the unknown variable, we can multiply (149) on both sides by the transformation matrix

$$T = \begin{bmatrix} T_c & 0 \\ 0 & I \end{bmatrix}$$

where $T_c = P_c^{-1}$. Then (149) becomes

$$\begin{bmatrix} Y_{11} & Y_2 \\ Y_2^T & Y_{22} \end{bmatrix} \leq 0 \tag{150}$$

where $Y_{11} = (A + H)T_c + T_c(A + H)^T - R^T B^T - BR$, $Y_{22} = A^T P_o + P_o A - C^T F^T - FC$, $Y_2 = BK + T_c H^T P_o$, $R = KT_c$, and $F = P_o L$.

Thus, the control and observer problems can be decoupled and solved by two step procedure. First, we compute the matrices $T_c = P_c^{-1}$ and R from the following LMI

$$(A + H)T_c + T_c(A + H)^T - R^T B^T - BR \leq 0 \tag{151}$$

and the control gain matrix is obtained from

$$K = RT_c^{-1} \tag{152}$$

Then, in the second step we solve the LMI (150) with respect to P_o and F which determines

$$L = P_o^{-1} F \tag{153}$$

6.2 PI-Observer Case

Let the PI-observer for each subsystem be described by

$$\begin{aligned} \dot{\hat{x}}_i(t) &= (A_i - L_{Pi}C_i)\hat{x}_i(t) + L_{Pi}y_i(t) + B_i u_i(t) + B_i v_i(t) \\ \hat{v}_i(t) &= L_{Ii}(C_i \hat{x}_i(t) - y_i(t)) \end{aligned} \tag{154}$$

The composite form of the above equations can be written as

$$\begin{aligned} \dot{\hat{x}}(t) &= (A - L_P C)\hat{x}(t) + L_P y(t) + Bu(t) + Bv(t) \\ \hat{v}(t) &= L_I(C\hat{x}(t) - y(t)) \end{aligned} \tag{155}$$

where $\hat{x} = \left[\hat{x}_1^T, \ldots, \hat{x}_N^T\right]^T$, $A = diag(A_i)$, $B = diag(B_i)$, $C = diag(C_i)$, $L_P = diag(L_{Pi})$ and $L_I = diag(L_{Ii})$ $\forall i = 1, \ldots, N$. When the control law $u(t) = -K\hat{x}(t)$ is applied to the system (139) and combined with the error dynamics of the system one can easily define the following augmented dynamics

$$\begin{bmatrix} \dot{x} \\ \dot{\psi} \end{bmatrix} = \begin{bmatrix} A + H - BK & BK_X \\ H_X & A_X - L_X C_X \end{bmatrix} \begin{bmatrix} x \\ \psi \end{bmatrix} + \begin{bmatrix} I \\ I_X \end{bmatrix} w \tag{156}$$

where

$$\psi = \begin{bmatrix} e \\ v \end{bmatrix}, A_X = \begin{bmatrix} A & B \\ 0 & 0 \end{bmatrix}, L_X = \begin{bmatrix} L_P \\ L_I \end{bmatrix}, I_X = \begin{bmatrix} I \\ 0 \end{bmatrix}$$

$$C_X = \begin{bmatrix} C & 0 \end{bmatrix}, K_X = \begin{bmatrix} K & 0 \end{bmatrix}$$

Now it is not difficult to employ the same Lyapunov based technique for PI-observer based control system in a similar fashion as we did in the case of P-observer via (145) which leads to a two-step procedure to obtain controller and observer gains by solving LMI's. This results can be summarized as the following theorem.

Theorem 3. *Consider linear interconnected system* (139). *The decentralized P-observer based controller* (141) *and* (143) *stabilizes the system if the following two steps have feasible solutions:*

Step 1:

Find R and T_c subject to $T_c > 0$ and (151).

Step 2:

Find F and P_o subject to $P_o > 0$ and (150) *in which R, T_c are obtained from step 1.*

whereby the controller and observer parameters are given by $K = RT_c^{-1}$, $L = P_o^{-1}F$.

Furthermore, a decentralized PI-observer based controller (143) and (155) stabilizes the system based on the two step rocedure as above with proper adjustment of parameter matrices by comparing (145) and (156).

Although we discussed the problem of stabilizing decentralized controller so far, it is possible to extend the results to the case of designing controllers that attenuate the disturbance while preserving the stability. This can be done easily with the robust control tool and obtain augmented LMI conditions that guarantee disturbance attenuation in L_2 gain sense.

Theorem 4. *Consider linear interconnected system* (139) *with performance measurement $z(t) = C_z x(t) + D_{zw} w(t)$. The decentralized P-observer based controller* (141) *and* (143) *stabilizes the system with disturbance attenuation less than γ in L_2 gain sense; i.e. $\|T_{zw}\|_2 \le \gamma$, if the following two steps have feasible solutions:*

Step 1:

Find R and T_c subject to $T_c > 0$ and (151)

Step 2:
Minimize γ
subject to $P_o > 0$ *and*

$$\begin{bmatrix} Y_{11} & Y_2 & C_z^T & T_c \\ Y_2^T & Y_{22} & 0 & I \\ C_z & 0 & -I & D_{zw} \\ T_c^T & I & D_{zw}^T & -\gamma^2 I \end{bmatrix} < 0$$

where R, T_c, *and* K *are obtained from step 1 and optimization variables* F *and* P_o.

Moreover, a decentralized PI-observer based controller (143) and (155) stabilizes the system and guarantees based on the two step procedure as above with proper adjustment of parameter matrices by comparing (145) and (156). The resulting LMI for the second step is given by

$$\begin{bmatrix} Y_{11} & BK_X + T_c H^T P_o & C_z^T & T_c \\ * & A_X^T P_o + P_o A_X - C_X^T F_X^T - F_X C_X & 0 & I \\ * & * & -I & D_{zw} \\ * & * & * & -\gamma^2 I \end{bmatrix} < 0$$

where all matrices with subscript X are in augmented format according to the PI-observer expressions as previously defined.

7 Conclusion

This chapter reviewed observer theory and its development by covering the most important results for state estimation, robust observer based controller design, disturbance estimation, and fault detection. The major emphasize was given to PI-observer and its applicability in various scenarios. The specific problem of designing a decentralized PI observer with prescribed degree of convergence for a set of interconnected systems has been considered in this chapter. When the interactions are linear, a direct design procedure for the PI observer is provided which can effectively be used in disturbance estimation and observer-based control design. We also extended the results to the case of designing controllers that attenuate the disturbance while preserving the stability. It was shown that the design can be formulated in terms of LMI which efficiently solve the problem.

References

1. Luenberger, D.: Observing the state of a linear system. IEEE Transactions on Military Electronics 8, 74–80 (1964)
2. Luenberger, D.: An Introduction to Observers. IEEE Transactions on Automatic Control 16, 596–602 (1971)
3. Chen, C.T.: Linear system theory and design. Oxford University Press, Inc. (1995)

4. O'Reily, J.: Observers for Linear Systems. Academic Press (1983)
5. Korovin, S., Fomichev, V.: State Observer for Linear Systems with uncertainty. de Gruyter Press, Berlin (2009)
6. Trinh, H., Fernando, T.: Functional Observers for Dynamical Systems. Springer, Berlin (2012)
7. Gupta, R., Fairman, F., Hinamoto, T.: A direct procedure for the design of single functional observers. IEEE Trans. Circuits and Systems CAS 28, 294–300 (1981)
8. Fortman, T., Williamson, D.: Design of a low order observer for linear feedback control laws. IEEE Transactions on Automatic Control 17, 301–308 (1972)
9. Trinh, H., Nahavandi, S., Tran, T.: Algorithms for designing reduced-order functional observers of linear systems. International Journal of Innovative Computing, Information and Control 4(2), 321–333 (2008)
10. Darouach, M.: Existence and design of functional observers for linear systems. IEEE Transactions on Automatic Control 45(5), 940–943 (2000)
11. Tsui, C.: A new algorithm for the design of multifunctional observers. IEEE Transactions on Automatic Control 30(1), 89–93 (1985)
12. Uddin, V., Shafai, B., Kawaji, S.: A simple method for the design of minimal order multifunctional observer. In: Proc. of 2nd IASTED International Conference on Control and Applications, Banff, Canada, pp. 107–112 (1999)
13. Shafai, B., Carroll, R.L.: Minimal-order observer designs for linear time-varying multivariable systems. IEEE Transactions on Automatic Control 31(8), 757–761 (1986)
14. Shafai, B.: Design of single-functional observers for linear time-varying multivariable systems. International Journal of Systems Science 20(11), 2227–2239 (1989)
15. Chai, W., Loh, N.K., Hu, H.: Observer design for time-varying systems. International Journal of Systems Science 22(7), 1177–1196 (1991)
16. Trumpf, J.: Observers for linear time-varying systems. Linear Algebra and its Applications 425(2), 303–312 (2007)
17. Rotella, F., Zambettakis, I.: On functional observers for linear time-varying systems. IEEE Transaction on Automatic Control 58(5), 1354–1360 (2013)
18. Bestle, D., Zeitz, M.: Canonical form observer design for non-linear time-variable systems. International Journal of Control 38(2), 419–431 (1983)
19. Gauthier, J.P., Hammouri, H., Othman, S.: A simple observer for nonlinear systems applications to bioreactors. IEEE Transactions on Automatic Control 37(6), 875–880 (1992)
20. Ciccarella, G., Dalla Mora, M., Germani, A.: A Luenberger-like observer for nonlinear systems. International Journal of Control 57(3), 537–556 (1993)
21. Fairman, F., Kumar, A.: Delayless observers for systems with delay. IEEE Transactions on Automatic Control 31(3), 258–259 (1986)
22. Hou, M., Zitek, P., Patton, R.: An observer design for linear time-delay systems. IEEE Transactions on Automatic Control 47(1), 121–125 (2002)
23. Trinh, H., Teh, P., Fernando, T.: Time-delay systems: Design of delay-free and low-order observers. IEEE Transactions on Automatic Control 55(10), 2434–2438 (2010)
24. Sadaka, H., Shafai, B., Sipahi, R.: Robust PI observer design for linear time-delay systems. In: Control Applications(CCA) & Intelligent Control(ISIC), pp. 1209–1213. IEEE (2009)
25. Shafai, B., Carroll, R.L.: Design of a minimal-order observer for singular systems. International Journal of Control 45(3), 1075–1081 (1987)
26. Dai, L.: Singular Control Systems. Springer, Berlin (1989)

27. Darouach, M., Boutayeb, M.: Design of observers for descriptor systems. IEEE transactions on Automatic Control 40(7), 1323–1327 (1995)
28. Hou, M., Muller, P.: Observer design for descriptor systems. IEEE Transactions on Automatic Control 44(1), 164–168 (1999)
29. Rami, M.A., Tadeo, F., Helmke, U.: Positive observers for linear positive systems, and their implications. International Journal of Control 84(4), 716–725 (2011)
30. Back, J., Astolfi, A.: Design of positive linear observers for positive linear systems via coordinate transformations and positive realizations. SIAM Journal on Control and Optimization 47(1), 345–373 (2008)
31. Shafai, B.: Parallel procedure for the design of observers in high-order multivariable systems. International Journal of Control 47(5), 1489–1496 (1988)
32. Jamshidi, M.: Large-scale systems: Modeling, control and fuzzy logic. Upper Saddle River. Prentice-Hall (1996)
33. Petkov, P.H., Christov, N.D., Konstantinov, M.M.: Computational methods for linear control systems. Prentice-Hall International (UK) Ltd (1991)
34. Saif, M., Guan, Y.: Decentralized state estimation in large-scale interconnected dynamical systems. Automatica 28(1), 215–219 (1992)
35. Doyle, J.C., Stein, G.: Robustness with Observers. IEEE Transactions on Automatic Control 24(4), 607–611 (1979)
36. Doyle, J.C., Stein, G.: Multivariable Feedback Design: concepts for a classical/modern synthesis. IEEE Transactions on Automatic Control 26, 4–16 (1981)
37. Athans, M.: A tutorial on the LQG/LTR method. In: Proc. of American Control Conference, Seattle, WA, pp. 1289–1296 (1986)
38. Shafai, B., Beale, S., Niemann, H.H., Stoustrup, J.L.: Modified structures for loop transfer recovery design. In: Proc. 32nd IEEE Conference on Decision and Control, San Antonio, TX, pp. 3341–3344 (1993)
39. Saberi, A., Chen, B.M., Sannuti, P.: Loop transfer recovery: Analysis and design. Springer, New York (1993)
40. Yao, Y.X., Zhang, Y.M., Kovacevic, R.: Loop Transfer recovery design with proportional integral observer based on H^∞ optimal observation. In: Proc. of American Control Conference, Albuquerque, New Mexico, vol. 1, pp. 786–790 (1997)
41. Niemann, H.H., Stoustrup, J., Shafai, B., Beale, S.: LTR design of proportional-integral observers. International Journal of Robust and Nonlinear Control 5(7), 671–693 (1995)
42. Shafai, B., Beale, S., Niemann, H.H., Stoustrup, J.: LTR design of discrete-time proportional-integral observers. IEEE Transactions on Automatic Control 41(7), 1056–1062 (1996)
43. Kim, B.K., Chung, W.K.: Advanced disturbance observer design for mechanical positioning systems. IEEE Transactions on Industrial Electronics 50(6), 1207–1216 (2003)
44. Guan, Y., Saif, M.: A novel approach to the design of unknown input observers. IEEE Transactions on Automatic Control 36(5), 632–635 (1991)
45. Kudva, P., Viswanadham, N., Ramakrishna, A.: Observers for linear systems with unknown inputs. IEEE Transactions on Automatic Control 25(1), 113–115 (1980)
46. Hou, M., Muller, P.: Design of observers for linear systems with unknown inputs. IEEE Transactions on Automatic Control 37, 871–875 (1992)
47. Busawon, K.K., Kabore, P.: Disturbance attenuation using proportional integral observers. International Journal of Control 74(6), 618–627 (2001)
48. Shafai, B., Pi, C.T., Nork, S.: Simultaneous disturbance attenuation and fault detection using proportional integral observers. In: Proc. of American Control Conference, Anchorage, Alaska, pp. 1647–1649 (2002)

49. Marx, B., Koenig, D., Georges, D.: Robust fault diagnosis for linear descriptor systems using proportional integral observers. In: Proc. of the 42nd IEEE Conference on Decision and Control, Maui, Hawaii, pp. 457–462 (2003)
50. Soffker, D., Yu, T.J., Muller, P.C.: State estimation of dynamical systems with nonlinearities by using proportional-integral observer. International Journal of Systems Science 26(9), 1571–1582 (1995)
51. Shafai, B., Pi, C.T., Nork, S., Linder, S.P.: Proportional integral adaptive observer for parameter and disturbance estimations. In: Proc. of the 41st IEEE Conference on Decision and Control, Las Vegas, Nevada, pp. 4694–4699 (2002)
52. Patton, R., Frank, P., Clark, R.: Fault diagnosis in dynamic systems: theory and application. Prentice-Hall, Upper Saddle River (1989)
53. Liu, Y.: Robust nonlinear control design with Proportional-Integral-observer technique, Diss. Universitat Duisburg-Essen, Fakultat für Ingenieurwissenschaften Maschinenbau und Verfahrenstechnik (2011)
54. Shafai, B., Ghadami, R., Saif, M.: Robust decentralized PI observer for linear interconnected systems. In: IEEE International Symposium on Computer-Aided Control System Design (CACSD), pp. 650–655 (2011)
55. Ghadami, R., Shafai, B.: Decentralized PI observer-based control of nonlinear interconnected systems with disturbance attenuation. In: Proc. 2011 American Control Conference, San Francisco, CA, pp. 4705–4710 (2011)

New Application of an Adaptive Controller Based on Robust Fixed Point Transformations

Imre J. Rudas and József K. Tar

Óbuda University, Budapest, Hungary
{rudas@,tar.jozsef@nik.}uni-obuda.hu

Abstract. In this paper the possible application of a novel adaptive solution, the Robust Fixed Point Transformations (RFPT) based adaptive controller was investigated in a complicated practical case: the control a caster-supported Wheeled Mobile Robot (WMR) with two actively driven wheels when the dynamic model of this equipment is only approximately known.

This control task is an interesting problem that is especially burdened from both its kinematic and dynamic aspects. The kinematic problems root in the practical need that we have to control both the orientation and the location of the WMR on a plain surface, i.e. we wish to keep under control three independent quantities (θ, x, y) while we have only two control agents (the torques of the drives of the active wheels), and also have to satisfy the non-holonomic constraints excluding free sliding, slipping or skidding over the plane. The dynamic problems are the formal consequences of the approximate model. While in the possession of a precise model it is possible to design a controller to track the mass center point therefore the equations of motion describing the acceleration of the mass center point and the spinning of the body around it become decoupled, in our case the appearance of the nonlinearly coupled equations is inevitable. Further problem arises from the interaction of the electromagnetic subsystem (the DC motors) and the mechanical one (the cart): for controlling the cart instantaneous torque values i.e. motor currents are needed, however, only the time-derivative of this current can be instantaneously set by the control voltages. Therefore in the practice we need some order reduction technique in the control of nonlinear systems.

It is shown via simulations that the RFPT-based adaptive controller simultaneously can solve these problems.

Keywords: robust fixed point transformations, adaptive control, nonlinear order reduction.

1 Introduction

For the control of Linear Time-Invraiant (LTI) systems efficient practically uniform design methodologies are available. Normally the aim is finding appropriate feedback gains that stabilize the controlled system that has to precisely track some prescribed trajectory. The general need for proceeding along these rails

A. El-Osery and J. Prevost (eds.), *Control and Systems Engineering*,
Studies in Systems, Decision and Control 27, DOI: 10.1007/978-3-319-14636-2_3

lead to the introduction of more general classes of systems that are more or less similar to the LTI system as Linear Parameter Varying (LPV), Quasi-Linear Parameter Varying (QLPV), and finally the big class of nonlinear systems.

Due to the fact that equations of motion of the nonlinear systems generally do not have closed analytical form the control design methods –while try to utilize the specialties of the problem classes– normally are based on Lyapunov's "direct" or "2nd" method [1, 2] that originally was developed for the investigation of the stability of the motion of dynamic systems in 1892. The essence of this method is that without revealing any subtle details of the motion under consideration its important property, namely its stability can be determined by relatively simple estimations. (For this purpose various stability definitions are in circulation.) In spite of its virtues this design approach has certain weak points as follows: a) its application in the practice is difficult, it needs innovative and skilled designers, b) normally it requires satisfying *satisfactory conditions* that are not inevitably necessary, c) it does not reveal subtle details of the motion that otherwise may appear as primary design intent, and d) the so obtained solutions normally contain ample number of arbitrary control parameters that are not optimized and somehow have to be set in the next step of the controller design (e.g. the application of Genetic Algorithm as in [3]). In spite of these difficulties the early adaptive control applications in the beginning of the nineties of the past century (e.g. [4–6]) successfully applied this design methodology.

The need for direct satisfaction of the primary design intents lead to seeking alternative adaptive design approach. The "Robust Fixed Point Transformations (RFPT)"-based design was introduced in 2009 [7]. This approach concentrates on the realization of prescribed details of the trajectory tracking that is achieved by an iterative learning process based on Stefan Banach's "Fixed Point Theorem" published in 1922 [8]. Though the original form of this approach guarantees only local stability because the iterative sequence it uses converges only within a bounded basin of attraction later it was improved with complementary parameter tuning that helped maintaining its convergence [9]. In this manner the requirement of "global stability" practically can met by the use of this simple approach for a wide class of physical systems. It worths noting that in many cases no complementary tuning is needed.

The control of caster-supported, DC motors driven WMR was intensively researched in the past years (e.g. [10–12]). These approaches used Lyapunov's technique that lead to complicated considerations.

The aim of this paper is to show that the RFPT-based method is applicable for adaptively controlling a WMR by tracking its arbitrary point that can differ from its mass center point. In the sequel at first the basic idea of order reduction in the case of nonlinear controllers is briefed. Following that the immediate antecedents are surveyed. In the next step a detailed dynamic model of the WMR is developed that was used in the simulations. Finally, following the simulation results the conclusions are drawn.

2 Nonlinear Order Reduction in Adaptive Control Based on Approximate Dynamic Model

Higher order control problems frequently arise whenever various physical subsystems are dynamically coupled. For instance if a mass-point can be moved through some elastic component that connects it to a directly forced rigid body, according to the laws of Classical Mechanics, the 2nd time-derivative of its Cartesian co-ordinates can be manipulated via forces. The necessary force can be exerted by the elastic component that needs appropriate deformation. The length of this component cannot be set instantaneously. Only its 2nd time-derivative can be set immediately by the force acting on the rigid body. Consequently, only higher order derivatives of the co-ordinates of the mass point are in direct relationship with the control forces.

If a Classical Mechanical System is driven by a Direct Current (DC) electric motor, for its acceleration appropriate motor torque is needed. This is proportional to the motor current that (due to the inductivity of its coils) cannot be modified in an abrupt manner. The control voltage directly concerns only the time-derivative of this current. Therefore the 3rd time-derivative of the co-ordinates of the mechanical system are in direct relationship with the control voltage. If we wish to evade the development of a 3rd order control we have to apply some order reduction technique.

In the realm of the Linear Time-Invariant (LTI) systems that can be described by some Transfer Function in the frequency domain the Padé approximation theory [13] can widely be used for order reduction. The main idea is very simple: the transfer function normally is a fractional expression with polynomials in the numerator and the denominator (rational function). This fractional expression can be replaced by a different one in which the order of the numerator and that of the denominator is less than in the original expression. If within the frequency range of practical interest this simple expression well approximates the original one it can be used for controller design. Technically the Taylor series expansion of the original function and the approximation are considered around a given point of the frequency domain and the approximation is built up by equating a few (normally low order) coefficients of these equations [14].

Padé's approximation theory can be widely used for order reduction even in the case of fractional order systems of long memory (e.g. [15]). However, in the case of nonlinear systems alternative approaches have to be chosen since they cannot be described by some transfer function.

In the case of DC motors driven electric carts we are just in such a situation. The suggested methodology assumes that the variation of the electromagnetic torque of the motor must be considerably faster than the variation of the force need of the mechanical components. In this case it seems to be a realistic approximation that during the cycle time of a dynamical controller the torque need of the mechanical parts is approximately constant and that during the consecutive control cycles the electromagnetic torque very quickly saturates at the desired level. However, this assumption is only an approximation. The basic idea in using the RFPT based adaptive approach is that –together with the effects of other modeling errors and certain unknown external disturbances– the consequences of this approximation can be compensated by the adaptive deformation applied by the controller.

In the sequel at first certain immediate antecedents will be considered regarding modeling and controlling WMRs.

3 Immediate Antecedents in WMR Control

The model-based control of WMRs was intensively researched in the past years. A special variant of these equipment has two actively driven wheels plus a freely rotating “caster wheel” having the role of keeping the vehicle in the (x, y) plane by supporting its plateau. The most “popular” dynamic model in circulation assumes that the mass center point of the WMR and the point the motion of which has to be tracked by the controller are identical. As a consequence the equations of motion are far less complicated than in a “general case” when the tracked point of the WMR is determined via plausible kinematic considerations and it is different to the mass center point.

If the tracked point is the mass center point the equations describing the rotation of the object around it and giving its acceleration with respect to an inertial system of reference become decoupled. This simple decoupled model form easies the controller design to a great extent.

Whenever the tracked point is different to the mass center point the equations describing its acceleration and the rotation of the WMR around that point become dynamically and nonlinearly coupled by terms that contain information on the motion of the mass center point, too. This form needs more sophisticated nonlinear control designs.

Though in principle it is always possible to track the mass center point, in the practice this option is not easily implementable. Whenever some payload is placed on the plateau of the WMR the mass center point of the WMR+payload system as a rigid body is shifted while the kinematically traceable point remains the same. Since the identification of the modified dynamical parameters of the system is a complicated task generally we cannot use decoupled equations of motion. For designing controllers for nonlinear systems generally Lyapunov's 2nd or “direct” method [1, 2] is used. For instance in 1993 Samson found an appropriate Lyapunov function in a nonlinear control design for car-like wheeled mobile robots [16].

Perhaps due to historical reasons the control community insists on using the terms rooting in the traditional linear control even in the case of nonlinear systems. The original and clear concept of the LTI system was generalized as the Linear Parameter-Varying (LPV) and the Quasi-Linear Parameter-Varying (QLPV) systems. Regarding model-based control this approach has the great advantage that by numerical operations a nonlinear model can be transformed into tensor product model form to which the use of the traditional linear design methods can be extended by solving linear matrix inequalities (e.g. [17, 18]). However, this approach assumes the existence of a static model that needs computation-greedy transformations before designing the controller. If the model is uncertain this approach seems to be less efficient.

The traditional linearization techniques do not work well in the case of even a simplified model of such WMR. In 2002 Oriolo, De Luca, and Vendittelli suggested the application of feedback linearization approach [10]. I their paper they stressed that on the basis of [19] the linearized model of the WMR at any point is not controllable therefore a linear controller cannot locally stabilize the posture of the vehicle. For such a stabilization either discontinuous and/or time-varying control rule is necessary. They applied dynamic feedback linearization.

In 2008 LeBel and Rodrigues developed a piecewise-affine parameter-varying control for a WMR [11]. They used the Lyapunov function technique to prove the stability of the various approximations.

In 2013 Ojleska applied switched-fuzzy and fuzzy-neural systems for controller design for a WMR [12]. These approaches intensively used Lyapunov's technique that lead to complicated considerations. In this approach the mechanical subsystem had the kinematic model as follows:

$$\begin{aligned} \dot{x} &= \tfrac{r_{\mathrm{w}}}{2}\left(\dot{q}_{\mathrm{r}}+\dot{q}_{\mathrm{l}}\right)\cos\theta, \\ \dot{y} &= \tfrac{r_{\mathrm{w}}}{2}\left(\dot{q}_{\mathrm{r}}+\dot{q}_{\mathrm{l}}\right)\sin\theta, \\ \dot{\theta} &= \tfrac{r_{\mathrm{w}}}{2D}\left(\dot{q}_{\mathrm{r}}-\dot{q}_{\mathrm{l}}\right), \end{aligned} \tag{1}$$

in which r_{w} denotes the radii of the wheels, D means the distances between the wheels and the center of the cart, that is assumed to be on the line connecting the centers of the wheels, and q_{r} and q_{l} denote the rotation angles of the wheels at the *right* and the *left sides* of the cart, respectively, and θ describes its orientation. The consequences for the 2nd time-derivatives of the coordinates are:

$$\begin{aligned} \ddot{x} &= \tfrac{r_{\mathrm{w}}}{2}\left(\ddot{q}_{\mathrm{r}}+\ddot{q}_{\mathrm{l}}\right)\cos\theta - \tfrac{r_{\mathrm{w}}}{2}\left(\dot{q}_{\mathrm{r}}+\dot{q}_{\mathrm{l}}\right)\sin\theta\dot{\theta}, \\ \ddot{y} &= \tfrac{r_{\mathrm{w}}}{2}\left(\ddot{q}_{\mathrm{r}}+\ddot{q}_{\mathrm{l}}\right)\sin\theta + \tfrac{r_{\mathrm{w}}}{2}\left(\dot{q}_{\mathrm{r}}+\dot{q}_{\mathrm{l}}\right)\cos\theta\dot{\theta}, \\ \ddot{\theta} &= \tfrac{r_{\mathrm{w}}}{2D}\left(\ddot{q}_{\mathrm{r}}-\ddot{q}_{\mathrm{l}}\right) \quad . \end{aligned} \tag{2}$$

The dynamic model again directly was deduced from the laws of Classical Mechanics: the acceleration of the mass center point of the cart with respect to the inertial road-system is proportional to the inertia of the whole system, M, and its acceleration. The angular acceleration of the rotation around the mass center point multiplied by the momentum of the system I must yield the rotary torque.

$$\begin{pmatrix} T_{\mathrm{r}} \\ T_{\mathrm{l}} \end{pmatrix} = \begin{pmatrix} \frac{Mr_{\mathrm{w}}^2}{4} + \frac{Ir_{\mathrm{w}}^2}{4D^2} & \frac{Mr_{\mathrm{w}}^2}{4} - \frac{Ir_{\mathrm{w}}^2}{4D^2} \\ \frac{Mr_{\mathrm{w}}^2}{4} - \frac{Ir_{\mathrm{w}}^2}{4D^2} & \frac{Mr_{\mathrm{w}}^2}{4} + \frac{Ir_{\mathrm{w}}^2}{4D^2} \end{pmatrix} \begin{pmatrix} \ddot{q}_{\mathrm{r}} \\ \ddot{q}_{\mathrm{l}} \end{pmatrix} \quad . \tag{3}$$

As it can be seen the dynamic model has a simple LTI structure and no any difficulty we had if our task would be to track some nominal trajectory prescribed in the terms $q_{\mathrm{r}}^{\mathrm{N}}(t)$, $q_{\mathrm{l}}^{\mathrm{N}}(t)$. Our problems are of kinematic nature because we should like to prescribe the nominal trajectory in the terms $\theta_{\mathrm{r}}^{\mathrm{N}}(t)$, $x^{\mathrm{N}}(t)$, and $y^{\mathrm{N}}(t)$, for which we have only two physical control actions T_{r}, and T_{l} while the non-holonomic constraints described by (1) and (2) have to be satisfied, too. This situation is further complicated if the tracked point and the mass center point of the WRM are not identical. In the next section this more complicated model will be developed.

4 The More General Model

In developing this model we revisited Newton's laws by directly computing the "Translational Momentum" and the "Angular Momentum" with respect to an inertial frame using an arbitrary point to be tracked.

4.1 The Physical Basis

Let the initial position of the cart with respect to this frame given by that of the elementary mass points i of mass $m^{(i)}$ and position $\hat{r}^{(i)}$. (Both $m^{(i)}$ and $\hat{r}^{(i)}$ are constant quantities.) Since the cart is approximated as a rigid body, the whole system's displacement can be described by a single rotation around the origin denoted by the Orthogonal Matrix $O(t) \in \mathrm{I\!R}^{3\times 3}$ and, following that, a displacement by vector $R(t) \in \mathrm{I\!R}^3$ where t denotes the time. Consequently the actual position of the mass-point $m^{(i)}$ and its velocity with respect to the workshop frame are given as

$$\begin{aligned} r^{(i)}(t) &= O(t)\hat{r}^{(i)} + R(t), \\ \dot{r}^{(i)}(t) &= \dot{O}(t)\hat{r}^{(i)} + \dot{R}(t). \end{aligned} \tag{4}$$

Since the orthogonal matrices form a Lie Group for which the matrix $G \stackrel{def}{=} \dot{O}(t)O^{-1}(t)$ (or $\dot{O} = GO$) can be geometrically interpreted as the actual rotational velocity it is expedient to transform the 2nd equation of (4) as follows

$$\begin{aligned} \dot{r}^{(i)}(t) &= \dot{O}(t)O^{-1}(t)O(t)\hat{r}^{(i)} + \dot{R}(t) \Longrightarrow \\ \dot{r}^{(i)}(t) &= G(t)O(t)\hat{r}^{(i)} + \dot{R}(t) \end{aligned} \tag{5}$$

since the actual rotational speed $G(t)$ can be used for control purposes by the driver of the cart. Following this preparation the complete "linear momentum" or "translational momentum" of the rigid body P can be defined as

$$\begin{aligned} P \stackrel{def}{=} \textstyle\sum_i m^{(i)}\dot{r}^{(i)} = G(t)O(t)M\hat{S} + M\dot{R}(t), \text{ where} \\ M \stackrel{def}{=} \textstyle\sum_i m^{(i)}, \text{ and } \hat{S} \stackrel{def}{=} \frac{\sum_i m^{(i)}\hat{r}^{(i)}}{M} \end{aligned} \tag{6}$$

i.e. the introduction of the mass of the whole system M, and the location of the mass center point in the initial position $\hat{S}$ can be introduced. (Naturally both M and $\hat{S}$ are constant quantities.)

The angular momentum of the whole body can be defined as the sum of the angular momentum of the individual mass-points as

$$\begin{aligned}\mathcal{M} &\stackrel{def}{=} \sum_i m^{(i)} \left[\dot{r}^{(i)}[r^{(i)} - R(t)]^T - [r^{(i)} - R(t)]\dot{r}^{(i)^T}\right] = \\ &= \sum_i m^{(i)} \left[\dot{r}^{(i)}\hat{r}^{(i)^T} O^T - O\hat{r}^{(i)}\dot{r}^{(i)^T}\right]\end{aligned} \tag{7}$$

that describes the spinning motion of the body around the point having the actual coordinates $R(t)$. (The notation $\mathcal{M}$ was intentionally used to distinguish between the concepts of full mas and rotary momentum.) The quantity $\dot{r}^{(i)}$ can be substituted into this equation by the use of $G(t)$, $O(t)$, and $\dot{R}(t)$ that are physically well-interpreted quantities:

$$\begin{aligned}\mathcal{M} = \sum_i m^{(i)} &\left[GO\hat{r}^{(i)}\hat{r}^{(i)^T} O^T + \dot{R}\hat{r}^{(i)^T} O^T \right. \\ &\left. - O\hat{r}^{(i)}\hat{r}^{(i)^T} O^T G^T - O\hat{r}^{(i)}\dot{R}^T\right] \ .\end{aligned} \tag{8}$$

In (8) the constant vector $M\hat{S}$ can be identified. This equation also suggests the introduction of another constant quantity, the "momentum of inertia" as

$$\hat{\Theta} \stackrel{def}{=} \sum_i m^{(i)}\hat{r}^{(i)}\hat{r}^{(i)^T} \tag{9}$$

that is a symmetric constant matrix. By the use of these quantities (8) can be rewritten as

$$\mathcal{M} = GO\hat{\Theta}O^T - O\hat{\Theta}O^T G^T + \dot{R}M\hat{S}^T O^T - OM\hat{S}\dot{R}^T \ . \tag{10}$$

The equations of motion of the whole system can be derived by considering the quantities $\dot{P}$ in (6) and $\dot{\mathcal{M}}$ in (7) in which the fragments $F^{(i)} \stackrel{def}{=} m^{(i)}\ddot{r}^{(i)}$ naturally appear and physically can be interpreted as the sum of all forces acting on the mass-point i.

The complicated details are omitted in this paper, however, this introduction substantiates why the quantities M, $\hat{S}$, $\hat{\Theta}$, as well as their transformed (i.e. rotated variants) $S(t) \stackrel{def}{=} O(t)\hat{S}$, and $\Theta(t) \stackrel{def}{=} O(t)\hat{\Theta}O^T(t)$ appear in them. In the calculations the special property of the rotational group that $O^{-1} \equiv O^T$, the symmetry of Θ and the skew-symmetry of G were also utilized.

4.2 The Sign Conventions and the Final Results

Further mathematical details also depend on the way chosen for the parametrization of the rotational group and choosing sign conventions for describing the rotation of the wheels of the cart (Fig. 1).

For describing a rotation with angle $q_{\mathrm{v}} \equiv \theta$ radians around axis $\hat{x}_3$ the orthogonal matrix $O(q_{\mathrm{v}})$ generated by the generator $\hat{G}^{(\mathrm{O})}$ was chosen as

$$O(q_{\mathrm{v}}) \stackrel{def}{=} \begin{pmatrix} \cos q_{\mathrm{v}} & -\sin q_{\mathrm{v}} & 0 \\ \sin q_{\mathrm{v}} & \cos q_{\mathrm{v}} & 0 \\ 0 & 0 & 1 \end{pmatrix}, \ \hat{G}^{(\mathrm{O})} \stackrel{def}{=} \begin{pmatrix} 0 & -1 & 0 \\ 1 & 0 & 0 \\ 0 & 0 & 0 \end{pmatrix} \ . \tag{11}$$

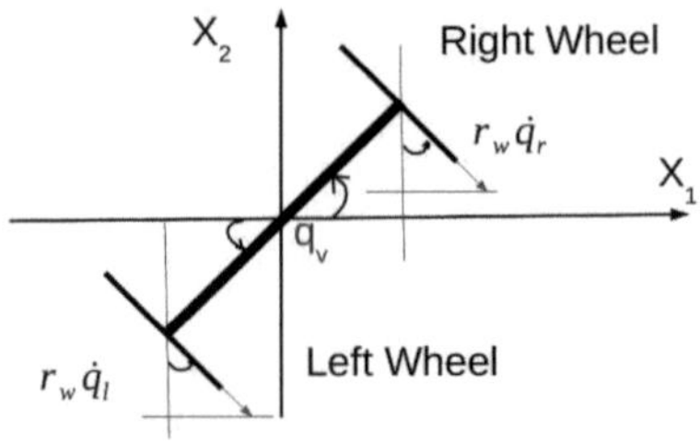

Fig. 1. The sketch of the non-holonomic cart also defining the sign conventions used in the development of the model

This operation is possible, since it moves the whole cart on the surface of the horizontal plain. By the use of the above conventions we arrive at the final result for the non-holonomic kinematic constraints as follows:

$$\begin{aligned}\dot{R}_1 &= r_{\mathrm{w}}\tfrac{\dot{q}_{\mathrm{r}}+\dot{q}_{\mathrm{l}}}{2}\sin q_{\mathrm{v}}\\ \dot{R}_2 &= -r_{\mathrm{w}}\tfrac{\dot{q}_{\mathrm{r}}+\dot{q}_{\mathrm{l}}}{2}\cos q_{\mathrm{v}}\\ \dot{q}_{\mathrm{v}} &= -\tfrac{r_{\mathrm{w}}}{D}\tfrac{\dot{q}_{\mathrm{r}}-\dot{q}_{\mathrm{l}}}{2}\quad .\end{aligned} \tag{12}$$

The kinematically allowable 2nd time-derivatives for the skidding/sliding-free case can be derived from (12) as follows:

$$\begin{aligned}\ddot{R}_1 &= r_{\mathrm{w}}\tfrac{\ddot{q}_{\mathrm{r}}+\ddot{q}_{\mathrm{l}}}{2}\sin q_{\mathrm{v}} + r_{\mathrm{w}}\tfrac{\dot{q}_{\mathrm{r}}+\dot{q}_{\mathrm{l}}}{2}\cos q_{\mathrm{v}}\dot{q}_{\mathrm{v}}\\ \ddot{R}_2 &= -r_{\mathrm{w}}\tfrac{\ddot{q}_{\mathrm{r}}+\ddot{q}_{\mathrm{l}}}{2}\cos q_{\mathrm{v}} + r_{\mathrm{w}}\tfrac{\dot{q}_{\mathrm{r}}+\dot{q}_{\mathrm{l}}}{2}\sin q_{\mathrm{v}}\dot{q}_{\mathrm{v}}\\ \ddot{q}_{\mathrm{v}} &= -\tfrac{r_{\mathrm{w}}}{D}\tfrac{\ddot{q}_{\mathrm{r}}-\ddot{q}_{\mathrm{l}}}{2}\quad .\end{aligned} \tag{13}$$

Regarding the dynamical equations it can be observed that generally the rotation effects of the external and contact forces around the actual x_1 and x_2 axes as well as the force components in the direction of x_3 can compensate each other, so we can concentrate only on the translation within the horizontal plain and the rotation around x_3. If no horizontal external force is acting on the system (14) is obtained in which the contact forces at the driven wheels are orthogonal to the wheels' axles and directly are related to the torques acting on the axles:

$$\begin{aligned}&\left(\begin{array}{c|c}\frac{r_{\mathrm{w}}}{2D}(\Theta_{11}+\Theta_{22}) & \frac{Mr_{\mathrm{w}}}{2}(S_2\sin q_{\mathrm{v}}+S_1\cos q_{\mathrm{v}})\\ \hline \frac{Mr_{\mathrm{w}}}{2D}(S_2\sin q_{\mathrm{v}}+S_1\cos q_{\mathrm{v}}) & \frac{Mr_{\mathrm{w}}}{2}\end{array}\right)\begin{pmatrix}\ddot{q}_{\mathrm{r}}-\ddot{q}_{\mathrm{l}}\\ \ddot{q}_{\mathrm{r}}+\ddot{q}_{\mathrm{l}}\end{pmatrix}+\\ &+\begin{pmatrix}\frac{Mr_{\mathrm{w}}}{2}(\dot{q}_{\mathrm{r}}+\dot{q}_{\mathrm{l}})\dot{q}_{\mathrm{v}}(S_2\cos q_{\mathrm{v}}-S_1\sin q_{\mathrm{v}})\\ M\dot{q}_{\mathrm{v}}^2(S_2\cos q_{\mathrm{v}}-S_1\sin q_{\mathrm{v}})\end{pmatrix}=\begin{pmatrix}\frac{D(T_{\mathrm{l}}-T_{\mathrm{r}})}{r_{\mathrm{w}}}\\ \frac{(T_{\mathrm{l}}+T_{\mathrm{r}})}{r_{\mathrm{w}}}\end{pmatrix}\quad .\end{aligned} \tag{14}$$

In the next step the dynamic model of the DC motor has to be given and it must be combined with the kinematic and dynamic model of the cart.

4.3 Combination of the Dynamic Models of the DC Motor and the Cart

It is assumed that the right and left wheels are driven by identical motors. The model of the electrical part of the motor was taken from [20] as follows:

$$\dot{Q}_{\mathrm{e}} = \frac{-RQ_{\mathrm{e}} - K^2\dot{q} + KU}{L} \tag{15}$$

where

- q denotes the rotational angle of the motor's axle,
- Q_{e} is the torque of electromagnetic origin exerted on the motor's axle,
- R is the Ohmic resistance of the motor's coil system,
- L is its inductivity,
- K is the motor's torque coefficient,
- U denotes the control voltage of the motor.

This model can be completed by mechanical parts as the rotational inertia momentum of the rotary part of the motor Θ^{Mot}, a viscous friction coefficient b determining the viscous torque curbing the rotation of the motor axle, and by the gear reduction ratio ν that is valid for the mechanical connection of the cart and motor axles.

In order to make it possible to combine (14) with the motor model of the drives as it was done in (15) it is expedient to rearrange (14) for obtaining $\ddot{q}_{\mathrm{r}}$ and $\ddot{q}_{\mathrm{l}}$ and T_{r} and T_{l} separately, and following that we can add the additional torque contributions at the motor axle level as

$$\begin{gathered}
\left[\frac{\Theta^{\mathrm{Mot}}}{\nu} I + \frac{-\nu r_{\mathrm{w}}^2}{2D} \begin{pmatrix} \frac{1}{r_{\mathrm{w}}} & -\frac{D}{r_{\mathrm{w}}} \\ -\frac{1}{r_{\mathrm{w}}} & -\frac{D}{r_{\mathrm{w}}} \end{pmatrix} \times \right. \\
\left. \left(\begin{array}{c|c} \frac{r_{\mathrm{w}}}{2D}(\Theta_{11}+\Theta_{22}) + \frac{Mr_{\mathrm{w}}}{2}(S_2 sq_{\mathrm{v}} + S_1 cq_{\mathrm{v}}) & -\frac{r_{\mathrm{w}}}{2D}(\Theta_{11}+\Theta_{22}) + \frac{Mr_{\mathrm{w}}}{2}(S_2 sq_{\mathrm{v}} + S_1 cq_{\mathrm{v}}) \\ \hline \frac{Mr_{\mathrm{w}}}{2D}(S_2 sq_{\mathrm{v}} + S_1 cq_{\mathrm{v}}) + \frac{Mr_{\mathrm{w}}}{2} & \frac{Mr_{\mathrm{w}}}{2} - \frac{Mr_{\mathrm{w}}}{2D}(S_2 sq_{\mathrm{v}} + S_1 cq_{\mathrm{v}}) \end{array} \right) \right] \times \\
\begin{pmatrix} \ddot{q}_{\mathrm{r}} \\ \ddot{q}_{\mathrm{l}} \end{pmatrix} + \frac{b}{\nu} \begin{pmatrix} \dot{q}_{\mathrm{r}} \\ \dot{q}_{\mathrm{l}} \end{pmatrix} + \frac{-\nu r_{\mathrm{w}}^2}{2D} \begin{pmatrix} \frac{1}{r_{\mathrm{w}}} & -\frac{D}{r_{\mathrm{w}}} \\ -\frac{1}{r_{\mathrm{w}}} & -\frac{D}{r_{\mathrm{w}}} \end{pmatrix} \begin{pmatrix} \frac{Mr_{\mathrm{w}}}{2}(\dot{q}_{\mathrm{r}} + \dot{q}_{\mathrm{l}})\dot{q}_{\mathrm{v}}(S_2 cq_{\mathrm{v}} - S_1 sq_{\mathrm{v}}) \\ M\dot{q}_{\mathrm{v}}^2 (S_2 cq_{\mathrm{v}} - S_1 sq_{\mathrm{v}}) \end{pmatrix} = \begin{pmatrix} Q_{\mathrm{r}}^{\mathrm{eMot}} \\ Q_{\mathrm{l}}^{\mathrm{eMot}} \end{pmatrix}
\end{gathered} \tag{16}$$

in which $Q_{\mathrm{r}}^{\mathrm{eMot}}$ and $Q_{\mathrm{l}}^{\mathrm{eMot}}$ describe the actual electromagnetic torque of the DC motor drives, $I \in \mathrm{I\!R}^{2\times 2}$ is the unit matrix, and $sq_{\mathrm{v}} \equiv \sin q_{\mathrm{v}}$, and $cq_{\mathrm{v}} \equiv \cos q_{\mathrm{v}}$ were introduced only for reducing the size of (16).

The most important conclusion is that (16) does not correspond to an LTI system if the tracked point of the cart is not placed in the mass center point, i.e. whenever $\hat{S} \neq 0$ and consequently $S \neq 0$. Besides the nonlinearities of the $\sin q_{\mathrm{v}}$ and $\cos q_{\mathrm{v}}$ function quadratic terms appear in $\dot{q}_{\mathrm{r}}$ and $\dot{q}_{\mathrm{l}}$ if $\hat{S} \neq 0$. This more complicated model and the assumption that we are in the possession of only approximate model parameters substantiated the idea that instead finding some model-based feedback gains that normally can be done by the use of Lyapunov's direct method it would be simpler to apply some RFPT-based adaptive deformation in the dynamic control of the system. We also had to keep it in mind that that the appropriate combination of (16) with the dynamical constraint equations (12) and (13) is necessary.

4.4 Anisotropic and Dynamically Varying Resolution of the Non-Holonomic Constraints

Assume that the controller has to track a nominal route defined by the coordinates $\xi^{\mathrm{N}} \stackrel{def}{=} \left(q_{\mathrm{v}}^{\mathrm{N}}(t), R_1^{\mathrm{N}}(t), R_2^{\mathrm{N}}(t)\right)^T$. By the use of a time-constant $\Lambda > 0$ a purely kinematically designed PID-type tracking error relaxation can be prescribed as

$$h_{\mathrm{int}}(t) \stackrel{def}{=} \int_{t_0}^{t} \left(\xi^{\mathrm{N}}(\zeta) - \xi(\zeta)\right) d\zeta, \; \left(\frac{d}{dt} + \Lambda\right)^3 h_{\mathrm{int}}(t) \equiv 0 \tag{17}$$

where $\xi \stackrel{def}{=} \left(q_{\mathrm{v}}(t), R_1(t), R_2(t)\right)^T$. This idea works because the solution of the differential equation $\left(\frac{d}{dt} + \Lambda\right) f(t) \equiv 0$ is $f(t) = f(t_0)\exp(-\Lambda(t-t_0)) \longrightarrow 0$ as $t \longrightarrow \infty$. Equation (17) simply leads to the desired 2nd time-derivatives as

$$\ddot{\xi}^{\mathrm{Des}}(t) = \Lambda^3 h_{\mathrm{int}}(t) + 3\Lambda^2 \xi(t) + 3\Lambda\dot{\xi}(t) + \ddot{\xi}^{\mathrm{N}}(t) \ . \tag{18}$$

If we need only a PD-type tracking policy by similar considerations we can arrive to

$$\ddot{\xi}^{\mathrm{Des}}(t) = 2\Lambda^2 \xi(t) + 2\Lambda\dot{\xi}(t) + \ddot{\xi}^{\mathrm{N}}(t) \ . \tag{19}$$

The kinematically designed (18) or (19) tracking prescriptions cannot be exactly realized due to the anti-skidding constraints given in (12) and (13).

Since in the practice precise tracking of each component of $\xi^{\mathrm{N}}(t)$ is important we cannot afford to permanently neglect one of them as it was done in [12] where the Author concentrated only on tracking the coordinate y. Instead of that the tracking error somehow must be distributed between these components. If we apply a digital controller it seems to be an appropriate tackling of this problem to minimize the following weighted sum at each control cycle according to $\ddot{q}_{\mathrm{r}}$ and $\ddot{q}_{\mathrm{l}}$:

$$\Phi(\ddot{q}_{\mathrm{r}}, \ddot{q}_{\mathrm{l}}) := \kappa_1(\ddot{\xi}_1(t) - \ddot{\xi}_1^{\mathrm{Des}}(t))^2 + \kappa_2(\ddot{\xi}_2(t) - \ddot{\xi}_2^{\mathrm{Des}}(t))^2 + \kappa_3(\ddot{\xi}_3(t) - \ddot{\xi}_3^{\mathrm{Des}}(t))^2 \tag{20}$$

where the weights κ_1, κ_2, $\kappa_3 > 0$ determine the relative significance of the orientation error and the positioning error. If $\kappa_2 \neq \kappa_3$ we can speak about non-isotropic satisfaction of the constraints over the plane (x, y). This minimum can be found easily as

$$\begin{gathered}
0 = \frac{\partial\Phi}{\partial\ddot{q}_{\mathrm{r}}} = \\
2\kappa_1(\ddot{\xi}_1(t) - \ddot{\xi}_1^{\mathrm{Des}}(t))\frac{\partial\ddot{\xi}_1}{\partial\ddot{q}_{\mathrm{r}}} + 2\kappa_2(\ddot{\xi}_2(t) - \ddot{\xi}_2^{\mathrm{Des}}(t))\frac{\partial\ddot{\xi}_2}{\partial\ddot{q}_{\mathrm{r}}} + \\
2\kappa_3(\ddot{\xi}_3(t) - \ddot{\xi}_3^{\mathrm{Des}}(t))\frac{\partial\ddot{\xi}_3}{\partial\ddot{q}_{\mathrm{r}}} \\
0 = \frac{\partial\Phi}{\partial\ddot{q}_{\mathrm{l}}} = \\
2\kappa_1(\ddot{\xi}_1(t) - \ddot{\xi}_1^{\mathrm{Des}}(t))\frac{\partial\ddot{\xi}_1}{\partial\ddot{q}_{\mathrm{l}}} + 2\kappa_2(\ddot{\xi}_2(t) - \ddot{\xi}_2^{\mathrm{Des}}(t))\frac{\partial\ddot{\xi}_2}{\partial\ddot{q}_{\mathrm{l}}} + \\
2\kappa_3(\ddot{\xi}_3(t) - \ddot{\xi}_3^{\mathrm{Des}}(t))\frac{\partial\ddot{\xi}_3}{\partial\ddot{q}_{\mathrm{l}}}
\end{gathered} \tag{21}$$

in which according to (13)

$$
\begin{aligned}
&\frac{\partial \ddot{\xi}_1}{\partial \ddot{q}_r} = \frac{\partial \ddot{q}_v}{\partial \ddot{q}_r} = -\frac{r_w}{2D}, \frac{\partial \ddot{\xi}_2}{\partial \ddot{q}_r} = \frac{\partial \ddot{R}_1}{\partial \ddot{q}_r} = \frac{r_w \sin q_v}{2}, \frac{\partial \ddot{\xi}_3}{\partial \ddot{q}_r} = \frac{\partial \ddot{R}_2}{\partial \ddot{q}_r} = -\frac{r_w \cos q_v}{2}, \\
&\frac{\partial \ddot{\xi}_1}{\partial \ddot{q}_l} = \frac{\partial \ddot{q}_v}{\partial \ddot{q}_l} = \frac{r_w}{2D}, \frac{\partial \ddot{\xi}_2}{\partial \ddot{q}_l} = \frac{\partial \ddot{R}_1}{\partial \ddot{q}_l} = \frac{r_w \sin q_v}{2}, \frac{\partial \ddot{\xi}_3}{\partial \ddot{q}_l} = \frac{\partial \ddot{R}_2}{\partial \ddot{q}_l} = -\frac{r_w \cos q_v}{2},
\end{aligned}
\tag{22}
$$

that is

$$
\begin{aligned}
0 = &-\kappa_1(\ddot{\xi}_1(t) - \ddot{\xi}_1^{\mathrm{Des}}(t))\frac{r_w}{2D} + \kappa_2(\ddot{\xi}_2(t) - \ddot{\xi}_2^{\mathrm{Des}}(t))\frac{r_w \sin q_v}{2} \\
&-\kappa_3(\ddot{\xi}_3(t) - \ddot{\xi}_3^{\mathrm{Des}}(t))\frac{r_w \cos q_v}{2} \\
0 = &\kappa_1(\ddot{\xi}_1(t) - \ddot{\xi}_1^{\mathrm{Des}}(t))\frac{r_w}{2D} + \kappa_2(\ddot{\xi}_2(t) - \ddot{\xi}_2^{\mathrm{Des}}(t))\frac{r_w \sin q_v}{2} \\
&-\kappa_3(\ddot{\xi}_3(t) - \ddot{\xi}_3^{\mathrm{Des}}(t))\frac{r_w \cos q_v}{2}.
\end{aligned}
\tag{23}
$$

According to (13) these expressions are linear in $\ddot{q}_r$ and $\ddot{q}_l$, therefore the solution simply can be obtained:

$$
\begin{aligned}
0 = &-\kappa_1\left(-\frac{r_w}{D}\frac{\ddot{q}_r-\ddot{q}_l}{2} - \ddot{\xi}_1^{\mathrm{Des}}(t)\right)\frac{r_w}{2D} + \\
&\kappa_2\left(r_w\frac{\ddot{q}_r+\ddot{q}_l}{2}\sin q_v + r_w\frac{\dot{q}_r+\dot{q}_l}{2}\cos q_v \dot{q}_v - \ddot{\xi}_2^{\mathrm{Des}}(t)\right)\frac{r_w \sin q_v}{2} \\
&-\kappa_3\left(-r_w\frac{\ddot{q}_r+\ddot{q}_l}{2}\cos q_v + r_w\frac{\dot{q}_r+\dot{q}_l}{2}\sin q_v \dot{q}_v - \ddot{\xi}_3^{\mathrm{Des}}(t)\right)\frac{r_w \cos q_v}{2}, \\
0 = &\kappa_1\left(-\frac{r_w}{D}\frac{\ddot{q}_r-\ddot{q}_l}{2} - \ddot{\xi}_1^{\mathrm{Des}}(t)\right)\frac{r_w}{2D} + \\
&\kappa_2\left(r_w\frac{\ddot{q}_r+\ddot{q}_l}{2}\sin q_v + r_w\frac{\dot{q}_r+\dot{q}_l}{2}\cos q_v \dot{q}_v - \ddot{\xi}_2^{\mathrm{Des}}(t)\right)\frac{r_w \sin q_v}{2} \\
&-\kappa_3\left(-r_w\frac{\ddot{q}_r+\ddot{q}_l}{2}\cos q_v + r_w\frac{\dot{q}_r+\dot{q}_l}{2}\sin q_v \dot{q}_v - \ddot{\xi}_3^{\mathrm{Des}}(t)\right)\frac{r_w \cos q_v}{2}.
\end{aligned}
\tag{24}
$$

This system of linear equations conveniently can be solved for $\ddot{q}_- \stackrel{def}{=} (\ddot{q}_r - \ddot{q}_l)$ and $\ddot{q}_+ \stackrel{def}{=} (\ddot{q}_r + \ddot{q}_l)$:

$$
\begin{aligned}
&\begin{pmatrix} \frac{\kappa_1 r_w^2}{4D^2} & \frac{r_w^2(\kappa_2 \sin^2 q_v + \kappa_3 \cos^2 q_v)}{4} \\ -\frac{\kappa_1 r_w^2}{4D^2} & \frac{r_w^2(\kappa_2 \sin^2 q_v + \kappa_3 \cos^2 q_v)}{4} \end{pmatrix} \begin{pmatrix} \ddot{q}_- \\ \ddot{q}_+ \end{pmatrix} = \\
&\begin{pmatrix} -\kappa_1 \ddot{\xi}_1^{\mathrm{Des}} \frac{r_w}{2D} + \kappa_2 \ddot{\xi}_2^{\mathrm{Des}} \frac{r_w \sin q_v}{2} - \kappa_3 \ddot{\xi}_3^{\mathrm{Des}} \frac{r_w \cos q_v}{2} + (\kappa_3 - \kappa_2) r_w^2 \frac{\dot{q}_r + \dot{q}_l}{4} \cos q_v \sin q_v \dot{q}_v \\ \kappa_1 \ddot{\xi}_1^{\mathrm{Des}} \frac{r_w}{2D} + \kappa_2 \ddot{\xi}_2^{\mathrm{Des}} \frac{r_w \sin q_v}{2} - \kappa_3 \ddot{\xi}_3^{\mathrm{Des}} \frac{r_w \cos q_v}{2} + (\kappa_3 - \kappa_2) r_w^2 \frac{\dot{q}_r + \dot{q}_l}{4} \cos q_v \sin q_v \dot{q}_v \end{pmatrix}.
\end{aligned}
\tag{25}
$$

By adding the two equations it is obtained that

$$
\ddot{q}_+^{\mathrm{Des}} = \frac{2\left(\kappa_2 \ddot{\xi}_2^{\mathrm{Des}} \sin q_v - \kappa_3 \ddot{\xi}_3^{\mathrm{Des}} \cos q_v + (\kappa_3 - \kappa_2) r_w^2 \frac{\dot{q}_r + \dot{q}_l}{2} \cos q_v \sin q_v \dot{q}_v\right)}{r_w^2(\kappa_2 \sin^2 q_v + \kappa_3 \cos^2 q_v)}
\tag{26}
$$

Via subtraction the 2nd equation from the 1st one we obtain that

$$
\ddot{q}_-^{\mathrm{Des}} = -\frac{2D^2}{\kappa_1 r_w^2}\frac{\kappa_1 r_w}{D}\ddot{\xi}_1^{\mathrm{Des}} = -\frac{2D}{r_w}\ddot{\xi}_1^{\mathrm{Des}}
\tag{27}
$$

from which it follows that

$$
\ddot{q}_r^{\mathrm{Des}} = \frac{\ddot{q}_+^{\mathrm{Des}} + \ddot{q}_-^{\mathrm{Des}}}{2}, \text{ and } \ddot{q}_l^{\mathrm{Des}} = \frac{\ddot{q}_+^{\mathrm{Des}} - \ddot{q}_-^{\mathrm{Des}}}{2}.
\tag{28}
$$

It is interesting that κ_1 is canceled but κ_2 and κ_3 remain within the solution.

On the basis of physical considerations one would be apt to insist on isotropic constraint satisfaction by prescribing the condition $\kappa_2 = \kappa_3$. However, this idea is not necessarily reasonable in the case of a double wheeled cart under nonholonomic constraints. To explain that consider Fig. 1 and the constraint equation (13). If $q_{\mathrm{v}} = 0$ and $q_{\mathrm{v}} = \pi$ $\ddot{q}_{\mathrm{r}}$ and $\ddot{q}_{\mathrm{l}}$ has no any effect on $\ddot{R}_1$. In similar manner whenever $q_{\mathrm{v}} = \pm\frac{\pi}{2}$ $\ddot{q}_{\mathrm{r}}$ and $\ddot{q}_{\mathrm{l}}$ has no any effect on $\ddot{R}_2$. In the vicinity of these special points the kinematically prescribed $\ddot{R}_1^{\mathrm{Des}}$ or $\ddot{R}_2^{\mathrm{Des}}$ values have only very little effect in the solution obtained for $\ddot{q}_+$. These effects can be well tracked in (26), too, and physically can be interpreted and further utilized in a very simple manner: if some effect is very small, e.g. there is no way to significantly modify say $\ddot{x}$ it is better to completely abandon it and let us freely correct $\ddot{y}$, and vice versa. This strategy is well approximated by the dynamically varying parameters as $\kappa_2 \overset{def}{=} \kappa_{20}\left(1-\cos^2 q_{\mathrm{v}}\right)$ and $\kappa_3 \overset{def}{=} \kappa_{30}\left(1-\sin^2 q_{\mathrm{v}}\right)$. The choice $\kappa_{20} = \kappa_{30}$ may be reasonable, too. In the simulations this choice will be applied. In the sequel the basics of the RFPT-based adaptivity is summarized.

5 The RFPT-Based Adaptivity

In the in control step n of the digital controller according to (26), (27), and (28) the kinematically formulated "desired acceleration" $\ddot{\xi}^{\mathrm{Des}}(n)$ can be calculated then its best realizable approximation $\ddot{q}_{\mathrm{r}}^{\mathrm{DesR}}(n)$ and $\ddot{q}_{\mathrm{l}}^{\mathrm{DesR}}(n)$ can be taken according to the non-holonomic constraints. In the next step this signal is adaptively deformed as $\ddot{q}_{\mathrm{r}}^{\mathrm{Req}}(n)$ and $\ddot{q}_{\mathrm{l}}^{\mathrm{Req}}(n)$ and it is substituted into the approximate dynamic model of the system as it was suggested in [7]. The motor torques are calculated according to the approximate model and are exerted on the controlled system the actual response of which is assumed to be observable. In this manner a sequence of control signals is generated as given in (29):

$$\ddot{q}_{\mathrm{rl}}^{\mathrm{Req}}(n) = G(\ddot{q}_{\mathrm{rl}}^{\mathrm{Req}}(n-1), \ddot{q}_{\mathrm{rl}}^{\mathrm{DesR}}(n), \ddot{q}_{\mathrm{rl}}(n-1)) \tag{29}$$

in which the function G has three adaptive parameters as A_{c}, K_{c}, and B_{c}, and $\ddot{q}_{\mathrm{rl}}(n-1)$ denotes the observable quantities at cycle n. These parameters must be so set that the mapping defined in (29) for constant (that is practically slowly varying) $\ddot{q}_{\mathrm{rl}}^{\mathrm{DesR}}$ becomes contractive for the sequence $\ddot{q}_{\mathrm{rl}}^{\mathrm{Req}}(n)$. Then, according to Banach's Fixed Point Theorem [8] this sequence is contractive and converges to the solution of the control task $\ddot{q}_{\mathrm{rl}}^{\star}$ for which the observable quantity equals to the desired one, i.e. $\ddot{q}_{\mathrm{rl}}(n) = \ddot{q}_{\mathrm{rl}}^{\mathrm{DesR}}$. If the quantity $\ddot{q}_{\mathrm{rl}}^{\mathrm{DesR}}$ slowly varies in time the fixed point drifts appropriately and the sequence $\ddot{q}_{\mathrm{rl}}^{\mathrm{Req}}(n)$ remains in the close vicinity of the actual solution of the control task. The G function is realized by program code in "Scilab" as follows:

```
function [ki]=szigmoid(x)
  ki=x/(1+abs(x));
endfunction
function [Req]=G(desired,past_exerted,past_observed)
```

```
    global Ac Bc Kc
    error=norm(past_observed-desired,'fro');
    if error>1e-4 then
        e_direction=(past_observed-desired)/error;
        B_factor=Bc*szigmoid(Ac*error);
        Req=(1+B_factor)*past_exerted+B_factor*Kc*e_direction;
    else
        Req=past_exerted;
    end
endfunction
```

Regarding the physical interpretation of this solution a nonlinear mapping is applied in the direction of the "response error" $e_{\text{direction}} \in \mathbb{R}^2$.

For appropriate convergence usually the settings $B_c = \pm 1$, $|K_c|$ is a big positive number, and A_c a small positive number used to work well. These quantities can be set by simulations via considering the operation of a simple non-adaptive PID controller and observing the order of magnitude of the occurring $\ddot{q}$ signals. Then $|K_c|$ must be significantly greater (about 10 or 100 times greater) than the observed maximum. The parameter A_c normally used to be about $0.1 \cdot |K_c|$, and B_c must be set to either 1 or -1 in order to obtain convergence.

Regarding the order reduction by the use of (16) for the rough estimation of the required electromagnetic motor torques (30) was used

$$\begin{aligned} Q_e^{\text{Req}} &= \Theta \ddot{q}^{\text{Mot}} - \nu T^{\text{Req}} + b\dot{q}^{\text{Mot}}; \\ U^{\text{Req}} &\stackrel{def}{=} \frac{R}{K} Q_e^{\text{Req}} + K\dot{q}^{\text{Mot}} \quad . \end{aligned} \tag{30}$$

In the simulations the exact system model was used to calculate the realized system response $\ddot{q}_{\text{rl}}$. In the sequel simulation results are presented.

6 Simulation Results

Various solutions were investigated as examples that needed various control parameter settings. The common parameters belong to the exact and approximate system models as follows:

Parameters of the Exact System Model. These parameters were set as follows:

- The significant inertia momentum matrix elements in the "initial position" $\hat{\Theta}_{11}^{\text{Exact}} = 5\,\text{kg}\,\text{m}^2$, $\hat{\Theta}_{22}^{\text{Exact}} = 6\,\text{kg}\,\text{m}^2$;
- The offset of the mass center point in the "initial position" $\hat{S}_1^{\text{Exact}} = 0.3\,\text{m}$, and $\hat{S}_2^{\text{Exact}} = 0.4\,\text{m}$;
- The mass of the whole cart $M^{\text{Exact}} = 2\,\text{kg}$;
- The viscous friction coefficient of the motor's axle $b^{\text{Exact}} = 0.1\,\text{N}\,\text{m}\,\text{s}\,\text{rad}^{-1}$;

- The gear reduction ratio $\nu^{\mathrm{Exact}} = 0.1$ non – dimenional;
- The Ohmic resistance of the motor's coil $R^{\mathrm{Exact}} = 1\,\mathrm{Ohm}$;
- The inductivity of the motor's coil $L^{\mathrm{Exact}} = 0.5\,\mathrm{H}$;
- The motor's torque coefficient $K^{\mathrm{Exact}} = 0.01\,\mathrm{N\,m\,A^{-1}}$ (Normally, the torque coefficient K_{t} relates the motor current to the motor torque as $T = K_{\mathrm{t}} i$ from which its dimension is derived. The back EMF is proportional to the angular velocity of the shaft by a constant factor K_{e} as $e = K_{\mathrm{e}} \dot{q}^{\mathrm{Mot}}$. In SI units, the motor torque and back emf constants are equal, that is, $K_{\mathrm{t}} = K_{\mathrm{e}}$, therefore, simply the common quantity K can be used to represent both the motor torque constant and the back EMF constant [20].)
- The inertia momentum of the motor's rotary part $\Theta^{\mathrm{MotExact}} = 0.01\,\mathrm{kg\,m^2}$;
- The common radius of the wheels $r_{\mathrm{w}}^{\mathrm{Exact}} = 0.1\,\mathrm{m}$;
- The half distance between the wheels $D^{\mathrm{Exact}} = 1\,\mathrm{m}$;

Parameters of the Approximate System Model. These parameters were set as follows:

- The significant inertia momentum matrix elements in the "initial position" $\hat{\Theta}_{11}^{\mathrm{Approx}} = 7\,\mathrm{kg\,m^2}$, $\hat{\Theta}_{22}^{\mathrm{Approx}} = 8\,\mathrm{kg\,m^2}$;
- The offset of the mass center point in the "initial position" $\hat{S}_1^{\mathrm{Approx}} = 0\,\mathrm{m}$, and $\hat{S}_2^{\mathrm{Approx}} = 0\,\mathrm{m}$;
- The mass of the whole cart $M^{\mathrm{Approx}} = 2.5\,\mathrm{kg}$;
- The viscous friction coefficient of the motor's axle $b^{\mathrm{Approx}} = 0.05\,\mathrm{N\,m\,s\,rad^{-1}}$;
- The gear reduction ratio $\nu^{\mathrm{Approx}} = \nu^{\mathrm{Exact}}$;
- The Ohmic resistance of the motor's coil $R^{\mathrm{Approx}} = 1.5\,\mathrm{Ohm}$;
- The inductivity of the motor's coil $L^{\mathrm{Approx}} = 0.55\,\mathrm{H}$;
- The motor's torque coefficient $K^{\mathrm{Approx}} = 0.02\,\mathrm{N\,m\,A^{-1}}$;
- The inertia momentum of the motor's rotary part $\Theta^{\mathrm{MotApprox}} = \Theta^{\mathrm{MotExact}}$;
- The common radius of the wheels $r_{\mathrm{w}}^{\mathrm{Approx}} = r_{\mathrm{w}}^{\mathrm{Exact}}$;
- The half distance between the wheels $D^{\mathrm{Approx}} = D^{\mathrm{Ecaxt}}$;

In the simulation the cycle time of the controller was 1 ms. In the sequel various examples will be considered.

Example 1 (PID-type control with anisotropic dynamic constraint satisfaction). In this example a variant of the kinematically prescribed PID-type tracking policy defined in (18) was applied as

$$\dddot{\xi}^{\mathrm{Des}}(t) = \Lambda^3 w \tanh\left(\frac{h_{\mathrm{int}}(t)}{w}\right) + 3\Lambda^2 \xi(t) + 3\Lambda \dot{\xi}(t) + \dddot{\xi}^{\mathrm{N}}(t) \qquad (31)$$

since it was observed that because this strategy cannot be exactly realized without some truncation determined by the parameter w the integrated term may cause high oscillation in the tracking error. In the case of a PID-type control

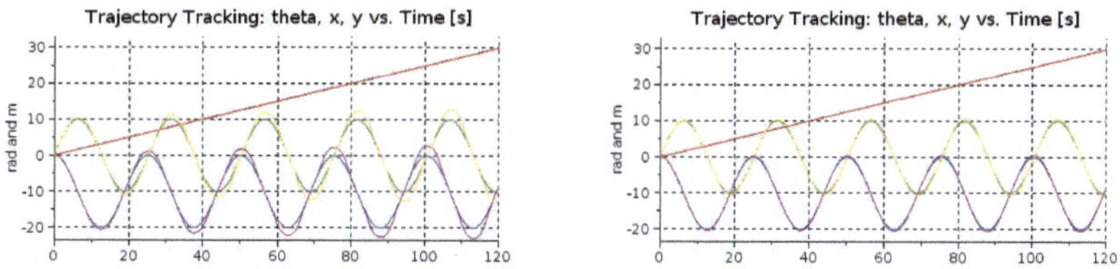

Fig. 2. The trajectory tracking of the PID-type controller: non-adaptive control at LHS, adaptive control at RHS (q_v^N: black, q_v: red, x^N: blue, x: purple, y^N: green, y: ocher)

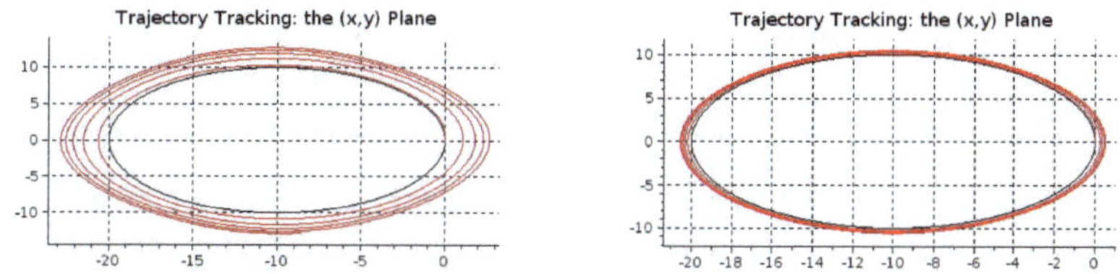

Fig. 3. The trajectory tracking of the PID-type controller on the (x, y) plane: non-adaptive control at LHS, adaptive control at RHS (nominal trajectory: black, simulated: red)

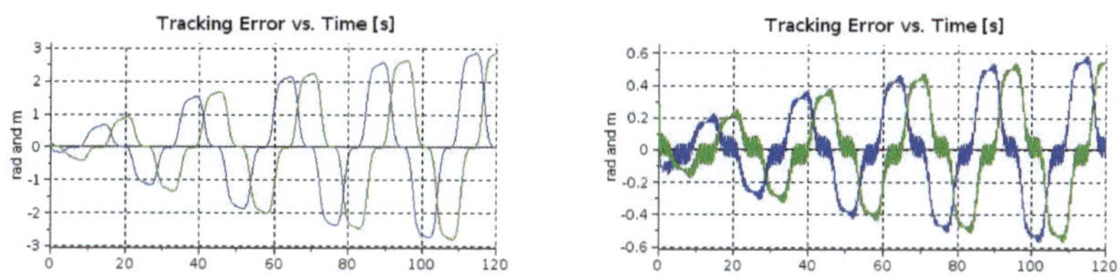

Fig. 4. The trajectory tracking error of the PID-type controller: non-adaptive control at LHS, adaptive control at RHS (for q_v: black, for x: blue, for y: green)

relatively small $\Lambda = 5\,\mathrm{s}^{-1}$ worked well for the motion considered with $w = 0.06\,\mathrm{rad\,s}$ or $w = 0.06\,\mathrm{m\,s}$ depending on the components describing the orientation or the displacement over the plane. The adaptive control parameters were $B_c = -1$, $K_c = 2000\,\mathrm{rad\,s}^{-2}$, and $A_c = 5 \times 10^{-5}\,\mathrm{rad}^{-1}\,\mathrm{s}^2$. The anisotropy parameters were $\kappa_{20} = \kappa_{30} = 1$ i.e. only the dynamic variation of κ_2 and κ_3 caused anisotropy in the trajectory tracking. Figures 2, 3, and 4 reveal the effect of adaptivity that considerably improved the tracking accuracy.

To reveal operation of the controllers the rotational speeds of the cart axles are described in Fig. 5. It is easy to give a qualitative interpretation of the achieved result. Whenever one wishes to take a narrow parking place normally an "iterative" strategy is applied in which at first the car's orientation is set at the cost of corrupting the position. In the next step the position is corrected that corrupts the orientation, etc. The fast corrections in $\dot{q}_r$ and $\dot{q}_l$ play similar role in the adaptive control, and as a consequence, the tracking accuracy is improved. The non-adaptive controller was not able to make such fast correcting movements. In the chart of the control voltages of the adaptive controller (Fig. 6) one

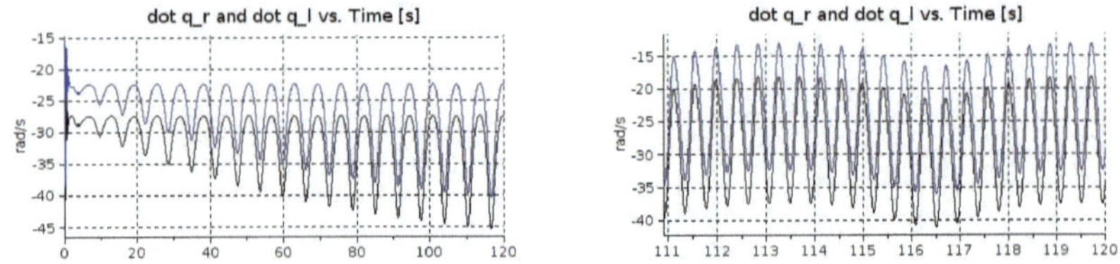

Fig. 5. The rotational speeds of the motor axles for the PID-type controller: non-adaptive control at LHS, adaptive control at RHS (a zoomed-in picture to reveal details) ($\ddot{q}_{\mathrm{r}}$: black, $\ddot{q}_{\mathrm{l}}$: blue)

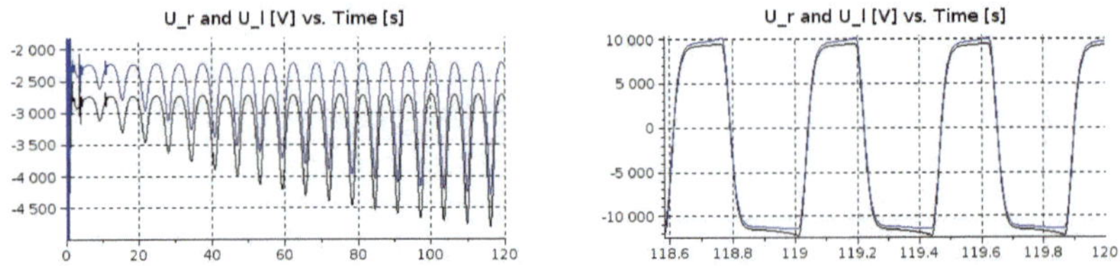

Fig. 6. The control voltages of the motors for the PID-type controller: non-adaptive control at LHS, adaptive control at RHS (a zoomed-in picture to reveal details) (U_{r}: black, U_{l}: blue)

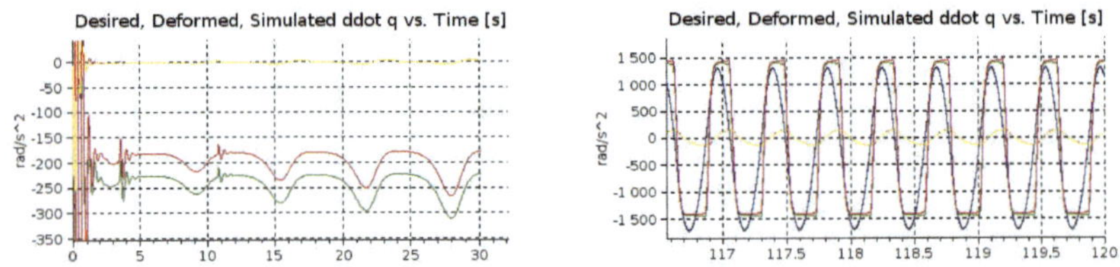

Fig. 7. The "Desired", the "Adaptively Deformed", and the "Simulated" $\ddot{q}_{\mathrm{r}}$ and $\ddot{q}_{\mathrm{l}}$ values for the PID-type controller: non-adaptive control at LHS, adaptive control at RHS (zoomed-in pictures to reveal details) ($\ddot{q}_{\mathrm{r}}^{\mathrm{Des}}$: black, $\ddot{q}_{\mathrm{r}}^{\mathrm{Req}}$: green, $\ddot{q}_{\mathrm{r}}$: purple, $\ddot{q}_{\mathrm{l}}^{\mathrm{Des}}$: blue, $\ddot{q}_{\mathrm{l}}^{\mathrm{Req}}$: red, $\ddot{q}_{\mathrm{l}}$: ocher; in the non-adaptive case the "Desired" and the "Required" values are exactly identical due to the lack of any deformation, therefore the black line is *exactly covered* by the green one, and the red line exactly hides the blue one)

can distinguish between the sessions of fast variation and slow fine-corrections. Such sessions are missing in the non-adaptive case. This experience substantiates the qualitative expectations related to the operation of the "nonlinear order reduction" achieved by the adaptive controller.

To give an insight into the operation of the RFPT-based adaptivity in Fig. 7 the kinematically realizable "Desired", the adaptively deformed "Required", and the "Simulated" $\ddot{q}$ values are described. It can well be seen that in the non-adaptive case the "Simulated" values cannot approach the "Desired" ones, there are order of magnitude differences between them. However, in the adaptive case simulated values better approach the desired ones, and this situation evidently is achieved by the considerable adaptive deformation achieved.

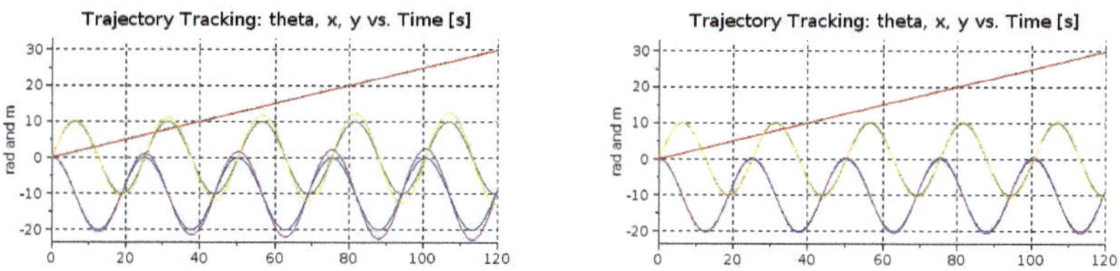

Fig. 8. The trajectory tracking of the PID-type controller: non-adaptive control at LHS, adaptive control at RHS (q_v^N: black, q_v: red, x^N: blue, x: purple, y^N: green, y: ocher)

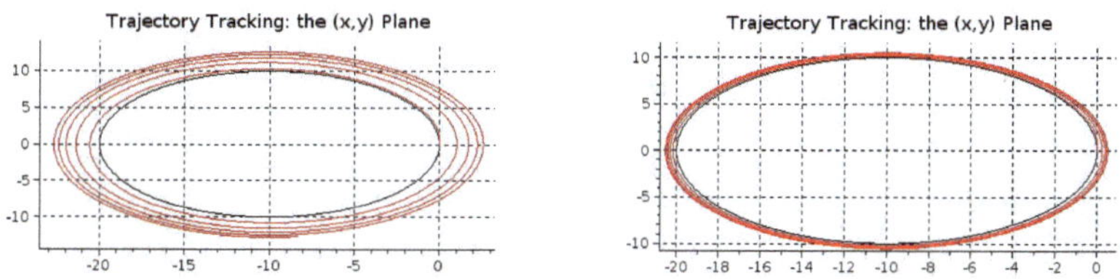

Fig. 9. The trajectory tracking of the PID-type controller on the (x, y) plane: non-adaptive control at LHS, adaptive control at RHS (nominal trajectory: black, simulated: red)

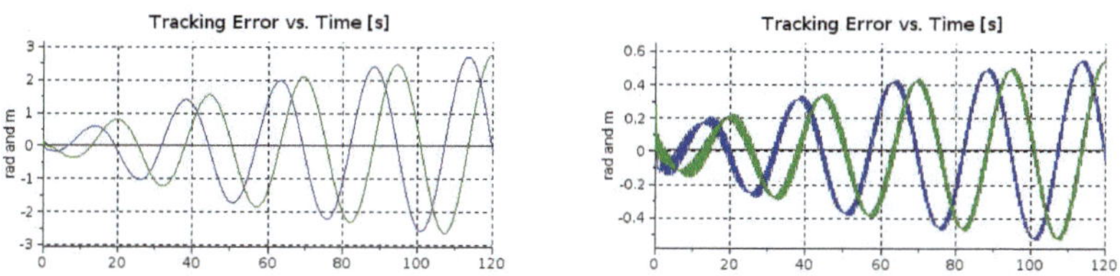

Fig. 10. The trajectory tracking error of the PID-type controller: non-adaptive control at LHS, adaptive control at RHS (for q_v: black, for x: blue, for y: green)

To make comparisons with the case using the "static" κ_2 and κ_3 parameters the next example has to be considered.

Example 2 (PID-type control with isotropic and static constraint satisfaction). In this case the $\kappa_2 = \kappa_3 \equiv 1$ constant settings were applied. The figures 8, 9, and 10 testify similar improvement achieved by the adaptivity. The precision of the trajectory tracking in the average is similar to that of the non-isotropic dynamic case, however, in little details significant differences can be observed.

Figure 11 testifies that the improved precision was achieved again by fast corrections in the $\dot{q}_r$ and $\dot{q}_l$ rotational speeds. According to Fig. 12 it can be stated that the idea of order reduction can be realized in similar manner as it was done in the case of the non-isotropic dynamic constraint satisfaction.

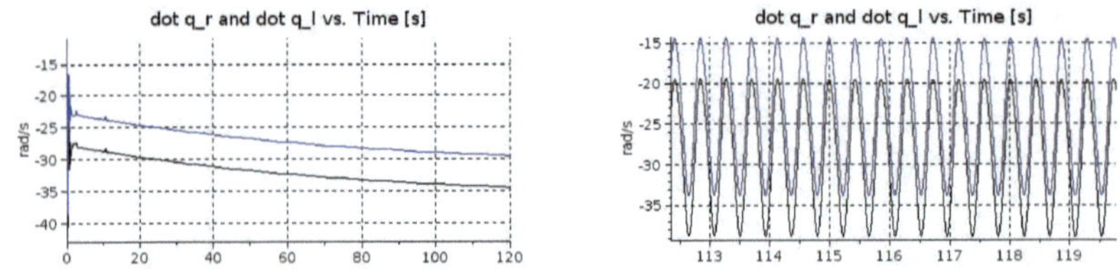

Fig. 11. The rotational speeds of the motor axles for the PID-type controller: non-adaptive control at LHS, adaptive control at RHS (a zoomed-in picture to reveal details) ($\ddot{q}_{\rm r}$: black, $\ddot{q}_{\rm l}$: blue)

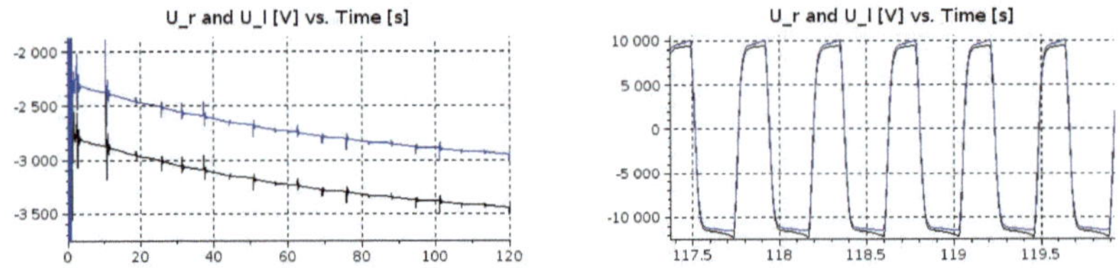

Fig. 12. The control voltages of the motors for the PID-type controller: non-adaptive control at LHS, adaptive control at RHS (a zoomed-in picture to reveal details) ($U_{\rm r}$: black, $U_{\rm l}$: blue)

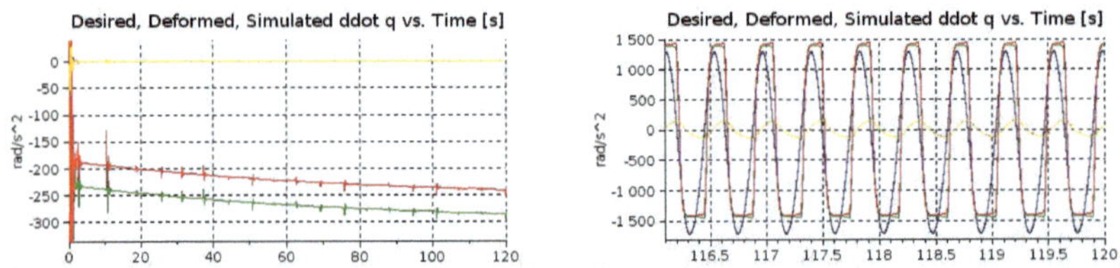

Fig. 13. The "Desired", the "Adaptively Deformed", and the "Simulated" $\ddot{q}_{\rm r}$ and $\ddot{q}_{\rm l}$ values for the PID-type controller: non-adaptive control at LHS, adaptive control at RHS (zoomed-in pictures to reveal details) ($\ddot{q}_{\rm r}^{\rm Des}$: black, $\ddot{q}_{\rm r}^{\rm Req}$: green, $\ddot{q}_{\rm r}$: purple, $\ddot{q}_{\rm l}^{\rm Des}$: blue, $\ddot{q}_{\rm l}^{\rm Req}$: red, $\ddot{q}_{\rm l}$: ocher; in the non-adaptive case the "Desired" and the "Required" values are exactly identical due to the lack of any deformation, therefore the black line is *exactly covered* by the green one, and the red line exactly hides the blue one)

The figures 13 and 14 again give some insight into the realization of the adaptive order reduction.

As a final example we consider a PD-type control with static, isotropic constraint resolution.

Example 3 (PD-type control with isotropic and static constraint satisfaction). In this case (19) was used but an increased value $\Lambda = 12\,{\rm s}^{-1}$ had to be chosen. The results are given in Figs. 20–15. It is evident that the PID-type controller is more advantageous because it results in much higher precision with considerably smaller feedback term determined by Λ.

Fig. 14. The "Desired", the "Adaptively Deformed", and the "Simulated" $\ddot{q}_r$ and $\ddot{q}_l$ values for the adaptive PID-type controller (zoomed-in picture to reveal details) ($\ddot{q}_r^{Des}$: black, $\ddot{q}_r^{Req}$: green, $\ddot{q}_r$: purple, $\ddot{q}_l^{Des}$: blue, $\ddot{q}_l^{Req}$: red, $\ddot{q}_l$: ocher; in the non-adaptive case the "Desired" and the "Required" values are exactly identical due to the lack of any deformation, therefore the black line is *exactly covered* by the green one, and the red line exactly hides the blue one)

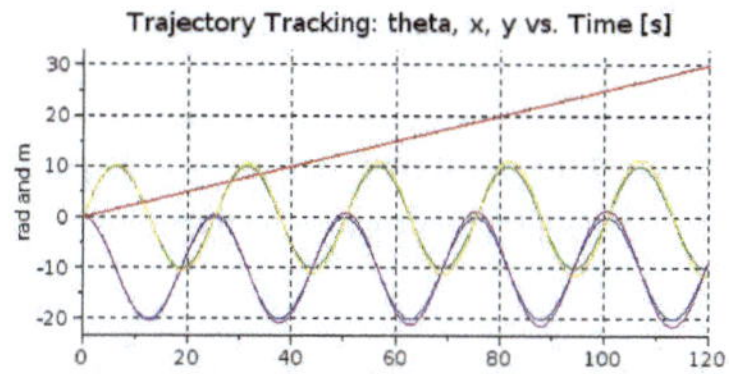

Fig. 15. The trajectory tracking of the adaptive PD-type controller (q_v^N: black, q_v: red, x^N: blue, x: purple, y^N: green, y: ocher)

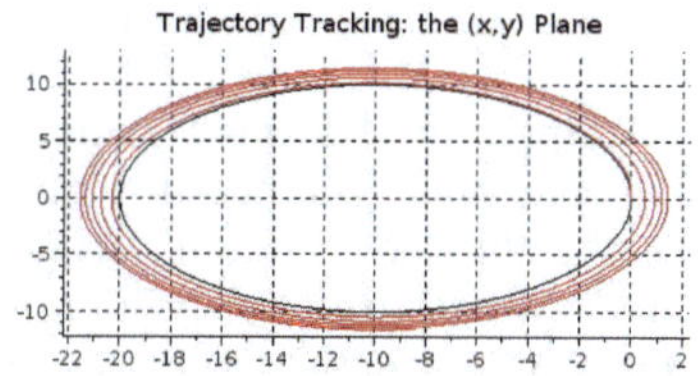

Fig. 16. The trajectory tracking of the adaptive PID-type controller on the (x, y) plane: (nominal trajectory: black, simulated: red)

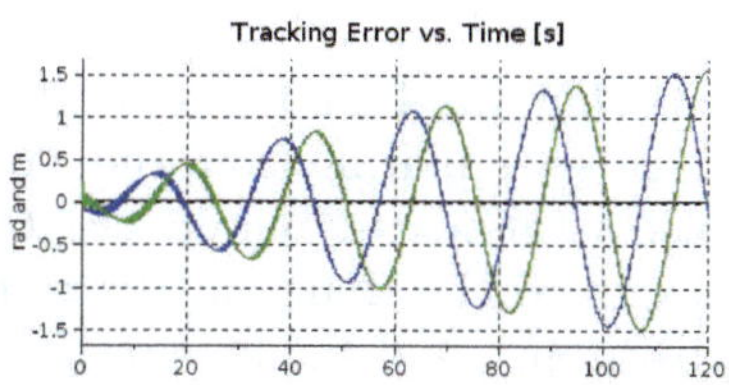

Fig. 17. The trajectory tracking error of the adaptive PD-type controller: (for q_v: black, for x: blue, for y: green)

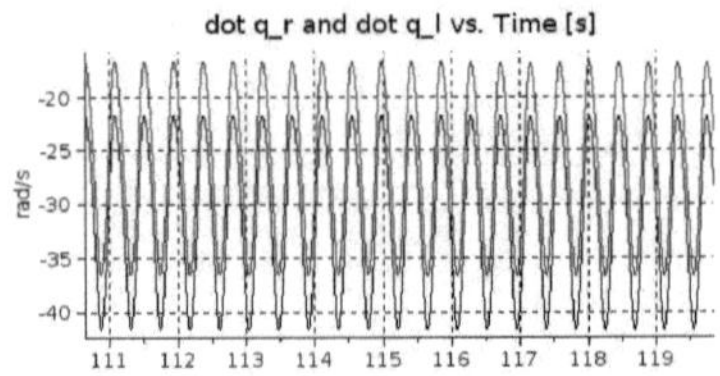

Fig. 18. The rotational speeds of the motor axles for the adaptive PD-type controller: (a zoomed-in picture to reveal details) ($\ddot{q}_\mathrm{r}$: black, $\ddot{q}_\mathrm{l}$: blue)

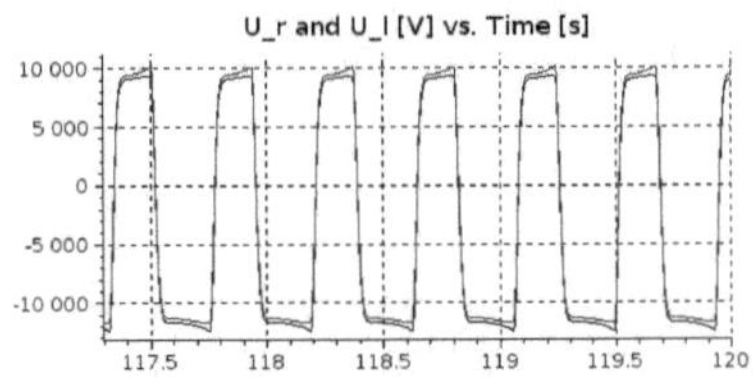

Fig. 19. The control voltages of the motors for the adaptive PD-type controller: (a zoomed-in picture to reveal details) (U_r: black, U_l: blue)

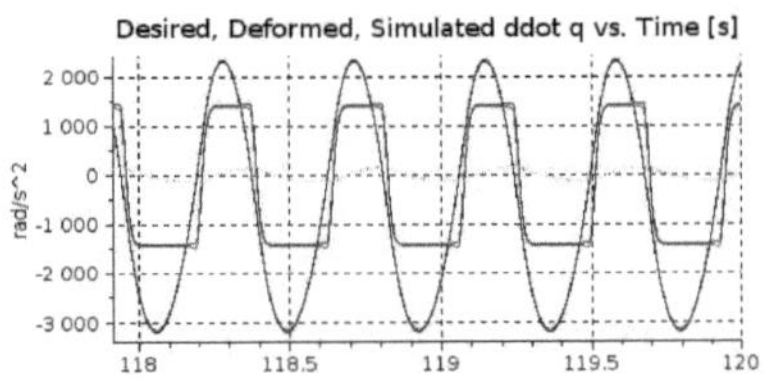

Fig. 20. The "Desired", the "Adaptively Deformed", and the "Simulated" $\ddot{q}_\mathrm{r}$ and $\ddot{q}_\mathrm{l}$ values for the adaptive PD-type controller (zoomed-in picture to reveal details) ($\ddot{q}_\mathrm{r}^\mathrm{Des}$: black, $\ddot{q}_\mathrm{r}^\mathrm{Req}$: green, $\ddot{q}_\mathrm{r}$: purple, $\ddot{q}_\mathrm{l}^\mathrm{Des}$: blue, $\ddot{q}_\mathrm{l}^\mathrm{Req}$: red, $\ddot{q}_\mathrm{l}$: ocher; in the non-adaptive case the "Desired" and the "Required" values are exactly identical due to the lack of any deformation, therefore the black line is *exactly covered* by the green one, and the red line exactly hides the blue one)

7 Conclusions

It was shown via simulations that the RFPT-based adaptive controller can simultaneously solve the problems originating from the modeling imprecisions necessitating the use of coupled nonlinear dynamic equations, the the necessary order reduction that arises due to the dynamic interaction of the mechanical and the electromagnetic subsystems, and the kinematic limitations that are the consequences of the non-holonomic constraints in the case of caster supported WMRs having only two actively driven wheels. The suggested approach is far simpler than any other solution that uses Lyapunov's direct method.

In the paper both PID and PD-type approaches were considered. It can be concluded that the PID-type solution with truncated integrated feedback is more advantageous than the PD-type because it can work with considerably smaller feedback gains and results in more precise trajectory tracking. It worths noting that in each case considered a slow increasing drift can be observed in the tracking error. This small drift may be the consequence of the incomplete realizability of the kinematically prescribed tracking policy.

An important feature of the RFPT-based design that it completely separates the design of the kinematically prescribed trajectory tracking and its realization by adaptively dealing with the dynamics of the system. It can be expected that via prescribing more "greedy" kinematic tracking policy this drift can be avoided.

Acknowledgment. This works has been partially supported by the Hungarian Scientific Research Fund OTKA K-106392 and the grant provided by the Project TÁMOP-4.2.2.A-11/1/KONV-2012-0012: Basic research for the development of hybrid and electric vehicles – The Project is supported by the Hungarian Government and co-financed by the European Social Fund.

References

1. Lyapunov, A.M.: A general task about the stability of motion (in Russian). Ph.D. Thesis, University of Kazan, Russia (1892)
2. Lyapunov, A.M.: Stability of motion. Academic Press, New-York (1966)
3. Sekaj, I., Veselý, V.: Robust output feedback controller design: Genetic Algorithm approach. IMA J Math Control Info 22(3), 257–265 (2005)
4. Slotine, J.-J.E., Li, W.: Applied Nonlinear Control. Prentice Hall International, Inc., Englewood Cliffs (1991)
5. Isermann, R., Lachmann, K.H., Matko, D.: Adaptive Control Systems. PPrentice-Hall, New York (1992)
6. Nguyen, C.C., Antrazi, S.S., Zhen-Lei, Z., Jr. Campbell, C.E.: Adaptive control of a Stewart platform-based manipulator. Journal of Robotic Systems 10(5), 657–687 (1993)
7. Tar, J.K., Bitó, J.F., Nádai, L., Tenreiro Machado, J.A.: Robust Fixed Point Transformations in Adaptive Control Using Local Basin of Attraction. Acta Polytechnica Hungarica 6, 21–37 (2009)
8. Banach, S.: Sur les opérations dans les ensembles abstraits et leur application aux équations intégrales (About the Operations in the Abstract Sets and Their Application to Integral Equations). Fund. Math. 3, 133–181 (1922)
9. Kósi, K., Tar, J.K., Rudas, I.J.: Improvement of the Stability of RFPT-based Adaptive Controllers by Observing "Precursor Oscillations". In: Proc. of the 9 the IEEE Intl. Conf. on Computational Cybernetics, Tihany, Hungary, pp. 267–272 (2013)
10. Oriolo, G., De Luca, A., Vendittelli, M.: WMR control via dynamic feedback linearization: design, implementation, and experimental validation. IEEE Transactions on Control Systems Technology 10(6), 835–852 (2002)
11. LeBel, S., Rodrigues, L.: Piecewise-affine parameter-varying control of wheeled mobile robots. In: Proc. of the American Control Conference, Seattle, WA, USA (WeA05.3), pp. 195–200 (2008)
12. Ojleska, V.M.: Control and computational synergy of switched-fuzzy and fuzzy-neural nonlinear systems. Ph.D. Thesis, Ss. Cyril and Methodius University in Skopje, Faculty of Electrical Engineering and Information Technologies, Republic of Macedonia (2013)

13. Padé, H.: Sur la répresentation approchée d'une fonction par des fractions rationelles (Thesis). Ann. École Nor. 9(3), 1–93 (1892)
14. Van Assche, W.: Padé and Hermite–Padé Approximation and Orthogonality. Surveys in Approximation Theory 2, 61–91 (2006), http://www.emis.de/journals/SAT/papers/6/6.pdf
15. Barbosa, R.S., Tenreiro Machado, J.A.: Implementation of discrete-time fractional-order controllers based on LS Approximations. Acta Polytechnica Hungarica 3(4), 5–22 (2006)
16. Samson, C.: Time-varying feedback stabilization of car-like wheeled mobile robots. Int. J. Robot. Res. 12(1), 55–64 (1993)
17. Baranyi, P.: Tensor product model-based control of two-dimensional aeroelastic system. Journal of Guidance, Control, and Dynamics 29, 391–400 (2006)
18. Takarics, B.: TP model transformation based sliding mode control and friction compensation (Ph.D. dissertation). Budapest University of Technology and Economics (2012)
19. Brockett, R.W.: Asymptotic stability and feedback stabilization. In: Brockett, R.W., Millman, R.S., Sussmann, H.J. (eds.) Differential Geometric Control Theory, pp. 181–191. Birkhäuser, Boston (1983)
20. http://ctms.engin.umich.edu/CTMS/index.php?example=MotorSpeed&section=SystemModeling (last time chechked February 1, 2014)

Hybrid Functions Approach for Variational Problems and Optimal Control of Delay Systems

Mohsen Razzaghi

Department of Mathematics and Statistics Mississippi State University,
Mississippi State, MS 39762, USA
razzaghi@math.msstate.edu

Abstract. In this chapter, a numerical method for solving variational problems and optimal control of delay systems is presented. The method is based upon hybrid functions approximation. The properties of hybrid functions consisting of block-pulse functions and Bernoulli polynomials are presented. The operational matrices of integration, product, delay and the integration of the cross product of two hybrid functions of block-pulse and Bernoulli polynomials vectors are given. These matrices are then utilized to reduce the solution of variational problems, delay systems and the optimal control of delay systems to the solution of algebraic equations. Illustrative examples are included to demonstrate the validity and applicability of the technique.

Keywords: Delay systems, optimal control, variational problems, hybrid, block-pulse, Bernoulli polynomials.

1 Introduction

The available sets of orthogonal functions can be divided into three classes. The first class includes sets of piecewise constant basis functions (e.g., block-pulse, Haar, Walsh, etc.). The second class consists of sets of orthogonal polynomials (e.g., Chebyshev, Laguerre, Legendre, etc.). The third class is the set of sine-cosine functions in the Fourier series. While orthogonal polynomials and sine-cosine functions together form a class of continuous basis functions, piecewise constant basis functions have inherent discontinuities or jumps. It is worth noting that approximating a continuous function with piecewise constant basis functions results in an approximation that is piecewise constant. On the other hand, if a discontinuous function is approximated with continuous basis functions, the resulting approximation is continuous and cannot properly model the discontinuities.

Orthogonal functions have been used when dealing with various problems of the dynamical systems. The main advantage of using orthogonal functions is that they reduce the dynamical system problems to those of solving a system of algebraic equations. The approach is based on converting the underlying

A. El-Osery and J. Prevost (eds.), *Control and Systems Engineering*,
Studies in Systems, Decision and Control 27, DOI: 10.1007/978-3-319-14636-2_4

differential equation into an integral equation through integration, approximating various signals involved in the equation by truncated orthogonal functions, and using the operational matrix of integration P to eliminate the integral operations. Special attention has been given to applications of Walsh functions [1], Chebyshev polynomials [2], Laguerre polynomials [3], Legendre polynomials [4] and Fourier series [5]. Among orthogonal polynomials, the shifted Legendre polynomials $p_m(t)$, $m = 0, 1, 2, \ldots$, where $0 \le t \le 1$, are computationally more effective [6]. The Bernoulli polynomials and Taylor series are not based on orthogonal functions. Nevertheless, they possess the operational matrix of integration. However, since the integration of the cross product of two Taylor series vectors is given in terms of a Hilbert matrix [7], which is known to be ill conditioned, the applications of Taylor series are limited.

For approximating an arbitrary time function, the advantages of Bernoulli polynomials $\beta_m(t)$, $m = 0, 1, 2, \ldots$, where $0 \le t \le 1$, over shifted Legendre polynomials are:

a) the operational matrix P in Bernoulli polynomials has less errors than P for shifted Legendre polynomials for $1 < m < 10$. This is because for P in $\beta_m(t)$ we ignore the term $\frac{\beta_{m+1}(t)}{m+1}$ while for P in $p_m(t)$ we ignore the term $\frac{p_{m+1}(t)}{2(2m+1)}$;

b) the Bernoulli polynomials have fewer terms than shifted Legendre polynomials. For example $\beta_6(t)$ has 5 terms while $p_6(t)$ has 7 terms, and this difference will increase by increasing m. Hence, for approximating an arbitrary function, we use less CPU time by applying Bernoulli polynomials as compared to shifted Legendre polynomials;

c) the coefficient of individual terms in Bernoulli polynomials $\beta_m(t)$ is smaller than the coefficient of individual terms in the shifted Legendre polynomials $p_m(t)$. Since the computational errors in the product are related to the coefficients of individual terms, the computational errors are lessened by using Bernoulli polynomials.

In recent years, the hybrid functions consisting of the combination of block-pulse functions with Chebyshev polynomials [8,9], Legendre polynomials [10,11], or Taylor series [12,13] have been shown to be mathematical power tools for discretization of selected problems. Among these three hybrid functions, the hybrid functions of block-pulse and Legendre polynomials have been shown to be computationally more effective.

In the large number of problems arising in analysis, mechanics, geometry, etc., it is necessary to determine the maximal and minimal of a certain functional. Such problems are called variational problems [14]. The direct method of Ritz and Galerkin in solving variational problems has been of considerable concern and is well covered in many textbooks [15,16].

Delays occur frequently in biological, chemical, electronic and transportation systems [17]. Time-delay systems are, therefore, a very important class of systems whose control and optimization have been of interest to many investigators [18].

Although the optimal control problem for linear systems without delays was solved in the 1960s [19], the optimal control problem for linear systems with delays is still open [20], depending on the delay type, specific system equation, criterion, etc. The application of Pontryagin's maximum principle to the optimization of control systems with time-delays, as outlined by Kharatishvili [21], results in a system of coupled two-point boundary-value (TPBV) problems involving both delay and advance terms whose exact solution, except in very special cases, is very difficult. Therefore, the main object of all computational aspects of optimal time-delay systems has been to devise a methodology to avoid the solution of the mentioned (TPBV) problems. Inoue et al. [22] have proposed a sensitivity approach to obtain the suboptimal control for linear systems with a small delay in the state. They expanded the control in a Maclaurin's series in the delay and obtained the series coefficients from the solution of simple (TPBV) problems. The methods presented in [23] is also a sensitivity approach in which the original system is embedded in a class of non-delay systems using an appropriate parameter. A reference book in the area ([24] and references therein), discussing the Pontryagin's maximum principle or the dynamic programming method for systems with delays, states that finding a particular explicit form of the criterion, must also be taken into account. The studies mostly focused on the time-optimal criterion for quadratic systems [20,25,26].

In this chapter, we focus on the following problems:

(a) The variational problems:

$$J[x(t)] = \int_0^1 F(t, x(t), \dot{x}(t), \ldots, x^{(n)}(t))dt, \tag{1}$$

with the boundary conditions

$$x(0) = a_0, \;\; \dot{x}(0) = a_1, \ldots, x^{(n-1)}(0) = a_{n-1}, \tag{2}$$

$$x(1) = b_0, \;\; \dot{x}(1) = b_1, \ldots, x^{(n-1)}(1) = b_{n-1}. \tag{3}$$

The problem is to find the extremum of Eq. (1), subject to boundary conditions in Eqs. (2) and (3).

(b) The linear time-varying delay system:

$$\dot{X}(t) = E(t)X(t) + F(t)X(t-\tau) + G(t)U(t), \qquad 0 \le t \le 1, \tag{4}$$

$$X(0) = X_0, \tag{5}$$

$$X(t) = \phi(t), \;\; -\tau \le t < 0, \tag{6}$$

where $X(t) \in \mathbb{R}^l$, $U(t) \in \mathbb{R}^q$, $E(t), F(t)$ and $G(t)$ are matrices of appropriate dimensions, X_0 is a constant specified vector, and $\phi(t)$ is an arbitrary known function. The problem is to find the optimal control $U(t)$ and the corresponding state trajectory $X(t)$, $0 \leq t \leq 1$, satisfying Eqs. (4)-(6) while minimizing (or maximizing) the quadratic performance index

$$J = \frac{1}{2}X^T(1)SX(1) + \frac{1}{2}\int_0^1 [X^T(t)Q(t)X(t) + U^T(t)R(t)U(t)]dt, \quad (7)$$

where T denotes transposition, S, $Q(t)$ and $R(t)$ are matrices of appropriate dimensions with S and $Q(t)$ symmetric positive semi-definite matrices and $R(t)$ a symmetric positive definite matrix. The method of this chapter can also be extended to case with delay in both state and control.

In the present work, we introduce a direct computational method to solve problems in (a) and (b). The method consists of reducing problems (a) and (b) to a set of algebraic equations by expanding the candidate function as the hybrid functions with unknown coefficients. These hybrid functions, which consist of block-pulse functions and Bernoulli polynomials, are introduced. The operational matrices of integration, product and delay for the hybrid of block-pulse and Bernoulli polynomials are given. It will be seen that the operational matrix of integration for the hybrid of block-pulse and Bernoulli polynomials has many zero elements and is more sparse than the operational matrix of integration for the hybrid of block-pulse and Legendre polynomials.

The outline of this chapter is as follows: In section 2, we describe the basic formulation of the hybrid functions required for our subsequent development. In section 3, the numerical methods for problems (a) and (b) are given and, in section 4, we report our numerical findings and demonstrate the accuracy of the proposed numerical scheme by considering six numerical examples.

2 Properties of Hybrid Functions

2.1 Hybrid of Block-Pulse and Bernoulli Polynomials

Hybrid functions $b_{nm}(t)$, $n = 1, 2, \ldots, N$, $m = 0, 1, \ldots, M$ are defined on the interval $[0, t_f)$ as

$$b_{nm}(t) = \begin{cases} \beta_m(\frac{N}{t_f}t - n + 1), & t \in [\frac{n-1}{N}t_f, \frac{n}{N}t_f), \\ 0, & \text{otherwise}, \end{cases} \quad (8)$$

where n and m are the order of block-pulse functions and Bernoulli polynomials, respectively. In Eq. (8), $\beta_m(t)$, $m = 0, 1, 2, \ldots$ are the Bernoulli polynomials of order m, which can be defined by [27]

$$\beta_m(t) = \sum_{k=0}^{m} \binom{m}{k} \alpha_k t^{m-k},$$

where $\alpha_k,\ \ k = 0, 1, ..., m$ are Bernoulli numbers. These numbers are a sequence of signed rational numbers which arise in the series expansion of trigonometric functions [28] and can be defined by the identity

$$\frac{t}{e^t - 1} = \sum_{n=0}^{\infty} \alpha_n \frac{t^n}{n!}.$$

The first few Bernoulli numbers are

$$\alpha_0 = 1, \quad \alpha_1 = \frac{-1}{2}, \quad \alpha_2 = \frac{1}{6}, \quad \alpha_4 = \frac{-1}{30},$$

with $\alpha_{2k+1} = 0,\ k = 1, 2, 3, \ldots.$

The first few Bernoulli polynomials are

$$\beta_0(t) = 1, \quad \beta_1(t) = t - \frac{1}{2}, \quad \beta_2(t) = t^2 - t + \frac{1}{6}, \quad \beta_3(t) = t^3 - \frac{3}{2}t^2 + \frac{1}{2}t.$$

These polynomials satisfy the following formula [28]

$$\beta_m(0) = \alpha_m, \qquad m \geq 0, \tag{9}$$

$$\int_a^x \beta_m(t)dt = \frac{\beta_{m+1}(x) - \beta_{m+1}(a)}{m+1}, \tag{10}$$

$$\int_0^1 \beta_n(t)\beta_m(t)dt = (-1)^{n-1}\frac{m!n!}{(m+n)!}\alpha_{n+m}, \qquad m, n \geq 1. \tag{11}$$

According to [29], Bernoulli polynomials form a complete basis over the interval [0,1].

2.2 Function Approximation

Suppose that $H = L^2[0,1]$ and $\{b_{10}(t), b_{20}(t), ..., b_{NM}(t)\} \subset H$ be the set of hybrid of block-pulse and Bernoulli polynomials and

$$Y = span\{b_{10}(t), b_{20}(t), ..., b_{N0}(t), b_{11}(t), b_{21}(t), ..., \\ b_{N1}(t), ..., b_{1M}(t), b_{2M}(t), ..., b_{NM}(t)\},$$

and f be an arbitrary element in H. Since Y is a finite dimensional vector space, f has the unique best approximation out of Y, such as $f_0 \in Y$, that is

$$\forall y \in Y, \| f - f_0 \| \leq \| f - y \| .$$

Since $f_0 \in Y$, there exists the unique coefficients $c_{10}, c_{20}, ..., c_{NM}$ such that

$$f \simeq f_0 = \sum_{m=0}^{M}\sum_{n=1}^{N} c_{nm}b_{nm}(t) = C^T B(t), \tag{12}$$

where

$$B^T(t) = [b_{10}(t), b_{20}(t), ..., b_{N0}(t), b_{11}(t), b_{21}(t), ..., \\ b_{N1}(t), ..., b_{1M}(t), b_{2M}(t), ..., b_{NM}(t)], \tag{13}$$

and

$$C^T = [c_{10}, c_{20}, ..., c_{N0}, c_{11}, c_{21}, ..., c_{N1}, ..., c_{1M}, c_{2M}, ..., c_{NM}]. \tag{14}$$

2.3 Integration of $\mathbf{B(t)B^T(t)}$

By using Eq. (12), we obtain

$$f_{ij} =< \sum_{m=0}^{M}\sum_{n=1}^{N} c_{nm}b_{nm}(t), b_{ij}(t) >= \sum_{m=0}^{M}\sum_{n=1}^{N} c_{nm}k_{nm}^{ij},$$

$$i = 1, 2, ..., N, \quad j = 0, 1, ..., M,$$

where $f_{ij} =< f, b_{ij}(t) >$, $k_{nm}^{ij} =< b_{nm}(t), b_{ij}(t) >$, and $<, >$ denotes inner product. Therefore,

$$f_{ij} = C^T [k_{10}^{ij}, k_{20}^{ij}, ..., k_{N0}^{ij}, k_{11}^{ij}, k_{21}^{ij}, ..., k_{N1}^{ij}, ..., k_{1M}^{ij}, k_{2M}^{ij}, ..., k_{NM}^{ij}]^T, \\ i = 1, 2, ..., N, \quad j = 0, 1, ..., M,$$

so we get

$$\Phi = K^T C,$$

with

$$\Phi = [f_{10}, f_{20}, \ldots, f_{N0}, f_{11}, f_{21}, \ldots, f_{N1}, \ldots, f_{1M}, f_{2M}, \ldots, f_{NM}]^T, \quad K = [k_{nm}^{ij}],$$

where K is a matrix of order $N(M+1) \times N(M+1)$ and is given by

$$K = \int_0^1 B(t)B^T(t)dt. \tag{15}$$

By using Eq. (11) in each interval $n = 1, 2, ..., N$, we can get matrix K. For example, with $N = 2$ and $M = 3$, K is

$$K = \frac{1}{2}\begin{bmatrix} 1 & 0 & 0 & 0 & 0 & 0 & 0 & 0 \\ 0 & 1 & 0 & 0 & 0 & 0 & 0 & 0 \\ 0 & 0 & \frac{1}{12} & 0 & 0 & 0 & \frac{-1}{120} & 0 \\ 0 & 0 & 0 & \frac{1}{12} & 0 & 0 & 0 & \frac{-1}{120} \\ 0 & 0 & 0 & 0 & \frac{1}{180} & 0 & 0 & 0 \\ 0 & 0 & 0 & 0 & 0 & \frac{1}{180} & 0 & 0 \\ 0 & 0 & \frac{-1}{120} & 0 & 0 & 0 & \frac{1}{840} & 0 \\ 0 & 0 & 0 & \frac{-1}{120} & 0 & 0 & 0 & \frac{1}{840} \end{bmatrix}.$$

It is seen that the matrix K is a sparse matrix. Furthermore, if we choose large values of M and N, the non zero elements of K will tend to zero.

2.4 Operational Matrix of Integration

The integration of the $B(t)$ defined in Eq. (13) is given by

$$\int_0^t B(t')dt' \simeq PB(t), \tag{16}$$

where P is the $N(M+1) \times N(M+1)$ operational matrix of integration and is given by [30,31]

$$P = \frac{t_f}{N}\begin{bmatrix} P_0 & I & O & \dots & O \\ \frac{-1}{2}\alpha_2 I & O & \frac{1}{2}I & \dots & O \\ \vdots & \vdots & \vdots & \ddots & \vdots \\ \frac{-1}{M}\alpha_M I & O & O & \dots & \frac{1}{M}I \\ \frac{-1}{M+1}\alpha_{M+1} I & O & O & \dots & O \end{bmatrix},$$

where I and O are $N \times N$ identity and zero matrices, respectively, and

$$P_0 = \begin{bmatrix} -\alpha_1 & 1 & \dots & 1 & 1 \\ 0 & -\alpha_1 & \dots & 1 & 1 \\ \vdots & \vdots & \ddots & \vdots & \vdots \\ 0 & 0 & \dots & -\alpha_1 & 1 \\ 0 & 0 & \dots & 0 & -\alpha_1 \end{bmatrix}.$$

It is seen that P is a sparse matrix.

2.5 The Operational Matrix of Product

The following property of the product of two hybrid function vectors will also be used.
Let

$$B(t)B^T(t)C \simeq \tilde{C}B(t), \tag{17}$$

where $\tilde{C}$ is a $N(M+1) \times N(M+1)$ product operational matrix and $M=2$ and $N=3$ is given by [31]

$$\tilde{C} = \begin{bmatrix} \tilde{C}_0 & \tilde{C}_1 & \tilde{C}_2 \\ \frac{1}{12}\tilde{C}_1 & \tilde{C}_0 + \frac{1}{6}\tilde{C}_2 & \tilde{C}_1 \\ \frac{1}{180}\tilde{C}_2 & \frac{1}{6}\tilde{C}_1 & \tilde{C}_0 + \frac{1}{3}\tilde{C}_2 \end{bmatrix},$$

where $\tilde{C}_i, \;\; i = 0, 1, 2$ are 3×3 matrices given by

$$\tilde{C}_i = \begin{bmatrix} c_{1i} & 0 & 0 \\ 0 & c_{2i} & 0 \\ 0 & 0 & c_{3i} \end{bmatrix},$$

similarly, for other values of M and N, the product operational matrix $\tilde{C}$ in Eq. (17) can be obtained.

2.6 The Operational Matrix of Delay

The delay function $B(t-\tau)$ is the shift of the function $B(t)$ defined in Eq. (13) along the time axis by τ. The general expression is given by

$$B(t-\tau) = DB(t), \quad t > \tau, \ \ 0 \le t \le 1, \tag{18}$$

where D is the operational matrix of delay. To find D, we first choose N in the following manner:

$$N = \begin{cases} \frac{1}{\tau}, & \frac{1}{\tau} \in \mathbb{Z}, \\ \lfloor \frac{1}{\tau} \rfloor + 1, & \text{otherwise}, \end{cases} \tag{19}$$

where $\lfloor \frac{1}{\tau} \rfloor$ denotes the greatest integer value less than or equal to $\frac{1}{\tau}$. With the aid of Eq. (8), it is noted that for the case $\tau \le t \le 2\tau$ the only terms with non-zero values are $b_{1m}(t-\tau)$ for $m = 0, 1, 2, \ldots, M$. If we expand $b_{1m}(t-\tau)$ in terms of $b_{2m}(t)$, since $b_{1m}(t-\tau) = b_{2m}(t)$, then the element of the delay matrix is the $N \times N$ matrix Ψ given by

$$\Psi = \begin{bmatrix} 0 & 1 & 0 & \cdots & 0 \\ 0 & 0 & 1 & \cdots & 0 \\ \vdots & \vdots & \vdots & \ddots & \vdots \\ 0 & 0 & 0 & \cdots & 1 \\ 0 & 0 & 0 & \cdots & 0 \end{bmatrix}. \tag{20}$$

In a similar manner, for $2\tau \le t \le 3\tau$, only $b_{2m}(t-\tau)$ for $m = 0, 1, 2, \ldots, M$ has nonzero values. If we expand $b_{2m}(t-\tau)$ in terms of $b_{3m}(t)$, since $b_{2m}(t-\tau) = b_{3m}(t)$, then the element of the delay matrix is the $N \times N$ matrix Ψ given in Eq. (20). Thus, if we expand $B(t-\tau)$ in terms of $B(t)$, we find $N(M+1) \times N(M+1)$ matrix D as

$$D = diag(\Psi, \Psi, \ldots, \Psi).$$

3 The Numerical Methods

In this section, we consider problems (a) and (b) given in Eqs. (1)-(3) and (4)-(7), respectively.

3.1 Problem (a)

By expanding $x^{(n)}(t)$ in the hybrid of block-pulse and Bernoulli polynomials, we have

$$x^{(n)}(t) = X^T B(t). \tag{21}$$

where X is a vector of order $N \times (M+1)$ given by

$$X = [x_{10}, x_{20}, ..., x_{N0}, x_{11}, x_{21}, ..., x_{N1}, ..., x_{1M}, x_{2M}, ..., x_{NM}]^T,$$

By integrating Eq. (21) from 0 to t, we get

$$x^{(n-1)}(t) - x^{(n-1)}(0) = \int_0^t X^T B(t')dt' = X^T PB(t),$$

where P is operational matrix of integration given in Eq. (16). By using Eq. (2), we get

$$x^{(n-1)}(t) = a_{n-1} + X^T PB(t). \tag{22}$$

By $n-1$ times integrating Eq. (22) from 0 to t and using boundary conditions given in Eq. (2), we have

$$x^{(n-2)}(t) = a_{n-2} + a_{n-1}t + X^T P^2 B(t), \tag{23}$$

$$\vdots$$

$$\dot{x}(t) = a_1 + a_2 t + \frac{a_3}{2!}t^2 + \cdots + \frac{a_{n-1}}{(n-2)!}t^{n-2} + X^T P^{n-1} B(t), \tag{24}$$

$$x(t) = a_0 + a_1 t + \frac{a_2}{2!}t^2 + \cdots + \frac{a_{n-1}}{(n-1)!}t^{n-1} + X^T P^n B(t). \tag{25}$$

Assume that each of t^i, $i = 1, 2, \ldots, n-1$, and each of a_{n-i}, $i = 1, 2, \ldots, n$, can be written in terms of hybrid functions as

$$t^i = d_i^T B(t), \quad i = 1, 2, \ldots, n-1 \tag{26}$$

$$a_{n-i} = a_{n-i} E^T B(t), \quad i = 1, 2, \ldots, n, \tag{27}$$

where

$$E^T = [\underbrace{1, 1, ..., 1}_{N}, \underbrace{0, 0, ..., 0}_{NM}].$$

By substituting Eqs. (26) and (27) in Eqs. (21)-(25), we obtain

$$x^{(n)}(t) = X^T B(t),$$
$$x^{(n-1)}(t) = (a_{n-1}E^T + X^T P)B(t),$$
$$\vdots$$
$$\dot{x}(t) = (a_1 E^T + a_2 d_1^T + \frac{a_3}{2!}d_2^T + \cdots + \frac{a_{n-1}}{(n-2)!}d_{n-2}^T + X^T P^{n-1})B(t),$$
$$x(t) = (a_0 E^T + a_1 d_1^T + \frac{a_2}{2!}d_2^T + \cdots + \frac{a_{n-1}}{(n-1)!}d_{n-1}^T + X^T P^n)B(t).$$

By substituting above equations in Eq. (1), we get

$$J[x(t)] = J[X], \tag{28}$$

the boundary conditions in Eq. (3) can be expressed as

$$q_k = x^{(k)}(1) - b_k = 0, \quad k = 0, ..., n-1. \tag{29}$$

We now find the extremum of Eq. (28) subject to Eq. (29) using the Lagrange multiplier technique. Let

$$J^*[X, \lambda] = J[X] + \lambda Q,$$

where the vector λ represents the unknown Lagrange multipliers and

$$Q = [q_0, q_1, \ldots, q_{n-1}]^T.$$

The necessary conditions are given by

$$\begin{cases} \frac{\partial}{\partial X} J^*[X, \lambda] = 0, \\ \frac{\partial}{\partial \lambda} J^*[X, \lambda] = 0. \end{cases}$$

By solving the above equations, we can obtain X. Next, we solve the problem in (b).

3.2 Problem (b)

Approximation of the System Dynamics

We approximate Eq. (4) as follows:
Let

$$X(t) = [x_1(t), x_2(t), \ldots, x_l(t)]^T, \quad U(t) = [u_1(t), u_2(t), \ldots, u_q(t)]^T, \tag{30}$$

$$\hat{B}(t) = I_l \otimes B(t), \quad \hat{B}_1(t) = I_q \otimes B(t), \tag{31}$$

where I_l and I_q are the l and q dimensional identity matrices and $\otimes$ denotes Kronecker product [32]. Also, $\hat{B}(t)$ and $\hat{B}_1(t)$ are $l(M+1)N \times l$ and $q(M+1)N \times q$ matrices, respectively. By using Eq. (12), each of $x_i(t)$ for $i = 1, 2, \ldots, l$ and each of $u_j(t)$ for $j = 1, 2, \ldots, q$, can be written in terms of hybrid functions as

$$x_i(t) = B^T(t) X_i, \quad u_j(t) = B^T(t) U_j.$$

From Eqs. (30) and (31), we get

$$X(t) = \hat{B}^T(t) X, \quad U(t) = \hat{B_1^T}(t) U, \tag{32}$$

where X and U are vectors of order $l(M+1)N \times 1$ and $q(M+1)N \times 1$, respectively, given by

$$X = [X_1, X_2, \ldots, X_l]^T, \quad U = [U_1, U_2, \ldots, U_q]^T.$$

Similarly, we have

$$X(0) = \hat{B}^T(t)d, \quad \phi(t-\tau) = \hat{B}^T(t)W, \tag{33}$$

where d and W are vectors of order $l(M+1)N \times 1$. We expand $E(t), F(t)$ and $G(t)$ by hybrid functions as follows:

$$E(t) = E^T\hat{B}(t), \quad F(t) = F^T\hat{B}(t), \quad G(t) = G^T\hat{B}_1(t),$$

where E^T, F^T and G^T are of dimensions $l \times l(M+1)N$, $l \times l(M+1)N$, and $l \times q(M+1)N$, respectively. We can write $X(t-\tau)$ in terms of hybrid functions as

$$X(t-\tau) = \begin{cases} \hat{B}^T(t)W, & 0 \le t < \tau, \\ \hat{B}^T(t)\hat{D}^T X, & \tau \le t \le 1, \end{cases}$$

where

$$\hat{D} = I_l \otimes D,$$

and D is the operational matrix of delay given in Eq. (18). Also, we have

$$E(t)X(t) = E^T\hat{B}(t)\hat{B}^T(t)X = \hat{B}^T(t)\tilde{E}^T X, \tag{34}$$

$$G(t)U(t) = G^T\hat{B}_1(t)\hat{B}_1^T(t)U = \hat{B}_1{}^T(t)\tilde{G}^T U, \tag{35}$$

$$\int_0^t \hat{B}^T(t')dt' = (I_l \otimes B^T(t))(I_l \otimes P^T) = \hat{B}^T(t)\hat{P}^T, \tag{36}$$

$$\int_0^t F(t')X(t'-\tau)dt' = \begin{cases} \hat{B}^T(t)\hat{P}^T\tilde{F}^T W, & 0 \le t \le \tau, \\ \hat{B}^T(t)Z\tilde{F}^T W + \hat{B}^T(t)\hat{P}^T\tilde{F}^T\hat{D}^T X, & \tau \le t \le 1, \end{cases} \tag{37}$$

where $\tilde{E}$, $\tilde{F}$,and $\tilde{G}$ can be calculated similarly to matrix $\tilde{C}$ in Eq. (17) and P is the operational matrix of integration given in Eq. (16). Also,

$$\int_0^\tau \hat{B}^T(t)dt = \hat{B}^T(t)Z,$$

where Z is a constant matrix of order $l(M+1)N \times l(M+1)N$.
By integrating Eq. (4) from 0 to t and using Eqs. (5), (6) and (32)-(37), we have

$$\begin{aligned} \hat{B}^T(t)X - \hat{B}^T(t)d &= \hat{B}^T(t)\hat{P}^T\tilde{E}^T X + \hat{B}^T(t)\hat{P}^T\tilde{F}^T W \\ &+ \hat{B}^T(t)Z\tilde{F}^T W + \hat{B}^T(t)\hat{P}^T\tilde{F}^T\hat{D}^T X + \hat{B}^T(t)\hat{P}^T\tilde{G}^T U. \end{aligned} \tag{38}$$

Using Eq. (38), we get

$$A^* = (\hat{P}^T\tilde{E}^T + \hat{P}^T\tilde{F}^T\hat{D}^T - I)X + \hat{P}^T\tilde{F}^T W + Z\tilde{F}^T W + \hat{P}^T\tilde{G}^T U + d = 0. \tag{39}$$

The Performance Index Approximation

By using Eqs. (7) and (32), we have

$$\begin{aligned} J = &\tfrac{1}{2}X^T\hat{B}(1)S\hat{B}^T(1)X + \tfrac{1}{2}X^T[\textstyle\int_0^1 \hat{B}(t)Q(t)\hat{B}^T(t)dt]X \\ &+\tfrac{1}{2}U^T[\textstyle\int_0^1 \hat{B}_1^T(t)R(t)\hat{B}_1^T(t)dt]U. \end{aligned} \tag{40}$$

Eq. (40) can be computed more efficiently by writing J as

$$\begin{aligned} J = &\tfrac{1}{2}X^T(B(1)B^T(1)\otimes S)X + \tfrac{1}{2}X^T[\textstyle\int_0^1 B(t)B^T(t)\otimes Q(t)dt]X \\ &+\tfrac{1}{2}U^T[\textstyle\int_0^1 B(t)B^T(t)\otimes R(t)dt]U. \end{aligned} \tag{41}$$

For problems with time-varying performance index, $Q(t)$ and $R(t)$ are functions of time, and

$$\int_0^1 B(t)B^T(t)\otimes Q(t)dt, \qquad \int_0^1 B(t)B^T(t)\otimes R(t)dt$$

can be evaluated numerically. For time-invariant problems, $Q(t)$ and $R(t)$ are constant matrices and can be removed from the integrals. In this case, Eq. (41) can be rewritten as

$$J = \frac{1}{2}X^T(B(1)B^T(1)\otimes S)X + \frac{1}{2}X^T(K\otimes Q)X + \frac{1}{2}U^T(K\otimes R)U,$$

where K is the integration of the cross product of two hybrid functions of block-pulse and Bernoulli polynomials given in Eq. (15).

Solution of the Optimization Problem

The optimal control problem has been reduced to a parameter optimization problem, which can be stated as follows. Find X and U so that $J(X, U)$ is minimized (or maximized), subject to the constraint in Eq. (39).
Let

$$J^*(X, U, \lambda) = J(X, U) + \lambda^T A^*, \tag{42}$$

where the vector λ represents the unknown constant Lagrange multipliers; then the necessary conditions for minimum (or maximum) of J^* are given by

$$\frac{\partial}{\partial X}J^*(X, U, \lambda) = 0, \quad \frac{\partial}{\partial U}J^*(X, U, \lambda) = 0, \quad \frac{\partial}{\partial \lambda}J^*(X, U, \lambda) = 0. \tag{43}$$

4 Main Feature of the Method

It will be shown in Examples (1) and (2) that the exact solutions of these variational problems are different functions on the distinct subintervals. Also, due to the nature of time-delay systems, the exact solutions of these systems are also different functions on the distinct subintervals. In such situations, neither the continuous basis functions nor piecewise constant basis functions taken alone would form an efficient basis in the representation of such solutions. Datta and Mohan [33] have correctly pointed out that, in general, the computed response of the delay systems via continuous or piecewise constant basis functions is not in good agreement with the exact response of the system. To meet these situations, we choose a suitable hybrid system of basis functions inherently possessing the required features of the solutions corresponding to these situations. For the delay systems, in the proposed method, with the aid of Eq. (19), we determine the appropriate value for N, the order of block-pulse functions. To select M, for the solution of delay systems, we first choose an arbitrary number depending on the problem. If the exact solution is a polynomial, we increase the value of M by 1 until two consecutive results are the same. When the exact solution is not a polynomial, we evaluate the results for two consecutive M for different t in $[0, 1]$ until the results are similar up to a required number of decimal places. To select M for the optimal control problems, we first choose an arbitrary number depending on the problem and evaluate J, the value of the performance index, then increase the value of M by 1 until two consecutive values for J are similar to a required number of decimal places.

5 Illustrative Examples

In this section, we use six examples to demonstrate the applicability, efficiency and accuracy of our proposed method.
Examples (1) and (2) are variational problems. The exact solutions are different functions on distinct subintervals. For these examples we get the exact solutions by using the present method. Examples (3) and (4) are solution of delay systems for which we can apply the present method and Examples (5) and (6) are optimal control of delay systems.

5.1 Example 1

Consider the extremization of [1]

$$J = \int_0^1 [\frac{1}{2}\dot{x}^2 - xg(t)]dt, \tag{44}$$

where

$$g(t) = \begin{cases} -1, 0 \leq t \leq \frac{1}{4}, \ \frac{1}{2} \leq t \leq 1, \\ 3, \ \ \frac{1}{4} \leq t \leq \frac{1}{2}, \end{cases} \tag{45}$$

with the boundary conditions

$$x(0)=0, \quad \dot{x}(0)=0, \quad \dot{x}(1)=0. \tag{46}$$

Schechter [34] gave a physical interpretation of this problem by noting an application in heat conduction. The exact solution is

$$x(t)=\begin{cases} \frac{1}{2}t^2, & 0\le t\le \frac{1}{4},\\ \frac{3}{2}t^2+t-\frac{1}{8}, & \frac{1}{4}\le t\le \frac{1}{2},\\ \frac{1}{2}t^2-t+\frac{3}{8}, & \frac{1}{2}\le t\le 1. \end{cases}$$

Here, we solve this problem by using the hybrid of block-pulse functions and Bernoulli polynomials. We assume

$$\dot{x}(t)=C^T B(t),$$

in view of Eq. (45), we write Eq. (44) as

$$J=\frac{1}{2}\int_0^1 \dot{x}^2(t)dt+4\int_0^{\frac{1}{4}} x(t)dt-4\int_0^{\frac{1}{2}} x(t)dt+\int_0^1 x(t)dt,$$

or

$$J=\frac{1}{2}\int_0^1 C^T B(t)B^T(t)Cdt+4C^T P\int_0^{\frac{1}{4}} B(t)dt-$$
$$4C^T P\int_0^{\frac{1}{2}} B(t)dt+C^T P\int_0^1 B(t)dt.$$

Let

$$V(t)=\int_0^t B(t')dt', \tag{47}$$

using Eq. (15), we get

$$J=\frac{1}{2}C^T KC+C^T P[4V(\frac{1}{4})-4V(\frac{1}{2})+V(1)], \tag{48}$$

the boundary conditions in Eq. (46) can be expressed in terms of hybrid functions as

$$C^T B(0)=0, \quad C^T B(1)=0. \tag{49}$$

We now find the extremum of Eq. (48) subject to Eq. (49) using the Lagrange multiplier technique. Suppose

$$J^*=J+\lambda_1 C^T B(0)+\lambda_2 C^T B(1),$$

where λ_1 and λ_2 are two multipliers. Then, the necessary condition is given by

$$\frac{\partial}{\partial C}J^*=DC+P[4V(\frac{1}{4})-4V(\frac{1}{2})+V(1)]+\lambda_1 B(0)+\lambda_2 B(1)=0. \tag{50}$$

Equations (49) and (50) define a set of simultaneous linear algebraic equations from which the vector C and the multipliers λ_1 and λ_2 can be found. By solving the above equations with $M=2$ and $N=4$, we get the exact solution.

5.2 Example 2

Consider the problem of finding the extremum of the functional [35]

$$J[x(t)] = \int_0^1 (\dot{x}^2(t)f(t))dt, \tag{51}$$

where

$$f(t) = \begin{cases} -1, 0 \leq t < \frac{1}{4}, \\ 1, \ \frac{1}{4} < t \leq 1, \end{cases} \tag{52}$$

with the boundary conditions

$$x(0) = 0, \qquad x(1) = 1. \tag{53}$$

Suppose

$$\dot{x}(t) = C^T B(t),$$

in view of Eq. (52), we write Eq. (51) as

$$J = -2\int_0^{\frac{1}{4}} \dot{x}^2(t)dt + \int_0^1 \dot{x}^2(t)dt,$$

or

$$J = -2\int_0^{\frac{1}{4}} C^T B(t)B^T(t)Cdt + \int_0^1 C^T B(t)B^T(t)Cdt.$$

By using Eqs. (15) and (17), we have

$$J = -2\int_0^{\frac{1}{4}} C^T \tilde{C}B(t)dt + C^T DC,$$

by applying Eq. (47), we get

$$J = -2C^T\tilde{C}V(\frac{1}{4}) + C^T DC, \tag{54}$$

the boundary condition in Eq. (53) can be expressed in terms of hybrid functions as

$$C^T PB(1) = 1. \tag{55}$$

We now find the extremum of Eq. (54) subject to Eq. (55) using the Lagrange multiplier technique. Suppose

$$J^* = J + \lambda(C^T PB(1) - 1),$$

where λ is multiplier. Then, the necessary conditions are given by

$$\begin{cases} \frac{\partial}{\partial C}J^* = 0, \\ \\ \frac{\partial}{\partial \lambda}J^* = 0. \end{cases}$$

By solving the above equations with $M = 1$ and $N = 4$, we get

$$x(t) = \begin{cases} -2t, & 0 \leq t \leq \frac{1}{4}, \\ 2t - 1, & \frac{1}{4} \leq t \leq 1, \end{cases}$$

which is the exact solution.

5.3 Example 3

Consider the time-varying delay systems described by [36]

$$\begin{pmatrix} \dot{x}_1(t) \\ \dot{x}_2(t) \end{pmatrix} = \begin{pmatrix} 0 & 1 \\ -25 & -5t \end{pmatrix} \begin{pmatrix} x_1(t-\frac{1}{4}) \\ x_2(t-\frac{1}{4}) \end{pmatrix} + \begin{pmatrix} 0 \\ 1 \end{pmatrix}, \tag{56}$$

with

$$\begin{pmatrix} x_1(t) \\ x_2(t) \end{pmatrix} = \begin{pmatrix} 0 \\ 0 \end{pmatrix}, \qquad t \in [-\frac{1}{4}, 0]. \tag{57}$$

The exact solutions are

$$x_1(t) = \begin{cases} 0, & 0 \le t \le \frac{1}{4}, \\ \frac{1}{32} - \frac{1}{4}t + \frac{1}{2}t^2, & \frac{1}{4} \le t \le \frac{1}{2}, \\ \frac{1}{32} - \frac{19}{96}t + \frac{3}{16}t^2 + \frac{5}{8}t^3 - \frac{5}{12}t^4, & \frac{1}{2} \le t \le \frac{3}{4}, \\ -\frac{9641}{32768} + \frac{37391}{24576}t - \frac{3183}{1024}t^2 + \frac{785}{256}t^3 - \frac{45}{128}t^4 - \frac{85}{96}t^5 + \frac{5}{18}t^6, & \frac{3}{4} \le t \le 1, \end{cases}$$

and

$$x_2(t) = \begin{cases} t, & 0 \le t \le \frac{1}{4}, \\ -\frac{5}{384} + t + \frac{5}{8}t^2 - \frac{5}{3}t^3, & \frac{1}{4} \le t \le \frac{1}{2}, \\ \frac{775}{1536} - \frac{17}{8}t + \frac{1295}{192}t^2 - \frac{115}{24}t^3 - \frac{75}{32}t^4 + \frac{5}{3}t^5, & \frac{1}{2} \le t \le \frac{3}{4}, \\ \frac{3666575}{5505024} - \frac{1051}{1024}t - \frac{95755}{49152}t^2 + \frac{21515}{1536}t^3 - \\ \quad \frac{55325}{3072}t^4 + \frac{335}{96}t^5 + \frac{2125}{576}t^6 - \frac{25}{21}t^7, & \frac{3}{4} \le t \le 1. \end{cases}$$

We solve this problem by choosing $N = 4$ and $M = 7$. Let

$$x_1(t) = X_1^T B(t), \tag{58}$$

$$x_2(t) = X_2^T B(t), \tag{59}$$

where X_1, X_2 and $B(t)$ can be obtained similarly to Eqs. (13) and (14). By expanding $-5t$ in terms of hybrid functions, we get

$$-5t = [\frac{-5}{8}, \frac{-15}{8}, \frac{-25}{8}, \frac{-35}{8}, \frac{-5}{4}, \frac{-5}{4}, \frac{-5}{4}, \frac{-5}{4}, \underbrace{0, \ldots, 0}_{24}]B(t) = W^T B(t). \tag{60}$$

We also have

$$-5tx_2(t-\frac{1}{4}) = X_2^T D B(t) B^T(t) W = X_2^T D \tilde{W} B(t), \tag{61}$$

where $\tilde{W}$ can be calculated similarly to Eq. (17) and D is the delay operational matrix. By using Eqs. (57)-(61), we get

$$X_1^T = X_2^T DP, \tag{62}$$

$$X_2^T = -25X_1^T DP + X_2^T D\tilde{K}P - \frac{1}{5}K^T. \tag{63}$$

By solving Eqs. (62) and (63), we obtain the exact value of $x_1(t)$ and $x_2(t)$.

5.4 Example 4

Consider the following delay system with delay in both control and state [37]

$$\dot{x}(t) = -x(t) - 2x(t - \frac{1}{4}) + 2u(t - \frac{1}{4}), \qquad u(t) = 1, \qquad t > 0,$$

$$x(t) = u(t) = 0, \qquad -\frac{1}{4} \leq t < 0.$$

The exact solution is

$$x(t) = \begin{cases} 0, & 0 \leq t \leq \frac{1}{4}, \\ 2 - 2\exp[-(t - \frac{1}{4})], & \frac{1}{4} \leq t \leq \frac{1}{2}, \\ -2 - 2\exp[-(t - \frac{1}{4})] + (2 + 4t)\exp[-(t - \frac{1}{2})], & \frac{1}{2} \leq t \leq \frac{3}{4}, \\ 6 - 2\exp[-(t - \frac{1}{4})] + (2 + 4t)\exp[-(t - \frac{1}{2})]- \\ (\frac{17}{4} + 2t + 4t^2)exp[-(t - \frac{3}{4})], & \frac{3}{4} \leq t < 1. \end{cases}$$

We solve this problem by choosing $N = 4$ and $M = 6$. In Table 1, we compare the solution obtained using the proposed method by choosing $N = 4$ and $M = 6$ with the hybrid of block-pulse and Legendre polynomials with $N = 4$ and $M_1 = 7$ in [37] and the exact solution of $x(t)$ for $\frac{1}{4} \leq t \leq 1$. The approximate value of $x(t)$ on $[0, \frac{1}{4}]$ is equal to zero, which is the same as the exact solution. In this table, M_1 is the order of Legendre polynomials.

Table 1. Estimated Values of $x(t)$

t	block-pulse and Legendre with $N = 4$ and $M_1 = 7$	present method with $N = 4$ and $M = 6$	Exact value
0.25	0	0	0
0.40	0.27858404	0.278584047	0.278584047
0.55	0.51352714	0.513527141	0.513527141
0.70	0.65465130	0.654651311	0.654651311
0.85	0.70892964	0.708929636	0.708929636
1.00	0.71174280	0.711742826	0.711742826

5.5 Example 5

This example was studied in [20]. Consider the optimal control problem of a time-varying delay system described by

$$\dot{x}(t) = tx(t) + x(t-1) + u(t), \quad 0 \le t \le 2, \tag{64}$$

$$x(t) = 1, \quad -1 \le t \le 0, \tag{65}$$

with the performance index

$$J = \int_0^2 [x^2(t) + u^2(t)]dt.$$

The problem is to find the optimal control $u(t)$, which minimizes J subject to Eqs. (64) and (65). By rescaling the time scale into $[0, 1]$, we have

$$\dot{x}(t) = 4tx(t) + 2x(t - \tfrac{1}{2}) + 2u(t), \quad 0 \le t \le 1, \tag{66}$$

$$x(t) = 1, \quad -\tfrac{1}{2} \le t \le 0, \tag{67}$$

with the performance index

$$J = 2\textstyle\int_0^1 [x^2(t) + u^2(t)]dt. \tag{68}$$

Here, we solve this problem by choosing $N = 2$ and $M = 2$. Let

$$x(t) = C^T B(t), \quad u(t) = U^T B(t). \tag{69}$$

By expanding $x(0)$ in terms of hybrid functions, we have

$$x(0) = [1, 1, 0, 0, 0, 0]B(t) = e_1^T B(t). \tag{70}$$

Also,

$$tx(t) = C^T B(t) B^T(t) d_1 = C^T \tilde{d}_1 B(t), \tag{71}$$

where

$$d_1 = \left[\frac{1}{4}, \frac{3}{4}, \frac{1}{2}, \frac{1}{2}, 0, 0\right]^T,$$

and $\tilde{d}_1$ can be calculated similarly to matrix $\tilde{C}$ in Eq. (17).
We also have

$$\int_0^t x(t' - \frac{1}{2})dt' = \begin{cases} e_2^T B(t), & 0 \le t \le \frac{1}{2}, \\ e_3^T B(t) + C^T DPB(t), & \frac{1}{2} \le t \le 1, \end{cases} \tag{72}$$

where

$$e_2 = \left[\frac{1}{4}, 0, \frac{1}{2}, 0, 0, 0\right]^T, \tag{73}$$

$$e_3 = \left[0, \frac{1}{2}, 0, 0, 0, 0\right]^T. \tag{74}$$

Integrating Eq. (66) from 0 to t and using Eqs. (69)-(74), we have

$$C^T[4\tilde{d_1}P + 2DP - I_6] + 2U^T P + e^T = 0,$$

where I_6 is a six dimensional identity matrix and

$$e = e_1 + 2e_2 + 2e_3.$$

By using Eqs. (68) and (69), we get

$$J = 2C^T K_1 C + 2U^T K_1 U, \tag{75}$$

where K_1 can be calculated similarly to Eq. (15).

In Table 2, a comparison is made between the value of J obtained by the present method with $N = 2$ and $M = 2$, together with the value of J reported in [20] by using 8 basis functions of linear Legendre Multiwavelets.

Table 2. Estimated values of J

Method of [20]	J
with 8 basis functions	5.1713
Present Method	
$N = 2$ and $M = 2$	4.7407

5.6 Example 6

Consider the following optimal control problem with different delays in state and control [38]

$$\dot{x}(t) = -x(t) + x(t - \frac{1}{3}) + u(t) - \frac{1}{2}u(t - \frac{2}{3}), \qquad 0 \leq t \leq 1, \tag{76}$$

with

$$x(t) = 1, \qquad -\frac{1}{3} \leq t \leq 0, \tag{77}$$

$$u(t) = 0, \qquad -\frac{2}{3} \leq t \leq 0, \tag{78}$$

and the performance index

$$J = \frac{1}{2}\int_0^1 [x^2(t) + \frac{1}{2}u^2(t)]dt. \tag{79}$$

The problem is to find the optimal control $u(t)$ which minimizes Eq. (79) subject to Eqs. (76)-(78).
Here, we solve this problem by choosing $N = 3$ and $M = 2$. Let

$$x(t) = A^T B(t), \qquad u(t) = U^T B(t), \tag{80}$$

where A, U and $B(t)$ can be obtained similarly to Eqs. (13) and (14). By expanding $x(0)$ in terms of hybrid functions, we have

$$x(0) = [1,1,1,0,0,0,0,0,0]B(t) = e_1^T B(t). \tag{81}$$

We also have

$$\int_0^t x(t' - \frac{1}{3})dt' = \begin{cases} e_2^T B(t), & 0 \le t \le \frac{1}{3}, \\ e_3^T B(t) + A^T D_1 P B(t), & \frac{1}{3} \le t \le 1, \end{cases} \tag{82}$$

$$\int_0^t u(t' - \frac{2}{3})dt' = \begin{cases} 0, & 0 \le t \le \frac{2}{3}, \\ U^T D_2 P B(t), & \frac{2}{3} \le t \le 1, \end{cases} \tag{83}$$

where

$$e_2 = \left[\frac{1}{6}, 0, 0, \frac{1}{3}, 0, 0, 0, 0, 0\right]^T, \quad e_3 = \left[0, \frac{1}{3}, \frac{1}{3}, 0, 0, 0, 0, 0, 0\right]^T. \tag{84}$$

In Eqs. (82) and (83), P is the operational matrix of integration, and D_1 and D_2 are 9×9 operational matrices of delay given by

$$D_1 = diag(\Psi_1, \Psi_1, \Psi_1), \qquad D_2 = diag(\Psi_2, \Psi_2, \Psi_2),$$

where

$$\Psi_1 = \begin{bmatrix} 0\,1\,0 \\ 0\,0\,1 \\ 0\,0\,0 \end{bmatrix}, \qquad \Psi_2 = \begin{bmatrix} 0\,0\,1 \\ 0\,0\,0 \\ 0\,0\,0 \end{bmatrix}.$$

Integrating Eq. (76) from 0 to t and using Eqs. (80)-(84), we have

$$A^T[I_9 + P - D_1 P] - U^T[P - \frac{1}{2}D_2 P] - e^T = 0, \tag{85}$$

where I_9 is a nine dimensional identity matrix and

$$e = e_1 + e_2 + e_3.$$

By using Eqs. (79) and (80), we get

$$J = \frac{1}{2}[C^T K_2 C + \frac{1}{2}U^T K_2 U],$$

where K_2 can be calculated similarly to Eq. (15).
In Table 3, a comparison is made between the values of J obtained by the present method with $N = 3$ and $M = 2, 4$, and 6 together with the value of J with the hybrid of block-pulse and Legendre polynomials $b_{n_1 m_1}(t)$, $n_1 = 1, 2, ..., N_1$, $m_1 = 0, 1, ..., M_1 - 1$ with $N_1 = 3$ and $M_1 = 4, 5$ and 6 reported in [38].

Table 3. Estimated values of J

Method of [38] with $N_1 = 3$ and	J
$M_1 = 4$	0.37311293
$M_1 = 5$	0.37311253
$M_1 = 6$	0.37311241
Present Method with $N = 3$ and	
$M = 2$	0.37310834
$M = 4$	0.37207693
$M = 6$	0.37207461

6 Conclusion

In the present work, the hybrid of block-pulse and Bernoulli polynomials are used to solve variational problems and optimal control of delay systems. The problem has been reduced to a problem of solving a system of algebraic equations. The matrices K, P, and $\tilde{C}$ in Eqs. (15), (16) and (17) have large numbers of zero elements, and they are sparse. Hence, the present method is very attractive and reduces the CPU time and the computer memory. Illustrative examples are given to demonstrate the validity and applicability of the proposed method.

References

1. Chen, C.F., Hsiao, C.H.: A Walsh series direct method for solving variational problems, J. Franklin. Instit. 300, 265–280 (1975)
2. Horng, I.R., Chou, J.H.: Shifted Chebyshev direct method for solving variational problems, Int. J. Syst. Sci. 16, 855–861 (1985)
3. Hwang, C., Shih, Y.P.: Laguerre series direct method for variational problems, J. Optim. Theory. Appl. 39, 143-149 (1983)
4. Chang, R.Y., Wang, M.L.: Shifted Legendre direct method for variational problems, J. Optim. Theory. Appl. 39, 299–307 (1983)
5. Razzaghi, M., Razzaghi, M.: Fourier series direct method for variational problems, Int. J. Control. 48 887–895 (1988)
6. Razzaghi, M., Elnagar, G.: Linear quadratic optimal control problems via shifted Legendre state parametrization, Int. J. Syst. Sci. 25, 393–399 (1994)
7. Razzaghi, M., Razzaghi, M.: Instabilities in the solution of a heat conduction problem using Taylor series and alternative approaches, J. Franklin Instit. 326, 683–690 (1989)
8. Razzaghi, M., Marzban, H.R.: Direct method for variational problems via hybrid of block-pulse and Chebyshev functions, Math. Prob. Eng. 6, 85–97 (2000)
9. Wang, X.T., Li, Y.M.: Numerical solutions of integro differential systems by hybrid of general block-pulse functions and the second Chebyshev, polynomials, Appl. Math. Comput. 209, 266–272 (2009)
10. Razzaghi, M., Marzban, H.R.: A hybrid analysis direct method in the calculus of variations, Int. J. Comput. Math. 75, 259–269 (2000)
11. Marzban, H.R., Razzaghi, M.: Hybrid functions approach for linearly constrained quadratic optimal control problems, Appl. Math. Model. 27, 471–485 (2003)
12. Marzban, H.R., Razzaghi, M.: Analysis of time-delay systems via hybrid of block-pulse functions and Taylor series, J. Vib. Control. 11, 1455–1468 (2005)
13. Marzban, H.R., Razzaghi, M.: Solution of multi-delay systems using hybrid of block-pulse functions and Taylor series, J. Sound. Vib. 292, 954–963 (2006)

14. Tikhomirov, V.M.: Stories about Maxima and Minima. American Mathematica Society, Providence, RI (1990)
15. Elgolic, L.E.: Calculus of variations. Pergamon Press, Oxford (1962)
16. Gelfand I.M., Fomin, S.V.: Calculus of variations. Prentice-Hall, Englewood Cliffs, NJ (1963)
17. Elsgolts, L.: Differential equations and the calculus of variations, translated from the Russian by G. Yankovsky, Mir Publisher, Moscow (1977)
18. Jamshidi, M., Wang, C.M.: A computational algorithm for large-scale nonlinear time-delay systems. IEEE Transactions Systems Man Cybern. 14, 2–9 (1984)
19. Kwakernaak, H., Sivan, R.: Linear Optimal Control Systems, Wiley-Inter-science, New York (1972)
20. Khellat, F.: Optimal control of linear time-delayed systems by linear Legendre multiwavelets, J. Optim. Theory. Appl. 143, 107–121 (2009)
21. Kharatishvili, G.L.: The maximum principle in the theory of optimal process with time-lags. Dokl. Akad. Nauk SSSR 136, 39–42 (1961)
22. Inoue, K., Akashi, H., Ogino, K., Sawaragi, Y.: Sensitivity approaches to optimization of linear systems with time-delay, Automatica 17, 671–676 (1971)
23. Jamshidi, M., Razzaghi, M.: Optimization of linear systems with input time-delay, Kybernetika. 11, 375–384 (1975)
24. Malek-Zavarei, M., Jamshidi, M.: Time-Delay Systems: Analysis, Optimization and Applications. North-Holland, Amsterdam (1978)
25. Delfour, M.C.: The linear quadratic control problem with delays in state and control variables: A state space approach. SIAM J. Control. Optim. 24, 835–883 (1986)
26. Uchida, K., Shimemura, E., Kubo, T., Abe, N.: The linear-quadratic optimal control approach to feedback control design for systems with delay, Automatica. 24, 773–780 (1988)
27. Costabile, F., Dellaccio, F., Gualtieri, M.I.: A new approach to Bernoulli polynomials, Rendiconti di Matematica, Serie VII. 26 1-12 (2006)
28. Arfken, G.: Mathematical Methods for Physicists, Third edition, Academic Press, San Diego (1985)
29. Kreyszig, E.: Introductory Functional Analysis with Applications, John Wiley and Sons Press, New York (1978)
30. Mashayekhi, S., Ordokhani, Y., Razzaghi, M.: Hybrid functions approach for nonlinear constrained optimal control problems, Commun. Nonlinear. Sci. Numer. Simulat. 17, 1831–1843 (2012)
31. Mashayekhi, S., Ordokhani, Y., Razzaghi, M.: Hybrid functions approach for optimal control of systems described by integro-differential equations, Appl. Math. Model. 37, 3355–3368 (2013)
32. Lancaster, P.: Theory of Matrixes. Academic Press, New York (1969)
33. Datta, K.B., Mohan, B.M.: Orthogonal functions in systems and control. World Scientific, Singapore (1995)
34. Schechter, R.S.: The Variation Method in Engineering. McGraw-Hill, New York (1967)
35. Russak, I.B.: Calculus of Variations. Ma 4311 Lecture Notes, Monterey, CA (2002)
36. Hwang C., Chen, M.Y.: Analysis of time-delay systems using the Galerkin method, Int. J. Control. 44, 847–866 (1986)
37. Marzban H.R., Razzaghi, M.: Solution of time-varying delay systems by hybrid functions, Math. Comput. Simulat. 64, 597–607 (2004)
38. Marzban, H.R., Razzaghi, M., Optimal control of linear delay systems via hybrid of block-pulse and Legendre polynomials, J. Franklin. Instit. 341, 279-293 (2004)

Punctuated Anytime Learning for Autonomous Agent Control

Gary Parker

Department of Computer Science, Connecticut College, USA
parker@conncoll.edu

Abstract. In this chapter we will discuss Punctuated Anytime Learning (PAL), which uses Evolutionary Computation running offline to learn online control programs for autonomous agents. Two methods of PAL will be discussed and considering their strengths and weaknesses they will be employed to learn control programs for autonomous agents operating in two distinct environments. One is a hexapod robot that needs an appropriate gait for its circumstances and the other is a virtual agent operating in Xpilot, the 2D space combat game.

Keywords: co-evolution of model parameters, fitness biasing, autonomous agent learning, evolutionary robotics, hexapod, real-time learning, video game learning, Xpilot-AI, cyclic genetic algorithm, evolutionary computation.

1 Introduction

Two of the main advantages of learning control programs for autonomous agents are that the programmer does not need to tweak the program for the intricacies of the controller and the system has the potential to adapt to changes in the environment and/or the capabilities of the agent. Evolutionary Computation (EC) has been a very successful method of learning that uses an agent's overall fitness as opposed to immediate feedback after each move to perform learning. This is a significant advantage because the negative feedback for a poor move may not be received until several moves later. However, most forms of EC have a drawback in that they require numerous trials of potential control programs to produce better programs. If these trials are done on the agent, the results should be good, but they will overwhelm the agent's time, and in the case where the agent is a robot, will result in undue wear and tear during training. This is not much of a problem if these trials are done in a simulation, but then there is an issue of making sure the simulation matches the actual robot. In this chapter, we will discuss Punctuated Anytime Learning (PAL), which bridges the gap between the simulation and the actual agent during EC.

The previously established techniques for evolving control programs for autonomous agents can be divided into three general approaches. The first is to do all of the training offline on a simulation and transfer the results to the actual agent when it is complete [1, 2, 3]. Significant attention must be paid to the simulation to ensure that it accurately models the actual agent in order to assure the usefulness of the results. The

A. El-Osery and J. Prevost (eds.), *Control and Systems Engineering*,
Studies in Systems, Decision and Control 27, DOI: 10.1007/978-3-319-14636-2_5

time and effort put into the model can sometimes exceed the work that would be required to program the controller by hand. In the second method, all of the training is done on the actual agent [4, 5, 6]. If the task can be completed and the fitness accurately judged in minimal time, this is a viable option. It precludes the need for any simulation. The problem is that training using evolutionary computation takes *num-generations* × *num-individuals* × *time-to-complete-task* time to complete the training. Assuming the *num-generations* and *num-individuals* are set by the nature and complexity of the problem, the *time-to-complete-task* is the most probable factor to influence the time for training. In most cases the *time-to-complete-task* is significantly longer when the tests are done on the actual agent (especially when it's a robot), which precludes this method except for very simple tasks. The third method is to start the training off line and transfer training to the actual agent for the remaining generations [7, 8]. This does not require the same level of simulation accuracy, although a model is still required. In addition, since the EC learning curve typically starts out steep dealing with a starting population that is filled with randomly generated solutions, this method passes through the initial phase with faster training on the simulation preserving time for more accurate training on the actual agent as the system tweaks the solution. Nevertheless, several generations are still required on the agent, which can take significant time for all but the simplest tasks.

The problem with all of these techniques is that either time has to be to put into the simulation or into the training on the actual agent. In addition, none of them are designed to be able to refine the solution while the actual agent is in operation unless it is only doing the task being learned. What is needed is a method of coupling the simulator to the actual autonomous agent [9, 10, 11]. The research reported in this chapter describes punctuated anytime learning, a method for making the simulated/actual agent link. This method is distinct from the previously established techniques in that it uses periodic tests on the actual agent to improve the accuracy of training done in simulation.

After introducing punctuated anytime learning and its two methods, we will discuss specific problems where it has been successfully used for autonomous agent learning, and consider the positive and negative attributes of each of the methods.

2 Punctuated Anytime Learning

Anytime algorithms are algorithms that can be interrupted at any time and return a solution [12]. Grefenstette and Ramsey [13] used the term *anytime learning* to refer to a method of learning that has the characteristics of an anytime algorithm, except that it never stops; it continually gives a meaningful answer any time the solution is improved. When used as part of a robot control system, anytime learning integrates the execution (control) module and the learning module. The control module directs what actions to take in response to the input sensors. The learning module uses an internal simulation of the robot and its environment to learn a better control program. Direct feedback from the control module to the learning module can be used to update the internal representation in the learning module.

Anytime learning allows the learning module to continually compute a best control solution while the robot operated in the environment is using the latest best solution. This system worked well for a robot that could carry both the operations and learning components on board and could detect specific changes in the environment, such as changes in the opponent's capabilities in the case of the work done by Grefenstette and Ramsey. An extension of anytime learning was needed for robots that are simple enough that their learning component cannot be carried on board. This is a reasonable configuration for teams of expendable robots. They need to have enough computing power to provide autonomous control, but do not want to carry the significantly more equipment on board to support learning (especially learning with EC).

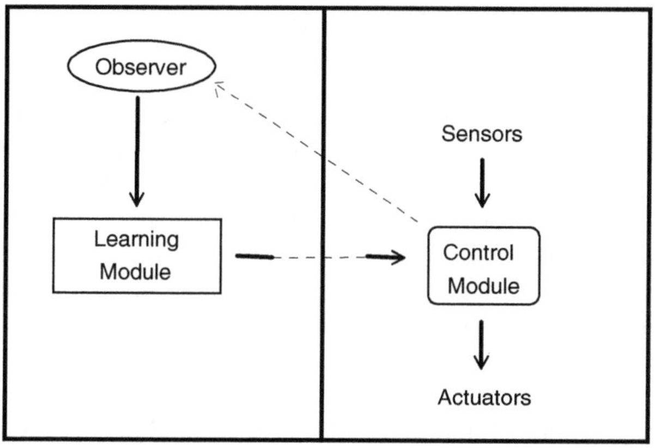

Fig. 1. Punctuated anytime learning system where the learning module is not on the robot. The Control Module runs the autonomous robot. The Observer can be separate from the robot and it evaluates the robot's performance. The Learning Module, which is using EC to learn on a simulation of the robot/environment, uses information from the Observer to alter the learning process.

Punctuated anytime learning (PAL) is a solution for the needed modification to anytime learning and the model verses actual agent issue discussed in the Introduction (Figure 1). Training with EC takes place offline on a simple model while periodic checks on the actual agent help improve the learning system's output. This type of anytime learning is referred to as punctuated because the learning, although it is continuous, has break points where it is working on newly updated information about the actual agent. Grefenstette's system could adapt quickly to changes in the robot or the environment by having the robot's sensors continually update the status of the environment and the robot's capabilities in the learning component's simulation. This is where a difficulty develops when this methodology is applied to remote learning, where the learning system can only depend on global observation, instead of feedback from precise sensors, to determine the robot's capabilities. PAL compensates for the lack of quick feedback, precise sensors, and high computational power on agent robots.

In a punctuated anytime learning system the learning module is not part of the agent, so it no longer has direct input to the control module. Updated control programs are sent via some form of external communication. In addition, feedback is no longer sent via a direct link. It can be transmitted, in the same way as the control program, if there is an evaluation unit on the robot. Alternatively, it may be provided to the learning module from an outside observer (such as an external overhead camera). The second option becomes more likely as the complexity of the robot goes down since it would carry a less sophisticated evaluation unit and communications transmitter. There are two versions of punctuated anytime learning: the co-evolution of model parameters and fitness biasing.

2.1 The Co-Evolution of Model Parameters

The co-evolution of model parameters [14, 15] is a PAL method that dynamically links the model to the actual agent during EC. Periodic checks on the actual robot alter the model (simulation) in an attempt to improve the result of training. This involves the co-evolution of the model parameters and the controller program. After each n generations of controller program evolution, three controller solutions are selected to be tested on the actual agent. The three solutions include the best, plus either two random or one random and the worst. The actual controller fitnesses of these three controller solutions are used to determine the model fitnesses. The individuals in the model parameter population are made up of a set number of genes, each of which represents a corresponding field in the model. These individuals are evolved to produce models that corresponded in performance to the actual agent. The model fitness of each individual is determined by comparing the solution fitness run on the actual agent with the solution fitness run on the model. The model fitness can be computed in different ways, but the point is to have a model that best corresponds to the actual agent when running all three controller solutions. The population of model parameters can start out either as randomly generated individuals or as a combination of perturbations (to varying degrees) of the original model parameters. The most accurate model parameter is used for the controller evolution; the population of model parameters will continue to evolve until interrupted by updated actual test information. This solution requires three actual tests every n generations of solution evolution.

2.2 Fitness Biasing

Fitness biasing [16, 17] allows the system to modify the EC based on the performance of the robot. The model is not affected; its parameters were set before anytime learning began. The periodic checks on the actual robot can alter the processing within the EC in an attempt to improve the result of training. This involves the biasing of evolved solution fitnesses by comparing their performance on the model to their performance on the actual robot. During EC, probability for selection is determined by computing each individual's fitness on the agent model. This fitness is computed for each individual in the population. After each n generations all solutions of the EC population are tested on the actual robot. These measurements are used to bias the fitnesses found on the model to make them equal to the actual robot fitnesses. A bias for each solution of the population is computed using the algorithm in Figure 2.

```
Model-Fitness = Compute-Model-Fitness(Solution)
if absolute-value(Model-Fitness) < 1
   Bias = 1
else
   Actual-Fitness = Test-on-Robot(Solution)
   Bias = Actual-Fitness / Model-Fitness
```

Fig. 2. Algorithm to compute bias

This bias is stored with its corresponding solution. It is used in subsequent generations of the EC to alter the fitness of the solution computed on the model of the robot. This is done by multiplying the fitness computed on the model by the bias:

$$\textit{Corrected-Fitness} = \textit{Model-Fitness} * \textit{Bias} \tag{1}$$

Pairs of individuals are stochastically selected for reproduction using the *Corrected-Fitnesses*. The two individuals produce a single offspring for the next generation; combining their attributes by crossover with possible a small chance of mutation. The new offspring's bias is computed by averaging the biases of its parents. With fitness biasing, solutions that are good on the actual agent but poor on the model have boosted fitnesses, which results in their production of more offspring. This solution requires *population-size* actual tests every *n* generations.

3 Using the Co-Evolution of Model Parameters for Hexapod Robot Gait Development

Hexapod gait development presents an interesting problem for learning algorithms since it usually requires sequential solutions where a series of actions is continually repeated. Although learning to actuate a single leg for a single step is not difficult, learning to walk requires all legs to repeat steps in the correct sequence, and transition smoothly from one step to the next, which increases in difficulty when more legs are added and an optimal gait is required. Additionally, unpredictable changes in robot capability make adaptivity a requirement. This dynamic system suggests that some form of EC should serve well as the learning algorithm. Gait generation for six-legged robots, has been addressed with several approaches. Donner [18] developed control strategies for the SSA six-legged robot using decomposition of the task's elements and locality of control. Brooks [19] used subsumption architecture in the development of his robots. Beer and Gallagher [20] used a neural net to control a simulated cockroach and an actual hexapod robot [21]. These solutions proved to be successful for the robots used but we wanted a solution that could work on the most primitive of robots and that required less a priori knowledge of how to walk. Spencer [22] used genetic programming to develop gaits for a simulated robot using minimal knowledge about the intricacies of walking. Although his results were promising in that the system consistently learned gaits which generated sustained forward movement, we wanted a system that would learn the optimal gait for our robot which had a very primitive control interface and no means of feedback. In addition, we wanted it to be continually adaptive to robot capabilities and be a quickly converging algorithm suitable for PAL.

3.1 Robot and Model

The robot used for these tests was the ServoBot (developed by David Braun), which is an inexpensive hexapod robot that has two degrees of freedom per leg. Twelve servos, two per leg, provide thrust and vertical movement. A control sequence is transmitted to an onboard controller (BASIC Stamp II) from a workstation. The onboard controller can store and execute (repeating the designated section) the sequence of primitive instructions downloaded. Each instruction corresponds directly to an activation that is sent to the servomotor every 30ms.

An input to the ServoBot servos is a 12 bit number where each bit represents a servo. A signal of 1 moves the leg back if it is a horizontal servo and up if it is a vertical servo. A signal of 0 moves it in the opposite direction. The activation can be thought of as 6 pairs of actuations. Each pair is for a single leg with the first bit of the pair being that leg's vertical activation and the second being that leg's horizontal activation. The legs are numbered 0 to 5 with 0,2,4 being on the right from front to back and 1,3,5 being the left legs from front to back. The activation 100101101001 results in one phase of the classic tripod gait, which is considered to be the optimal gait for speed in this hexapod robot when all its actuators are fully functioning. Legs 1,2,5 are providing thrust while legs 0,3,4 are being lifted. Some sequence of these activations will result in a pattern of leg movement that will produce a viable gait.

The model was a simple data structure that held each leg's capabilities and current state (Figure 3). Input activations were used with these capabilities to determine how much to change the current state of the model during training.

Fields specific for each leg:
current up -- current vertical position of the leg.
max up -- posit off the ground when completely up.
current back -- current horizontal position of the leg.
max back -- posit relative to completely forward completely back.

Fields applicable to all legs:
rate up/down -- rate of vertical movement when servo activated.
rate back/forward -- rate of horizontal movement when servo activated.

Fig. 3. ServoBot Model Data Structure

Measurements to fill the fields for each leg's capabilities were taken before training. Each leg's vertical position was measured relative to this initial zero position, which was on the ground. Each leg's horizontal position was measured relative to its at rest full forward position. Measurements to fill the maximum position fields of the model were taken by activating each control on the actual robot while recording the leg's maximum throw. An average rate per activation was calculated for horizontal and vertical movement by dividing the maximum throw by the minimum number of activations required to attain it.

3.2 Cyclic Genetic Algorithm

The cyclic genetic algorithm (CGA) was developed [23] to allow for the representation of a cycle of actions in the chromosome. It differs from the standard GA in that the genes of the chromosome represent tasks that are to be completed in a predetermined segment of time. The tasks can be as simple as primitives (activations) or they can be as complicated as cyclic sub-chromosomes that can be trained separately by a CGA. For our purposes, the genes represent a set of servo activations that are to be sent to the servos every 30 msec. The trained chromosome will contain the cycle of these primitive instructions that will be continually repeated by our robot's controller to produce a gait.

Each chromosome was made up of three sections: coordinators, inhibitors, and a set of 12 activation genes. The genes of the chromosome had two parts. The activations part was a 12 bit number that contained the encoding required to activate two possible primitives per leg. A bit setting of 1 meant the servo was moving the leg back, a bit setting of 0 meant it was moving the leg forward. When the servo reached the full throw of the leg it would hold its full back position. Similarly, when the servo ran continuously forward the leg would stay at its full forward position. The activation for up and down worked in the same way. The repetition part was an 8 bit number that designated the number of times to repeat the activations part. This repetition part was what gave the CGA the ability to vary the length of the sequence of primitives being sent to the robot in each cycle.

The CGA chromosome can also incorporate parameters for control in its structure. For gait generation, coordinators were 12 bit numbers that directed the coordination of individual leg movement. Inhibitors prevented pairs of legs from moving back at the same time. The inhibitors for the set of legs were stored in a single 15 bit number.

The effectiveness of the CGA has been tested on fully functional [24] and partially disabled robots [25]. In both cases, training was done on a model of the robot with tests on the actual robot. CGAs were very successful at evolving what appeared to be optimal gaits for the robots tested.

3.3 Using the Co-Evolution of Model Parameters to Learn Gaits

The CGA was used to evolve the control program, while a standard genetic algorithm (GA) was used to evolve the model parameters [15]. The control program chromosome was described in the previous section. The model parameter chromosome was made up of the *max up* and *max back* parameters of each leg (Figure 3). The chromosome had 12 genes, with each gene 6 bits. The starting population was made up of randomly generated chromosomes. After every 10 generations of CGA training on the model, three solutions (best and two random) are tested on the actual robot. Each of these three is run on the actual robot to gain a true fitness, which was the distance (measured in centimeters) traveled by the robot after 200 servo pulses (approximately 5 seconds). The three gait solutions were also tested on each of the individuals in the model population.

A comparison of each gait solution's fitness on the actual robot verses the model was used to determine the fitness of each individual of the model population (Figure 4). A model that performed similarly to the robot on all three tests was assumed to have accurate parameters and was given a high fitness. After the model parameters were checked at the 10th generation, the best was used to continue evolving gait solutions. The fitnesses applied to the individuals in the population of model parameters were used in the co-evolution of the model parameters, which continues to evolve until interrupted after another n generations of gait learning.

```
R1 = Test best solution on the robot
R2 = Test a random solution on the robot
R3 = Test a second random solution on the robot
For each individual in the model population
   M1 = Get model fitness using best solution
   M2 = Get model fitness using first random solution
   M3 = Get model fitness using second random solution
   Order-Same = True if the fitness order of R1, R2, R3 is the same as M1, M2, M3
   Signs-Same = True if Rx & Mx have the same sign for x = 1 to 3
   Fit-Comp = Fitness of the robot to model comparison
   If neither Order-Same or Sign-Same are True
     Return .001
   If either Order-Same or Sign-Same are True
     Return .001 x Fit_Comp
   If both are True
     Return Fit_Comp
```

Fig. 4. Algorithm to compute the fitness of each model's set of parameters in a population

M1 is the model's fitness when run using the best gait from the CGA, M2 and M3 are the model's fitnesses when run using the random gaits. The R1, R2, and R3 are the robot's fitnesses (or distance) traveled using the corresponding gaits. The value for Fit-Comp is computed using the following equation:

$$\frac{1}{\left\| \left|\frac{M1-R1}{R1}\right| - \left|\frac{M2-R2}{R2}\right| \right\|} + \frac{1}{\left\| \left|\frac{M1-R1}{R1}\right| - \left|\frac{M3-R3}{R3}\right| \right\|} \tag{2}$$

R1 and M1 are changed to equal 1 if their absolute value is less than 1 and if the denominator on either side of the plus sign is less than .01 the division results in a value of 100. This equation can be replaced with alternatives that are appropriate to the task.

In the case of gait generation, the purpose of the equation is to compute how closely the robot and model match while disregarding the differences in rates. The idea is that the model is highly fit if its computed distances moved vary in the same way as the robot as they are tested on differing gaits. M1 - M2 should equal R1 - R2 if the model is perfect. The divisions by M1 and R1 allow for differing rates in the model and robot by normalizing the gait distance traveled differences. For example, if the robot's rate is 3 cm of movement per 50 ms and the max throw is 45, the corresponding model can have a rate fixed at 2 cm per 50 ms and adjust its max throw through co-evolution to equal 30. In each case, 15 activations will move the leg full throw. A robot and model differing only in rates could have values of R1 = 3 and M1 = 2 for the best gait and R2 = 1.5 and M2 = 1 for the random gait. Using these values to compute the left side of the equation above results in a denominator of 0 yielding a fitness of 100. If the R3 and M3 correspond this nicely, the model will attain a maximum fitness of 200.

For some problems, a genetic algorithm will not be the best optimization technique for finding the model parameters. It may be hill climbing or any other form of greedy algorithm. While the CGA is learning the cycle for the gaits, or whatever else is being learned, the model parameters can be learned at the same time, whether it be by some form of evolutionary computation or through some other means of learning. In either case, the concept of co-learning model parameters is the same.

3.4 Results

Two tests were performed, each on five gait populations. In the ½ capability test, the initial populations were trained on models that had all of their max back capabilities reduced by ½. The max back capabilities used to make this computation were taken from the measurements made in the original gait generation experiments described in Section 3. The five solution populations were then used with the learning system employing the co-evolution of model parameters to generate a gait for the actual robot. The double capability test was similar except that the initial population of solutions was for a robot with max back capabilities doubled.

Figure 5 shows the results of the ½ capability tests done on the actual robot. The x axis shows the number of trials on the robot. Each mark (starting at 6 trials on the robot) also represents 10 CGA generations and 3 model parameter generations of evolution. The y axis shows the distance moved in centimeters. The results of each of the five tests are plotted with the fitness (distance moved) of the best solution tested on the robot shown. The average of the five tests is shown as a bold line. As can be seen, the initial best solution was at around 30 cm. By the end of 30 trials on the robot, all the solutions improved to result in forward movement of over 40 cm. A majority of the tests showed improvement within 6 actual trials on the robot, probably indicating that the elements of improved solution gaits were already in the population (one of the strengths of genetic algorithms). The baseline shows the actual robot performance without anytime learning; there is no improvement over time, because the learning system does not change the model of the robot that it is using for evolution.

The initial solution (without PAL) resulted in a tripod gait, but the strides were short to accommodate the reduced max back measurements in the model. This gait does not take full advantage of each leg's thrust capability because the short, choppy steps do not allow the servo to continue providing thrust in the middle of its rotation. After PAL training, the resultant gaits are still tripod, but much improved such that the legs are not lifted until they have moved nearly to their back limit.

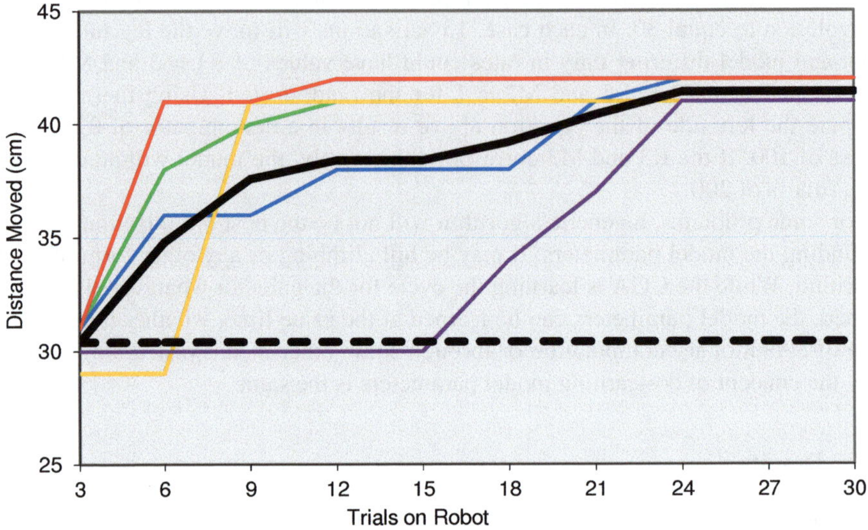

Fig. 5. Results of the ½ Capability Tests. The baseline is a bold dashed line, the 5 runs are thin colored lines, and the average of these 5 is a bold solid line.

Figure 6 shows the results of the double capability tests done on the actual robot. The graph is set up in the same way as in Figure 5 except that there are 81 generations showed along the x axis. This problem was significantly more difficult for the learning system to solve. The initial solutions resulted in an average of only 20 cm covered. Two out of the five tests did not produce solutions that resulted in more than 35 cm covered within 51 trials on the actual robot; one of these did not make it to 35 cm by 78 trials. Nevertheless, all the tests produced solutions that showed at least a 50% improvement over the baseline.

The initial solution (without PAL) resulted in a tripod gait, but it had very long strides. Once the legs hit their stops (went full swing), the robot would hold until the next phase began. This resulted in time wasted without forward movement. The final solution gait was similar to the gait produced in the ½ capability experiment except that its strides were a little longer. This was probably because the long stride solutions were more similar to the initial gaits in this test and because the robot's capabilities had changed during training.

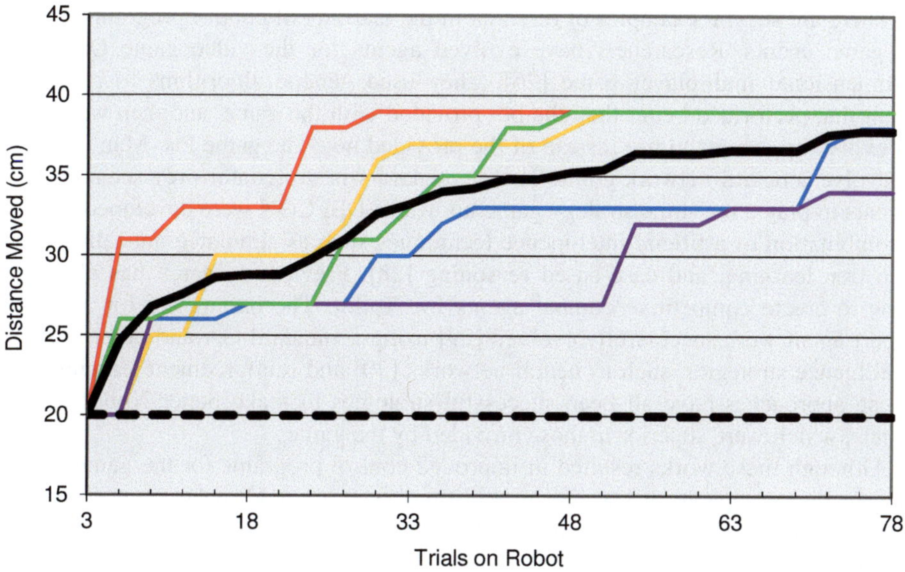

Fig. 6. Results of the Double Capability Tests. The baseline is a bold dashed line, the 5 runs are thin solid lines, and the average of these 5 is a bold solid line.

Tests show that the co-evolution of model parameters can adapt to inaccuracies in the model or changes in the robot's capabilities to provide a system of anytime learning. This system requires only external observation of the robot's performance and is an effective means of coupling the model to the robot during control learning using EC.

4 Using Fitness Biasing for Xpilot Game Agent Control

In this section we discuss the application of fitness biasing for real-time learning of the control program for an autonomous agent in the Xpilot-AI, a 2D combat-based video game [17]. Video games are popular for entertainment and some can be used to test autonomous agent control programs. Creating competitive non-player agents for these games is often difficult, especially if it's desirable to have the agents adapt during game play. One method of learning that can create distinct competitors and has the potential for adapting to different players is evolutionary computation (EC). This learning method is also important in learning robot control so its successful use for autonomous agents in video games is applicable to robotics research as well. The video game used for the research reported in this paper, Xpilot, is an excellent environment due to its low computational requirements and the Xpilot-AI add-on that allows users to create learning programs for game agents.

There are several examples of research in the learning of control programs for video game agents. Researchers have evolved agents for the video game Quake3, a 3-dimensional multiplayer game [26]. They used genetic algorithms to evolve an agent that performed better than the bot provided with the game, and then went on to co-evolve opponents to use instead of the provided bot. The game Pac-Man was used to evolve a neural network controller for predators in a predator-prey scenario [27]. Agents to play a real-time strategy game known as DEFCON were developed through a combination of artificial intelligence techniques such as simulated annealing, decision tree learning, and case-based reasoning [28]. Previous research has also been done to create competitive combat agents for Xpilot. The parameters for a combat expert agent were successfully evolved [29] using a standard GA and other artificial intelligence strategies, such as neural networks [30] and reinforcement learning [31]. These approaches have all been successful strategies to make better Xpilot combat agents, which were superior to those provided by the game.

Although these works resulted in improved control programs for the game agents, they were not capable of learning in real-time, which would make them capable of learning and adapting to the manner of play of others in the game. Creating artificially intelligent systems that are capable of doing this in real-time is an important, yet one of the most difficult aspects, in game development [27]. Others applied evolutionary techniques to neural networks by starting with simple networks then slowly adding nodes and connections while the game is running to make the agent learn increasingly complex behavior in real-time [32]. While both of these do learn in real-time, they also both rely on past knowledge and pre-defined courses of action. In previous Xpilot-AI research, dynamic programming-based reinforcement learning techniques, such as Q-learning, were used to produce an Xpilot agent capable of real-time learning, but the controllers learned were for a very simple environment. The Q-learning method that was implemented requires an accurate model to be successful. Since the Xpilot combat environment is very complicated, this method was applied only to a single agent in a simple environment with no opponents [31]. Though it was a successful implementation of real-time learning, it was determined to not be as scalable as desired. As the complexity of the environment increases, such a system's ability to cope rapidly deteriorates. In another attempt to demonstrate real-time learning in Xpilot, evolutionary strategies were used to learn agent controllers [33]. Though capable of real-time learning, their reliance on mutations of a chromosome to evolve led to slower learning and a lack of effectiveness.

Although GAs have been used successfully to generate controllers for interactive game environments, they have not been used in a real-time learning system since each individual of the population needs to be tested. In order to be able to adapt to an opponent's play in real-time, these tests would have to be done by playing the opponent. In addition, the GA typically starts with a random population of solutions. This would make particularly poor play for someone playing the game. This is a problem area where PAL is particularly appropriate. Learning can take place in a simulation of the game with periodic checks to make sure the simulation represents actual play.

4.1 Xpilot-AI

Xpilot-AI is a modification of the computer game Xpilot, a 2-dimensional space game (Figure 7). Xpilot has server and client components. The server handles the game-wide configuration and is the central source for all information about each ship and object in the game. The client gets this information from the server, accepts the commands of the user and relays them to the server. This allows the user to turn, thrust, and shoot, among other possibilities. Xpilot-AI works in between the client and the server, so that the information about the game is available to a programmer. A typical Xpilot server runs at 16 frames per second. Between each frame, the server and client exchange information and Xpilot-AI intercepts the communications to obtain necessary information for the user. The programmer can write programs that get info from functions Xpilot-AI provides, use this information to determine what actions to take depending on the situation, and set variables describing the actions that the client sends back to the server.

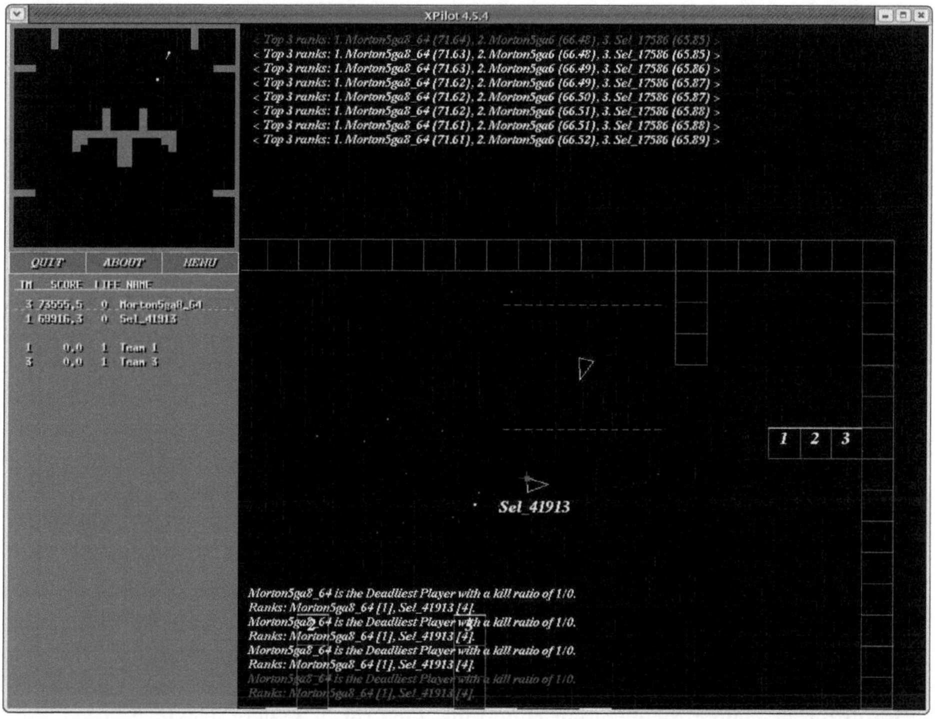

Fig. 7. The Xpilot environment. Evolved agent Morton has just attempted to shoot the enemy ship Sel, but the shot went behind its target. The blue blocks at the top are walls.

Xpilot-AI provides many functions to the programmer that give information about the current state of the game. For example, the speed of the ship and its location are available, as well as the speeds and locations of any other ships currently playing in the map. The programmer also has the ability to get the location and heading of the bullets that are in play, and the ability to determine if there is a wall between two points. The programmer can make use of this information in a control program that determines when to determine when to turn, shoot, and thrust. These control programs can take many forms, but one means of writing a control program is to write a rule-based system, using rules such as "if there is an enemy close by, then turn toward the enemy and shoot at the enemy." This rule would check the functions provided by Xpilot-AI to determine if the agent is close and then call functions that cause the agent to turn and to shoot.

4.2 Fitness Biasing Applied to Xpilot-AI

An additional benefit of using Xpilot for testing autonomous agent learning systems is that it is possible to specify the number of frames per second (FPS) when starting a server. Increasing the FPS causes the game to run faster, allowing more information to be sent between the server and the connected clients. This has the potential to significantly increase training time, but due to communication and computing limitations, running the system at high FPS causes abnormal behavior in the game, resulting in reduced learning capabilities. This is a problem for real-time learning. It would be nice to run a background GA system learning at 128 FPS so that it would be learning 8 times faster than normal play at 16 FPS. However, since this is flawed simulation of the actual game, the learning is limited. Tests showed that although 128 FPS caused issues for learning, it still showed consistent aspects of continued learning. In order to run the background GA at high speeds, fitness biasing was needed to periodically help in the simulated learning process.

The agent we used was controlled by an expert system and a standard GA was used to learn the best values for parameters within the rules. The expert system was written by the developer with best guesses for the many parameters such as angle of wall feelers and the turn to be made in case of an approaching wall. The resultant agent was good, but it was difficult to find the correct set of parameters to make it competitive. Twenty-two parameters that needed to be adjusted and their appropriate ranges were identified. It was determined that they could be set using a range of 0 to 15, so each was represented by a 4 bit gene. Some were multiplied, some were divided into plus or minus possibilities, while still others were added to previous parameters to set their parameter. Some of these parameters have to do with general use functions, such as "same", which was used to define what should be considered equal in reference to distances from the ship. Some of the parameters were for specific situations. The ship uses feelers to determine the distance to walls at angles set relative to its track. There were different feelers for different situations. One set of feelers was to be used early in the case statement, which signifies that these were for higher priority situations. The second set was only used when none of the prior cases were true. Some of the parameters were used to determine when a case would be considered true (such as those dealing with feelers), others were used to determine what the action would be when the case was true (such as those dealing with an angle required for the ship to turn to or an angle required before the ship would thrust or shoot).

For fitness biasing, the previously created GA-agent was modified into two forms: one to act as the simulation and one to act as the agent. The simulation would run just as the normal GA did with two key additions: (1) while calculating fitnesses, it would bias them as previously described. (2) Every 15 generations, it would connect to the agent and calculate new biases. While doing this, the GA would halt itself and discontinue learning until the new biases were received. The actual agent would run separately but simultaneously. Upon receiving a new population from the learning module, it would run each chromosome of the population on the agent, record their fitnesses, calculate their biases, and then send the new information back to the simulation. At this point it would continue to play using the best of the chromosomes for its controller.

The fitness was calculated after two minutes of game play. The agent's opponent was the best known hand-coded expert agent named Sel. During the two minutes of game play, the agent would receive one point of fitness for every frame it was alive, plus 1000 points of fitness for every time it killed its opponent, and lose 20 seconds off its total time available for each death it experienced. Upon receiving results from the agent, the simulation would overwrite its own collection of biases with the new information and continue learning until another 15 generations had passed. This process was repeated until stopped by the researcher or some predetermined number of generations had passed. The GA used single point crossover for recombination and a mutation rate of 1:100 per bit.

4.3 Results

Five populations of 128 randomly generated individuals were generated and run in three tests for 115 generations each. These starting populations were used in three different tests. The 16FPS test was a standard GA running at 16 FPS. This sets an upper boundary on the best we could hope to achieve. The 128FPS test was a standard GA running at 128 FPS. This would be the agent's fitness if it was using 128 FPS for learning without fitness biasing. The PAL test was the same starting five populations using fitness biasing for the learning system. It tested the current population on the agent to generate new biases every 15 generations. Figure 8 shows the average fitness of the five populations for each of the three tests at each 5 generations. A look at 16FPS versus 128FPS shows both the superiority of the 16FPS method and that the 128FPS method is less effective, but does make progress. Considering the learning curves of all three tests, it is clear that the fitness biasing produced favorable results. In the beginning, the fitness averaged at approximately 350 and rose to an average of approximately 3840 by the end of training. In addition to being highly effective, it is also important to note that the fitness biasing system only needed to engage the actual opponent every 15 generations, whereas the 16FPS GA was playing against the opponent every generation. If we compared fitness after equal numbers of games played against the opponent, we'd need to compare 16FPS generation 5 with PAL generation 60. As can be seen, 16FPS is at about 2100 at this point, whereas PAL is at 3300.

The above empirical data can also be confirmed by observation. At the start of the learning process, the agent regularly made clear mistakes in combat. For example, some control programs would not turn sharply enough to avoid a bullet or a wall, not thrust to avoid a bullet, or aim incorrectly at its opponent when firing. However, towards the end, these mistakes were corrected in the majority of the learned control programs. The agent would adeptly dodge enemy bullets while responding quicker and firing back with greater accuracy. This increase in optimal behavior can be attributed to the learning that took place as a result of the Fitness Biasing system. By the end of training, all five trials produced individuals with intelligent behavior that were able to consistently beat Sel both in fitness and by the scoring system Xpilot uses to measure player kills and deaths. Visual observation of these PAL evolved agents revealed that they were highly competitive and learned what appeared to be complex combat behavior.

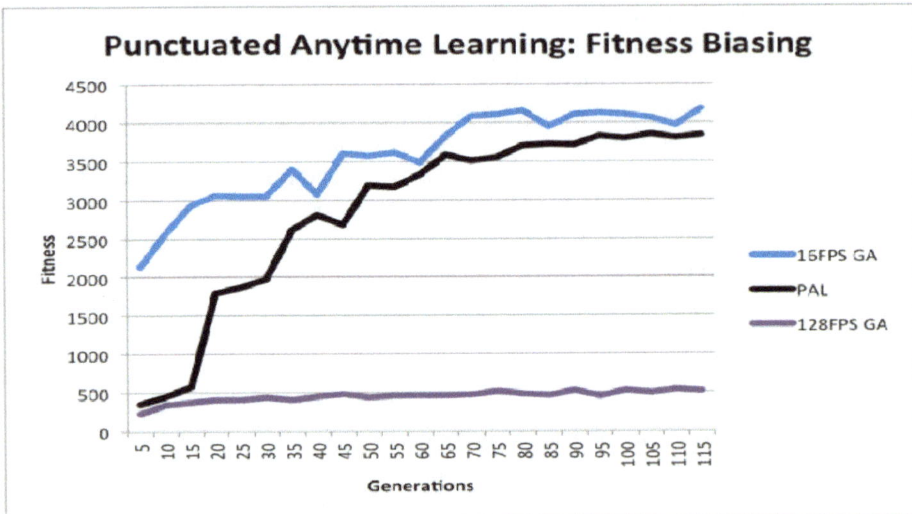

Fig. 8. Fitness biased agent (PAL) compared to a standard GA at 16 FPS and at 128 FPS.

5 Discussion

Fitness biasing takes minimal effort by the programmer to set up and can be applied to any evolutionary computation (EC) problem that is running on a model of the actual robot. The only provisions the programmer has to make are to provide a means of making the actual tests, performing the biasing, and determining how to pass an individual's bias on to its offspring. Any time these three steps can be taken, anytime learning with fitness biasing is an option. The difficulty with fitness biasing is that, since it does not change the model and training continues on an inaccurate model, it can only go so far. It cannot compensate for very inaccurate models. Another problem with fitness biasing is that it requires that all individuals in the population be tested. This can be very time consuming and disruptive to the actions of the robot if the control sequence being optimized takes long to execute. Nevertheless, fitness biasing is an improvement over doing all testing on the robot because less online testing is required and it is an improvement over doing all testing on the model since the model accuracy does not have to be painfully precise. It is even an improvement over doing the initial training on the model and then transferring to the actual robot because it allows the system to be continually adaptive throughout training.

The co-evolution of model parameters addresses the two major issues that are short-comings of fitness biasing. Since it actually changes the model, it has more potential to compensate for very inaccurate models by adjusting the model to be more like the actual agent. The other benefit of the co-evolution of model parameters is that it does not require that a test be done on all individuals of the population. It only requires three tests: the best individual and two others. Using the information from these three tests alone, it co-evolves the model parameters and in this way changes the model to be more accurate. In addition, it shares with fitness biasing the benefit that only a simple model is required. The model does not have to be extremely accurate in the initial phase so the programmer does not need to put in as much time as he would if all the training took place on the model. There are, however, some disadvantages to this form of PAL. One of the major disadvantages is that it is not applicable for all problems. Some problems are such that it is very difficult to set up the model parameter co-evolution. There are those situations where we cannot specifically define the model parameters that can be co-evolved and there are those situations where the model parameters are too numerous to be co-evolved. Another difficulty with the co-evolution of model parameters is that it is possible to get a solution that is not as good as the one produced without anytime learning. Since it is changing the model, subsequent changes in the environment could make the evolved model further from reality than the original inaccurate model.

The control learning problems addressed in this chapter demonstrated the types of problems appropriate for each method of PAL. In the hexapod gait learning problem, the model was easily accessible and had a reasonable number of parameters that needed to be evolved. In addition, the problem required minimal tests on the actual robot since with multiple tests it would take too long to learn and would result in undue wear to the robot. This played to the strengths of the co-evolution of model parameters, which was successful in adapting the gaits to be effective for the robot despite the errors in the model. In the Xpilot agent controller generation problem, there were no defined model parameters, the GA learned using a preprogrammed opponent representing a human opponent. Determining parameters to evolve would be very difficult. However, multiple tests on the actual agent were not as detrimental. The agent did not experience wear and multiple tests actually have an advantage in that they present the opponent with some variety in game play. Both methods of PAL are very effective, but the right one needs to be applied to the problem.

6 Conclusion

Tests done on these two distinct problems showed that both the co-evolution of model parameters and fitness biasing greatly improve the ability of an EC learning system to adapt to inaccuracies in the simulation. These methods of PAL, which require only external observation of the agent's performance, can be an effective means of coupling the learning system to the actual agent during EC. It is best to use the co-evolution of model parameters when one can easily define the model parameters in such a way that they can be co-learned along with the evolution of the solution that we are attempting to attain. In addition, it is good to use co-evolution of model parameters when the problem that we are trying to learn takes a long time to test each of the possible solutions. Fitness biasing is better in situations where one cannot define the parameters in such a way that they can be co-learned along with the solution and when the solution being tested is not so time consuming on the actual agent that it significantly disrupts its mission.

References

1. Jakobi, N.: Half-baked, Ad-Hoc and Noisy: Minimal Simulations for Evolutionary Robotics. In: Advances in Artificial Life: Proceedings of the 4th European Conference on Artificial Life (1997)
2. Schultz, A., Grefenstette, J., Adams, W.: RoboShepherd: Learning a Complex Behavior. In: Proceedings of the Robots and Learning Workshop, RoboLearn (1996)
3. Lee, W.-P., Hallam, J., Lund, H.: Applying Genetic Programming to Evolve Behavior Primitives and Arbitrators for Mobile Robots. In: Proceedings of IEEE 4th International Conference on Evolutionary Computation (1997)
4. Harvey, I.: Artificial Evolution and Real Robots. In: Proceedings of International Symposium on Artificial Life and Robotics (AROB), pp. 138–141 (1996)
5. Mondada, F., Floreano, D.: Evolution of Neural Control Structures: Some Experiments on Mobile Robots. Robotics and Autonomous Systems 16 (1995)
6. Nordin, P., Banzhaf, W.: Genetic Programming Controlling a Miniature Robot. In: AAAI Fall Symposium Series 1995, Symposium on Genetic Programming (1995)
7. Lund, H., Miglino, O.: From Simulated to Real Robots. In: Proceedings of IEEE 3rd International Conference on Evolutionary Computation (1996)
8. Miglino, O., Lund, H., Nolfi, S.: Evolving Mobile Robots in Simulated and Real Environments. In: Technical Report, Institute of Psychology, CNR, Rome (1995)
9. Husbands, P., Harvey, I., Cliff, D., Miller, G.: Artificial Evolution: A New Path for Artificial Intelligence? Brain and Cognition 34, 130–159 (1997)
10. Mataric, M., Cliff, D.: Challenges in Evolving Controllers for Physical Robots. In: Evolutional Robotics, special issue of Robotics and Autonomous Systems, vol. 19(1), pp. 67–83 (1996)
11. Nolfi, S., Florano, D., Miglino, O., Mondada, F.: How to Evolve Autonomous Robots: Different Approaches in Evolutionary Robotics. In: Proceedings of the International Workshop on the Synthesis and Simulation of Living Systems (1994)
12. Dean, T., Boddy, M.: An Analysis of Time-Dependent Planning. In: Proceedings of the Seventh National Conference on AI, AAAI 1988 (1988)
13. Grefenstette, J., Ramsey, C.: An Approach to Anytime Learning. In: Proceedings of the Ninth International Conference on Machine Learning, pp. 189–195 (1992)
14. Parker, G.: The Co-Evolution of Model Parameters and Control Programs in Evolutionary Robotics. In: Proceedings of the 1999 IEEE International Symposium on Computational Intelligence in Robotics and Automation (CIRA 1999), pp. 162–167 (1999)

15. Parker, G.: Co-Evolving Model Parameters for Anytime Learning in Evolutionary Robotics. Robotics and Autonomous Systems 33(1), 13–30 (2000)
16. Parker, G., Mills, J.: Adaptive Hexapod Gait Control Using Anytime Learning with Fitness Biasing. In: Proceedings of the Genetic and Evolutionary Computation Conference (GECCO 1999), pp. 519–524 (1999)
17. Parker, G., Fritzsche, P.: Fitness Biasing for Evolving an Xpilot Combat Agent. In: Proceedings of the 2011 IEEE Congress on Evolutionary Computation, CEC (2011)
18. Donner, M.D.: Real-Time Control of Walking. Birkhauser, Boston (1986)
19. Brooks, R.A.: A Robot That Walks: Emergent Behaviors from a Carefully Evolved Network. Neural Computation, 254–262 (1989)
20. Beer, R.D., Gallagher, J.C.: Evolving Dynamical Neural Networks for Adaptive Behavior. In: Adaptive Behavior, vol. 1, pp. 91–122. MIT Press, Cambridge (1992)
21. Gallagher, J.C., Beer, R.D.: Application of Evolved Locomotion Controllers to a Hexapod Robot. In: Technical Report CES-94-7, Department of Computer Engineering and Science, Case Western Reserve University (1994)
22. Spencer, G.: Automatic Generation of Programs for Crawling and Walking. In: Kinnear Jr., K. (ed.) Advances in Genetic Programming, pp. 335–353. MIT Press, Cambridge (1994)
23. Parker, G., Rawlins, G.: Cyclic Genetic Algorithms for the Locomotion of Hexapod Robots. In: Proceedings of the World Automation Congress (WAC 1996). Robotic and Manufacturing Systems, vol. 3, pp. 617–622 (1996)
24. Parker, G., Braun, D., Cyliax, I.: Evolving Hexapod Gaits Using a Cyclic Genetic Algorithm. In: Proceedings of the IASTED International Conference on Artificial Intelli-gence and Soft Computing (ASC 1997), pp. 141–144 (1997)
25. Parker, G., Cyliax, I.: Locomotion Control Cycles Adapted for Disabilities in Hexapod Robots. In: Proceedings of the World Automation Congress (WAC 1998). Robotic and Manufacturing Systems, vol. 7, pp. 359–364 (1998)
26. Priesterjahn, S., Kramer, O., Weimer, A., Goebels, A.: Evolution of Human-Competitive Agents in Modern Computer Games. In: Proceedings of the 2006 IEEE Congress on Evolutionary Computation, CEC (2006)
27. Yannakakis, G., Hallam, J.: Evolving Opponents for Interesting Interactive Computer Games. In: Proceedings of the 8th International Conference on the Simulation of Adaptive Behavior (SAB 2004), pp. 499–508 (2004)
28. Baumgarten, R., Colton, S., Morris, M.: Combining AI Methods for Learning Bots in a Real-Time Strategy Game. International Journal of Computer Games Technology (2009)
29. Parker, G., Parker, M.: Evolving Parameters for Xpilot Combat Agents. In: Proceedings of the 2007 IEEE Symposium on Computational Intelligence in Games, CIG (2007)
30. Parker, G., Parker, M.: The Evolution of Multi-Layer Neural Networks for the Control of Xpilot Agents. In: Proceedings of the 2007 IEEE Symposium on Computational Intelligence in Games, CIG (2007)
31. Allen, M., Dirmaier, K., Parker, G.: Real-Time AI in Xpilot Using Reinforcement Learning. In: Proceedings of the 2010 World Automation Congress International Sympo-sium on Intelligent Automation and Control, ISIAC (2010)
32. Stanley, K., Bryant, B., Karpov, I., Miikkulainen, R.: Real-Time Evolution of Neural Networks in the NERO Video Game. In: AAAI-2006, pp. 1671–1674 (2006)
33. Parker, G., Probst, M.: Using Evolutionary Strategies for the Real-Time Learning of Controllers for Autonomous Agents in Xpilot-AI. In: Proceedings of the 2010 IEEE Congress on Evolutionary Computation, CEC (2010)

Big Data Analytic: Cases for Communications Systems Modeling and Renewable Energy Forecast

Yashar Sahraei Manjili[1] and Maryam Niknamfar[2]

[1] ACE Laboratory, Department of Electrical and Computer Engineering
University of Texas at San Antonio, San Antonio, Texas, USA
yashar.sahraeimanjili@utsa.edu
[2] Photonics Research Lab., Department of Electrical and Computer Engineering
University of Texas at San Antonio, San Antonio, Texas, USA
mniknamfar@gmail.com

Abstract. Data analytic-based approaches are proposed for different cases of communications systems modeling based on qualitative methods and solar energy forecasting based on artificial intelligence in order to represent the spread of areas to which such techniques are applicable.

The first case uses qualitative methods for modeling of communications systems. Radio over fiber (RoF) communications system is considered for cable television (CATV) channels over wavelength division multiplexing (WDM) network using an optical direct modulator (DM). The link performance is studied for different combinations of AC electrical power and Module to Bias ratio. The best setting options are proposed for 30 CATV channels with 64-QAM constellation transmitted over the 4-WDM link.

Moreover, the communications system is modeled by Fuzzy logic using dataset obtained from simulation. Combination of Fuzzy logic and genetic algorithms (GA) is implemented to provide the user with reasonable estimates of electrical power required to achieve desired performance. Fuzzy-GA modeling capabilities are illustrated and discussed.

The second example considers application of data mining and artificial intelligence in forecasting of expected solar energy. The models are developed and validated utilizing a raw large data base from the National Renewable Energy Laboratory (NREL) archive sampled at 1-minute intervals during 8 years (2004 – 2012) at the NREL site in Golden, Colorado.

The methodology uses an integrated serial time-domain analysis coupled with multivariate analysis for pre-processing of the data. The resulting enhanced data set is then used for training of the forecast core engine core implemented using an artificial neural networks (ANN). Characteristics of the proposed data analytic-based approach are illustrated and compared to the physics-based methods. Performance of the proposed framework is compared to those of the current state-of-the-art in 24-hour ahead solar energy forecasting in order to provide a clear understanding of its capabilities.

Keywords: Direct modulator (DM), Fuzzy-GA estimation, Fuzzy reverse modeling, radio over fiber (ROF), adaptive systems, data analytic, solar forecast, artificial neural networks, weather synoptic systems detection.

A. El-Osery and J. Prevost (eds.), *Control and Systems Engineering*,
Studies in Systems, Decision and Control 27, DOI: 10.1007/978-3-319-14636-2_6

1 Cost-Efficient Approach to ROF Communications Systems Design for CATV Channels over WDM Network and Fuzzy-GA Estimation[1]

1.1 Introduction

In order to accommodate the increasing demand of transmitting multiple data channels simultaneously through a fiber link, efficient and economic network design is valuable. Different optical modulation methods can be used for transferring CATV signals from electrical domain to the optical [1]. Only some modulation methods are cost-effective and reliable. We have considered various electrical M-QAM schemes for transmitting 30 CATV channels over a 4-WDM RoF network. The cost effective optical modulation method for a WDM network with four or less routes is utilization of DM for each route [2]. However, bandwidth of RF signals, laser clipping, chirping and nonlinearity effects can limit the quality of the link performance [3]. Using DM laser is cheaper than using the external modulator. But in return, a DM laser with a higher bandwidth will be needed to support the RF signal spectra. The higher the bandwidth of the DM laser, the higher the cost. On the other hand, increasing the number of bits per symbol will decrease the required laser bandwidth and will consequently reduce the cost; higher M-QAM brings more symbol error rate (SER) in the symbol detection stage. Hence, there should be a tradeoff between using laser with higher bandwidth and utilizing higher M-QAM scheme in order to achieve the optimum link SER while reducing cost of the network. Furthermore, different system adjustments in order to achieve a linear performance of the direct modulator laser are important for data transmission. For instance, an appropriate laser biasing prevents undesirable effects of the laser clipping [4].

In this section, a radio over fiber (RoF) communication system is considered for cable television (CATV) channels over wavelength division multiplexing (WDM) network using an optical direct modulator (DM) [Niknamfar 2013]. Since the DM bandwidth is limited, increasing bandwidth efficiency of the system using M-ary quadrature amplitude modulation (MQAM) schemes will improve the link SER performance. Radio frequency (RF) signals with constellations of 16-64-256 QAM are simulated for CATV channels and the effect of using higher bit per symbol on the link performance was studied in order to reduce the cost of the network through utilizing financially feasible optical modulator [Niknamfar 2013]. The 64-QAM scheme which is the most appropriate is used for the system of CATV channels. The link performance is studied for different combinations of effective parameters such as AC electrical power and Module to Bias ratio, which are crucial to be set properly,

[1] Parts of this section are based on the article "Cost effective ROF communication system for CATV channels over WDM network and Fuzzy modeling of the system" by Niknamfar et. al., In proceeding of: International Conference on Computing, Networking and Communications (ICNC), 2013 DOI: 10.1109/ICCNC.2013.6504053

and SER is obtained at the front end of the receiver. The best setting options are proposed for 30 CATV channels with 64-QAM constellation with direct modulation transmitted over the 4-WDM link [Niknamfar 2013]. The communication system is also modeled using Fuzzy logic considering the data-set obtained from simulation of the original system in VPI® tool. Furthermore, reverse modeling of the system is also implemented using a combination of Fuzzy logic and Genetic Algorithms (GA) in order to provide reasonable estimates of electrical power required to achieve a desired SER. Fuzzy modeling capabilities are illustrated using error measures and discussed.

1.2 Optimized System Design to Overcome DM Laser Limitations

Increasing the number of bits per symbol is an approach which is popularly used in communications in order to increase the efficiency of a system's bandwidth. Equation (1) gives the mathematical representation of a system's the bandwidth efficiency [6]:

$$\frac{R}{W} = \frac{K}{WT_S} = \frac{1}{WT_b} \quad bits/s/Hz \quad ; \quad K = \log_2 M \tag{1}$$

Where R is the bit rate, W is the bandwidth, Ts, Tb, K and M represent symbol duration, bit duration, number of bits per symbol and symbol alphabet size, respectively. The drawback associated with growing K is the increase in bit error rate (BER) which in case of M-QAM signals up to some level is not as noticeable as the other M-ary schemes and is represented as follows [6]:

$$P_B \approx \frac{2(1-L^{-1})}{\log_2 L} Q\left[\sqrt{\left(\frac{3\log_2 L}{L^2-1}\right)\frac{2E_b}{N_0}}\right] \quad ; \quad L = \sqrt{M} \tag{2}$$

Where P_B is the QAM probability of bit error and Q(x) is complementary error function. Consequently, there is a limit for increasing the bandwidth efficiency due to the rise in detection error probability.

On the other hand, increasing the bias level of the laser will result in lower laser chirping that will impose higher power penalty due to lower extinction ratio of the laser [7]. Therefore, there should also be a tradeoff between decreasing laser chirping and avoiding large power penalty.

1.3 System Analysis

Figure 1 show the system structure which is simulated in VPI® software package. This system includes four different wavelength-routes with channel spacing of 100 GHz. Each route contains six QAM blocks each of which has five M-QAM channels with bit rate of 30 (Mb/s) for every channel, resulting in total bit rate of 900 (Mb/s) optically modulated on each wavelength-route. For the purposes of this study, however, only the results for one of the four routes, i.e. route No 2 as represented in figure 1, is presented. At the Power Control block, the electrical power of the 30

CATV channels will be set in order to sweep different power values in dBm to give the simulation results of the electrical power effect on the link performance by observing the SER at the front end of the QAM demodulator block. RF signals are then superimposed on a fixed DC bias current to intensity modulate the optical carrier [8]. The next block is optical direct modulator which is simulated based on numerical solutions of the single-mode rate equations. This block can show the nearly single-mode Fabry-Perot (FP) laser and can be modeled in more details by the laser equations in normalized form as follows [9]:

$$\frac{dE}{dt} = \frac{1}{2\tau_{ph}}\left[\tilde{G}(N,|E|^2)|E|^2 - 1\right]E + F_E(t) \tag{3}$$

$$\frac{dN}{dt} = \frac{1}{\tau_e}\left[\frac{J_{bias}+J_{mod}(t)}{J_{th}} - N - G(N,|E|^2)|E|^2\right] + F_N(t) \tag{4}$$

Where, N and $\tilde{G}$ are the carrier density and the laser gain, respectively, both normalized to their values at the laser threshold. $J_{bias}, J_{mod}, J_{th}$ and E, stand for the laser bias current, modulation current, threshold current and the electric field, respectively. τ_{ph} and τ_e are the photon lifetime and the carrier lifetime, respectively.

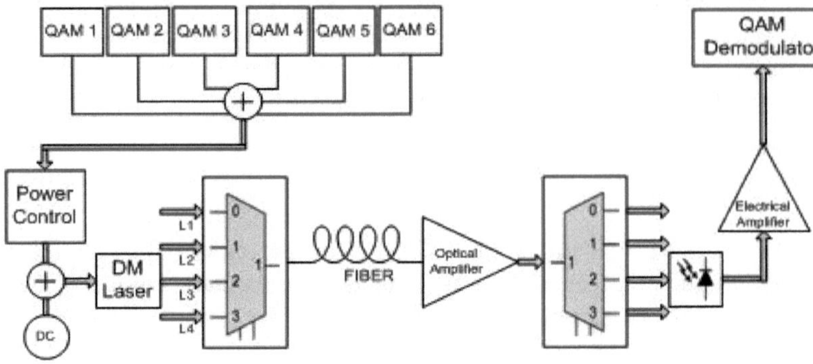

Fig. 1. Block Diagram of RoF System Multiplexed on 4-WDM

A 4-WDM is used for multiplexing 4 routes and RF signals are routed on a single mode fiber and then are being optically amplified. At the end point of the system, after the de-multiplexer and the electrical amplifier an M-QAM detector is utilized in order to study performance of one of the channel frequencies.

1.4 Simulation Results

For transmitting 30 QAM CATV channels with bit rate of 30 (Mbits/sec) per channel over 40 (km) single mode fiber a set of NTSC-M standard carrier frequencies (301.25 MHz-475.25 MHz) are selected. At the receiver side, one of the CATV channels with

397.25 (MHz) is studied. Since, for RF transmission through fiber the 1550 - 1555 nm wavelength range is the optimum range [10], laser center wavelength is set to 1553 nm. Moreover, according to the equation 4, increasing (J_{bias}/J_{th}) and ($J_{mod}(t)/J_{th}$) will affect the DM laser performance. In the VPI model of DM laser two coefficients can be set as(J_{bias}/J_{th}), (J_{mod}/J_{th}). If we define a new ratio of these two coefficients ($m = J_{mod}/J_{bias}$) called Module-Bias ratio, by fixing (J_{mod}/J_{th}) = 1, and changing the (J_{bias}/J_{th}) the optical output power will change due to different values of "m". The actual AC electrical power in the system is influenced by both the ratio (J_{mod}/J_{th}) and the AC power value being swept in the power control block. Therefore, for simplicity of simulation, we manually set the ratio (J_{mod}/J_{th}) to 1 so that there remains only one variable - i.e. the AC power value determined by power control block - to study as the AC electrical power. The smaller the "m" the larger the laser biasing level. Appropriate adjustments of this ratio "m" and RF power have a vital role in obtaining an acceptable SER for the transmission of M-QAM CATV channels over the fiber. Figure 2 shows the SER performance of the system when applying different sizes of M-ary QAM modulation for different values of "m" with the AC electrical power fixed at -5 dBm. From Figure 2 it can be inferred that 64-QAM modulation will result in the least SER for all values of "m". This shows that for k=6, tradeoff between more bandwidth efficiency and less symbol detection error is well set for 64-QAM. The 16-QAM ends up not to a large enough bandwidth efficiency and therefore will have worse SER compared to 64-QAM due to the bandwidth limitation of the DM laser. Although, 256-QAM has the largest bandwidth efficiency, it results in an unacceptable SER due to larger symbol alphabet size and closer received symbol detection threshold compared to that of 64-QAM.

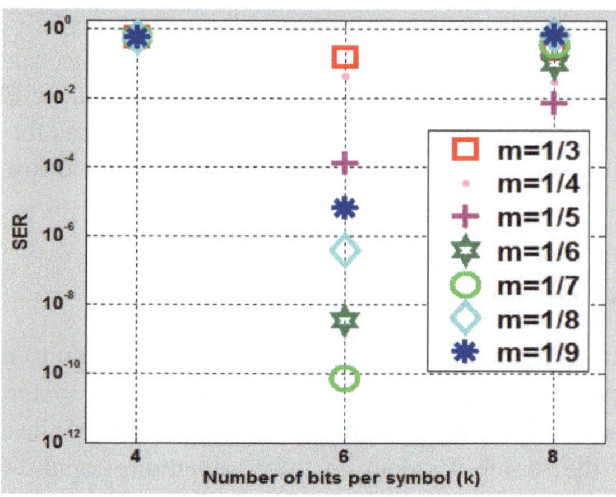

Fig. 2. SER for 2^K-QAM channels vs. K; ac electrical power at -5dBm

In Figure 2 the values of SER in case of K=6 is enhanced through decreasing “m” up to m=1/7 and then for m<1/7 it is degraded. This shows that increasing (J_{bias}/J_{th}) up to some level will add the modulation bandwidth of the DM laser and will reduce chirping effect. However, for (J_{bias}/J_{th}) >7, the DM performance will be degraded due to either nonlinearity or extinction ratio power penalty.

Electrical power of the RF signals is being controlled by power control module as shown in fig. 1 and is swept from -20 dBm to 3 dBm. The 3 dimensional diagram of 64-QAM CATV channels SER performance for different values of AC electrical power and different values of “m” is represented in Figure 3.

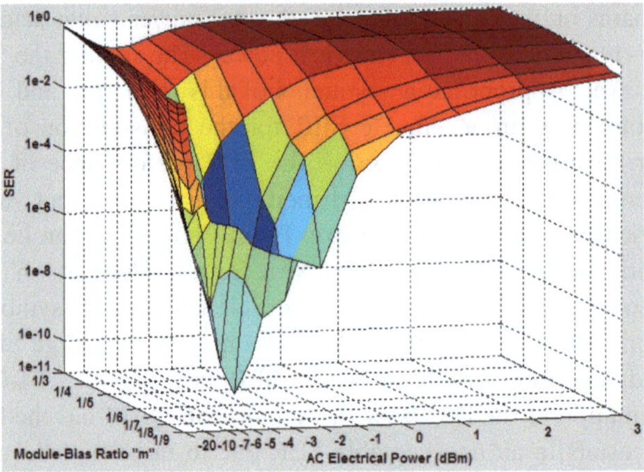

Fig. 3. SER vs. “m” and ac electrical power for received 64-QAM channel with carrier frequency= 397.25 (MHz)

Figure 3 shows that increasing signal electrical power up to some level will result in better Signal to Noise Ratio (SNR) and therefore lowers the SER. However, after some threshold nonlinearity due to higher power level degrades the SER performance of the link. There should be a tradeoff between higher SNR and lower nonlinearity for the system.

1.5 Fuzzy-GA Estimation

Fuzzy modeling uses the input-output data of the actual system to imitate its behavior [11]. Fuzzy reverse modeling, however, deals with a set of system variables, which may include some or all of the system outputs and some of the input variables, in order to estimate the required values of other remaining inputs. In other words, in Fuzzy-GA modeling the processes of fuzzification, defuzzification, inference, feeding the inputs to the Fuzzy system and getting the outputs from it are similar to those in Fuzzy modeling. The major difference is that in the reverse modeling scenario the Fuzzy reverse model is fed by Module-Bias ratio, i.e. "m" and the desired SER, while the estimated electrical power, P in dBm, required for achieving the desired SER will be the output of the Fuzzy reverse model. In reverse scenario, one of the inputs, here Module-Bias ratio, along with the output of the actual system, here SER, will be fed to the inference engine in order for the required value of the other input, in this case the electrical power P, to be estimated. This can be helpful specifically when the system model is significantly large and sophisticated and a relatively huge amount of time and effort is required to adjust the variables and feed the inputs to the system and run time-consuming simulations repeatedly in order to find the appropriate settings to achieve the desired SER. In such cases, one smart approach is to run a limited number of simulations, record the data and form the Fuzzy reverse model. The Fuzzy input data should be suitably chosen in order to cover the whole area of interest so that the reverse system can be mimicked accurately. After the Fuzzy reverse model is formed, the next step is to simply feed the Module-Bias ratio being used for the system and the desired SER to the reverse model. The output will be an estimate of the lowest amount of electrical power required to achieve the desired performance for the adjusted value of Module-Bias ratio. Also, based on the data set obtained by the communication system, an accurate Fuzzy reverse model can represent if the desired SER is not achievable for some setting of Module-Bias ratio due to non-linearity of the link. Rules of the inference engine in reverse Fuzzy modeling are determined based on human reasoning so that the Fuzzy inference system (FIS) will be able to provide the user with accurate estimations of the required electrical power.

As represented in Equation 7, ratio of absolute difference between output of the Fuzzy model, i.e. $\hat{X}$, and the required power settings of the actual communication system, i.e. X, to the actual system setting determines the error, e ,which has to be reduced in order for the Fuzzy reverse model to estimate the required system power setting precisely:

$$e = \frac{1}{p \times r} \sum_{i=1}^{p} \sum_{j=1}^{r} w_{ij} \frac{|X_{ij} - \hat{X}_{ij}|}{X_{ij}} \tag{5}$$

where p and r represent the number of different electrical power settings and number of different module-bias ratio settings used for reverse modeling respectively, and w_{ij} is the weight factor used to emphasize the importance or sensitivity of some settings compared to the others. Using w_{ij} it is possible to eliminate regions which are of lower interest for communication purposes such as points where higher electrical power are required or areas that the setting of module-bias ratio is typically not feasible for communication system settings, e.g. extremely low values or values greater than $\frac{1}{4}$.

Simulation results of the communication system for electrical power setting between -20 dBm to 3 dBm, with step size of 1 dBm, and module-bias ratio setting from $\frac{1}{9}$ to $\frac{1}{3}$ were obtained using VPI simulation tool and were used for the purpose of Fuzzy modeling in MATLAB; In other words, p and r are 24 and 7 respectively. Region of interest for this case study includes results obtained for electrical power between -12 dBm and -2 dBm, and module-bias ratio between $\frac{1}{8}$ and $\frac{1}{5}$ for which the average value of relative error is computed using equation 6 and its reciprocal is regarded as the fitness function for smart search in genetic algorithm.

Two sample inference rules for fuzzy reverse modeling are as follows:

-- **IF** the *Module-Bias ratio* is ***Very Low***, **AND** the *Desired SER* is ***Medium***, **THEN** the *Electrical Power* is estimated to be ***Very Low***.
-- **IF** the *Module-Bias ratio* is ***Low***, **AND** the *Desired SER* is ***Very Good***, **THEN** the *Electrical Power* is estimated to be ***Medium***.

where the medium value for SER means values in the range 10-5 while SER values in the range 10-9 are considered very good depending on the application requirements and structure of the system. In the same fashion, as mentioned before, the electrical power values range from around -11 dBm as very low to -1 dBm as very high. For m, very low refers to values around $\frac{1}{9}$ to very high as a representative of values close to one-third.

A total of 35 rules are defined for the Fuzzy reverse model with 5 membership functions for Module-Bias ratio, 7 membership functions for SER. Output variable of the Fuzzy reverse model, i.e. the estimate of the required electrical power, also had 7 membership functions in this model.

Genetic algorithm was deployed on top of the Fuzzy reverse model with the same adjustments mentioned for Fuzzy modeling while in this case each chromosome of the input variable SER and also each chromosome of the output variable electrical power had 21 cells since these two variables had 7 membership functions which result in 21 edges for each.

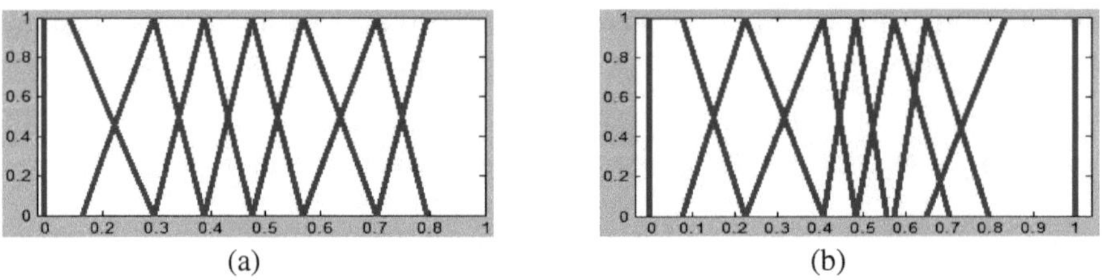

Fig. 4. Suboptimal fuzzy sets obtained by genetic algorithms for Fuzzy reverse model; (a) Input variable; SER (b) Output variable; power

Figure 4 represents the normalized final fuzzy sets obtained by the genetics algorithm for the input variable SER and the output variable, i.e. electrical power.

Two-dimensional view for the simulation results including the output of the Fuzzy reverse model and the error associated with estimation of the required electrical power are represented in figure 5a and figure 5b respectively.

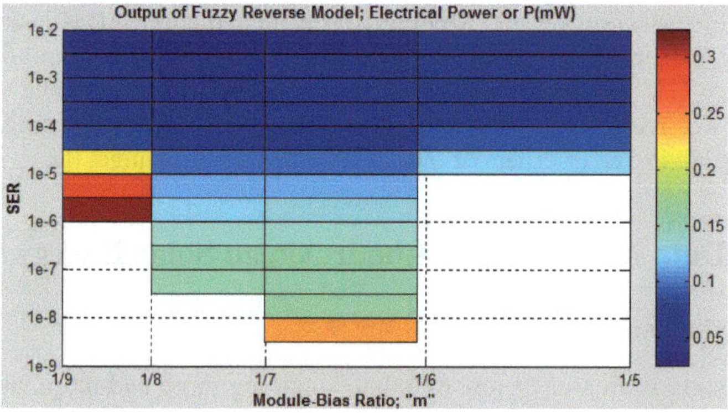

(a)

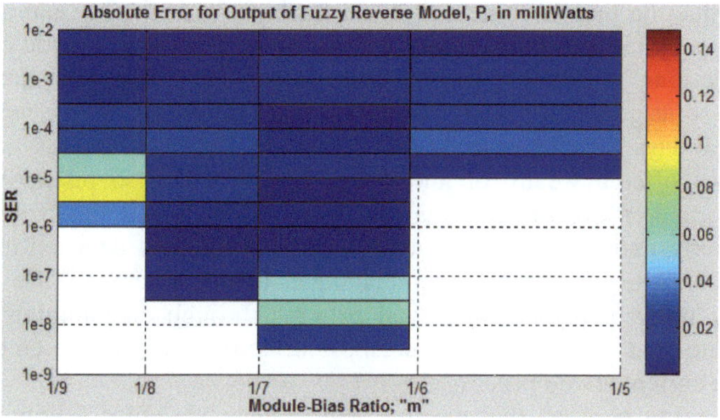

(b)

Fig. 5. Fuzzy reverse model results; (a) Output, estimated electrical power (mW)(b) Absolute error for electrical power estimation (mW)

Figure 5b represents the absolute value of the actual error for electrical power estimation using Fuzzy reverse model in milliWatts. Within the region of interest, the estimation error is far less than 0.02 mW in almost a 2 mW range for AC electrical power, which means error percentage or the estimation tolerance is less than 1%. This implies that the Fuzzy reverse model has a satisfactory performance within the region of interest. This must also be noted that in figure 5a, the points where no power estimation is made show actually the adjustments for which the desired SER could not be achieved at all regardless of how much electrical power was available.

In overall, it can be said that for transmitting CATV channels over a 4-WDM optical fiber communication system utilizing the DM laser is the cost effective design. DM laser bandwidth limitation can be overcome by increasing bandwidth efficiency. 16-QAM scheme will not increase bandwidth efficiency enough while 256-QAM exhibits an unacceptable SER due to its large symbol alphabet size and closer detection thresholds. 64-QAM scheme will result in an acceptable SER for transmitting 30 CATV channels with bit rate of 30 (Mb/s) per channel when an optimized adjustment of RF electrical power and laser biasing is considered. A proper selection of electrical power and Module-Bias ratio can control DM laser nonlinearity and chirping effect. Fuzzy modeling help obtain results for a continuous range of inputs. Fuzzy-GA combination proved useful in order to estimate the value of electrical power required to achieve a desired SER for different settings of Module-Bias ratio by having error orders of 0.02 mW or 1% for the range.

2 Big Data Analytic for 24-hour Ahead Solar Energy Forecast[2]

2.1 Introduction

Solar forecasting is an area of research that is gaining more and more visibility due to the increasing penetration of solar energy into conventional electricity grids. Once the amount of solar penetration reaches a certain threshold, the variability of the solar energy production may become a problem for both grid stability and reliable bidding of electricity by electric utilities. Solar forecasting solutions provide electric utilities with predictions of power output from large-scale solar installations or from distributed solar generation with a temporal scale ranging from the next few minutes up to several days ahead forecast [12].

Databases offer a wealth of knowledge but are often so restrictively large that information and/or knowledge cannot be extracted, analyzed and synthesized quickly and efficiently enough for daily, hourly, and intra-hourly decision-making. This is where data analytic and data mining techniques can be leveraged. Data analytic seek to discover relevant data features and patterns through data mining processes and communicate that information in the most effective way to the end-user; either a human or a computerized process [13-15].

The changing patterns that weather and climate variables follow will affect the cloud formation in the sky which requires specific conditions for pressure, temperature and relative humidity with respect to the dew point and air particles present in the sky. The cloud formation consequently influences the amount of sun rays that reach the ground negatively affecting the electrical energy generated by different approaches such as photovoltaic (PV) solar cells, concentrated solar power (CSP), etc.

There are a number of approaches that focus on direct estimation of patterns of cloud formation, their movement and spread in the sky such as methods based on sky-imaging which are high fidelity approaches for intra-hourly solar energy forecast. Other methods such as numerical weather prediction (NWP) and satellite-based approaches are used for recognition and prediction of cloud formation patterns in large temporal and spatial resolution.

NWP discretize space and time at the mesoscale/regional and national levels and solve the equations of motion, thermodynamics and mass transfer at pre-defined time and space intervals [16]. The problem with NWP is that it cannot be considered site-specific to any plant but based on regional commingled datasets and interpolation techniques. Such commingled datasets propagate error from node to node and from time step to time step and its uncertainties are still too large, in the order of mean

---

[2] Parts of this section are based on patent "Method and System for Hours-ahead and Day-ahead Forecasting of Solar Energy", Document No: 292007-8100/2014.028.UTSA.

absolute error (MAE) of 20%-40% and root mean square error (RMSE) of 30%-50% for day-ahead predictions [16-20]. More recently, there has been an increased interest in smart modeling of weather and climate events through training of systems based on artificial intelligence techniques [12,14,15,21-23]. The results of such modeling approaches can be extensively used in solar forecasting.

The proposed data analytic-based solution is not dependent on theories defined for physical processes that constrain natural processes to a set of equations. If we compare NWP to the solution of a textbook, our solution using data analytic-based artificial neural networks (ANN) can be compared to the experience of a veteran in weather who is able to predict climate conditions accurately because he/she knows the local micro-meteorological features and has a robust memory to recall historical anatomies of weather. Besides, this method can be modified to adapt itself to abrupt and long-term changes in climate and environmental conditions and to modify its forecasts accordingly [24].

The correlation analysis, weather synoptic events detection, feature extraction, dynamic performance evaluation, and the artificial neural networks all together enable the proposed framework to analyze and recognize patterns in meteorological and solar data inputs so that the ANN understand how different variables are related to each other and what would be the climate-related outcome of certain changing patterns in some variables with respect to others.

The data analytic-based approach introduced in this section is demonstrated for the prediction of global horizontal irradiance of the next day using the irradiance information and weather sensory data of current day for up to 24 hours ahead of time. Predicting irradiance on a point in space is more difficult than energy over larger areas since there are no distributed local features to average out [20,25]. Therefore, by developing the site-specific irradiance solar forecast tool the methodology addresses the forecast of energy over larger areas more accurately when aggregating the predictions from the micro-climate of each location.

2.2 Problem Statement and Solar Forecasting State-of-the-art

A report performed for the Regents of the University of California titled "Current State of the Art in Solar Forecasting" [26] best summarizes current status of this field of research. In the report Glassley et al., state that satellite and NWP are the preferred methods for longer duration solar forecasting (one hour to a few days ahead) [20,26]. As for intra-hour forecasts, sky imagery based forecasting methods are most common. However, several basic assumptions regarding cloud shape and linear cloud movement vectors reduce the potential accuracy of this type of forecasting [27,28].

Making intra-hourly or site-specific forecasts using satellite data is also not very common due to the infrequent sampling interval (30 minutes) and the low image resolution [20]. This problem is only increased in the NWP method due to a larger sampling interval and lower cloud imaging precision. The limitations of NWP models do not allow for shorter time-scale and smaller spatial sampling to be accounted for [16,20,27]. Some of the error associated with the aforementioned problems can be corrected through a method known as Modeled Output Statistics (MOS), which determines statistical correlations in observed weather data and related imagery (satellite, sky imagery, etc.) [16,17,20,29].

Practical methods for day-ahead up to week-ahead solar forecasting are mostly based on NWP and statistical approaches as mentioned earlier [17,20,26,30]. A number of global models are available for this purpose, among them the Global Forecasting Service (GFS) and the European Center for Medium Range Weather Forecasting (ECMWF) [31] both of which are considered the state-of-the-art of global forecast models. However, in order to increase spatial and temporal resolution of these models, other models have been developed which are generally called national or mesoscale models among which, the North American Model (NAM), the High Resolution Limited Area Model (HIRLAM) and the Weather Research and Forecasting (WRF) are widely used by different communities. A broad range of in-depth expertise is needed in order to obtain accurate results when running these models, due to the wide variety of parameters that can be configured. Also, sophisticated techniques such as data assimilation or statistical post-processing have been needed in order to calibrate the estimations or obtain a probabilistic point of view of the accuracy of the output. Usually, techniques that mix outputs of different models are used for this reason, which finally result in better estimates of those variables along with a degree of uncertainty associated with the forecast results [28].

Recently, artificial intelligence has gained much more interest in forecasting applications and a number of approaches are developed specifically for solar forecasting based on neural networks, support vector machines, etc. [19,22,23,32] which still need to improve the accuracy of their forecast engine, better stabilize and further fine-tune their forecast results in order to be comparable with already well-established forecasting methods that have been in the market for the past decades such as NWP-based approaches mentioned above.

NWP and satellite imagery-based forecast techniques are the two most versatile approaches for longer forecast horizons, i.e. have largest temporal capacity. The mentioned approaches also have the largest coverage, i.e. biggest spatial capacity, in solar energy forecast.

A database of instrumentation characteristics is necessary in order to implement MOS [29]. Unfortunately, the instrument characteristics database by itself does not translate to improved accuracies because the MOS solar forecasting approach would need accurate correlations between measured weather variables and the expected irradiance outputs, and these would be delegated to a statistical model, which often includes outlier data that do not improve forecasting results [17,26,30]. Furthermore, a statistical model does not learn from the decision-making and relationship policies developed by its operators over time.

2.3 Framework Development; Proposed Approach to Solar Energy Forecasting

The forecasting framework is implemented based on correlation analysis and feature-extraction of weather data and solar irradiance in combination with dynamic performance evaluation and a synoptic events detection algorithm which are used to determine the appropriate training set for the ANN in real time. This creates a knowledge-based library that is ever-growing in size and increasing in intelligence. The content of the mentioned knowledge-based library are information regarding the best sets of data chosen for training of the ANN and the forecast performance for each and every target day in the past. This information enables the forecasting framework to modify its decision rules accordingly to choose the best training subsets for upcoming target days so that the level of accuracy is maintained for the future forecasts.

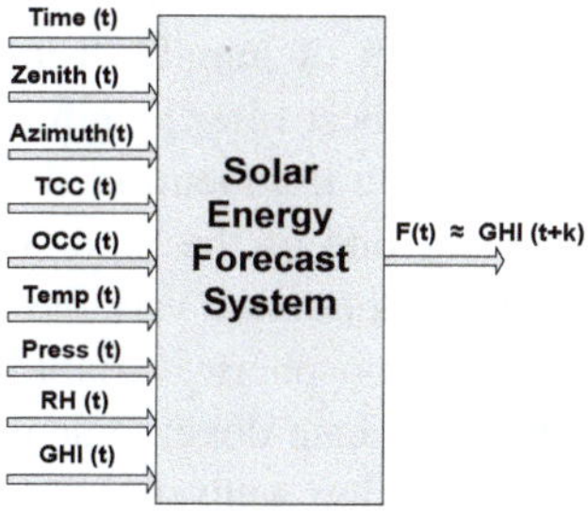

Fig. 6. Basic Solar Forecasting Input/Output Structure

Figure 6 represents the basic Input/Output (I/O) setup of the proposed approach used to develop the forecast engine. The proposed forecast framework, as represented in Figure 6, is capable of giving estimations of solar irradiance of the next "k" sample times based on current time data and the extracted feature-matrix of the past data.

Figure 7 represents the block diagram of the proposed solar forecast framework. This block diagram can also be regarded as a flow chart of the forecast process. As it can be seen in the block diagram of Figure 7 the data coming from the sensors includes the weather, timing data and the solar plant output, i.e. is the amount of DC power generated by solar array. The sensory data is continuously recorded added to the existing database.

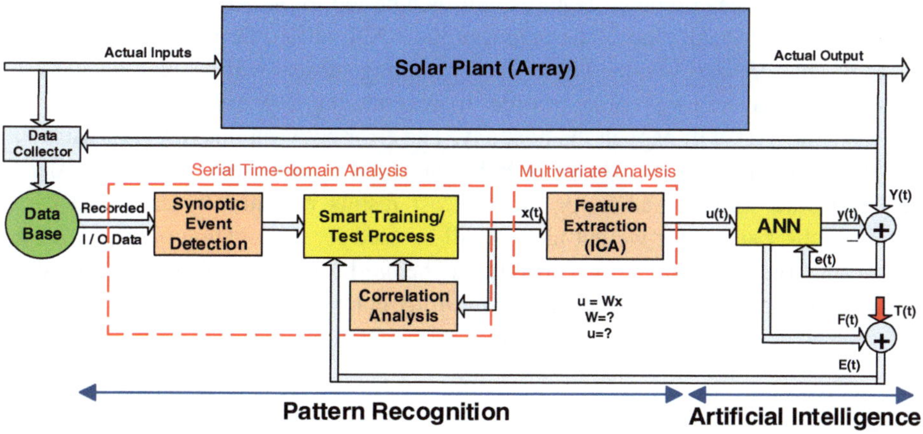

Fig. 7. Block Diagram of the Adaptive Data Analytic-based Solar Energy Forecasting Framework

A number of training subsets with different lengths and coverage of training windows are formed, considering the results of synoptic events detection algorithm, and a consecutive training and test process will take place over the target forecast horizon using each and every one of the training subsets. The process goes on by passing each training subset through a correlation analysis where the correlation function between temperature, pressure, relative humidity and GHI input with the plant output, i.e. the generated electrical power or the on-site measured GHI, is computed after using low-pass filters of lengths from 1 minute to 3 hours on the mentioned input variables. After the best filter length and shift values, i.e. the values that maximize the correlation between the mentioned variables and the plant output, are determined for each and every one of the input variables mentioned above the rates of change (ROC) for the aforementioned variables will also be computed. These ROC's represent gradients in the atmospheric variables and constitute precursors of changes in other variables which indicate development or dissipation of clouds. The best fit variables obtained by correlation analysis are fed together with the ROC's to the feature extraction block which transforms the data into a new domain using ICA in order to find the underlying independent features of those variables. These independent components or features will then be used for training of the ANN. After the ANN training is converged the test phase, i.e. the forecasting process, begins. As the first step in this phase, the real-time temperature, pressure, relative humidity and GHI obtained from the plant are shifted and filtered using the best set of filter length and shift value, obtained from correlation analysis, for each training subset. Then, the ROC's of those weather variables are computed. This will follow by computing the features of the enhanced data set using the separation matrix, i.e. W, obtained in the training phase. The data input to the feature computation unit is transformed into the independent components' domain using the separation matrix W(t) obtained for each training subset, by applying the ICA approach to the same training subset, in the training phase. The computed features will then be fed to the ANN for the prediction of solar energy to take place. The output of the ANN will go through the aggregation process. The forecasts obtained by different training subsets will be aggregated based on a weighted averaging regime in order to generate the forecast result, i.e. F(t). The target signal, T(t), will happen the next day. The error between this raw forecast result and the target signal, E(t), will be used by the smart training/test process in order to determine the weight factors for the forecast results obtained by each and every training subset and those weight factors will be used for aggregation purposes for the following target day. The forecast result may be used for electric energy bid decisions by the energy market operations (EMO) of utility companies.

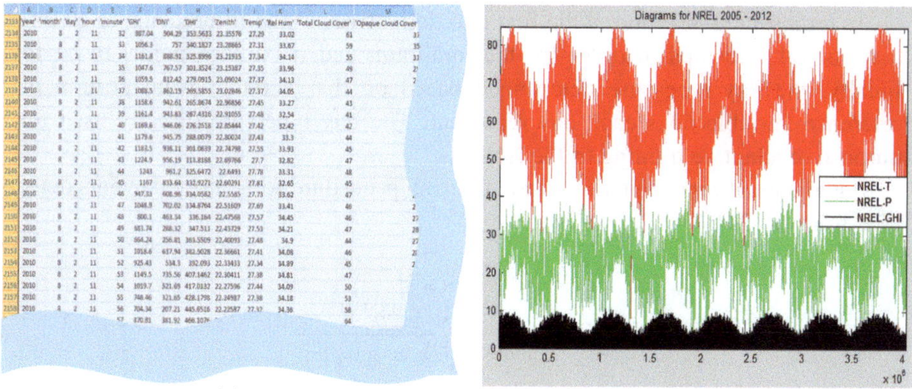

(a) (b)

Fig. 8. Removing the Undesired NREL Data for Solar Energy Forecast; (a) Snapshot of NREL library in the form of a spread sheet; 250+ variables sampled every 1 minute for 8 years (b) Sample scaled data for 8 years; Temperature – RED, Pressure – GREEN, and GHI – BLACK

Figure 8.a represents a shot of the data set, obtained from National Renewable Energy Laboratory (NREL), in the form of a spread sheet including more than 250 variables for the duration of 2004 to 2012 sampled at 1 minute intervals. Figure 8.b shows diagrams of the three variables atmospheric temperature, atmospheric pressure and the GHI over the period of eight years. In figure 8.b values of temperature, pressure and GHI are scaled to that the fit within the same range in the one diagram solely for representation purposes.

Weather synoptic events occur in a periodic fashion and since the atmosphere is a fluid, much like in turbulence of fluid dynamics, synoptic events are regarded as stochastic processes. These events may seem random when analyzed by a historical statistical approach. However, when studied in a time-dependent manner, the time-varying coherence of those synoptic systems and the effects of individual variables and their features on atmospheric cloud formation and eventually their influence on the solar irradiance can be established.

The procedure for detection of weather synoptic events is based on locating the low pressure weather systems which happen in between two consecutive points of high pressure that meet certain conditions. Two basic criteria were originally recommended to distinguish a fully developed synoptic system from a weak one [33]. The extended version of the algorithm introduced in [33] will be given in this study that involves MIN, MAX, Lo and Hi points.

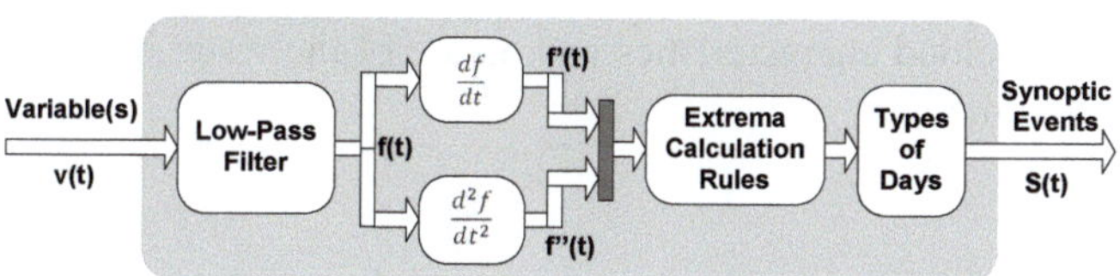

Fig. 9. Weather Synoptic Events Detection Process Block Diagram

As represented in Figure 9 the first two derivatives of the filtered signal go through a set of relationships in order for the minimum and maximum points to be found. Finally, the day types are determined based on the fact that what portion of the reference variable does each day represent. The day types are directly used for the purpose of adaptive training of the ANN.

Mathematical representation of the filtering procedure for synoptic events detection is as follows [21]:

$$f(t_i) = \frac{\sum_{j=\max(1,i-\left[\frac{L}{2}\right])}^{\min(N,i+\left[\frac{L}{2}\right])} v(t_j)}{\min\left(N,i+\left[\frac{L}{2}\right]\right)-\max\left(1,i-\left[\frac{L}{2}\right]\right)} \tag{6}$$

where $i = 1,2,\ldots,N$ is the sample number. N and L , respectively, represent the number of samples for the input variable, $v(t)$, and the low-pass filter length which works like an averaging window. The relationships used to determine the points of maximum and minimum values for the input variable are as follows [24]:

- IF $f'(t_i) = 0$ AND $f''(t_i) < 0$ *THEN* t_i is a MAX
- IF $f'(t_i) = 0$ AND $f''(t_i) > 0$ *THEN* t_i is a MIN

where f' and f'' represent the first and the second derivaties of the filtered function $f(t_i)$, respectively. After the MAX and MIN points are found the high, i.e. Hi, and low, i.e. Lo, points are determined based on the following criteria:

(a) If the difference in value of a MAX with its adjacent MIN is less than 1 unit the MAX point is neglected.
(b) If any two consecutive MAX points are less than 36 hours away the largest is defined as Hi, and the other local MAX is neglected.
(c) The smallest MIN point between each two consecutive Hi's is considered as a Lo.

Subsequently, two different signals are used for extracting relationships of climate anomalies, these in turn define the training window later. The first signal is the atmospheric pressure which is used for synoptic events detection. The second signal is calculated based on the ratio of the three variables of atmospheric pressure, temperature and relative humidity, as $\frac{P\times T}{RH}$. The reason to choose this second signal is that it is more likely for clouds to form at low pressure, low temperature, and high RH conditions which translate to the local minimum points of the $\frac{P\times T}{RH}$ reference signal and vice versa due to cloud characteristics and nature of the water droplets or icy crystals that form them [34,35].

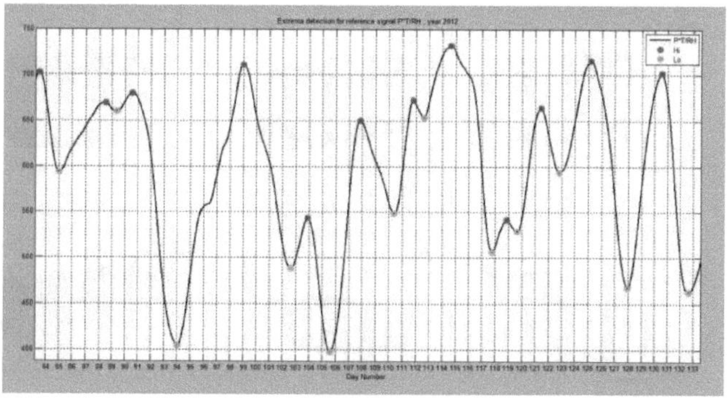

Fig. 10. Detection of Hi's and Lo's for Atmospheric Reference signal (P*T/RH)

Figure 10 represents the detection of extreme points obtained by the synoptic event detection algorithm using reference signal, $\frac{P \times T}{RH}$, for a portion of year 2012.

Types of days are determined based on the extreme point detection results obtained for the two reference signals of P and P*T/RH. The day-types are used to generate the adaptive training rules. The following conditions are considered in order to determine type of each day:

- IF day d has a t_i = MAX AND t_i meets the criteria (a), (b), (c) THEN d is a Hi
- IF day d has a t_i = MIN AND t_i meets the criteria (a), (b), (c) THEN d is a Lo
- IF day d is after a Lo AND d is NOT a Hi THEN d is a Rise
- IF day d is after a Hi AND d is NOT a Lo THEN d is a Fall

Correlation analysis and an algorithm for weather synoptic systems detection are the two first steps towards adding intelligence to the proposed solar forecasting solution. The correlation analysis is implemented in time domain and consists of filtering, shifting and forming the correlation function of temperature, pressure, relative humidity, and the GHI with the plant output.

Figure 11 represents the correlation functions for nine variables with the GHI. Weather variables influence the cloud formation significantly, which in turn affects the amount of solar irradiance that reaches the ground. For instance, atmospheric pressure works as the driver for the weather synoptic system so that a high pressure system is regarded as the sign of a stable climate and a low pressure system can be followed by an unstable, stormy weather or overcast days. Hence, by computing the correlation functions between each of those variables with GHI, over the training window, in a continuous manner for each consecutive target day, we can not only obtain the refined training dataset with respect to optimally correlated variables but also use the information obtained by the climatic variables in order to detect the weather synoptic events' beginnings and ends.

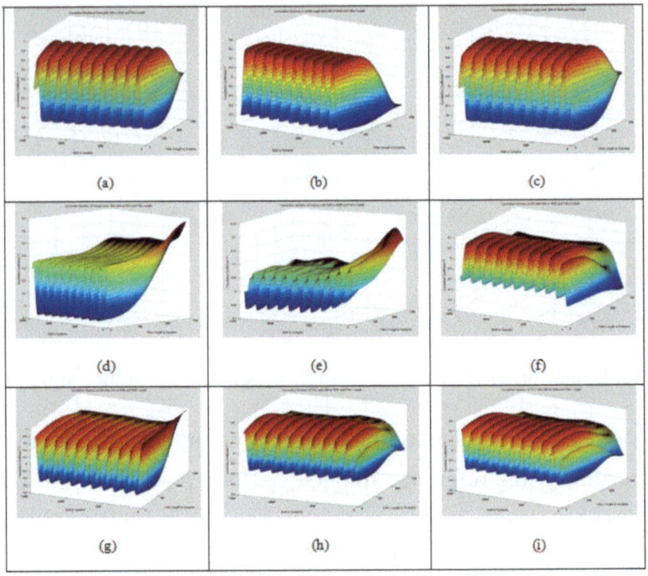

Fig. 11. Cross Correlation Functions of Nine Input Variables with Measured GHI Using Different Filters; (a) Time (b) Zenith (c) Azimuth (d) Temperature (e) Pressure (f) Relative Humidity (g) GHI (h) OCC (i) TCC

The process of correlation analysis takes place in real-time for each target day in order to find the low-pass filter length and the shift amount, both in number of samples, which result in the peak cross-correlation between inputs and the plant output over the training window. This process is mathematically as follows [21]:

$$r_L^{T_{sh}} = \mathrm{corr}\left(f(t - T_{fh} - T_{sh}), I(t)\right) = \frac{\sum_{i=N-N_d^\tau}^{N} \left(f(t_i - T_{fh} - T_{sh}) - \bar{f}\right)\left(I\left(t_{i-N+N_d^\tau+1}\right) - \bar{I}\right)}{\sqrt{\sum_{i=N-N_d^\tau}^{N} \left(f(t_i - T_{fh} - T_{sh}) - \bar{f}\right)^2 \left(I\left(t_{i-N+N_d^\tau+1}\right) - \bar{I}\right)^2}} \tag{7}$$

$$\bar{f} = \frac{\sum_{i=N-N_d^\tau}^{N} f(t_i - T_{fh} - T_{sh})}{N_d^\tau} \quad , \quad \bar{I} = \frac{\sum_{i=N-N_d^\tau}^{N} I\left(t_{i-N+N_d^\tau+1}\right)}{N_d^\tau} \tag{8}$$

$$f(t_i) = \frac{\sum_{j=\max\left(1, i-\left[\frac{L}{2}\right]\right)}^{\min\left(N, i+\left[\frac{L}{2}\right]\right)} v(t_j)}{\min\left(N, i+\left[\frac{L}{2}\right]\right) - \max\left(1, i-\left[\frac{L}{2}\right]\right)} \tag{9}$$

where T_{sh} and L, respectively, represent the shift value and the length of low-pass filter which is applied to the signal for correlation analysis and T_{fh} stands for the forecast horizon which is the default shift value between the variable $v(t)$ and the irradiance $I(t)$. Parameter N represents the total number of samples that exist in the vector $v(t)$ while N_d^τ is a positive integer showing the number of samples in the training subset $\boldsymbol{\tau}$ for target day d, i.e. S_d^τ, which is equal to the number of samples in

I(t) itself. Term T_{sh} ranges from 0 to N_d^τ while the filter length, L, changes between 1 and 1440 samples, i.e. in minutes.

The FastICA algorithm, based on independent component analysis, is used to extract the underlying features of input data in the proposed framework which will be later used to train the neural network for further pattern recognition [36-38].

The predefined candid lengths for the adaptive training windows are stored in a vector called training specifications vector, $\varphi = [5,8,10,12,15,20]$. For each candid training window the days that will constitute the training subset will be determined based on training relationships. In order to form the training subset, the following relationships consider the day-types obtained by the synoptic event detection algorithm on the pressure signal:

- IF current day is "Lo" of "*P*" THEN include days "Lo", "Rise" and "Hi"
- IF current day is "Rise" of "*P*" THEN include days "Lo", "Rise" and "Hi"
- IF current day is "Hi" of "*P*" THEN include days "Hi", "Fall" and "Lo"
- IF current day is "Fall" of "*P*" THEN include days "Hi", "Fall" and "Lo"

Dynamic evaluation of performance helps with the adaptive training and test. Let's denote the training subsets by $S_{d-1}^\tau, \tau = 1,2,\ldots,\boldsymbol{\tau}$ where τ and $\boldsymbol{\tau}$ respectively represent the training set number and the total number of sets considered for adaptive training and d stands for current target day.

The performance function used in the process of dynamic performance evaluation is defined as the ratio of the correlation coefficient between forecast and target for each target day d to the forecast RMSE of the same day. The function is represented in the following:

$$\gamma_d^\tau = \frac{r_d^\tau}{\mathrm{RMSE}_d^\tau} \tag{10}$$

where τ refers to the training subset used for training of the forecast engine for the specific target day d. Performances of all training subsets for each target day will be stored in a vector called performance vector. The performance vector for each target day will be added to the end the performance matrix as a new row.

The aggregation coefficients are determined as follows and are used to compute the forecast result for the target day:

$$c_d^\tau = \begin{cases} \frac{1}{\boldsymbol{\tau}} & d = 1 \\ \frac{v_d^\tau}{\sum_{\tau=1}^{\boldsymbol{\tau}} v_d^\tau}, \quad v_d^\tau = \frac{\alpha \cdot \gamma_d^\tau + \sum_{d=1}^{d-1} \gamma_d^t/(d-1)}{(\alpha+1)} & d > 1 \end{cases} \tag{11}$$

where c_d^τ is the aggregation coefficient, i.e. the weight factor, for the forecast result of target day d obtained using the training subset number $\boldsymbol{\tau}$, i.e. S_d^τ, to train the ANN. This forecasting result is denoted by y_d^τ. Total number of training subsets is $\boldsymbol{\tau}$. The symbol $\boldsymbol{\alpha}$ stands for a coefficient which is used to give more dominance to the performance value of the last target day, i.e. target day $d-1$, with respect to those of the previous target days. In our forecasts $\boldsymbol{\alpha}$ is set to 10.

Assuming that y_d^{τ} is the forecast obtained for target day d by training set τ the aggregated forecast can then be represented as follows:

$$Y_d = \frac{\sum_{\tau=1}^{T} y_d^{\tau} \cdot c_d^{\tau}}{\sum_{\tau=1}^{T} c_d^{\tau}} \tag{12}$$

where Y_d is the aggregated forecast for target day d and is used for performance evaluation purposes of this study.

2.4 Performance Evaluation Metrics

Performance of the proposed method is evaluated for the hourly average values of day-time target and forecast GHI of the year 2012. The error indicators are:

Mean Error (ME) or Bias Error (BE)

$$BE = \frac{1}{N}\sum_{i=1}^{N}(T_i - F_i) \tag{13}$$

Mean Absolute Error (MAE)

$$MAE = \frac{1}{N}\sum_{i=1}^{N}(|T_i - F_i|) \tag{14}$$

Root Mean Square Error (RMSE)

$$RMSE = \sqrt{\frac{1}{N}\sum_{i=1}^{N}(T_i - F_i)^2} \tag{15}$$

where N represents the total number of points for which the error values are calculated. The number of day-time samples in year 2012 for hourly average solar energy is approximately $N = 4500$. Relative error measures are also introduced which consider the ratio of error to the average value of the target:

- Relative Mean Error (ME_{Rel}) or Bias Error (BE_{Rel})

$$BE_{Rel} = \frac{\frac{1}{n}\sum_{i=1}^{n}(T_i - F_i)}{\bar{T}} \tag{16}$$

- Relative Mean Absolute Error (MAE_{Rel})

$$MAE_{Rel} = \frac{\frac{1}{n}\sum_{i=1}^{n}(|T_i - F_i|)}{\bar{T}} \tag{17}$$

- Relative Root Mean Squared Error ($RMSE_{Rel}$)

$$RMSE_{Rel} = \frac{\sqrt{\frac{1}{n}\sum_{i=1}^{n}(T_i - F_i)^2}}{\bar{T}} \tag{18}$$

where $\overline{T}$ is the mean value of solar energy that reaches the ground during the day-time for the whole year 2012 and is computed as follows:

$$\overline{T} = \frac{1}{N}\sum_{i=1}^{N}(T_i) \ , \ T_i = \frac{1}{n}\sum_{j=(i-1)*n+1}^{i*n}(GHI)_j > 0 \tag{19}$$

Percent Error is also considered for performance evaluation:

$$e_i^{\%} = 100 \times \frac{(T_i - F_i)}{C_i} \ , \ T_i = \frac{1}{n}\sum_{j=(i-1)*n+1}^{i*n}(GHI)_j > 0 \tag{20}$$

where T_i and F_i are the ith sample of the target and forecast respectively and the C_i represents the maximum solar energy capacity, i.e. the maximum amount of solar energy that may reach the ground, at time instance i. Mean value and the standard deviation of percent error serve as additional performance measures:

$$\mu^{e\%} = \frac{1}{N}\sum_{i=1}^{N} e_i^{\%} \tag{21}$$

$$\sigma^{e\%} = \sqrt{\frac{1}{N-1}\sum_{i=1}^{N}\left(e_i^{\%} - \mu^{e\%}\right)^2} \tag{22}$$

The correlation coefficient between the forecast and the target is also computed which represents the strength of the linear association between them. In other words, it tells us how well is the forecast tracking the actual upcoming deviations in the solar irradiance curve. The correlation coefficient is computed as follows:

$$r = \frac{1}{N-1}\sum_{i=1}^{N}\left(\frac{T_i - \overline{T}}{\sigma_T}\right)\left(\frac{F_i - \overline{F}}{\sigma_F}\right) = \frac{\sum_{i=1}^{N}(T_i - \overline{T})(F_i - \overline{F})}{\sqrt{\sum_{i=1}^{N}(T_i - \overline{T})^2 (F_i - \overline{F})^2}} \ , \quad T_i = \frac{1}{n}\sum_{j=(i-1)*n+1}^{i*n}(GHI)_j > 0 \tag{23}$$

where σ_T and σ_F respectively stand for the standard deviation of the target samples and their associated forecast.

2.5 Forecast Results and Discussion

Forecast is done for the whole duration of the year 2012 and the forecast of the solar energy for the next day is obtained for the 366 days in that year and the results are compared to current state-of-the-art benchmarks in solar forecasting and to standard reference cases using the aforementioned performance measures.

Monthly analysis of performance is considered for the proposed approach and is compared to that of the persistence reference case in order to get a better understanding of the improvements that the adaptive framework brings to the solar irradiance forecasting [39]. Figure 12 represents the results of monthly forecast performance analysis. As shown in Figure 12, the RMSE values obtained by the proposed forecasting framework are consistently smaller than those of the persistence case for next day prediction. The performance of the proposed framework is comparable to some of the dominating benchmarked solutions for solar energy forecast in European Union which are mainly based on NWP.

Forecast methods presented by the University of Oldenburg, Meteotest, CENER, the statistical method of Blue Sky (BLUE), and the traditional synoptic method of the meteorologists of Blue Sky (SYNOP) are evaluated using Austrian data set for 2007 and 2008. The SYNOP method is spatially restricted and only available for Austria [20]. Actual and relative values for RMSE, MAE and BE are given in the Table I for the next day forecast in Austria for European NWP-based approaches and their persistence method for the years 2007 to 2008.

For the duration of the year 2012, forecast results obtained using the method proposed in this article, i.e. UTSA Method, and the persistence approach for the specific NREL site at Golden CO, USA are also included in table II for comparison. The actual improved RMSE with respect to persistence is the difference between the RMSE of persistence and that of the method while the relative improvement value is computed as follows:

$$\vartheta_{\mathrm{RMSE}} = \mathrm{RMSE}_{\mathrm{Persistence}} - \mathrm{RMSE}_{\mathrm{method}} \tag{24}$$

$$_{\mathrm{RMSE}} = \frac{\vartheta_{\mathrm{RMSE}}}{\mathrm{RMSE}_{\mathrm{Persistence}}} \times 100 \tag{25}$$

where $\vartheta_{\mathrm{RMSE}}$ and $_{\mathrm{RMSE}}$ stand for the actual RMSE improvement and the relative RMSE improvement over the persistence method, respectively.

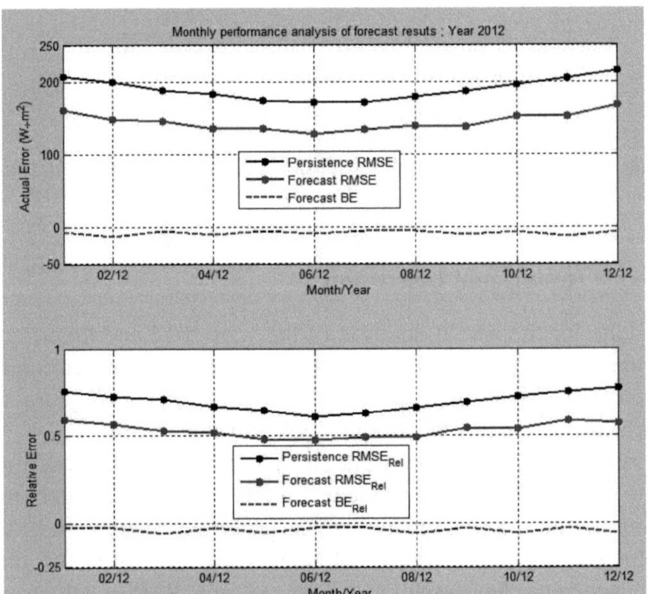

Fig. 12. Monthly performance analysis; Actual Error (TOP graph), Relative Error (BOTTOM graph), Persistence (BLACK), Forecast (BLUE)

Table 1. Actual and relative RMSE, MAE, and BE for next day forecast using Austrian dataset (2007 and 2008) and for the proposed method using Golden CO, USA dataset (2012).

	Approach	ϑ_{RMSE} (ϑ_{RMSE})	RMSE ($\frac{W}{m^2}$) ($RMSE_{Rel}$)	MAE ($\frac{W}{m^2}$) (MAE_{Rel})	BE ($\frac{W}{m^2}$) (BE_{Rel})
1	BLUE	43 (30.3 %)	99 (44.6%)	61 (27.6%)	1 (0.5%)
2	ECMWF-OL	41 (28.9%)	101 (45.6%)	65 (29.2%)	16 (7%)
3	**Proposed Method (NREL Golden CO)**	**45.26 (23.3%)**	**149.29 (41.69%)**	**98.09 (27.39%)**	**8.91 (2.49%)**
4	SYNOP	30 (21.1%)	112 (50.4%)	70 (31.5%)	9 (0.4%)
5	WRF-MT	19 (13.4%)	123 (55.4%)	77 (34.7%)	47 (21.0%)
6	CENER	13 (9.2%)	129 (58.1%)	87 (39.0%)	30 (13.6%)
	Austrian Persistence	0 (0.00%)	142 (64.3%)	91 (41.2%)	-14 (-6.4%)
	NREL Persistence	0 (0.00%)	194.55 (54.36%)	116.32 (32.5%)	-0.23 (-0.06%)

From Table 1 and Figure 12 it is seen that the proposed approach strongly outperforms the persistence for 24-hour ahead forecast based on the results obtained for year 2012. Considering the relative improved RMSE values the proposed UTSA Method stands in the third place after BLUE and ECMWF-OL. This means that the proposed framework for UTSA Method outperforms CENER, WRF-MT and SYNOP. Noticing that SYNOP is a real-time 24-hr ahead approach done by human meteorology experts it is inferred that the proposed framework for solar energy forecast outperforms the predictions suggested by the experts in the field. This can be of critical value to utility companies for microgrid energy management [40].

3 Conclusion and Remarks

A radio over fiber (RoF) communication system was developed for cable television (CATV) channels over wavelength division multiplexing (WDM) network using an optical direct modulator (DM). For transmitting CATV channels over a 4-WDM optical fiber communication system utilizing the DM laser is the cost effective design. DM laser bandwidth limitation can be overcome by increasing bandwidth efficiency. 16-QAM scheme will not increase bandwidth efficiency enough while 256-QAM exhibits an unacceptable SER due to its large symbol alphabet size and closer detection thresholds. 64-QAM scheme proved to result in an acceptable SER for transmitting 30 CATV channels with bit rate of 30 (Mb/s) per channel when an optimized adjustment of RF electrical power and laser biasing was considered. A proper selection of electrical power and Module-Bias ratio can control DM laser nonlinearity and chirping effect. Fuzzy modeling of the communication system helps obtain results for a continuous range of inputs. Fuzzy-GA combination also provided estimates of the electrical power that was required to achieve a desired SER for different settings of Module-Bias ratio with satisfactory accuracy and precision, in the 0.02 mW actual and 1% relative error orders, for the specific CATV application.

Adaptive site-specific framework for next-day solar energy forecast was developed based on a combination of data-analytic approaches and artificial intelligence techniques. The models are developed and validated utilizing a raw large dataset consisting 8 years of data from the NREL archive for a site in Golden, Colorado. The proposed method improves the RMSE by more than 23% and the MAE by more than 15% over the next-day persistence as the standard reference for the Golden, Colorado, USA considering prediction of the next day's solar energy over the year 2012. The capabilities of the proposed data analytic-based approach compared to physics-based models is illustrated with the fact that the proposed method compares better than some of the current state-of-the-art in the field and is very competitive with the leading approaches. Furthermore, the proposed method represents predictions of solar irradiance in a point in space. As the predictions are distributed in space over a larger area the forecasting framework introduced is expected to outperform others since patterns of weather and climate variables can be recognized on a distributed manner thus adding intelligence to the prediction which would not be possible otherwise.

References

1. Sabella, R., Paciotti, M., di Fonzo, A.: Impact of non-linear distortions on different modulation schemes in analog CATV distribution systems. In: IEEE Conf. (1996)
2. Chen, L., Wen, S.C., Li, Y., He, J., Wen, H., Shao, Y., Dong, Z., Pi, Y.: Optical Front-Ends to Generate Optical Millimeter-Wave Signal in Radio-Over-Fiber Systems With Different Architectures. J. of Lightwave Tech. 25(11), 3381–3387 (2007)
3. Schatz, R., Kjebon, O., Akram, M.N.: Design and evaluation of high speed DBR lasers for analog and digital transmission. In: IEEE Conf. (2003)
4. Lckinger, E.S.S.: Broadband Circuits for Optical Fiber Communication, ch. 7. John Wiley & Sons, Inc. (2005)
5. Niknamfar, M., Manjili, Y.S., Shadaram, M., Jamshidi, M.: Cost Effective ROF Communication System for CATV Channels over WDM Network and Fuzzy Modeling of the System. In: International Conference on Computing, Networking and Communications (ICNC), San Diego, CA, USA (2013)
6. Sklar, B.: Digital communications, ch. 9. Prentice-Hall (2001)
7. Keiser, G.: Optical Fiber Communications, ch. 8. McGraw-Hill (2010)
8. Jemison, W.D., Paolella, A.C.: RF and Microwave Applications and Systems, ch. 9. CRC press (2008)
9. VPI photonics manual help, version 8.7
10. Niknamfar, M., Barnes, R., Shadaram, M.: Comparative Study of the Per-formance of CATV Channels over DWDM Network Using Direct and External Modulator. In: IEEE Conf. ICTON (2012)
11. Zadeh, L.A.: The role of fuzzy logic in modeling, identification and control. MIC Journal 15(3), 191–203 (1994)

12. Nummikoski, J., Manjili, Y.S., Vega, R., Krishnaswami, H.: Adaptive rule generation for solar forecasting: Interfacing with a knowledge-base library. In: Proc. IEEE 39th Photovoltaic Specialists Conference (PVSC), Tampa, FL, USA, pp. 980–984 (2013)
13. Kohavi, R., Rothleder, N., Simoudis, E.: Emerging Trends in Business Analytics. Communications of the ACM 45(8), 45–48 (2002)
14. Fayyad, U., Piatetsky-Shapiro, G., Smyth, P.: From Data Mining to Knowledge Discovery in Databases. AI Magazine 17(3), 37–54 (1996)
15. Tannahill, B., Jamshidi, M.: Big Data Analytic Paradigms - From PCA to Deep Learning. In: Proc. AAAI Workshop, Stanford, CA, USA, March 24-25 (2014)
16. Mathiesen, P., Kleissl, J.: Evaluation of numerical weather prediction for intraday solar forecasting in the continental United States. Solar Energy 85(5), 967–977 (2011)
17. Perez, R., Moore, K., Wilcox, S., Renné, D., Zelenka, A.: Forecasting solar radiation–Preliminary evaluation of an approach based upon the national forecast database. Solar Energy 81(6), 809–812 (2007)
18. Bartholomy, O., Vargas, T., Simone, M.: Solar Forecasting Evaluation at SMUD. In: UVIG Variable Generation Forecasting Workshop, Tucson, AZ, Feb 25-26 (2014)
19. Marquez, R., Coimbra, C.: Forecasting of global and direct solar irradiance us-ing stochastic learning methods, ground experiments and the NWS database. Solar Energy 85, 746–756 (2011)
20. Lorenz, E., Remund, J., Müller, S., Traunmüller, W., Steinmaurer, G., Pozo, D., Ruiz-Arias, J., Fanego, V., Ramirez, L., Romeo, M., Kurz, C., Pomares, L., Guerrero, C.: Benchmarking of Different Approaches to Forecast Solar Irradiance. In: Proc. 24th European Photovoltaic Solar Energy Conference (EU PVSEC), Germany, pp. 4199–4208 (2009)
21. Manjili, Y.S., Vega, R.E., Jamshidi, M.: Method and System for Hours-ahead and Day-ahead Forecasting of Solar Energy, US patent, filing under process, UTSA Office of Commercialization and Innovation, Serial No: 292007-8100 / 2014.028.UTSA, San Antonio, TX, USA (2014)
22. Tannahill, B.K., Maute, C.E., Yetis, Y., Ezell, M.N., Jaimes, A., Rosas, R., Motaghi, A., Kaplan, H., Jamshidi, M.: Modeling of System of Systems via Data Analytics – Case for "Big Data" in SoS". In: Proceedings to 8th International Conf. on System of Systems Engineer-ing (SoSE), HI, USA (2013)
23. Cococcioni, M., D'Andrea, E., Lazzerini, B.: 24-hour-ahead forecasting of energy production in solar PV systems. In: IEEE 11th International Conference on Intelligent Systems Design and Applications (ISDA), Spain (2011)
24. Manjili, Y.S.: Data Analytic-based Adaptive Forecasting Framework for Smart Energy Management in Electrical Microgrids, PhD dissertation, University of Texas at San Antonio, San Antonio, TX (2014)
25. Banunarayanan, V., Brockway, A.: Solar Forecasting Metrics, Webinar by SunShot, US DOE (Feb. 2014)
26. Glassley, W., Kleissl, J., Shiu, H., Huang, J., Braun, G., Holland, R.: Current state of the art in solar forecasting. Final report, Appendix A, California Renewable Energy Forecasting, Resource Data and Mapping, California Institute for Energy and Environment (2010)
27. Sabburg, J., Long, C.N.: Improved sky imaging for studies of enhanced UV irradiance. Atmospheric Chemistry and Physics 4(11/12), 2543–2552 (2004), http://www.atmos-chem-phys.net/4/2543/2004/acp-4-2543-2004.pdf
28. Bacher, P., Madsen, H., Nielsen, H.A.: Online short-term solar power fore-casting. Solar Energy 83(10), 1772–1783 (2009), http://www.ge-min.de/book/solar

29. Diagne, H.M., David, M., Lauret, P., Boland, J.: Solar irradiation forecasting: state-of-the-art and proposition for future developments for small-scale insular grids. In: Proc. of the World Renewable Energy Forum (WREF), Denver, USA, pp. 4802–4809 (2012)
30. Coppin, P.: A Review of Solar Energy Forecasting Requirements, and a Proposed Approach for the Development of an Australian Solar Energy Forecasting System (ASEFS)., Report to Australian Solar Inst (2012)
31. European Centre for Medium-Range Weather Forecasts (ECMWF) (last visited: April 30, 2014), `http://www.ecmwf.int/`
32. Bing, J., Bartholomy, O., Krishnani, P.: Validation of Solar PV Power Forecasting Methods for High Penetration Grid Integration. IEEE Power and Energy Society, General Meeting, San Diego (2012)
33. Vega, R.E.: Wind Directionality: W Reliability-based Approach, PhD dissertation, Tex-as Tech University, Lubbock, TX (2008)
34. Pidwirny, M.: Cloud Formation Processes, Ch. 8 in Fundamentals of Physical Geography, 2nd edn (2006)
35. Elementary Meteorology Online, "Humidity, Saturation, and Stability", vsc.edu, (last visited: July 30, 2014)
36. Hyvarinen, A., Oja, E.: A fast fixed-point algorithm for independent component analysis. Neural Computation 9(7), 1483–1492 (1997)
37. Hyvarinen, A.: Fast and robust fixed-point algorithms for independent component analysis. IEEE Transactions on Neural Networks 10(3), 626–634 (1999a)
38. Hyvärinen, A., Oja, E.: Independent Component Analysis: Algorithms and Applications. Neural Networks 13(4-5), 411–430 (2000)
39. Manjili, Y.S., Vega, R.E., Jamshidi, M.: Data Analytic-based Adaptive Solar Energy Forecasting Framework. IEEE Systems Journal, Accepted for Publication, ISJ-RE-14-03213 (October 2014)
40. Manjili, Y.S., Vega, R., Jamshidi, M.: Cost-Efficient Environmentally-Friendly Control of Micro-Grids Using Intelligent Decision-Making for Storage Energy Manage-ment. Journal of Intelligent Automation and Soft Computing, Taylor & Francis Publication 19(4), 649–670 (2013), doi:10.1080/10798587.2013.842346.

Area Coverage in a Fixed-Obstacle Environment Using Mobile Sensor Networks

Hamid Mahboubi[1], Amir G. Aghdam[2], and Kamran Sayrafian-Pour[3]

[1] Department of Electrical & Computer Engineering, McGill University, 845 Rue Sherbrooke Ouest, Montréal, Québec H3A 0G4, Canada
hamid.mahboobi@mail.mcgill.ca

[2] Department of Electrical & Computer Engineering, Concordia University, 1455 de Maisonneuve Blvd. W., EV005.139, Montréal, Québec H3G 1M8 Canada
aghdam@ece.concordia.ca

[3] National Institute of Standards and Technology (NIST), 100 Bureau Drive, Stop 8910 Gaithersburg, MD 20899 USA
ksayrafian@nist.gov

Abstract. In this chapter, efficient deployment strategies for mobile sensor networks are proposed to improve the coverage area in target fields containing obstacles. The developed algorithms iteratively calculate and update the position of the sensors in order to improve the overall achievable coverage by the network. The *visibility-aware multiplicatively weighted Voronoi* (VMW-Voronoi) diagram is used to discover coverage gaps in networks that have sensors with different sensing capabilities. The sensors, then, reduce the size of the coverage gaps in the target field. The relocation strategy also considers possible existing obstacles in the field. Simulation results are provided to demonstrate the effectiveness of the proposed distributed deployment schemes.

1 Introduction

Autonomous mobile sensor networks (MSN) have received increasing attention in the past decade due to their many commercial and defense related applications such as health and environmental monitoring [1], [2], target tracking [3], [4], [5] and surveillance [6], [7]. Advances in Micro-Electro-Mechanical Systems (MEMS) are also helping in the low-cost and practical implementation of MSNs.

Field coverage is a major problem that arises in many of the MSN applications. In this problem, the mobile nodes are expected to provide sensory coverage of as much area as possible (ideally the entire area) in a given target field. Given the mobility feature of nodes in an MSN, it would be desirable to have autonomous networks that can achieve maximal field coverage. Therefore, intelligent and decentralized relocation algorithms are extremely important in mobile sensor network field coverage problems. Design of efficient relocation algorithms are quite challenging due to the limited energy resources in such networks. Constraints such as node's energy consumption (due to motion or data communication) as well as limited communication and sensing ranges of the sensors must be taken into account.

A lot of research has been focused in this area. For example, in [8], gradient descent algorithms are proposed for a class of utility functions that increase network coverage.

A. El-Osery and J. Prevost (eds.), *Control and Systems Engineering*,
Studies in Systems, Decision and Control 27, DOI: 10.1007/978-3-319-14636-2_7

In [9], a sensor deployment strategy is proposed to solve coverage problem on a complicated surface in 3D space. A collaborative coverage algorithm using a combination of mobile and static sensors has been discussed in [10], where mobile sensors try to compensate for possible coverage gaps between static sensors. In [11], distributed control laws are presented to achieve convex equi-partition configuration in mobile sensor networks. The field coverage strategy in [12] is based on a localized Voronoi diagram, where each sensor uses the local information of the neighboring sensors to construct its own Voronoi region. A coordination algorithm and performance analysis have been presented in [13] using a class of aggregate objective functions based on the geometry of the Voronoi partitions and proximity graphs.

A Delaunay graph has been used in [14] to propose distributed gradient-descent coverage algorithms. An optimal algorithm is proposed in [15] to monitor an environmental boundary with mobile agents, where the boundary is approximated by a polygon. The problem of sensor deployment in a network with non-uniform coverage priority has been considered in [16], [17]. Several efficient deployment algorithms have also been proposed in [18] to increase the coverage area based on an iterative method.

Using Voronoi diagram, three sensor deployment algorithms (i.e. VEC, VOR and Minimax) have been introduced in [19] to discover and reduce coverage gaps throughout the field. In [20], the LRV (Least Recently Visited) algorithm has been presented to simultaneously solve the problem of coverage, exploration and sensor deployment. The coverage problem in three-dimensional space has been addressed in [21]; and, a distributed algorithm has been proposed for achieving full field coverage. Efficient network coverage strategies that do not use simple sensing models or Voronoi partitions are presented in [22]-[26]. Distributed gradient-descent algorithms for optimal coverage control are presented in [22], [23]. These algorithms exploit cooperation among sensors and optimize a probabilistic detection metric.

In [24], distributed algorithms for optimal field coverage in the presence of polygonal obstacles and limited sensors' field-of-view are proposed. The problem of distributed convergence to a Nash equilibrium in mobile sensor networks is investigated in [25]. In [26], a field coverage strategy has been developed to maximize the probability of detecting events which occur in a given region. The approach prolongs the network lifetime by properly reducing the communication cost. Distributed control laws are provided in [27] for the disk-covering and sphere-packing problems using non-smooth gradient flows. In [28], an efficient procedure for sensor relocation is introduced to minimize the maximum error variance and extended prediction variance.

In this chapter, two different distributed algorithms are presented for deployment and relocation of sensors in an MSN. It is assumed that the nodes in the network have different sensing capabilities; and also, they are operating on a flat field containing obstacles. The approach in our proposed algorithms is based on using the *visibility-aware multiplicatively weighted Voronoi* (VMW-Voronoi) diagram. The effect of obstacles on sensors' sensing capabilities (i.e. visibility area of each sensor) is taken into consideration in the proposed algorithms. The VMW-Voronoi diagram is used for discovering coverage gaps, and sensing radii of sensors are used as node weights in the construction of this diagram. In this work, it is assumed that if the line-of-sight between a point and a sensor is blocked with an obstacle, then the sensor is not able to sense any object located at that point. This is an acceptable assumption and has been used in the literature previously [29], [30].

Using VMW-Voroni diagram, the following two algorithms are presented in this chapter: *Obstructed Farthest Point* (OFP) and *Obstructed Minmax Point* (OMP) algorithms.

By iterative application of these algorithms, a gradual improvement in the overall coverage can be obtained. At each iteration, a new candidate coordinate for sensor relocation is calculated based on the current position of each sensor and its VMW-Voronoi region. The decision to move each sensor to the calculated candidate coordinate is made after comparing the value of its local coverage at the current and candidate locations. If the sensor's local coverage cannot be improved, it will remain in its current position; otherwise it will move to the new calculated coordinate. The algorithm terminates when none of the sensors can improve their local coverage by moving to their respective candidate positions.

The rest of this chapter is organized as follows. Section 2 proposes visibility-aware multiplicatively weighted Voronoi diagram as an extension of the conventional Voronoi diagram. The proposed algorithms for sensor relocation are proposed in Section 3. Simulation results that demonstrate the effectiveness of our approach are provided in Section 4. Finally, conclusions and future works are discussed in Section 5.

2 Visibility-Aware Multiplicatively Weighted Voronoi Diagram

Let $\mathbf{F} \subset \mathbb{R}^2$ represent a 2D target field. Consider $\mathbf{S} = (S_1, w_1), (S_2, w_2), \ldots, (S_n, w_n)$ to be a set of n distinct weighted nodes within the field $\mathbf{F}$. $w_i > 0$ is the weighting factor associated with the node S_i, for any $i \in \mathbf{n} := \{1, 2, \ldots, n\}$. The *visible set* of an arbitrary point $Q \in \mathbf{F}$ is defined as the largest subset of $\mathbf{S}$ with a non-obstructed line-of-sight view from all of its elements to the point Q. Let $\mathbf{Indx}_Q$ represent the indices of the nodes in this subset. In the presence of obstacles in the field and depending on the location of point Q, the set $\mathbf{Indx}_Q$ may have between 0 to n elements. In particular, Q is called an *invisible point* if $\mathbf{Indx}_Q$ is an empty set, otherwise it is called a *visible point*. The set of all invisible points in the field $\mathbf{F}$ is called the *invisible region* and will be denoted by $\Theta_{\mathbf{S}}$. The invisible region is highly dependent on the positions of the nodes S_i and the obstacles in the field.

Definition 1. *The weighted distance of a point Q from a node (S_i, w_i), $i \in \mathbf{n}$ is defined as:*

$$d_w(Q, S_i) = \frac{d(Q, S_i)}{w_i}$$

where $d(Q, S_i)$ indicates the Euclidean distance between the point Q and the node S_i in the 2D field $\mathbf{F}$.

It is desired now to partition the visible area of the field $(\mathbf{F} \backslash \Theta_{\mathbf{S}})$ into n regions such that:

- Each region contains only one node, and
- the closest node, in the sense of weighted distance, to any point inside a region is the node assigned to that region.

The mathematical characterization of the invisible region and each visible region is respectively given by:

$$\Theta_{\mathbf{S}} = \{Q \in \mathbb{R}^2 \mid \mathbf{Indx}_Q = \varnothing\} \tag{1}$$

$$\Pi_i = \{Q \in \mathbb{R}^2 \mid i \in \mathbf{Indx}_Q,\ d_w(Q, S_i) \leq d_w(Q, S_j), \forall j \in \mathbf{Indx}_Q - \{i\}\} \tag{2}$$

The diagram obtained by partitioning the field $\mathbf{F}$ into the invisible region and the above-mentioned n regions is called the *visibility-aware multiplicatively weighted Voronoi* (VMW-Voronoi) diagram [31]. According to (2), any point Q in the i-th VMW-Voronoi region Π_i has the following property:

$$\frac{d(Q, S_i)}{d(Q, S_j)} \leq \frac{w_i}{w_j},\ \forall i, j \in \mathbf{Indx}_Q,\ i \neq j \tag{3}$$

Definition 2. *Similar to conventional Voronoi diagram [32], the nodes S_i and S_j ($i, j \in \mathbf{n}$, $i \neq j$) in a VMW-Voronoi diagram are called neighbors if $\Pi_i \cap \Pi_j \neq \varnothing$. The set of all neighbors of S_i, $i \in \mathbf{n}$, is denoted by $\mathbf{N}_i$ and is formulated as:*

$$\mathbf{N}_i = \{S_j \in \mathbf{S} \mid \Pi_i \cap \Pi_j \neq \varnothing,\ \forall j \in \mathbf{n} - \{i\}\} \tag{4}$$

The VMW-Voronoi diagram is the main tool for developing the sensor deployment strategies in this chapter. A circle is used to characterize the sensing area of each sensor. The size of these circles could be different for different sensors. Consider the position of each sensor in the field as a node with a weight equal to the sensor's sensing radius, and sketch the VMW-Voronoi region for each sensor; the resultant diagram, together with the invisible region, covers the entire field $\mathbf{F}$.

The following definitions and assumptions are borrowed from [33].

Definition 3. *Consider the sensor S_i with the sensing radius r_i and the corresponding VMW-Voronoi region Π_i, $i \in \mathbf{n}$, and let Q be an arbitrary point inside Π_i. The intersection of the region Π_i and a circle of radius r_i centered at Q is referred to as the* i-th coverage area w.r.t. Q, *and is denoted by $\beta_{\Pi_i}^{Q}$. The i-th coverage area w.r.t. the location of the sensor S_i is called the local coverage area of that sensor, and is denoted by β_{Π_i}.*

From the characterization of the VMW-Voronoi regions provided in (2), it can be easily shown that if a sensor cannot detect a point in its corresponding region, no other sensor can detect it either. This means that in order to find the coverage gaps (i.e., the undetectable points in $\mathbf{F}$), it is sufficient to compare the VMW-Voronoi region of every sensor with its local coverage area. A VMW-Voronoi diagram with 3 sensors S_1, S_2 and S_3 with the sensing radii 10m, 18m, and 18m, respectively, is depicted in Fig. 1.

Definition 4. *Consider an arbitrary point Q inside the VMW-Voronoi region Π_i, $i \in \mathbf{n}$. The area inside the VMW-Voronoi region Π_i which lies outside the i-th coverage area w.r.t. Q is referred to as the* i-th coverage gap w.r.t. Q, *and is denoted by $\theta_{\Pi_i}^{Q}$. The i-th coverage gap w.r.t. the location of the sensor S_i is called the local coverage gap of that sensor, and is denoted by θ_{Π_i}.*

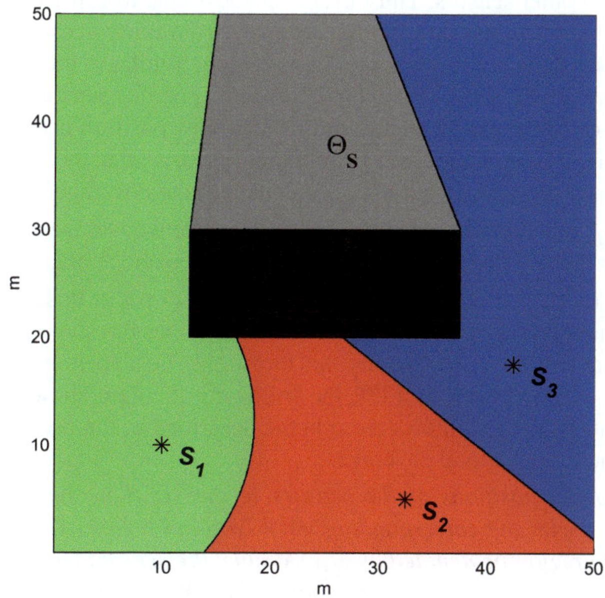

Fig. 1. An example of the VMW-Voronoi diagram for a group of 3 non-identical sensors in a field with obstacles [31]

Assumption 1. *The sensors are able to obtain their own positions with sufficient accuracy in the target field (using, for instance, the methods proposed in [34], [35]).*

Assumption 2. *The communication range of the sensors is limited (and possibly different for different nodes). This limitation could potentially prevent sensors from communicating with their neighbors, and can result in wrong VMW-Voronoi regions around some sensors. Consequently, such a limitation can have a negative impact on the detection of coverage gaps. Since the number of sensors in a mobile sensor network is typically large (or more precisely, there is a sufficient number of sensors per area unit) [36], [37], it is assumed that the graph representing sensors' communication topology is connected [38]. Hence, each sensor can obtain the information about the locations and sensing radii of the other sensors (and in particular its neighbors) through proper communication routes. Also it is assumed that the obstacles' locations are known by each sensor as a priori information. As a result, each sensor can calculate its VMW-Voronoi region accurately.*

3 Deployment Protocols

In this section, two different deployment protocols are developed for a network of non-identical sensors in the presence of obstacles. The proposed techniques are iterative, where in each iteration every sensor S_i, $i \in \mathbf{n}$, first broadcasts its sensing radius r_i

and position P_i to other sensors. Thus, every sensor is able to construct its own VMW-Voronoi region based on the information received from other sensors in the network. Then, every sensor detects coverage gaps in its region. Similar to the relocation scheme in [33], when a coverage gap is discovered, the corresponding sensor calculates its new position using one of the proposed algorithms. The new position is calculated with the aim of eliminating the coverage gap or reducing its size after the sensor relocates to that position. Once the new location $\acute{P}_i$ is calculated, the coverage area w.r.t. this new location, i.e., $\beta_{\Pi_i}^{\acute{P}_i}$ is evaluated and compared to the current coverage area, i.e., $\beta_{\Pi_i}^{P_i}$. The sensor moves to the new location only if the resultant coverage area is greater than the present value, i.e. $\beta_{\Pi_i}^{\acute{P}_i} > \beta_{\Pi_i}^{P_i}$; otherwise, it does not move in this iteration. In order to have a termination criterion for the algorithms, a proper threshold ϵ is defined; if no sensor can improve its coverage area by this threshold, the algorithm is terminated.

Analogously to Theorem 1 in [33], the following theorem shows that any sensor deployment strategy which follows the scheme described in the previous paragraph is guaranteed to increase the total coverage.

Theorem 1. *Let the positions of the sensors in the set* $\mathbf{S}$ *be represented by* $\mathbf{P} = \{P_1, P_2, \ldots, P_n\}$ *with the corresponding VMW-Voronoi regions* $\Pi_1, \Pi_2, \ldots, \Pi_n$. *Let also the invisible region be denoted by* $\Theta_{\mathbf{P}}$. *Assume the sensors move to new positions* $\acute{\mathbf{P}} = \{\acute{P}_1, \acute{P}_2, \ldots, \acute{P}_n\}$ *with the corresponding VMW-Voronoi regions* $\acute{\Pi}_1, \acute{\Pi}_2, \ldots, \acute{\Pi}_n$ *such that* $\acute{P}_i \neq P_i$ *for all* $i \in \mathbf{k}$, *where* $\mathbf{k}$ *is a non-empty subset of* $\mathbf{n}$. *If the* i*-th coverage area w.r.t.* $\acute{P}_i$ *in the previously constructed VMW-Voronoi region* Π_i *is greater than the previous* i*-th local coverage area (i.e.,* $\beta_{\Pi_i}^{\acute{P}_i} > \beta_{\Pi_i}^{P_i}$*) for all* $i \in \mathbf{k}$, *then the total coverage in the network increases.*

Proof. Denote by θ and $\acute{\theta}$ the total uncovered area (coverage gap) of the field when the sensors are located in $\mathbf{P}$ and $\acute{\mathbf{P}}$, respectively. It is deduced from the characterization of the VMW-Voronoi diagram that:

$$\theta = \Theta_{\mathbf{P}} + \sum_{i=1}^{n} \theta_{\Pi_i}^{P_i} \tag{5}$$

It can be easily shown that by increasing the coverage area in Π_i, $i \in \mathbf{k}$, the corresponding coverage gap will be decreased. Since it is assumed that the i-th coverage area w.r.t. $\acute{P}_i$ is greater than the i-th local coverage area for any $i \in \mathbf{k}$, one can conclude that:

$$\theta_{\Pi_i}^{\acute{P}_i} < \theta_{\Pi_i}^{P_i}, \quad \forall i \in \mathbf{k} \tag{6}$$

On the other hand, it is possible that some of the points in $\theta_{\Pi_i}^{\acute{P}_i}$ and $\Theta_{\mathbf{P}}$ are also covered by other mobile sensors at $\acute{\mathbf{P}}$. Hence:

$$\acute{\theta} \leq \Theta_{\mathbf{P}} + \sum_{i=1}^{n} \theta_{\Pi_i}^{\acute{P}_i} \tag{7}$$

Using the last two relations and noting that for any $i \in \mathbf{n} \backslash \mathbf{k}$ by definition $\theta_{\Pi_i}^{\acute{P}_i} = \theta_{\Pi_i}^{P_i}$, one can obtain the following inequality:

$$\acute{\theta} < \Theta_{\mathbf{P}} + \sum_{i=1}^{n} \theta_{\Pi_i}^{P_i} \tag{8}$$

Now, it is concluded from (5) and (8) that:

$$\acute{\theta} < \theta \tag{9}$$

which means that the total coverage area increases under this deployment scheme.

The above-mentioned procedure will be used in the next two subsections to develop two algorithms, namely, OFP and OMP.

3.1 Obstructed Farthest Point (OFP) Strategy

The main idea behind this algorithm is to move every sensor to the farthest point in its VMW-Voronoi region such that any existing coverage gap is covered. If a sensor S_i detects a coverage gap in its corresponding VMW-Voronoi region, it calculates the farthest point $X_{i,far}$ in that region, and moves toward it until this point is covered. Fig. 2 shows a sample VMW-Voronoi region constructed by the sensor S_1. The segments g and e are generated due to the two neighboring sensors with sensing radii equal to that of S_1. The segments a and h are obstacle edges, and the segment c is formed because of the field boundary. The edge b, in fact, is constructed by the line-of-sight of the sensor, and finally arcs d and f are formed by two neighboring sensors with larger and smaller sensing radii than that of S_1, respectively. As the figure illustrates, the OFP algorithm finds the farthest point to S_1, (i.e. $X_{1,far}$) as a candidate for the next location of the sensor. Since the local coverage of the sensor will increase by moving to S_1', the sensor moves toward $X_{1,far}$ until it is covered.

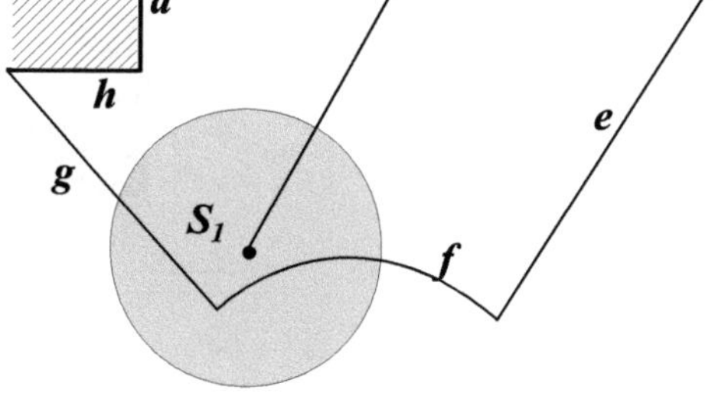

Fig. 2. A sample VMW-Voronoi region and a candidate point calculated using the OFP method

Fig. 3 shows an example of coverage improvement as a result of using the OFP Algorithm. In this example, 27 mobile sensors with varying sensing ranges are randomly deployed in a 2D field of size 50m × 50m. Among these 27 sensors, 15 have sensing range of 6m, 6 with a radius of 5m, 3 with a radius of 7m and the remaining 3 sensors with a radius of 9m. The communication range of each sensor is assumed to be greater than its sensing range by a factor of $10/3$. Three snapshots of the field coverage are shown in Fig. 3. The circles represent the sensing area of each sensor. As observed, the coverage is 68.44% initially, but it increases to 81.31% after the first iteration of the presented algorithm (Fig. 3(b)). The final coverage is 95.32% (Fig. 3(c)).

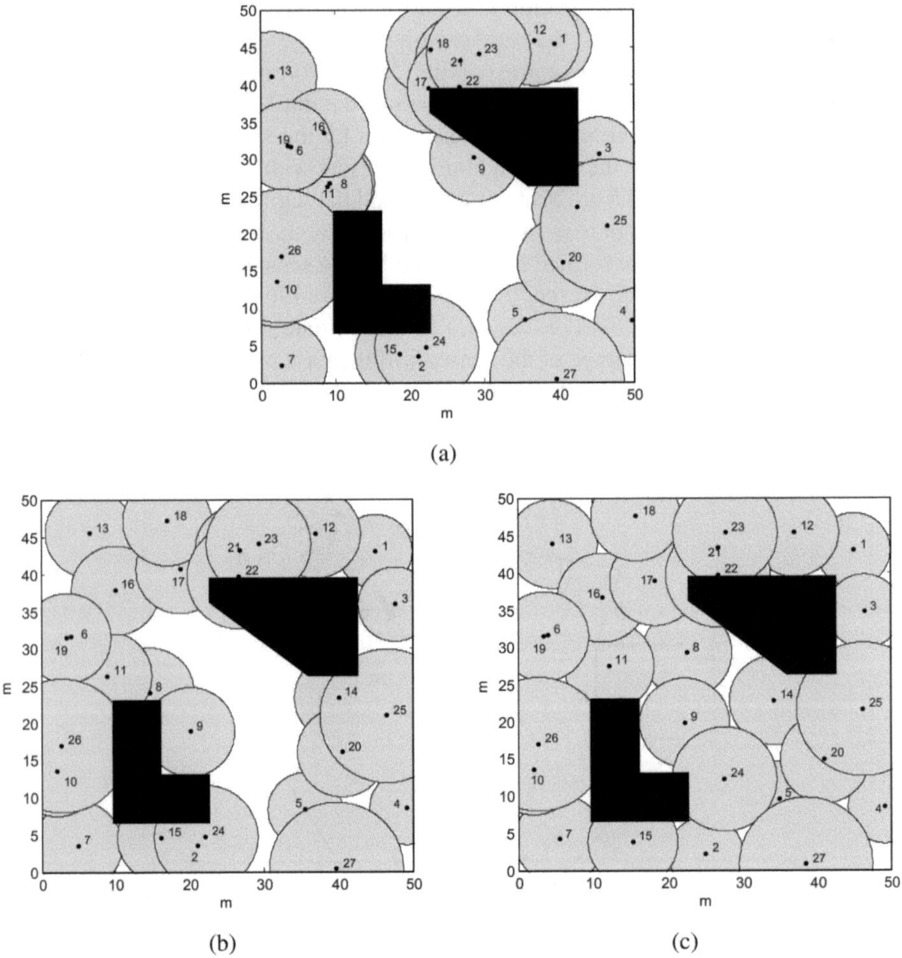

Fig. 3. Snapshots of the movement of the sensors under the OFP algorithm. (a) Initial coverage; (b) field coverage after the first round, and (c) final coverage.

3.2 Obstructed Minmax Point (OMP) Strategy

Although the OFP algorithm performed well in most simulated scenarios, there exist certain network setups and node configurations, where it might not be as effective. Fig. 4 shows such an example. The next candidate location for the sensor under the OFP algorithm does not lead to any improvement in the local coverage of the sensor. Thus, the mobile sensor remains in its previous location. However, there exit other potential positions for sensor relocation that can increase its local coverage. Another configuration for which the OFP algorithm is not as effective is when the calculated candidate point lies on a line-of-sight connecting the sensor to an obstacle (e.g., see Fig. 5). These candidate points are, by definition, on the line-of-sight; therefore, if the sensor moves to such a position, then there is a good chance that part of the sensor's sensing capability is blocked in its corresponding region.

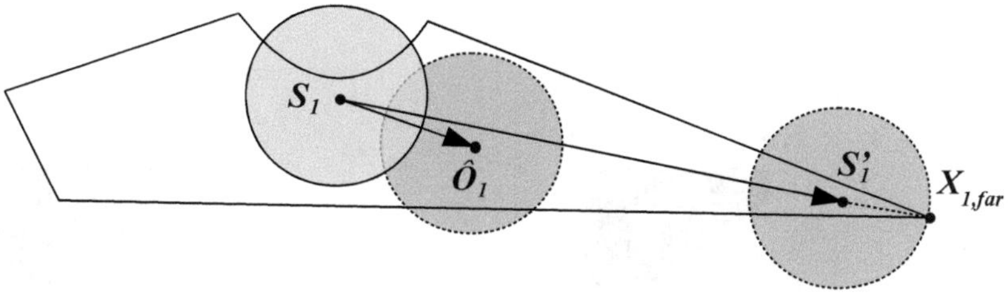

Fig. 4. A sample VMW-Voronoi region for which the OFP method performs poorly because of a narrow angle, while the OMP strategy finds a proper candidate location

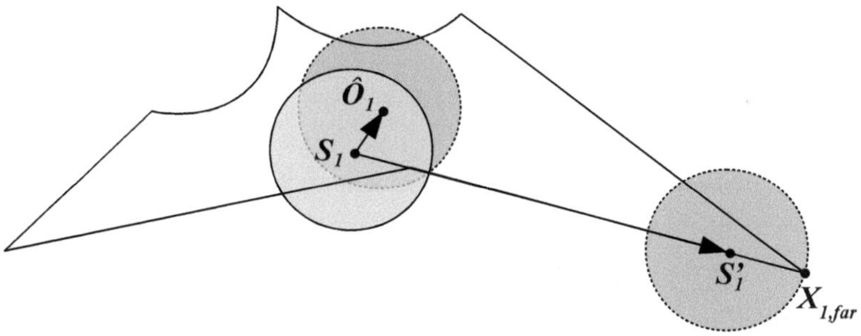

Fig. 5. A sample VMW-Voronoi region for which the OFP method performs poorly because of the sensor's line-of-sight, while the OMP strategy finds a proper candidate location

As it can be concluded from the above discussion, although the OFP algorithm is effective in many cases, one may find a proper location for the sensor in the special cases described above. The obstructed minmax point (OMP) strategy is proposed in the sequel to address this shortcoming of the OFP algorithm. The main idea behind the OMP strategy is that to achieve maximum coverage, no sensor should be too far

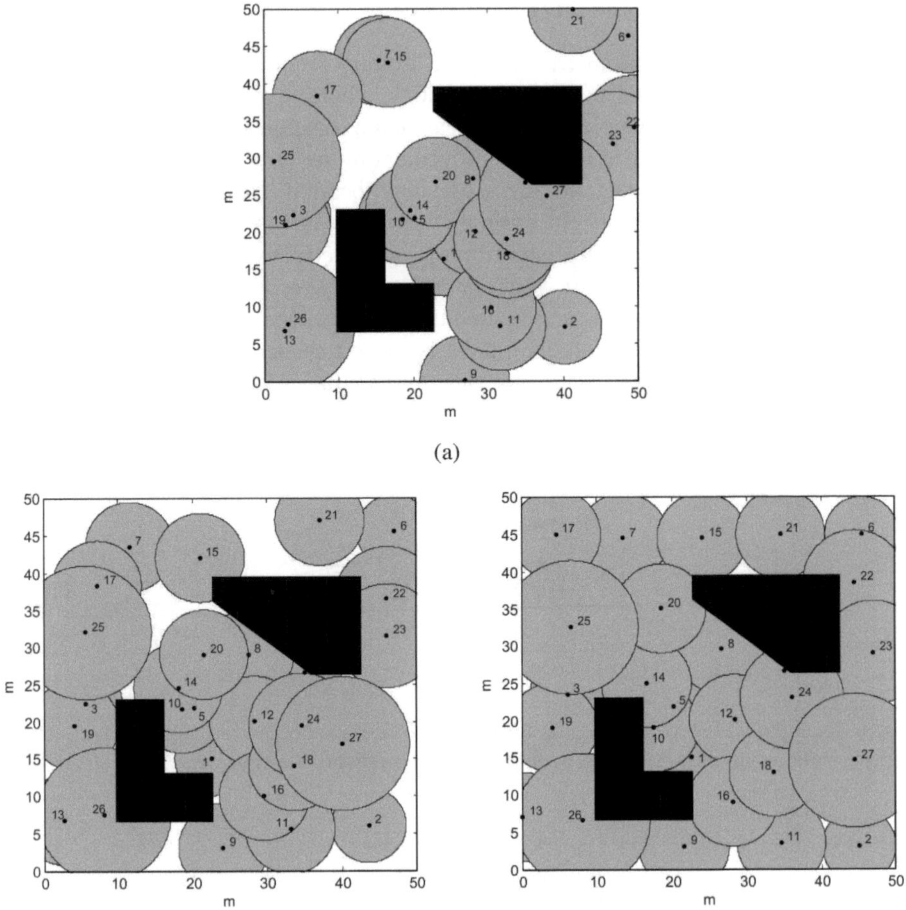

Fig. 6. Snapshots of the movement of the sensors under the OMP algorithm. (a) Initial coverage; (b) field coverage after the first round, and (c) final coverage.

from any point in its corresponding VMW-Voronoi region. The OMP strategy finds the location where its distance from the farthest point of the region is minimum and considers it as the candidate location for the sensor in the next step. This point is called the *OMP centroid*, and is denoted by $\hat{O}_i$ for the i-th region, $i \in \mathbf{n}$. It is clear that the candidate point $\hat{O}_1$ in Fig. 4 yield better coverage compared to the one obtained by using the OFP technique. Also, in Fig. 5, the OMP strategy performs more efficiently than the OFP algorithm due to the specific shape of the region which makes the farthest point lie on the line-of-sight of the sensor.

Consider the initial setting of Fig. 6(a), and let the OMP strategy be employed. The results are depicted in Figs. 6(b) and 6(c), where it is shown that after the first round the coverage increases from 67.09% to 84.94%, and that the final coverage is 97.62%.

Remark 1. One of the important properties of the VMW-Voronoi diagram is that first of all it partitions the field, and more importantly, there is exactly one sensor in each visible region. Since in the proposed algorithms the new location for each sensor is inside the corresponding VMW-Voronoi region and each sensor moves within its region, hence the sensors would not collide. Assume now that a sensor cannot communicate with some of its neighbors, and consequently it obtains a VMW-Voronoi region which some of its boundaries are different from the exact ones. As a result, the VMW-Voronoi regions do not necessarily partition the field in the sense that some of them overlap with each other. Such VMW-Voronoi regions can negatively affect the detection of coverage gaps. On the other hand, the overlapping of the VMW-Voronoi regions can lead to sensor collisions.

Remark 2. It is worth mentioning that the analytical solution to the optimal sensor deployment problem with respect to a prescribed coverage performance function is, in general, highly complex. Hence, to evaluate the performance of the algorithms proposed here, several simulations are performed using random settings. This method of performance assessment is typically used to evaluate the effectiveness of any sensor deployment strategy [19], [39], [10].

In the next section, the performance of the proposed algorithms in terms of the coverage area, energy consumption of the sensors, rate of convergence, and computational complexity are investigated.

4 Simulation Results

Consider a sensing field of size $50\text{m} \times 50\text{m}$ with two obstacles. Also consider a network of 36 mobile sensors with varying sensing radii i.e. out of these 36 mobile sensors, 20 sensors have a sensing radius of 6m, 8 have a radius of 5m, 4 have a radius of 7m, and the remaining 4 sensors have a sensing radius of 9m. The communication radius of each sensor is assumed to be greater than its sensing radius by a factor of $10/3$; e.g., a sensor with a sensing radius of 6m has a communication radius of 20m.

Define the *coverage factor* as the ratio of the covered area to the total area in the field. The simulation results presented in this section are the average values obtained from 20 different random initial sensor deployments. Coverage factor under both algorithms is shown in Fig. 7. While both algorithms provide satisfactory coverage, the OMP technique exhibits a better performance in this example.

To investigate the effect of the number of sensors on the performance of the algorithms, we considered four more setups: n=9, 18, 27, and 45, in addition to n=36 discussed above. It is assumed that the changes in the number of identical sensors in the new setups are proportional to the changes in the total number of sensors. For example, for n=27 there will be 15 sensors with sensing radius of 6m, 6 with sensing radius of 5m, 3 with sensing radius of 7m, and 3 with sensing radius of 9m. Fig. 8 shows the resulting final coverage versus number of sensors. Both algorithms yield satisfactory results with the OMP strategy still performing better.

An important factor in the performance evaluation of different deployment strategies is the time it takes for each strategy to reach the desired termination criterion. Assuming that both relocation strategies require the same deployment time in each round of

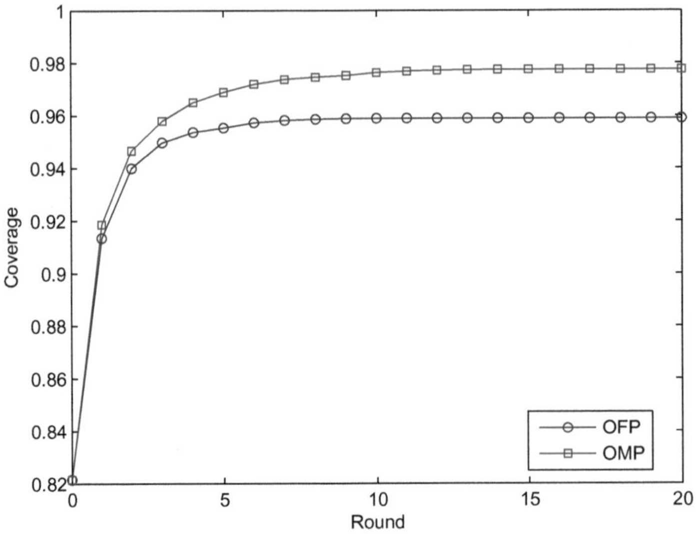

Fig. 7. Total normalized field coverage per round for 36 sensors

algorithm execution, then, the number of rounds to reach a predetermined termination criterion is a good measure of the deployment speed of each algorithm.

Fig. 9 shows that under both algorithms the number of rounds required to meet a certain termination condition increases with the number of sensors. This is the case up to certain number of sensors and then beyond that the number of required rounds decreases. The reason can be explained as follows. When the number of sensors is small, the sizes of their corresponding VMW-Voronoi regions are relatively larger than their coverage circles. Hence, it will be likely for some sensors that their entire coverage circles are enclosed within their VMW-Voronoi regions. Therefore, further relocation of each sensor in its region would not increase the coverage level. On the other hand, when the number of sensors is relatively large, the size of their corresponding VMW-Voronoi regions will be small. Thus, with a high likelihood, the coverage circles of most sensors enclose their VMW-Voronoi regions. This in turn implies that the termination criterion is satisfied in smaller number of rounds. Fig. 9 shows that the convergence rate of the OFP algorithm is faster than that of the OMP; therefore, it is a better candidate for field coverage if higher deployment speed is required.

Another important factor in performance evaluation of deployment algorithms in mobile sensor networks is energy-efficiency. Movement of a sensor, or more precisely, the distance it travels and also the number of times it stops (impact of static friction) are the dominant sources of energy consumption. Fig. 10 shows the average distance traveled by a sensor versus number of sensors in the network. As observed, the average traveled distance for a large number of sensors is small. For large number of sensors, the distance between each sensor's position and its candidate location in its corresponding

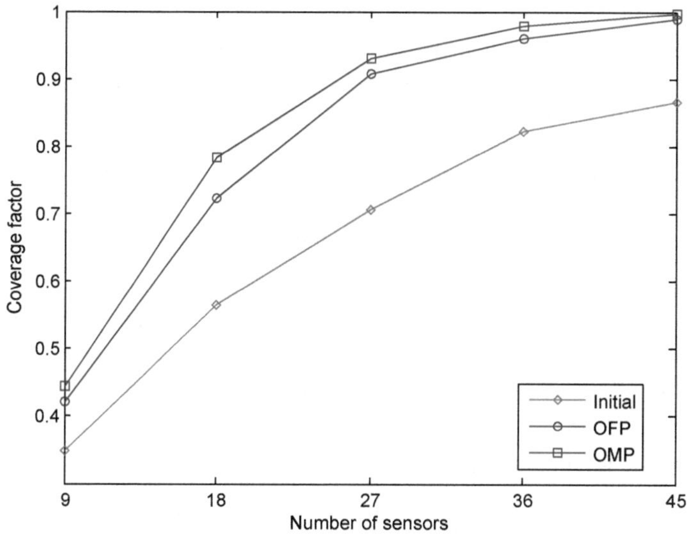

Fig. 8. The coverage factor versus number of sensors under the proposed algorithms

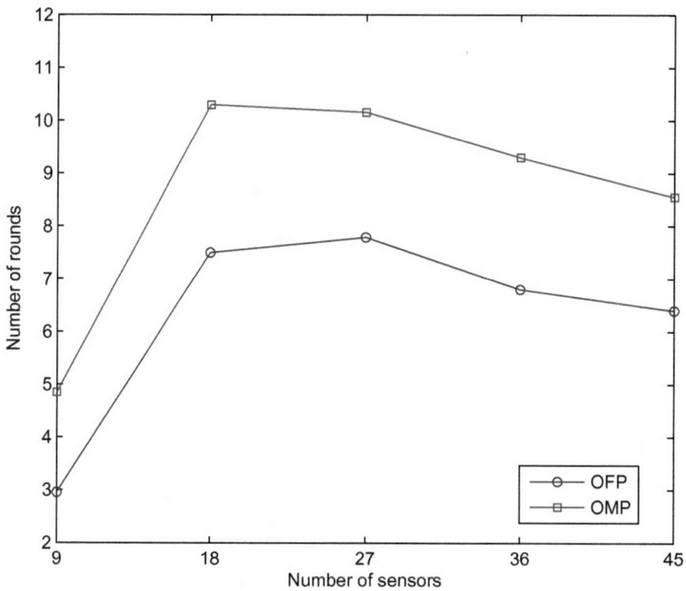

Fig. 9. Number of rounds needed to reach the termination criterion for different number of sensors using the proposed algorithms

VMW-Voronoi region decreases. Therefore, the average traveling distance required by a sensor decreases. This, in turn, leads to a reduction in energy consumption. Also, it can be seen from Fig. 10 that the OFP algorithm is more efficient than the OMP strategy for a larger number of sensors. Fig. 11 shows the number of relocations versus the number of sensors in the network. In both algorithms, and up to certain value, the number of relocations increases with the number of sensors, and then decreases after that. Again, this can be justified based on the relative sizes of the sensors' coverage circles versus their VMW-Voronoi regions. Figs. 10 and 11 clearly demonstrate that the OFP algorithm outperforms the OMP strategy in energy efficiency.

Remark 3. Note that the algorithms proposed in this chapter differ only in the way the new locations of the sensors are determined. Since the complexity of finding the new location of the i-th sensor in the OMP strategy is more than that in the OFP technique, hence the OFP algorithm outperforms the OMP algorithm as far as the computational complexity is concerned.

The above discussion is summarized below:

1. The OMP algorithm is more preferable as far as network coverage is concerned.
2. The OFP algorithm is more desirable when:
   - the deployment time is the main concern,
   - the energy consumption is the main concern,
   - the computational complexity is concerned.

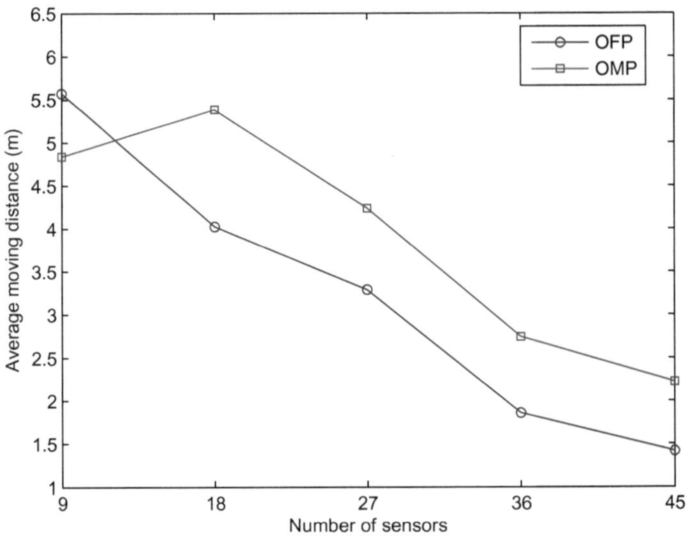

Fig. 10. Average traveled distance by each sensor versus number of sensors under proposed algorithms

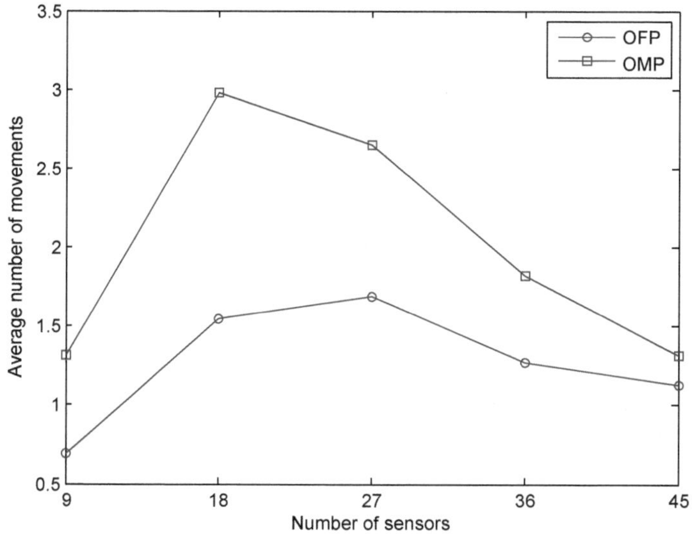

Fig. 11. Average number of movements required for different number of sensors, using the proposed algorithms

5 Conclusions and Future Works

In this chapter, two efficient distributed sensor relocation techniques are proposed to increase field coverage of mobile sensor networks. The algorithms are applicable to networks having non-identical mobile sensors and target coverage fields with obstacles. To account for the existence of obstacles, an extension of MW-Voronoi diagram, namely visibility-aware MW-Voronoi (VMW-Voronoi) diagram has been proposed as a tool to allow enhancement in sensor's local coverage. The iterative implementation of the algorithms provides gradual improvement in the overall network coverage. Simulation results confirm the effectiveness of the proposed techniques for different number of sensors. There are a number of related problems that can be investigated in the future. For example, adding static sensors to the field and proposing new deployment algorithms to maximize network coverage, and investigating the effect of communication delay and link failure on network coverage are some of the open problems in this area.

Acknowledgment. The authors would like to thank Dr. Kaveh Moezzi and Mr. Walid Masoudimansour for helpful discussions and suggestions.

References

1. Moh, M., Culpepper, B.J., Dung, L., Moh, T.-S., Hamada, T., Su, C.-F.: On data gathering protocols for in-body biomedical sensor networks. In: Proceedings of IEEE Global Communications Conference, pp. 2991–2996 (2005)

2. Talukder, A., Panangadan, A., Herrington, A.: Autonomous adaptive resource management in sensor network systems for environmental monitoring, pp. 1–9 (2008)
3. Mahboubi, H., Masoudimansour, W., Aghdam, A.G., Sayrafian-Pour, K.: Cost-efficient routing with controlled node mobility in sensor networks. In: Proceedings of IEEE Multiconference on Systems and Control, pp. 1238–1243 (2011)
4. Oka, A., Lampe, L.: Distributed scalable multi-target tracking with a wireless sensor network, p. 6 (2009)
5. Mahboubi, H., Momeni, A., Aghdam, A.G., Sayrafian-Pour, K., Marbukh, V.: Optimal target tracking strategy with controlled node mobility in mobile sensor networks. In: Proceedings of American Control Conference, pp. 2921–2928 (2010)
6. Olteanu, A., Xiao, Y., Wu, K., Du, X.: An optimal sensor network for intrusion detection (2009)
7. Huntwork, M., Goradia, A., Xi, N., Haffner, C., Klochko, C., Mutka, M.: Pervasive surveillance using a cooperative mobile sensor network, pp. 2099–2104 (2006)
8. Cortes, J., Martinez, S., Karatas, T., Bullo, F.: Coverage control for mobile sensing networks. IEEE Transactions on Robotics and Automation 2, 243–255 (2004)
9. Zhao, M.-C., Lei, J., Wu, M.-Y., Liu, Y., Shu, W.: Surface coverage in wireless sensor networks. In: Proceedings of 28th IEEE INFOCOM, pp. 109–117 (2009)
10. Wang, G., Cao, G., Berman, P., Porta, T.F.L.: A bidding protocol for deploying mobile sensors. IEEE Transactions on Mobile Computing 6(5), 563–576 (2007)
11. Pavone, M., Frazzoli, E., Bullo, F.: Distributed policies for equitable partitioning: Theory and applications. In: Proceedings of 47th IEEE Conference on Decision and Control, pp. 4191–4197 (2008)
12. Boukerche, A., Fei, X.: A Voronoi approach for coverage protocols in wireless sensor networks. In: Proceedings of IEEE Global Communications Conference, pp. 5190–5194 (2007)
13. Cortes, J., Martinez, S., Bullo, F.: Spatially-distributed coverage optimization and control with limited-range interactions. ESAIM. Control, Optimization and Calculus of Variations 11, 691–719 (2005)
14. Kwok, A., Martinez, S.: Unicycle coverage control via hybrid modeling. IEEE Transactions on Automatic Control 55(2), 528–532 (2010)
15. Susca, S., Bullo, F., Martinez, S.: Monitoring environmental boundaries with a robotic sensor network. IEEE Transactions on Control Systems Technology 16(2), 288–296 (2008)
16. Bullo, F., Cortes, J., Martinez, S.: Distributed Control of Robotic Networks. Princeton University Press (2009)
17. Cortes, J.: Coverage optimization and spatial load balancing by robotic sensor networks. IEEE Transactions on Automatic Control 55(3), 749–754 (2010)
18. Mahboubi, H., Moezzi, K., Aghdam, A.G., Sayrafian-Pour, K.: Self-deployment algorithms for field coverage in a network of nonidentical mobile sensors: Vertex-based approach. In: Proceedings of American Control Conference, pp. 3227–3232 (2011)
19. Wang, G., Cao, G., Porta, T.F.L.: Movement-assisted sensor deployment. IEEE Transactions on Mobile Computing 5(6), 640–652 (2006)
20. Batalin, M.A., Sukhatme, G.S.: The design and analysis of an efficient local algorithm for coverage and exploration based on sensor network deployment. IEEE Transactions on Robotics 23(4), 661–675 (2007)
21. Watfa, M.K., Commuri, S.: The 3-dimensional wireless sensor network coverage problem. In: Proceedings of IEEE International Conference on Networking, Sensing and Control, pp. 856–861 (2006)
22. Li, W., Cassandras, C.G.: Distributed cooperative coverage control of sensor networks. In: Proceedings of the IEEE Conference on Decision and Control, and the European Control Conference, vol. 2005, pp. 2542–2547 (2005)

23. Cassandras, C.G., Li, W.: Sensor networks and cooperative control. European Journal of Control 11(4-5), 436–463 (2005)
24. Zhong, M., Cassandras, C.G.: Distributed coverage control and data collection with mobile sensor networks. IEEE Transactions on Automatic Control 56(10), 2445–2455 (2011)
25. Stankovic, M.S., Johansson, K.H., Stipanovic, D.M.: Distributed seeking of nash equilibria in mobile sensor networks. In: Proceedings of the IEEE Conference on Decision and Control, pp. 5598–5603 (2010)
26. Zhong, M., Cassandras, C.G.: Asynchronous distributed optimization with event-driven communication. IEEE Transactions on Automatic Control 55(12), 2735–2750 (2010)
27. Cortes, J., Bullo, F.: Coordination and geometric optimization via distributed dynamical systems. SIAM Journal on Control and Optimization 44(5), 1543–1574 (2006)
28. Graham, R., Cortes, J.: Asymptotic optimality of multicenter voronoi configurations for random field estimation. IEEE Transactions on Automatic Control 54(1), 153–158 (2009)
29. Tan, H., Wang, Y., Hao, X., Hua, Q.-S., Lau, F.: Arbitrary obstacles constrained full coverage in wireless sensor networks. In: Proceedings of 5th International Conference of Wireless Algorithms, Systems and Applications, pp. 1–10 (2010)
30. Wang, Y.-C., Hu, C.-C., Tseng, Y.-C.: Efficient deployment algorithms for ensuring coverage and connectivity of wireless sensor networks. In: Proceedings of 1st International Conference on Wireless Internet, pp. 114–121 (2005)
31. Mahboubi, H., Sharifi, F., Aghdam, A.G., Zhang, Y.M.: Distributed coordination of multi-agent systems for coverage problem in presence of obstacles. In: Proceedings of American Control Conference, pp. 5252–5257 (2012)
32. Okabe, A., Boots, B., Sugihara, K., Chiu, S.N.: Spatial Tessellations: Concepts and Applications of Voronoi Diagrams. Wiley (2000)
33. Mahboubi, H., Moezzi, K., Aghdam, A.G., Sayrafian-Pour, K., Marbukh, V.: Distributed deployment algorithms for improved coverage in a network of wireless mobile sensors. IEEE Transactions on Industrial Informatics 10(1), 163–174 (2014)
34. Niculescu, D., Nath, B.: Ad hoc positioning system (APS) using AOA. In: Proceedings of IEEE INFOCOM. 22nd Annual Joint Conference of the IEEE Computer and Communications Societies, pp. 1734–1743 (2003)
35. Intanagonwiwat, C., Govindan, R., Estrin, D.: Directed diffusion: A scalable and robust communication paradigm for sensor networks. In: Proceedings of 6th Annual International Conference on Mobile Computing and Networking, pp. 56–67 (2000)
36. Wang, Y.-C., Tseng, Y.-C.: Distributed deployment schemes for mobile wireless sensor networks to ensure multilevel coverage. IEEE Transactions on Parallel and Distributed Systems 19(9), 1280–1294 (2008)
37. Yoon, S., Soysal, O., Demirbas, M., Qiao, C.: Coordinated locomotion and monitoring using autonomous mobile sensor nodes. IEEE Transactions on Parallel and Distributed Systems 22(10), 1742–1756 (2011)
38. Kwok, A., Martinez, S.: A distributed deterministic annealing algorithm for limited-range sensor coverage. IEEE Transactions on Control Systems Technology 19(4), 792–804 (2011)
39. Howard, A., Matarić, M.J., Sukhatme, G.S.: An incremental self-deployment algorithm for mobile sensor networks. Autonomous Robots 13(2), 113–126 (2002)

Energy Aware Load Prediction for Cloud Data Centers

John J. Prevost, Kranthimanoj Nagothu, Mo Jamshidi, and Brian Kelley

jeff.prevost@utsa.edu, {kranthimanoj,dr.brian.kelly}@gmail.com, moj@wacong.com

Abstract. Amazon recently estimated that the cost of energy for its datacenters reached 42% of the total cost of operation. Our previous research proposed an algorithm to predict how much cloud workload is expected at a specific time. This allows physical servers determined not to be needed to be placed in a low-power sleep state to save energy. If more system capacity is required, servers in a sleep state are transitioned back to an active state. In this paper, we extend our prior research by presenting both a stochastic model for state change as well as a new approach to determining the sampling frequency for performing the prediction of the expected capacity. The first result we show is that this allows the optimal prediction time horizon to be chosen. We next present a dynamic prediction quantization method to determine the optimal number of prediction calculation intervals. Both of these new algorithms allow us to predict future load within required Service Level Agreements while minimizing the number of prediction calculations. This effectively optimizes our ability to predict while minimizing the detrimental effect of additional calculations on our cloud resources. Finally, we test this model by simulating the stochastic time horizon and dynamic quantization algorithms and compare the results with three competing methods. We show that our model provides up to a 20% reduction in the number of calculations required while maintaining the given Service Level Agreement.

Keywords: Cloud computing, Energy conservation, Green design, Optimal control, Prediction algorithm.

1 Introduction

In 2006, it was estimated that the total energy consumption of data centers worldwide was equivalent to 1.5% of the energy consumed by the United States. This dollar amount represented over $4.5 billion, and was expected to double in a five year period [1]. For the operators of datacenters, the cost of energy is now over 42% of the total cost of operating the datacenter, and rising, according to Amazon [2]. This trend is accelerating due to the rapid adoption of cloud computing as the primary form of providing compute to users.

Cloud computing is an architectural transition occurring in datacenters across the world. The traditional approach to datacenter hosting is to provide access to each physical machine (PM) to only one client. Increasingly, the traditional model of datacenter hosting is giving way to a more efficient model of computing, cloud computing. Cloud computing offers the hosting provider higher efficiency of operation by allowing

A. El-Osery and J. Prevost (eds.), *Control and Systems Engineering*,
Studies in Systems, Decision and Control 27, DOI: 10.1007/978-3-319-14636-2_8

services to be stacked into virtual machines (VM), and the VMs to be stacked into physical machines at a ratio of up to 50 VMs or more, to one PM. This allows the cloud operator to utilize a higher average percentage of the physical infrastructure (CPU, MEM, HD, Net I/O), as compared with the traditional hosting model. Allocation of new VMs can also be done dynamically through a software interface which increases the speed and ease of deploying the new machines. Most of the modern cloud operating systems, such as, OpenStack, Amazon's Amazon Web Services and Microsoft's Azure allow the user to control the number and type of VM instances via a web portal. This reduces the personnel required by the cloud hosting provider therefore lowering the operational cost of running cloud datacenters (CDC). The consumer also directly benefits from this change in computing. Traditional hosting models force the consumer to own or rent computing equipment at a fixed monthly cost. The hosting provider shares varying degrees of maintenance responsibility of the computing equipment with the customer. Typically, the provider agrees to provide service level agreements (SLA) on the up-time of the network and the physical machines (e.g. CPU, RAM, HD, NIC, and power supply) and the base operating system (OS) and the customer is responsible for the applications running on top of the OS. Cloud computing separates the responsibility not at the OS of the physical machine, but at the virtual machine. The customer in an Infrastructure as a Service (IaaS) cloud computing model (such as Amazon's EC2, or Rackspace's OpenStack) provisions an operating system of choice inside a hypervisor running on the physical machine. The customer never sees the actual host OS nor has direct access to the physical machine's hardware. Discovered problems with a virtual machine instance can be solved by reloading a known good backup of the virtual machine from disk. Since the customer-facing OS is stored in a file, the same virtual machine can be instantiated on physical machines almost instantly, allowing for immediate scale-out of services to handle spikes in demand.

Another benefit that Cloud computing offers is economies of scale. This is because the hosting provider can spread the physical cost of the datacenter over more compute, storage and network nodes, which lowers the cost per node. This gives the hosting provider an incentive to build ever larger datacenters. As datacenters increase in size so does the amount of energy spent operating datacenters worldwide. Energy efficient methods of operating datacenters will increasingly become a primary focus of attention by the operators of CDCs.

1.1 Related Research

The problem of reducing the amount of energy required for the operation of CDCs has had much research [1-9]. One technique that is found in research literature is Dynamic Frequency and Voltage Scaling (DVFS) [3]. In this research the authors show a technique of creating power management policies for using processor and memory power reduction techniques to minimize the energy used across the data center. This is accomplished by changing the operating frequency of the CPU clock up or down, depending on the operational requirements. Another researched energy conservation method is to change the state of the compute nodes from high-power (active) to low power (off or sleep), when they are not needed. Cardosa, et al. [4] showed a technique for optimizing the power used in the data center by shunting workload to a minimum number of compute nodes, then switching off unused nodes. The algorithm used the utilization of system resources to help determine the proper mix of nodes left active and nodes switched off. The results showed that the reconfiguration of the system is time consuming due to the time required to switch off nodes and the time to process the optimization algorithm. Liu, et al. [5] proposed a system architecture that monitors the average CPU utilization of the compute nodes and performed live migration of the VMs to achieve a more optimal running configuration. In this approach, energy conservation for the data center was achieved by switching unused nodes off. One concern with this approach is that the authors did not take into account the time latency of cycling systems all the way to an off state, nor did they consider the thermal and mechanical degradation which results from thermal cycling of the system components. Optimization algorithms have been researched that do take the decrease of reliability due to thermal cycling into account. Deng, et al. [6], presented power-efficient resource management in CDCs with the use of a reliability aware VM consolidation strategy. In this research, utility models were used to optimize both the short term and the long term costs of making each consolidation decision. Parameters such as number of on-off thermal cycles, SLA performance violations and the cost of performing VM migration were used in a multi-objective genetic algorithm to determine the optimal placement decision for VMs. The approach fails to consider the time-varying nature of server state changes and VM migration periods.

1.2 Optimum Cloud Provisioning Method

The ideal state of operation for CDCs is for systems to be provisioned exactly when they will be needed to service the incoming user requests, and then taken offline when the request load diminishes.

Fig. 1 illustrates the desired CDC server provisioning policy for an arbitrary server load. The dashed line represents an actual expected capacity, or incoming workload. Traditional provisioning methods allocate new capacity based on a maximum predicted value, plus a buffer amount. This causes unused capacity to remain online, and therefore consuming more energy than actually required by the expected workload.

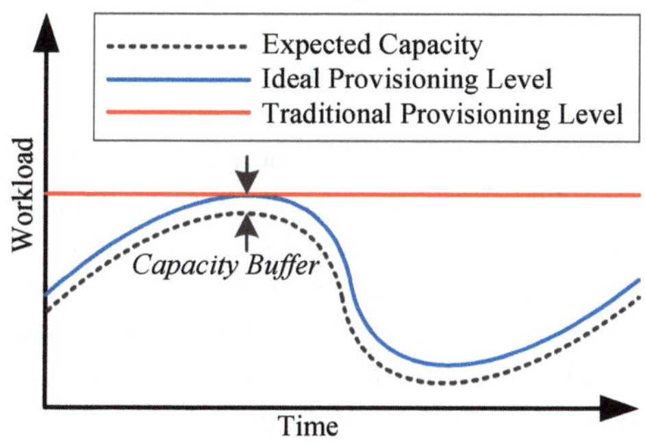

Fig. 1. Ideal cloud capacity provisioning

It is recognized that the process of calculating the optimal number of compute nodes required adds additional workload to the compute nodes. The determination of the optimal number of systems required should also be made in a manner that minimizes the impact to the system of performing the necessary calculations. The total impact on the system due to calculations is the product of the number of time the calculations are performed during a time interval, or time horizon (***T***), and the average time it takes for the calculations to complete. The time required for the required calculations to take place becomes the minimum time horizon due to the fact that performing the calculations more frequently would result in the calculations not completing in time to run the calculations for the next time horizon. The time horizon has another lower-bounds constraint relating to the amount of time required to change the compute nodes state from active to sleep and from sleep to active. If a time horizon is chosen that is smaller than the state change latency time, an inadequate number of compute nodes will potentially be available to properly handle the expected workload. The upper-bounds constraint on the time horizon is based on the dynamic characteristic of the expected workload. If the workload is relatively constant, larger prediction windows can be used without sacrificing accuracy. As workloads become more dynamic, a smaller prediction window is necessary to determine with accuracy the expected workload.

1.3 Research Contributions and Paper Outline

In this paper, we will present a new method of determining both the optimal prediction time horizon, ***T***, and the optimal number of prediction quantization levels to be used in the larger VM placement algorithm problem. Fig. 2 shows the overall VM placement energy-aware algorithm.

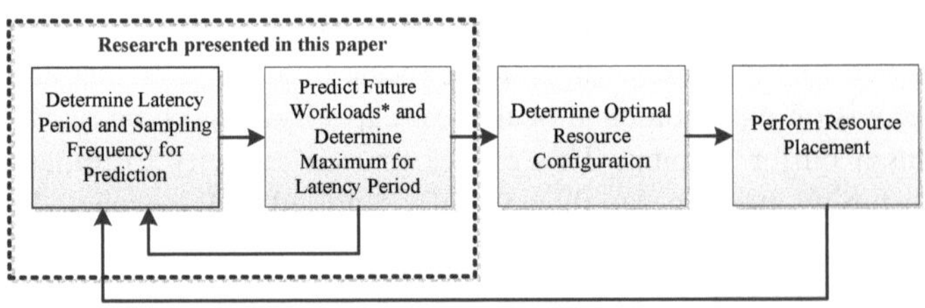

Fig. 2. Algorithm sequence for energy aware systems optimization

The red-dashed line indicates the research being presented in this paper. The first step in our algorithm is to determine the expected state change latency period. In order to properly understand the nature of server state-change times, an experiment will be presented that empirically measures the time it takes for servers to go from active to sleep and then from sleep back to active. From this experiment, a probability density function will be derived that models the experimental data. This model will be used in the later algorithm to provide a method of determining the optimal prediction time horizon. Utility functions which describe the costs of SLA violations, over and under allocation of system capacity and the excess of energy consumed by the overhead of calculating the expected load are used to determine the optimal frequency of prediction. Finally, a linear filter is used to perform the prediction for the prior determined sample intervals, and the maximum expected load is determined. This information will be feed into the VM placement optimization algorithm, which will be described in later research.

The rest of this paper is organized as follows. Section 2 addresses the nature of state change for compute nodes. This section presents experimental data that is used to determine a stochastic model for determining the proper time horizon, given a desired SLA. Section 3 describes a dynamic quantization model for determining the optimal number of times the prediction algorithm should be used to determine the maximum load expected in the next time horizon. Section 4 describes the use of an AR linear filter (Weiner filter) to predict the expected load. The results of all the simulations in the paper are presented in Section 5. The results of the paper are then summarized and the paper is concluded in Section 6.

2 Stochastic State Change Model

One of the core attributes affiliated with cloud data centers is the ability to respond to changes in the incoming requests via scaling. Cloud systems scale up when loads increase and scale down when loads decrease. This paper leverages this work profile, thereby allowing the introduction of a stochastic state-change latency model.

We start with a set of VMs, $\boldsymbol{V}=\{v_1,v_2,...v_M\}$ with M being the number VMs, and a set of PMs, $\boldsymbol{P}=\{p_1, p_2,...p_N\}$ with N being the number of PMs. An instance of a VM is represented as $\boldsymbol{v_i}$ and an instance of PM is represented as $\boldsymbol{p_j}$. The number of PMs under control is N and is also referred to as a "huddle". We let $\boldsymbol{t}$ represent the current time. The primary assumption is that each PM's state-change time can be represented as a probability distribution. The time horizon, $\boldsymbol{T}$, is chosen so the probability of the actual state change Prediction of the future workload occurs at a time horizon beyond the next period horizon.

The optimal VM placement algorithm starts with a determination of the proper time horizon, $\boldsymbol{T}$, in which to predict the maximum expected future load, $\boldsymbol{L}$. The time horizon is dependent on an understanding of the time it takes to allow servers to change from their original state **{Sleep, Active}** to a next desired state. If a chosen time horizon is less than the time it takes for servers to change state, then we will not have our expected number of servers online and ready to process the request load. If we make the time horizon too large, then we risk over allocating systems because we are not following the time-varying nature of the request load with our system allocation decisions. Fig. 3 shows the relationship between the actual load, the allocated capacity and the time horizon, $\boldsymbol{T}$. The graph shows that in each time horizon interval, the maximum predicted load (denoted by the red x) determines the future allocated capacity. It can be readily seen that as $\boldsymbol{T}$ approaches zero, our predicted load approaches our actual load. We therefore desire to reset our allocated capacity as frequently as the server state-change allows.

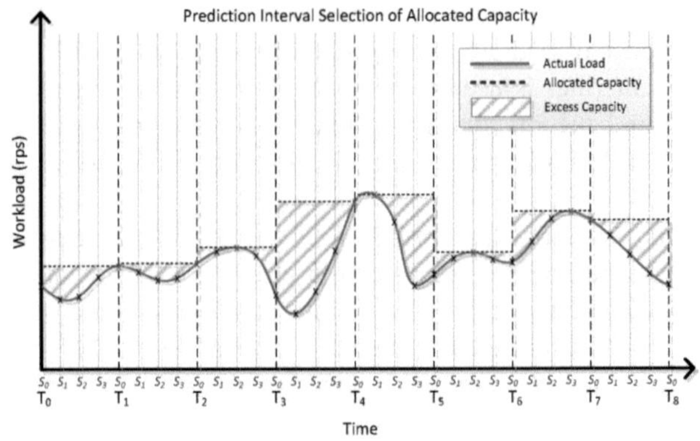

Fig. 3. Time horizon and prediction period relationship to prediction of the required capacity

We represent the server state change times as a random variable $\boldsymbol{X}$, with $\boldsymbol{x}_j$ representing the individual state change for the jth physical host. The physical state for a server is designated as k, and in our case is the set *{Active, Sleep}*. We can therefore represent a specific state change for an instance of a physical server as $\boldsymbol{x}_{jk}$.

The probability density function, $\boldsymbol{f_X(x)}$, for our random variable $\boldsymbol{X}$ needs to be determined in order to effectively model the state change time and therefore choose an appropriate new $\boldsymbol{T}$. We assume when server configurations are static, and have little or no request load, they will have minimal variance in the state-change times. The dynamic nature of the cloud allows for variance across servers and across time. As the needs of the cloud evolve to deal with the incoming request workloads, the services hosted on the PMs will change due to the need to host new VMs or to create VMs on a new PM.

To determine the stochastic model of state-change latency, we employ empirically observation to determine the times for changing servers from Active-to-Sleep and then from Sleep-to-Active.

2.1 Determination of Probability Distribution Function

We began with four Dell workstations, three were used as remote clients and one was used as the master controller. Each of the Dell workstations had a unique set of services configured (web server, email server and SQL database). The Ubuntu 11.10 Linux distribution was installed across all four systems. A Bash script was used to connect to each computer and issue Sleep and Wake commands in a cycle. The time taken between each command state was recorded in a local file. The script ran through 1000 Active-Sleep and Sleep-Active cycles on each machine. The data verified that, for the same system with no modification to the configuration, the state-change time remained static. However, each of the systems experienced different state-change times. We then reconfigured each system by adding a new service along with the existing service. The procedure was then performed again and the results recorded. This was repeated for ten times on each of the three client machines. The results indicated that as long as the configuration of the machine was modified, the time taken to go from active to sleep then sleep to active varied. Each sample set mean was then examined by plotting in a histogram. From the experimental results, the gamma probability distribution function, pdf, was determined to be the best fit pdf for the observed data and agreed with the theoretical models of server state change latency timings. Fig. 4 shows the histogram representing the experimental runs of waking an experimental server, Fig. 4a. Fig. 4b represents the resulting gamma pdf model created with the Matlab® gamfit function. The gamma function parameters used for the pdf model shown in Fig. 4b was α of 23.4543 and β of 0.1741. In production settings, these values would be updated based on actual observed state change times. It is noted that the state-time also increases as new services or VMs are added to the system, and correspondingly decreases as services or VMs are removed.

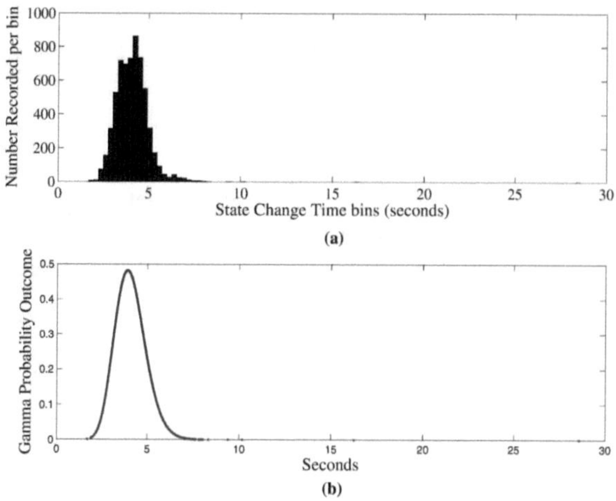

Fig. 4. Histogram of experimental state change time for servers (a) and the resulting gamma pdf model (b)

The gamma probability density function for $f_X(x)$ is presented below [7].

$$f_X(x) = \begin{cases} \frac{x^{\alpha-1}}{\Gamma(\alpha)\beta^{\alpha}} e^{-x/\beta}, \; x \geq 0 \\ 0, \; otherwise \end{cases} \quad (1)$$

Where Γ(α) represents the gamma function defined as

$$\Gamma(\alpha) = \int_0^{\infty} x^{\alpha-1} e^{-x} dx \quad (2)$$

2.2 Definition of the SLA Cost Function

With a stochastic model for determining the proper time horizon from the state-change statistics, we can now implement a cost function that will potentially modify our estimated time horizon with an appropriate buffer amount. CDC operators typically provide an uptime guarantee to their users that the VMs will be available for processing requests with a minimum probability defined as the SLA. We define α_{sla} as the maximum allowed failure probability for a given SLA, or the SLA failure threshold. We define $\boldsymbol{P_{TH}}$, as a condition on $\boldsymbol{T}$, as the probability that the actual maximum state change time for our systems exceeds the allowed time horizon.

$$T[n]=\hat{T}[n+1] \tag{3}$$

Where,

$$\begin{aligned} P_{TH} &= Probability\left[Max\left(\boldsymbol{x}_{jk}\right)>T[n]\right]\leq\alpha_{sla} \\ &\forall\boldsymbol{x}_j\in N,\ k=0,1 \end{aligned} \tag{4}$$

We now define a cost function that can be used to determine when and how we need to update our time horizon value. The cost function for the time horizon implements a monetary penalty, ***sla_$***, for failing to make the SLA failure threshold.

The cost function for time horizon violation, $\boldsymbol{C_{TH}}$, is defined as:

$$C_{TH}\left(\hat{T}\right)=\sum_{j=1}^{N}C_{TH}\left(p_j\right) \tag{5}$$

Where,

$$C_{TH}\left(p_j\right)=\begin{cases}0, & if\ P_{TH}\leq\alpha_{sla}\\ sla\_\$, & otherwise\end{cases} \tag{6}$$

The term $\boldsymbol{C_{TH}}$ will cause a buffer, $\boldsymbol{B_{TH}}$, to be added to the predicted time horizon from the state change model presented above. When the cost is zero, the amount of $\boldsymbol{B_{TH}}$ is zero, when the cost is greater than zero, the amount the buffer will add is $\boldsymbol{\beta_{TH}}$ of the estimated $\boldsymbol{T}$ value. In practice, the value of $\boldsymbol{\beta_{TH}}$ is determined by the CDC operator and is adjusted to keep the buffered value plus the estimated time horizon equal to the average state change time for the CDC resources. The buffer calculation is presented below.

$$B_{TH}=\begin{cases}\beta_{TH}*\hat{T}, & sla\_\$>0\\ 0, & otherwise\end{cases} \tag{7}$$

Where

$$0\leq\beta_{TH}\leq1$$

The estimated time horizon for the next period then becomes the updated time horizon plus the calculated buffer, B_{TH}.

$$\hat{T} = \hat{T} + B_{TH} \tag{8}$$

3 Dynamic Quantization Model

3.1 Determination of Cost Functions

This section continues the energy optimization algorithm by presenting two cost functions, one for the cost of the number of calculations performed and one for the difference in the estimated workload and the actual workload. Now that an estimate of the proper time horizon has been made, we can turn our attention to the determination of the number of quantization levels for prediction, η.

The time horizon under consideration is always the next future time horizon interval, which is estimated based on the state-change time statistics maintained for each of the PMs in the huddle as presented in Section 2. As illustrated in Fig. 3, the prediction interval, S, is defined as the prediction sample period and is determined by dividing the time horizon, T, by the number of quantization levels, η as shown in (9).

$$S = \frac{T[n]}{\eta} \tag{9}$$

Cost functions are used to perform the multi-objective minimization optimization of the time horizon and quantization level model. The two cost functions that are considered are number of calculations incurred, and the error in the estimated workload minus the actual workload required to service the user requests with the designated number of active PMs.

$$C_C(\hat{T}) \stackrel{\text{def}}{=} Cost\ function\ for\ calculations\ required$$
$$C_L(\hat{T}) \stackrel{\text{def}}{=} Cost\ function\ for\ workload\ error$$

We define the actual workload done by the system as L. The load prediction for the next time horizon is defined as $\hat{L}$. The workload prediction is performed with a Weiner Filter, which is discussed in detail in section 4. The workload prediction is made for the interval from $T[n+1]$ to $T[n+2]$. The system performs the prediction calculation η times during the prediction interval T, therefore as η increases so does the calculation overhead incurred by the system. The range for η is from 1 to the prediction time horizon divided by the minimum required processing time, ν, and is shown in (11).

$$1 \le \eta \le \frac{\widehat{T}}{\nu} \tag{10}$$

We can therefore define a linear cost function, C_c, that is based on the current value of η as,

$$C_C\left(p_j, \eta, T, v\right) = R_j * calc\_\$ \tag{11}$$

Where R is the linear function defined as,

$$R = \left(\frac{1}{T/v}\right)\eta - \left(\frac{1}{T/v}\right) \tag{12}$$

For example, when $\eta=1$, then $R=0$, meaning that when we are at the lower limit of η we have no calculation penalty. When $\eta=T/v$, then $R=1$, meaning when η is at the upper limit we incur the maximum calculation penalty. The basis of the calculation penalty is set to ***calc_$***, and is determined by the CDC operator.

The next cost function captures the error in the estimated workload estimated versus the actual workload. We define the workload error as the difference between the estimated workload and the actual workload.

$$\tilde{L} = \hat{L} - L \tag{13}$$

The probability that our predicted workload is at least as large as our expected workload, meaning we do not have under allocated our systems, is:

$$P_L \equiv \text{Prob}\left[\tilde{L} \geq 0\right] \leq \alpha_{SLA} \tag{14}$$

We can now define our cost function, C_L, as zero when we are not under allocated and ***load_$*** when we are under allocated.

$$C_L = \begin{cases} 0, & P_L \leq \alpha_{SLA} \\ load\_\$, & \text{otherwise} \end{cases} \tag{15}$$

3.2 Prediction Frequency Calculation

With the creation the cost functions we can now use fuzzy logic to determine the actual next value of η.

A fuzzy inference system was developed that uses two inputs and one output [8]. The fuzzy input sets defined are the expected cost of the number of calculations required, C_C, and the cost of workload prediction error, C_L. The output of the fuzzy inference engine is $\Delta\eta$. Fig. 5 shows the relationship between the input and output fuzzy sets.

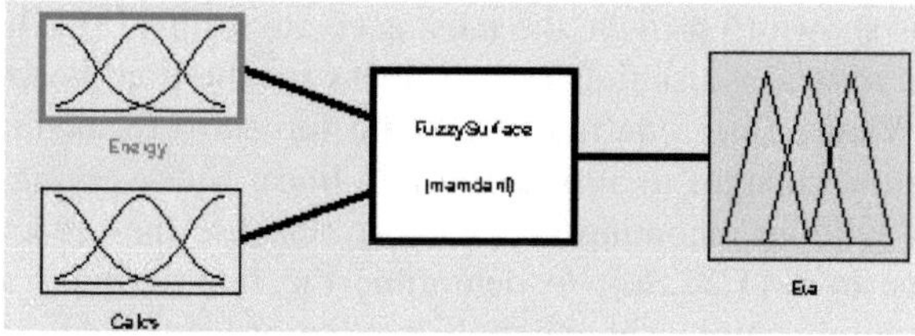

Fig. 5. Fuzzy input and output sets

Each fuzzy set is described by corresponding fuzzy variables, and the sets combined with fuzzy rules to create the fuzzy inference engine. The surface map for the fuzzy rules is shown in Fig. 6.

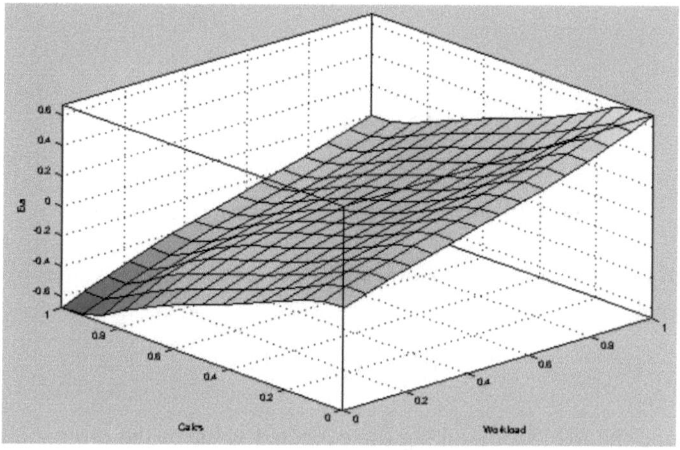

Fig. 6. Fuzzy rules surface map for determining $\Delta\eta$

The fuzzy inference engine outputs the updated value for $\boldsymbol{\Delta\eta}$ with each iteration of the optimization algorithm. The new value for $\boldsymbol{\eta}$ then is defined as the prior $\boldsymbol{\eta}$ plus the calculated change in $\boldsymbol{\eta}$.

$$\eta[n+1]=\eta[n]+\Delta\eta,\ 1\leq\eta\leq\frac{T}{v}. \tag{16}$$

4 Linear Prediction Model

With the values of $\boldsymbol{T}$ and $\boldsymbol{\eta}$ determined, we can now employ our prediction algorithm to determine the future workload at the determined time points.

Prediction is performed by implementing a discrete autoregressive Wiener Filter. This and other methods were examined in prior research by the author, and the

Weiner Filter was shown to provide the most accurate results [8]. The network traffic is assumed to be wide sense stationary and linear for the time horizons used for the prediction. The Wiener filter coefficients, ***a***, are forced to recalculate each time the system configuration changes to allow for a non-linear time-varying signal. The goal of the filter is to take an incoming signal, ***x[n]***, and use the predetermined Wiener Filter coefficients, $\boldsymbol{a_i}$, **i={1,2,..,K}**, to determine the future signal, ***x[n+θ]***, where θ represents the future time points at which the predictions are made.

$$\hat{x}[n+\theta]=\sum_{i=0}^{K} a_i x[n-i] \tag{17}$$

The coefficients of the filter are determined by minimizing the mean square (MMSE) error of the system shown in Fig. 7.

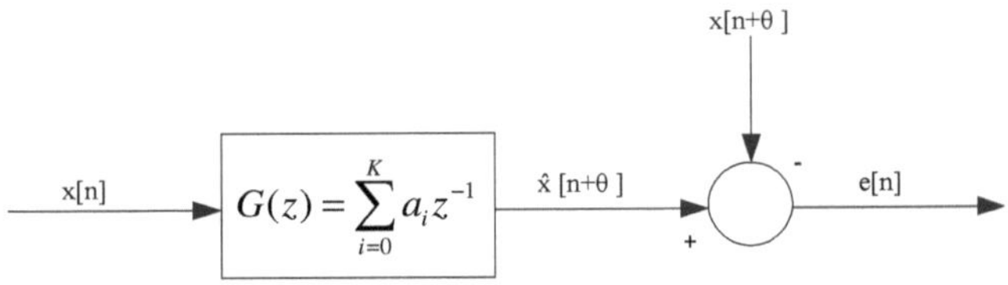

Fig. 7. Block diagram for linear filter load prediction

To simplify the variables, we can let **s[n]** represent the actual values of our predicted future signal.

$$s[n]=x[n+\theta] \tag{18}$$

We can find the MMSE by taking the expected value of the error squared, and expanding the terms.

$$E\{e^2[n]\}=E\{(s[n]-\hat{x}[n+\theta])^2\} \tag{19}$$

$$\begin{aligned} E\{e^2[n]\} &= E\{s^2[n]\}+E\{\hat{x}^2[x+\theta]\}- \\ &2E\{s[n]\hat{x}[x+\theta]\} \end{aligned} \tag{20}$$

$$\begin{aligned} E\{e^2[n]\} &= E\left\{\left(\sum_{i=0}^{K} a_i x[n-i]\right)^2\right\}+ \\ &E\{s^2[n]\}-2E\left\{\sum_{i=0}^{K} a_i x[n-i]s[n]\right\} \end{aligned} \tag{21}$$

In order to determine the correct set of $\boldsymbol{a_i}$, we will take the partial derivative with respect to $\boldsymbol{a_i}$, and set it to zero.

$$\frac{\partial}{\partial a_i} E\{e^2[n]\} = 2E\left\{\left(\sum_{j=0}^{K} a_j x[w-j]\right) x[n-i]\right\} - 2E\{s[n]x[n-i]\} \tag{22}$$

$$\frac{\partial}{\partial a_i} E\{e^2[n]\} = 2\sum_{j=0}^{K} E\{x[n-j]x[n-i]\} a_j - 2E\{x[n-i]s[n]\} \tag{23}$$

We recognize that if **x[n]** and **s[n]** are both jointly stationary (over our time horizon, $\boldsymbol{T}$), we can define an autocorrelation function, $\mathbf{R_x[m]}$, and cross-correlation function, $\mathbf{R_{xx}[m]}$, respectively as the following:

$$R_x[m] = E\{x[n]x[n+m]\} \tag{24}$$

$$R_{xs}[m] = E\{x[n]s[n+m]\} \tag{25}$$

Substituting into (23) gives us,

$$\frac{\partial}{\partial a_i} E\{e^2[n]\} = 2\sum_{j=0}^{K} R_x[j-i]a_j - 2R_{xs}[i] = 0 \tag{26}$$

which leads to the the Weiner-Hopf equation,

$$\sum_{j=0}^{K} R_x[j-i]a_j = R_{xs}[i]. \tag{27}$$

Equation (27) can now be rewritten in matrix form, **Xa=y** as:

$$\begin{bmatrix} R_x[0] & R_x[1] & \cdots & R_x[K] \\ R_x[1] & R_x[0] & \cdots & R_x[K-1] \\ \vdots & \vdots & \ddots & \vdots \\ R_x[K] & R_x[K-1] & \cdots & R_x[0] \end{bmatrix} \begin{bmatrix} a_0 \\ a_1 \\ \vdots \\ a_K \end{bmatrix} = \begin{bmatrix} R_{xs}[0] \\ R_{xs}[1] \\ \vdots \\ R_{xs}[K] \end{bmatrix} \tag{28}$$

The solution for the $\boldsymbol{a_i}$ coefficients then becomes $\boldsymbol{a=X^{-1}y}$, which can be solved using various techniques, depending on whether or not $\boldsymbol{R}$ is directly invertible (positive definite). In practice, we use a sample of known input-output exemplars to train our system and determine the coefficients. This process is repeated periodically to update our coefficients with new values, therefore allowing our system to change over time. [9] [10]

Once we have calculated the coefficients, we can use (28) to determine the future value of the signal given the current value. Prediction is performed η times, which each run S seconds apart. Each run produces a prediction value of the workload, $\hat{L}$. The final predicted workload then becomes the maximum of the η predictions taken during the prediction time horizon.

$$\begin{gathered}\hat{L}_{ij}\left[T\right]=\max\left(\hat{L}_{ij}\left(S_k\right)\right)\\ \forall v_i\in M,\ \forall p_j\in N,\ k=1\ to\ \eta\end{gathered}\tag{29}$$

A thee point moving average filter is applied to the predicted load. If the average predicted value is less than the actual value, an amount is added to the load prediction buffer, B_P. When the average predicted values fall within an acceptable threshold, the buffer amount is removed and the predicted value is used unaltered. This allows the system to respond to changes not captured in the predictive Wiener Filter to not have detrimental effects on the ability of the cloud to service incoming requests.

$$\hat{L}[T]=\begin{cases}\hat{L}[T]+B_P, & avg\left(\tilde{L}[t]\right)>0\\ \hat{L}[T], & otherwise\end{cases}\tag{30}$$

5 Simulation and Results

In order to simulate the model described in this paper, Matlab© was used to test input data taken from data published on a NASA web site. The data represents user requests for page access on a web server at NASA over a period of one month [11]. 10,000 data points were taken from the data set and used in the simulation.

The procedure started with an estimation of the Time Horizon. The gamma distribution presented in Section 2 was used to predict the next time horizon with a probability of success equal to an arbitrary input SLA. Fig. 8 shows the results of a statistical random number generator based on the gamma probability distribution model for server state-change latency times. This data was stored as an array for the duration of the experiment as the predicted state change times. The predicted state change times were then used for each run of the algorithm. This allowed a consistent time horizon to be used for each of the simulated models, ensuring a relevant comparison for each Monte Carlo run.

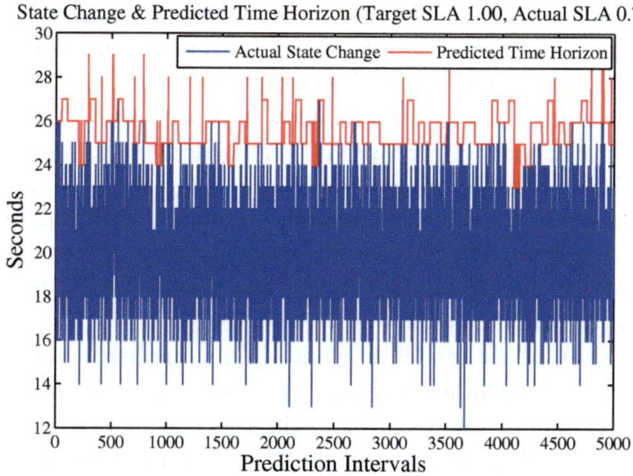

Fig. 8. Time Horizon Model Used During Simulation

If the predicted Time Horizon was found to be less than the actual Time Horizon, a buffer was added to the predicted value of the next Time Horizon. Fig. 8 shows the results of predicting the workload obtained from NASA using the stochastic time horizon presented in Fig. 7. For each time horizon, the simulation estimated the max workload using the prior value of η. The error signal was then generated and the percentage of SLA violations determined. The cost functions where then used with the fuzzy logic inference engine to determine the next value of η. This value was then stored and used as the η value for the next run.

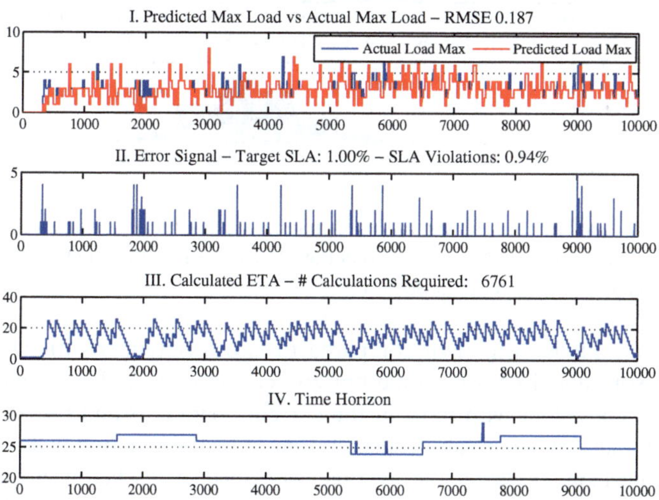

Fig. 9. Stochastic Time Horizon with Dynamic Prediction Frequency Model Results. Graph I shows the predicted vs. actual loads. Graph II displays the squared error. Graph III shows how η is recalculated to maintain the required SLA. Graph IV shows the predicted time horizon values used for the simulation.

The first sub-plot in Fig. 8 shows the actual maximum workload experienced during each time horizon compared with the predicted maximum workload for that time horizon. The second sub-plot in Fig. 8 shows the error signal, which was calculated as the actual max workload minus the predicted max workload. Sub-plot 3 of Fig. 9 displays how η (# of sampling intervals) is recalculated by the algorithm in response to the cost functions presented in section 3. The last graph shows the time horizon used during the simulation. For each run in the simulation, 10K data points were used from the NASA workload. The SLA used for each simulation was 99% success, 1% fail. Fig. 8 shows that the SLA failure rate was actually 0.94%. For this simulation run, it took 6,761 prediction calculations.

To test the stability of the model, a Monte-Carlo simulation was performed using 10,000 iterations. Each iteration simulated server state times by randomly generating a signal using a gamma distribution. The results of the Monte Carlo simulation for the model are presented in Fig. 10. The 10,000 simulations show that SLA violations occurred on average 1.0% of the time. This was equal to our target SLA. The number of calculations required had a mean value of 6,833.

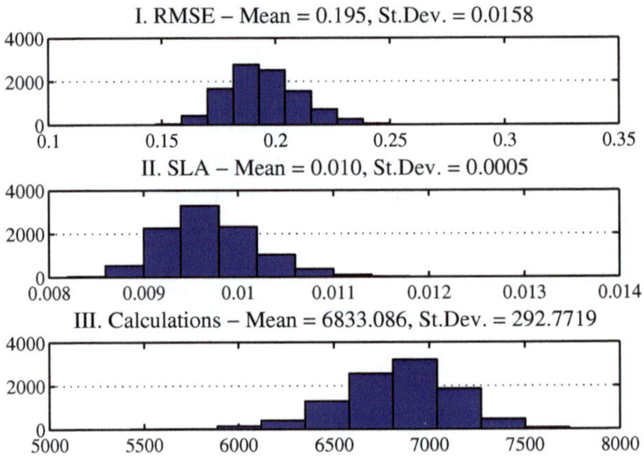

Fig. 10. Monte Carlo Simulation results for Dynamic Model. Graph I shows the RMSE values for the simulations. Graph II shows the percentage of SLA Violations incurred; Graph II shows the number of calculations needed based on η.

The dynamic model was then compared to two different traditional models. The first comparison model held the time horizon constant at the largest expected value. This model is consistent with most of the current operations and research seen in the cloud hosting industry where any consideration of load prediction is done with a fixed period time horizon, or when the average workload drops below a threshold value [12].The only deviation from industry norm was allowing η to be modified with the optimization algorithm in this paper. Fig. 11 presents the actual state change times of a cloud huddle and shows that the chosen static time horizon is the maximum of the state change times seen historically.

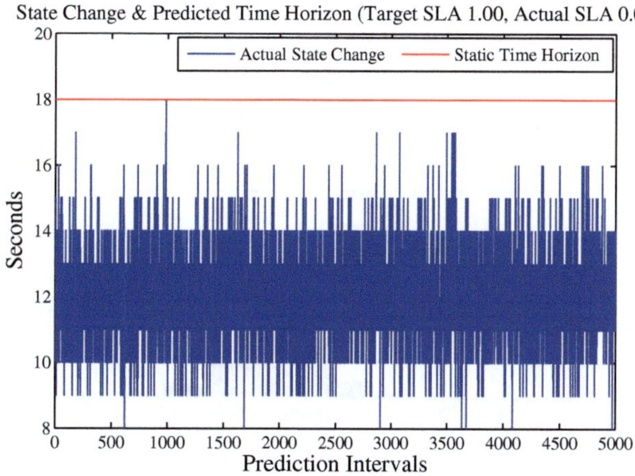

Fig. 11. Determination of the Static Time Horizon

The results of the initial static time horizon simulation are presented in Fig. 12. This simulation had a RMSE of 0.176, SLA violations occurred .86% of the time, and required 7,734 calculations.

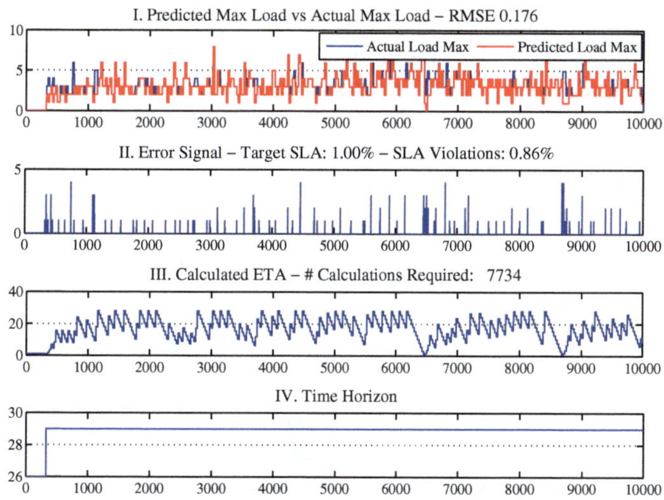

Fig. 12. Static Time Horizon Prediction Results. Graph I shows the predicted vs. actual loads. Graph II displays the squared error. Graph III shows how η is recalculated to maintain the required SLA. Graph IV shows the static time horizon used throughout the simulation.

A Monte Carlo simulation of 10,000 iterations based on the static time horizon approach yielded a mean SLA violation of .8% but required 8,200 calculations, on average. This represented an increase in the mean number of calculations by 20.0%. These results are presented in Fig. 13.

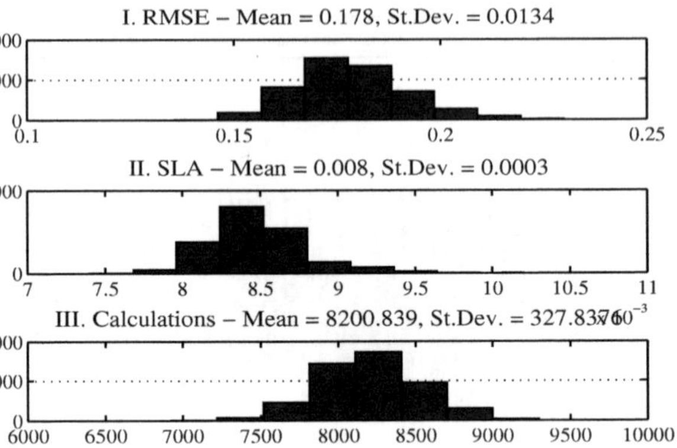

Fig. 13. Static Time Horizon with Dynamic η. Graph I shows the RMSE values for the simulations. Graph II shows the percentage of SLA Violations incurred; Graph II shows the number of calculations needed based on η.

The next model considered for comparison with the dynamic model presented in this paper keeps the Time Horizon static but fixes ***η*** at the maximum allowed value, which is ***T/v***, (time horizon/calculation time). Since v in our case is 1 second, the value of ***η*** becomes the time horizon. The effect of this is to perform prediction once a second for the duration of the time horizon. Since we are including all possible prediction intervals, we end up never under-predicting our workload estimate for the time horizon. This comes at the expense of overhead. The number of calculations required rises to 11,573 which are 69.3% higher than the fully dynamic model presented in the paper. Fig. 14 presents the results of the static time horizon using max ***η*** model.

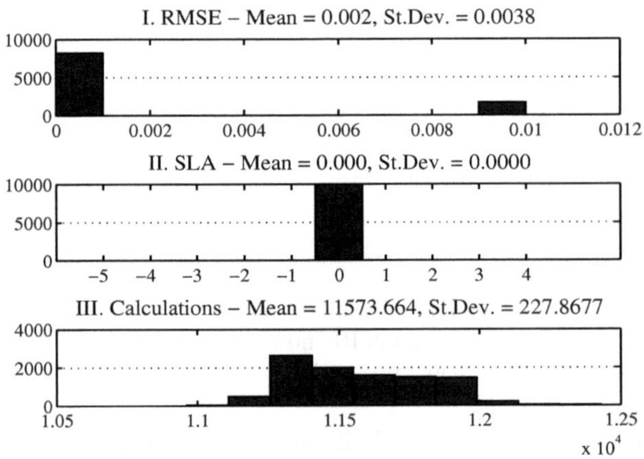

Fig. 14. Static Time Horizon with Max η. Graph I shows the RMSE values for the simulations. Graph II shows the percentage of SLA Violations incurred; Graph II shows the number of calculations needed based on η.

The last model considered for comparison uses a static time horizon but sets η to the minimum allowed value, which is 1. This approach calculates the predicted workload only once each Time Horizon, T. This results in the mean SLA violation of 3.4%, which are 2.4X the mean SLA violation achieved by the fully-dynamic model. The consolation is that we reduce our calculation overhead to a mean of 787. (Fig. 15).

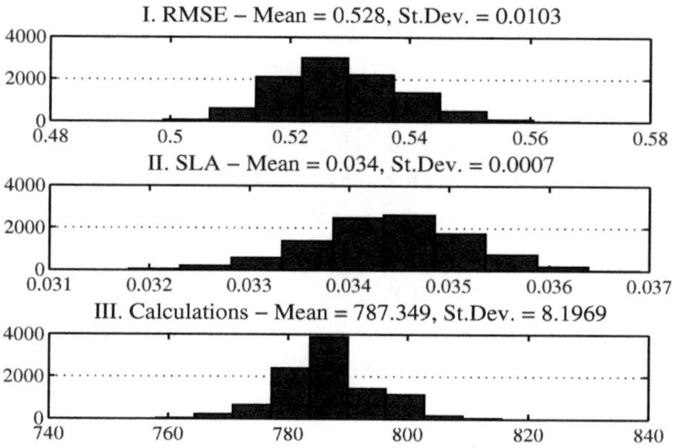

Fig. 15. Static Time Horizon with Min η. Graph I shows the RMSE values for the simulations. Graph II shows the percentage of SLA Violations incurred; Graph II shows the number of calculations needed based on η.

In summary, the fully-dynamic model presented in this paper predicted the workload with a RMSE value of 0.195 and had SLA violations of 1.0%, which was equal to the established target SLA. The number of calculations required to accomplish this was 6,833. These results were then compared to three other models.

The first model kept η dynamic, but set the Time Horizon to the maximum value seen of server state-change latency times. This model had a RMSE of 0.178 with SLA violations of .8%. The number of calculations required increased to 8,200, which represented a 20.0% rise.

The last two models also kept the time horizon fixed at the maximum value of the state-change latency but also fixed η. The first model set η to the maximum allowed value where the second model set η to the minimum allowed value. The first model had a RMSE of 0.002 and a mean SLA violation of effectively 0.0%. This came at a cost in increased calculations, which rose to a maximum seen value of 11,573. The second model saw an increase in mean RMSE to 0.528 and a corresponding increase in mean SLA to a value of 3.4%. Since we only perform prediction in this model once per time horizon, we saw the mean number of calculations drop to a lowest observed 787.

Table 1. Listing of Variables, Constants and Functions

Symbol	Description	Units
T	time horizon for prediction of expected workload	second (s)
L	workload of incoming requests for processing	requests/s
S	time interval for calculating workload prediction	s
X	server state change time latency random variable	s
P_{TH}	probability of predicted time horizon violating SLA terms	
V	set of available virtual machines	
P	set of available physical host machines	
α_{sla}	SLA violation threshold	
C_{TH}	cost of time horizon violating SLA terms	$
B_{TH}	buffer value added to estimated time horizon	s
β_{TH}	relative weight of buffer value as percentage of estimated time horizon	
η	number of predictions made of S length during each time horizon	
C_C	cost function for number of calculations required	$
C_L	cost function for under allocating the estimated load	$
ν	Minimum time required to perform one prediction calculation	s
R	function describing effect of η on C_C	
calc_$	constant used to determine value of calculation penalty	$
P_L	probability of under allocating estimated load	
load_$	constant used to apply penalty to under allocating load	$
a_i	Weiner filter coefficients	
x[n]	signal representing current time data	requests/s
x[n+θ]	signal representing future time data	requests/s
s[n]	variable change of x[n+θ] to simplify presentation	requests/s
θ	future time offset	s
e[n]	error signal, s[n] – x[n]	requests/s
$R_x[]$	autocorrelation function	
$R_{xs}[]$	cross-correlation function	
B_p	predicted load error buffer value	requests/s

6 Conclusion

The goal of this paper was to provide an energy aware algorithm for the prediction of cloud workloads. Our results showed that the algorithm presented in this paper required, on average, 6833.086 calculations to estimate the required workload within the tolerances of our SLA, which was 1.0% in the simulation. Using a static time horizon with a dynamically determined η required 8200.839 calculations to keep the SLA violations under our given SLA of 1.0%. This represents approximately a 20.0% increase in the number of calculations required. The third scenario held the time horizon constant but set η at the maximum allowed value. This resulted in the largest number of calculations seen, 11,573.664 on average. The benefit to this approach is that we achieved zero SLA violations. The forth simulated scenario resulted in the fewest number of calculations (787.349) but was rendered invalid due to having an SLA violation percentage of 3.4%, roughly 3X our maximum SLA violation threshold of 1.0%.

The optimal model was the use of a static time horizon along with dynamically determining the proper number of predictions to occur within each time horizon. This resulted in adding the fewest additional calculations onto our system while assuring that all required SLA conditions will be met.

Acknowledgments. This work was supported in part by the Lutcher Brown Endowed Chair at the University of Texas at San Antonio and the 80/20 Foundation. Special thanks to Jarret Hornsby who assisted the first author in obtaining the experimental data presented in Section 2.

References

[1] Koomey, J.G.: Estimating Total Power Consumption by Servers in the U.S. and the World. Lawrence Berkely National Laboratory, Berkley (2007)

[2] Hamilton, J.: Cooperative Expendable Micro-Slice Servers (CEMS): Low Cost, Low Power Servers for Internet-Scale Services. Amazon (2009)

[3] : D. Dharwar, S. S. Bhat, V. Srinivasan, D. Sarma and P. K. Banerjee. In: 2012 IEE International Conference on Cloud Computing in Emerging Markets (2012)

[4] Cardosa, M., Korupolu, M., Singh, A.: Shares and utilities based power consolidation in virtualized server environments. In: Proceedings of the 11th IFIP/IEEE Integrated Network Management, Long Island, NY (2009)

[5] Liu, L., Wnag, H., Liu, X., He, W., Wang, Q.: Green Cloud: a new architecture for a green data center. ACM, New York (2009)

[6] Deng, W., Liu, F., Jin, H., Liao, X., Liu, H., Chen, L.: Lifetime or energy: Consolidating Servers with Reliability Control in Virtualized Cloud Data Centers. In: 4th International Conference on Cloud Computing Technology and Science (CloudCom), Taipei (2012)

[7] Papoulis, A., Pillai, S.U.: Chapter 4 The Concept of a Random Variable. In: Probability, Random Variables and Stochastic Processes, 4th edn., pp. 72–119. McGraw-Hill, New York (2002)

[8] Jamshidi, M.: Chapter 7, in Large-Scale Systmes: Modeling, Control and Fuzzy Logic. Prentice Hall, Saddle River (1997)

[9] Moon, T., Stirling, W.: Wiener Filters, in Mathematical Methods and Algorithms. Prentice Hall, New Jersey (2000)

[10] Prevost, J.J., Nagothu, K., Kelley, B., Jamshidi, M.: Prediction of cloud data center networks loads using stochastic and neural models. In: 6th International Conference on System of Systems Engineering (SoSE), Albuquerque (2011)

[11] Arlett, M., Williamson, C.: NASA (May 1996), http://ita.ee.lbl.gov/html/contrib/NASA-HTTP.html (accessed April 17, 2013)
[12] Feller, E., Rohr, C., Margery, D., Morin, C.: Energy Management in IaaS Clouds: A Holistic Approach, Rennes, France. INRIA Centre Renes (2012)
[13] Jung, G., Hiltunen, M.A., Joshi, K.R., Schlichting, R.D., Pu, C.: Mistral: Dynamically Managing Power, Performance, and Adaption Cost in Cloud Infrastructures. In: Procedings of ICDCS, Genova (2010)

Behaving Nicely in a System of Systems – Harmonising with the Environment

Michael Henshaw, Carys Siemieniuch, and Murray Sinclair

Engineering Systems of Systems Group, Loughborough University, UK
{m.j.d.henshaw,c.e.siemieniuch,m.a.sinclair}@lboro.ac.uk

Abstract. Systems of Systems (SoS) Engineering has emerged as an important area of study; as the interconnectedness of systems rapidly increases, new challenges are arising that are of global impact. A major difficulty in SoS is that individual systems operators tend to maximize the performance of their own systems at the expense of the overall performance of the SoS. In this chapter a set of global drivers for encouraging "nice behavior" are identified. The results of a European project to identify a strategic research agenda in SoS are presented and the relationships of the research themes to the global drivers are explained.

1 Introduction

"It is less expensive to protect the planet now than to repair it later", said Mr. Barroso [1], President of the European Commission in March 2010. The rapid increase in connectivity between various systems creates the opportunity to damage the planet through increased energy demands and economically driven depletion of resources but it also provides the opportunity for better management of resources if engineers and others can exploit the potential of systems of systems. In this article the research agenda collaboratively created by Systems of Systems engineers in Europe and the USA will be described. Its contribution to understanding how technological systems can be better harmonized with the needs of the environment will be discussed. It will become apparent through this discussion that Systems of Systems Engineering (SoSE) must involve a wide range of disciplines and that attention to the social and political aspects are at least, and possibly more, important than the technical aspects. The quote from Mr. Barroso, above, indicates that research effort in SoSE cannot be to urgently undertaken.

The collaborative project T-AREA-SoS[1] addressed the following aim:

> *To increase European competitiveness in, and improve the societal impact of the development and management of large complex systems in a range of sectors through the creation of a commonly agreed EU-US systems of systems research agenda that will be used to support collaborative EU and joint EU-US research initiatives to be embedded in Horizon 2020. [2]*

[1] Trans-Atlantic Research and Education Agenda in Systems of Systems.

A. El-Osery and J. Prevost (eds.), *Control and Systems Engineering*,
Studies in Systems, Decision and Control 27, DOI: 10.1007/978-3-319-14636-2_9

Horizon 2020 is the financial instrument implementing the Innovation Union, a Europe 2020 flagship initiative aimed at securing Europe's global competitiveness [3]. The programme will run from 2014 until 2020 and is worth in excess of €70Bn.

The partners in the T-AREA-SoS project were Loughborough and Bournemouth universities from the UK and University of Texas at San Antonio and Purdue University in the US. The research agenda that it published [4] was developed through a consultative approach involving more than 100 SoS experts from industry, academia, and Government bodies.

2 Systems of Systems Engineering

2.1 Definition and Characterization of Systems of Systems

In recent years, there has been a convergence of views about what constitutes systems of systems. Jamshidi [5] has reviewed more than twenty definitions of SoS; many are very similar to each other, but frequently these definitions are rooted in the domain from which the author writes. In fact, by refining these multiple definitions and determining common features, the definition due to Jamshidi himself was considered by the T-AREA-SoS project to be the most appropriate from a general point of view:

A SoS is an integration of a finite number of constituent systems which are independent and operatable and which are networked together for a period of time to achieve a certain higher goal [5]

This definition emphasizes important features that are significance to the theme of this chapter; i.e. behaving nicely in a SoS. Firstly the independence of the constituent systems is of fundamental importance. Traditional systems engineering often implies an underlying assumption that the system in question is completely under the control of the developer or operator. In SoS, such control is lacking and whilst some aspects may be directly controlled, other mechanisms (e.g. influence, contracts, trust, incentives) are required in order for effective interoperation of the constituent systems. The higher goal is not achievable by any of the systems operating alone, and so the mechanisms are crucial for realization of the higher goal.

Secondly, this definition emphasizes the dynamic nature of SoS. The period of time might be short or long, but it implies that the SoS is evolutionary and, in some cases, it may be ad hoc, with constituent systems joining or leaving during the SoS operation. It has sometimes been suggested that, whereas the Systems Engineer is concerned with the detailed construction of complex systems, the SoS Engineer is concerned with making good choices of constituent systems to achieve appropriate interoperation (rather than the detailed internal workings of those constituent systems) [6].

Ackoff introduced the notion of SoS from a Systems Science perspective in the early 1970s [7] in a paper that considered the behaviors of systems interacting with each other and laid emphasis on the organizations as systems point of view. More recently, Maier has described the characteristics of SoS from a software perspective [8], but these have been generally accepted as applicable to SoS in general. The five characteristics he identified are:

- Operational independence of constituent systems, which means that the systems can be operated on their own and provide useful outcomes.
- Managerial independence of constituent systems, which implies that the systems are under the control of multiple organizations (or owners) and are not fully under the control of a single organization.
- Evolutionary development processes, which implies that the systems change over time but, perhaps more challengingly, an important implication is that systems built at different times (possibly to different standards) and not designed to work together from the outset must interoperate.
- Emergent behavior is, in fact, the property of all systems; i.e. a system is built from components that produce emergent behavior when they interact. However, in the case of SoS the implication is that certain emergent behavior is very difficult (even impossible) to predict in advance.
- Geographical distribution of constituent systems, which implies that the main concern is with the exchange of information between the systems. However, this particular characteristic is not usually a significant concern for many situations; it does become significant when the time delay for transfer of information impacts performance (e.g. satellite networks).

It is the first two characteristics that are greatest concern with regards to management of SoS and, particularly, behaving well. The problem with SoS has been perfectly described by George Rebovich:

> *From the single-system community's perspective, its part of the SoS capability represents additional obligations, constraints and complexities. Rarely is participation in a SoS seen as a net gain from the viewpoint of single-system stakeholders. [9]*

Essentially, individual system owners will strive to optimize their own systems (maybe for performance, or commercial profit, or robustness), but this may lead to sub-optimal or even undesirable behavior of the SoS.

2.2 Types of System of Systems

A significant contribution to understanding the nature of SoS was made by Dahmann and Baldwin in 2008 with the classification of SoS types [10].

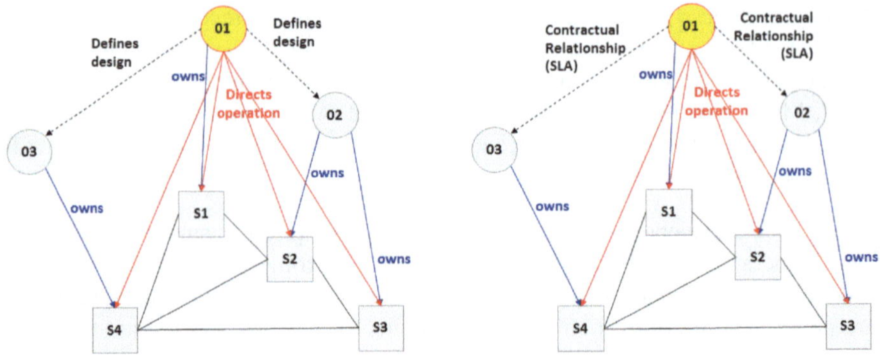

a) In a Directed SoS: operators O2 and O3 accept direction from O1 in terms of the specification and operation of the systems they own (O2 owns systems S2 and S3; O3 owns S4) This type of SoS is highly controlled by the central managing entity (O1).

b) In an Acknowledged SoS: O1 directs choice of systems and operation; O2 and O3 have a contractual relationship (e.g. Service Level Agreement) with O1. In this case, the central managing entity (O1) has less control over the systems owned by O2 and O3 (S2, S3, S4) and must rely more on influence.

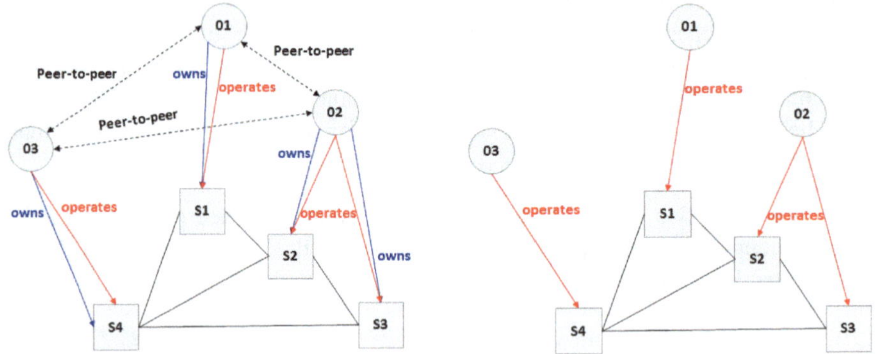

c) In a Collaborative SoS: there is mutual agreement to collaborate; usually covered by agreements of some form, but there is no overall managing entity; systems owners (O1, O2, O3) operate their own systems and collaborate with others to realize some shared benefit.

d) In a Virtual SoS: Owners (O1, O2, O3) access other systems through their own systems in order to realize individually sought benefits, though high level emergent behavior may still occur. There is no overall goal, no central management and interoperation is achieved by recognized protocols, or standards, not through individual agreements between pairs of systems.

Fig. 1. SoS types due to [10] as described in the T-AREA-SoS project [4]

These descriptions have been widely adopted by the SoS community and appear in the US DoD guide [11]. The classification postulates a number of operators of constituent systems (O1, O2…) and a set of known constituent systems (S1, S2, S3 …).

It should be noted that some operators may be responsible for several systems. There are four types: Directed, Acknowledged, Collaborative, and Virtual; the T-AREA-SoS project described these types as shown in Figure 1. Recently, [12] have suggested a further category of accidental, in which systems interoperate in a completely unplanned manner, coming together by accident as the name suggests.

The types described above are not intended to represent a specific classification for a particular SoS, but rather to describe the behavior and operation of SoS, which will generally contain more than one type. The hierarchy of Systems defined by [13] is helpful in this regard. The authors construct a conceptual hierarchy of Systems in which SoS are contained within higher level SoS. Decomposing the SoS, eventually one reaches single systems to which one might suggest traditional Systems Engineering is applicable. However, within the hierarchy of SoS it is plausible that different types (e.g. acknowledged and collaborative) are simultaneously present.

2.3 Interoperability

Clearly, the type of any particular SoS identified within the hierarchy will be determined by the nature of the interactions between the constituent systems of that particular SoS. Using a very broad interpretation of the term contract one can consider the SoS to be a network of contracts; these could be formal (legal) contracts, informal relationships in which behavior depends on levels of trust, or standards (e.g. for data transfer), etc.. Interoperability frameworks (e.g. NCOIC[2]) indicate that interoperation between systems occurs at different levels, such as mechanical interoperation, data exchange, semantic interoperability, process or policy alignment, so even within the context of interaction between two systems, there may be several different mechanisms (contracts) that are present. Indeed, it is frequently the case that problems arise when there is a failure at one particular level (e.g. the data exchange is satisfactory, but the lack of semantic interoperation means that information exchanged between two systems is not understood in the same way by the receiving system as the sending system). A formal approach to contracting in SoS, with a focus on evolving SoS, has been addressed by Mangeruca et. al. in the EU DANSE project [14].

2.4 Summary of SoS Basics

To summarize, the key difficulty with SoS is the independent management of the constituent systems and the level of independence of those systems provides a means through which SoS can be characterized from tightly controlled (directed) to little or no control (virtual and accidental). SoS are hierarchical and a high level SoS may contain different types of lower level SoS. Problems may arise in SoS because of interoperability failures between constituent systems, meaning that they do not interoperate as designed or expects; but, the chief difficulty is that individual system operators will generally optimize the operation of their own systems locally without

---

[2] Net Centric Operations Industry Consortium; for interoperability framework, visit `ncoic.org/technology/deliverables/nif/`

paying attention to the impact of this on the higher level SoS. In fact, it is often the case that operators will lack situation awareness of the behavior of the higher level SoS until a calamitous or, at any rate, significant unexpected behavior occurs.

3 Global Drivers for Improved SoS Design and Management

3.1 Increasing Interconnectivity, Increasing Complexity, and Increased Inflexibility

It is clear from the foregoing discussion that SoS have always existed (even if it is only in the last twenty years that the term System or Systems has gained currency). However, it is also clear that the complexity of SoS is continually increasing; as interconnection becomes easier, so the opportunities for unanticipated interoperation increase. Furthermore, the creation of highly heterogeneous SoS, in which sophisticated software intensive devices are mixed with simple systems either intentionally or unintentionally increases the vulnerability of the SoS to complex failures and simple attacks. It has been estimated that by the year 2020, there could be 50 billion devices connected [15]; six times the number of connected people! Setting aside the accidental relationships, it is apparent that within designed SoS the coupling between systems is creating complex dependencies, the behavior of which cannot be easily predicted.

Based on an extensive study of ancient civilizations, the archeologist, Joseph Tainter, has argued that social complexity, when faced with widespread change in economic and/or environmental conditions, causes that society to collapse [16]. The rapid increase in connectivity certainly increases complexity of the systems that support human endeavor; the local optimization of the contributing systems [9] causes increased rigidity in the SoS to which they contribute. Without wishing to be overly pessimistic or sensationalist, it is clear that the rapid development of SoS (outstripping the understanding of how to design and manage them most effectively) is contributing to the increase in social complexity and (ironically) lack of agility that contribute to societal collapse as discussed by Tainter.

3.2 A Perfect Storm for Europe

There are a number of key risks to human society as a whole; the analysis which follows is focused on the risks faced by Europe, but all will be present in some form across the globe. These are illustrated schematically in Figure 2. It is important to understand that the global drivers whilst independently articulated are highly overlapping and interrelated.

Between 2014 and 2050 the population of the world is likely to increase from 7 billion to about 9 billion [17]; this poses significant challenges to resources. However, in Europe the increase is very modest: from approximately 500 million to 526 million [18], but of greater concern is the demographics associated with these figures. The growth is likely to be uneven geographically, with countries like the UK seeing substantial population growth and others in Eastern Europe seeing a decline.

More significant still is the change in the age demographic, from an aged population (>65 years) of about 25% in 2014 to about 52% by 2050. This means that an increasingly large number of people will need to be supported by a decreasing working population.

The global population increase naturally implies challenges for provision of food. However, it should be noted that both rich and poor countries waste up to about half of the food produced (though the reasons for waste are different) [19]. Europe is affected by this, despite the modest population growth. Europe generally imports raw materials and exports finished products; thus the impact of growth in the global population is likely to reduce the availability of the raw materials (basic foodstuffs) that Europe is able to import, because it will be required by the burgeoning populations in other parts of the world.

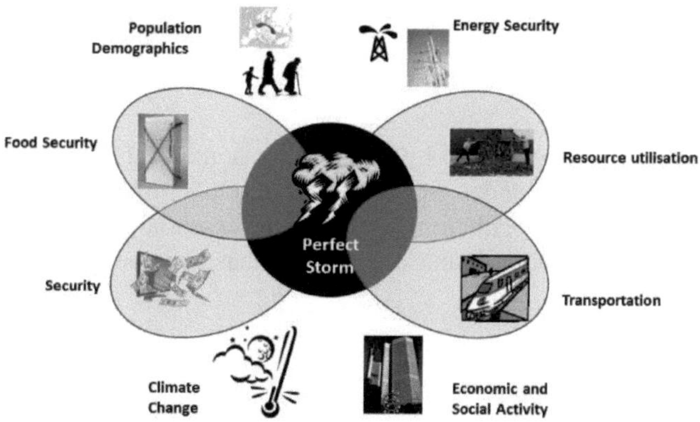

Fig. 2. A Perfect Storm for Europe – a symbolic representation of the threats faced by Europe that set the drivers for improved operation of SoS

Europe faces two simultaneous and reinforcing challenges with regard to energy. Firstly, the heavy dependence of Europe on external suppliers of fossil fuels provides both economic and energy security challenges. Indeed, during 2014 these dependencies were brought into sharp focus because of disputes in Eastern Europe [20]. Secondly, there is a pressing need to reduce carbon (and other damaging) emissions [21]. There is a strong focus, within the EU, on transportation and manufacturing as areas through which reduction in carbon emissions can be secured. Because the European Union has already invested considerable effort into reducing emissions and improving energy use through initiatives such as the European Climate Change Programme (ECCP), further reductions will be hard to achieve.

Depletion of mineral resources (such as iron, titanium, uranium, etc.) also poses problems for continued economic prosperity in Europe [22]. It is considered that significant changes in political and social thinking will be required to achieve the full benefits of recycling [23].

Although Europe has a good record on citizen security, there are significant developing threats to both physical and cyber security. As noted above, the highly interconnected nature of modern systems means that there are vulnerabilities to cyber-attack and especially greater exposure to privacy risks.

The EU has adopted a roadmap for transport [24] which has implications for energy management and transport availability across Europe. More effective networking of international transport systems will be an essential ingredient of strategy for energy.

More generally, globalization impacts economic and social concerns within Europe. Globalization can be considered to be the world trade in goods and services, and it has been brought about through the design and management of SoS.

3.3 Agility

Whether it concerns the SoS used by industry for somewhat narrow applications, or the highly complex SoS under the auspices of Government for delivery of the essential services to large populations, agility is the essential property of SoS that will perform well. The following definition of agility has been adopted by the T-AREA-SoS project:

Agility is the ability to successfully effect, cope with, and/or exploit changes in circumstances. [25]

Clearly, the dynamic nature of SoS implies that better design of SoS will yield improved agility. Perhaps this should more properly be the better design of constituent systems for their role in the SoS, because it will be the overall performance of the SoS, and its ability to reconfigure appropriately to cope with changes in circumstance, that will be critical for resilience and effectiveness.

Previous work in the area of Network Enabled Capability (NEC) for military purposes led to a model of agility [26], [27] in which the main influences are interoperability, affordability, dependability and availability (of a constituted system), as shown in Figure 3. This model also includes collaboration and knowledge management, which (human aspects) are considered to be an essential part of agility. These represent the way of working (collaborative) and the competence available within the extended enterprise. Although simplistic and derived for military applications, this model provides a useful general starting point for consideration of the incentives towards good behaviors for a SoS. Put simply, the constituent systems must be dependable, available, affordable, and interoperable with other systems. The operators must exhibit collaborative behavior and have sufficient knowledge to understand when reconfiguration is needed and what changes are required to maintain the fitness for purpose of the SoS.

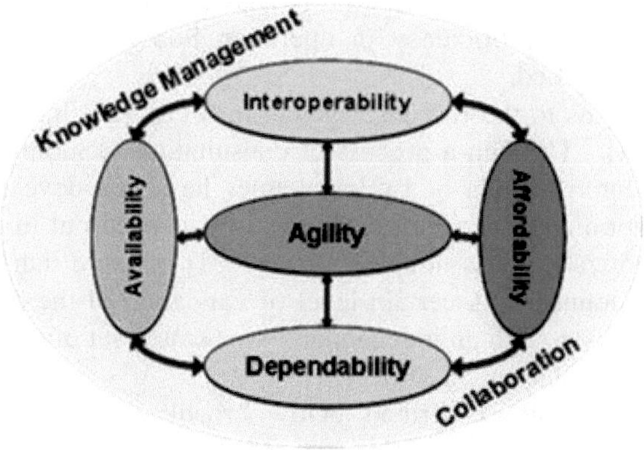

Fig. 3. The NECTISE NEC Themes [26]

A number of barriers to agility exist. Within supply chains, which can be considered to be enterprise SoS and likely include technological SoS, there are two main barriers to agility, as described in [28]. Firstly, there is the need for various commercial entities within the supply chain to protect their own Intellectual Property (IP); this means that there is only a partial sharing of information about the SoS operation and operators of one system will not have good situation awareness of the whole network because of a lack of information about specific constituent systems. It should be noted that in an endeavor to achieve better SoS working, the UK Ministry of Defence introduced nine Systems of Systems Approach principles, which includes (principle 8) the adoption of open standards, in order to improve awareness across the SoS enterprise [29].

Secondly, separately acquired information systems that are not compatible with each other, together with cultural differences between organizations in the SoS can lead to interoperability failures. This type of barrier is especially in evidence in disaster situations where, for example, a military organization is not able to work with civilian authorities due to both cultural differences (decision making) and communications devices incompatibility [30].

As implied above, the adoption of open specifications, open architecture, and open standards is viewed as an enabler for SoS agility. However, the business models for open systems architectures are yet to be firmly established.

4 The SoS Research Agenda for Europe

The foregoing sections have provided a characterization of the SoS concept and identified the main difficulties in achieving high performance for SoS operation; namely failures in interoperability and the lack of incentives for the single-system stakeholder to value participation in a SoS.

The risks associated with highly interconnected systems and the inevitability of their proliferation have been outlined. Whilst many aspects of SoS are important from a business perspective, it is clear that the impact of SoS on various global drivers is an important consideration. Agility has been implicitly identified as the key attribute through which proficiency in operating SoS may be judged. Barriers to agility have been identified.

Attention now turns to the research agenda in SoSE, developed by the T-AREA-SoS programme [4]. Through a process of consultation including a series of expert workshops, a research agenda of twelve themes has been developed, as described below. Identification and definition of themes was carried out in expert workshops, but the final prioritization was done by survey [31] of more than 100 experts from more than twelve countries. A certain level of validation of the priority themes was achieved by comparison with an independently generated set of priorities provided by the INCOSE[3] Working Group on SoSE [32].

The twelve themes are described below broadly in the order in which the community prioritized them, though many were of equal weight. A slightly surprising result was that characterization and theoretical foundations emerged as the top priorities among both academic and industrial stakeholders alike.

Although the prioritization of the themes is a statement of the research needs in SoSE in 2013 (when the T-AREA-SoS project completed), it is easily understood that the themes will remain a priority for many years to come. Furthermore, the themes will remain a priority if the global drivers identified above are to be satisfactorily addressed, and these drivers have timescales of decades associated with them.

4.1 Characterization and Description of SoS

As described above, there has been considerable convergence on agreeing the definition of the SoS concept. The five characteristics of Maier [8] together with the types of SoS [10] provide a satisfactory basis upon which to characterize the SoS concept. However, these descriptions are inadequate when trying to characterize a specific SoS and further detail is required to distinguish one SoS from another. Some progress has been made in this respect by Kinder [33] in which a set of nine characteristics are identified; these are related to each other and cannot be easily considered independently. The characteristics are: description of all components systems; the SoS lifecycle, probably including the variety of life cycles of the constituent systems; classification of the SoS (to the most closely resembling of directed, acknowledged, etc.); variability is used to describe the frequency at which various parts of the SoS change or reconfigure; the nature of all the interactions available for use between systems in this SoS; the nature of the relationships between systems (e.g. peer-to-peer); the functions that are performed within the SoS; the owners and operators of the constituent systems; the concept of employment or operation. Although this is a positive step towards detailed characterization, much work is still required on the definition of towns.

Specific characterization of SoS is important for understanding the types of SoS and how they operate in relation to the global drivers. There is insufficient knowledge about the direct impact of different SoS on the drivers to enable long-term prediction

[3] International Council for Systems Engineering.

of the effects of interventions. Better characterization is the starting point for a better understanding of how to address the global drivers and mitigate the negative effects of SoS.

4.2 Theoretical Foundations

Phenomenolgically SoS are understood, but there still lacks firm theoretical foundations upon which to develop predictive tools. This is fundamental research into the paradigm for SoS and SoSE. The types and characterization previously described will be a starting point for development of theory for SoS. The extension of this work would be to establish a set of SoSE principles and this would support SoS developers and researchers. The significance of theoretical foundations to the global drivers is the same as the characterization theme above.

4.3 Emergence

Arguably, most research in SoS is about emergent behavior and how to manage it. The research that is suggested for this theme concerns development of a fundamental understanding of emergence in terms of developing mathematical techniques or models to describe the interactions between two or more systems. Naturally, all themes are relevant to resolving the global drivers, but the theme of emergence is particularly relevant because the problems now faced by Europe in terms of the global drivers have all arisen through unpredicted emergent behavior resulting from technology interaction with the environment.

The global drivers, as described in the previous sections, are clearly emergent behaviors of the operation of large and complex SoS. It is sometimes considered that negative emergent behaviors are catastrophic events, but it is equally true that they can be very long term catastrophes that are only noticed when it is too late to do anything about them. Better prediction methods for emergence need to be able to predict the long-term behavior of SoS on drivers such as the climate.

4.4 Multi-level Modelling

Models may be used for prediction or for illustrating a concept for purposes of communication or education. In this case, the required modelling is very much motivated by the need for better predictive techniques. Much of the modelling effort to date has been in the area of architectural modelling, using developments of enterprise architecture to describe SoS and formal approaches to the use of such models in predicting likely behavior.

The term, multi-level, refers to two distinct approaches to modelling. The first is concerned with models that are consistent through different levels of abstraction and, clearly, there is an implicit assumption here that such models are possible. The second refers to the types of model already in use in which combinations of different models (from different disciplines) are combined in a simulation. There are considerable challenges in developing such complex models because validation is so difficult. A comprehensive survey of modelling techniques for SoS has been carried out by Kinder [34], in which fourteen techniques are compared for a case study on tactical data links.

Although concerned more with the analysis aspects than modelling per se, the emerging topic of Big Data is significant to the development of SoSE. The ability to deal with vast amounts of data, from multiple sources, provides the opportunity to extract meaningful information that can enlighten users about emergent behaviors [35]. This could be very important in early observation of trends that impact the global drivers.

4.5 Measurement and Metrics

The problem with measurement of useful parameters for SoS is two-fold. The first is a tendency to focus on performance metrics rather than on measures of effectiveness. For instance, measuring the number of solar panels installed in a year rather than the direct reduction in greenhouse gases due to such installations. In some ways, this is hardly surprising since the measures of performance often require a detailed understanding of relationships between outcomes and SoS features that may not be known with confidence. The second problem is related to the Rebovich difficulty [9] that no single user has access to all the data; individual organizations may restrict access to the systems under their direct control for commercial reasons or for reasons of security. This is an example of how incentivizing greater sharing of information within a SoS enterprise may yield a better collective understanding of the SoS behavior and, hence, a better understanding of how to achieve progress on the global drivers. It is to a large extent the absence of timely measures at the holistic (i.e. global) level that leads to unwanted impacts on the environment, even though local (constituent) systems are optimal.

As with the modelling theme the use of big data (or data analytics to give it, its proper name) could be a significant advance in understanding the emergent behaviors of SoS, provided suitable measures and metrics can be constructed. Research that leads to well-founded measures of effectiveness is a priority.

4.6 Evaluation of SoS

Although, on the face of it, this theme is very similar to the one above (metrics), in fact, the area of interest is more process related. Essentially, given suitable metrics, it is required to know how metrics may be related to decision making in the SoS. As described above, SoS are highly dynamic and so evaluating an SoS in order to then make decisions concerning reconfiguration and then, if reconfiguration is needed, determining how to reconfigure are problems to be addressed. A difficulty with evaluation is that a comparison will generally need to be made, but SoS are almost always one-offs, i.e. the constituent systems are networked for a period of time [5] in a unique way. Direct comparison in a controlled experimental way is simply not possible. So techniques for reliable evaluation remain troublesome.

4.7 Definition and Evolution of SoS Architecture

There has been a considerable effort over several years to develop systems architecting in such a way that risks associated with the integration or (more usually) the interoperation of complex heterogeneous networks may be achieved reliably. The motivation is clear: there are many examples (especially in the ICT world) of expensive failures, such as the National Programme for IT in the NHS (NPfIT) which was cancelled after spending £11.4 billion [36].

Enterprise Architecture has been used in the ICT domain for many years, since Zachmann created the framework for relating business transformation to ICT management [37], but there is still significant work needed for the use of such techniques in more general systems (even though they generally have substantial software included). Representing human action as part of the system is problematic, although there have been efforts to include human views in architecture frameworks [38].

Two European projects that began at the same time as T-AREA-SoS have focused on development of systems architecting techniques. In both cases architectural patterns have been an important consideration as these potentially enable consistent and, importantly, affordable approaches to the use of architectures. The COMPASS project has taken a formal methods approach and examined a number of use cases for effective SoS integration; the work has involved the creation of a modelling language (CML) to better describe the SoS. The work on SoS patterns is described in [39]. The DANSE project has also endeavored to apply rigorous approaches for modelling and architecting SoS; see, for instance [40].

The stakeholders in T-AREA-SoS noted that there are a great many architecting tools in existence and a good many frameworks for enterprise architectures. However, there is a need in most organizations to develop the architecting skills of their employees to adequately exploit the opportunities that systems architecting may offer.

4.8 Prototyping SoS

Largely driven by the manufacturing community, the theme of prototyping SoS has arisen because of the difficulty in capturing sufficient context and, indeed, sufficient of the overall SoS in a prototyping activity. Prototyping is standard practice for risk reduction with new products and systems, but in an evolutionary development, the scope may be rather limited. Because a SoS does not necessarily stop developing (unlike a product that has a design freeze and goes into production), then choice of when and what to prototype becomes more problematic. Of course, this theme is highly related to two above (metrics and evaluation). Its relationship to the global drivers is complex; whereas it is straightforward to understand the nature of prototyping as it applies to assessment of a commercial offering into the marketplace, the nature of prototyping as it applies to, say, climate change is not defined. Over what period should one prototype? What are the criteria for success or failure of the prototype?

The strategic agenda recommends the development of prototyping strategies, though it is anticipated that these will be sector specific.

4.9 Trade-Off

Trade studies is, of course, one of the tasks of systems engineers; with complex systems where many design features come together, it is necessary to understand the benefits of certain configurations over others balanced against the disadvantages that each introduces. Trade-off, because it specifically requires consideration of the environment in which a system operates is, by its nature, SoS-aware. However, the difficulty of drawing the systems boundary for SoS [33] makes construction of the trade-off problem space difficult. The trade space is, furthermore, very dense, making adequate coverage of possible options infeasible. As with prototyping, trade-off in the context of the global drivers is unreliable because of the lack of appropriate metrics and, especially, a lack of firm relational information between SoS features and the key parameters (the effects on) the global drivers. Concern about the global drivers is encouraging significant scientific research and the results of such research need to be incorporated appropriately into the models used by Systems Engineers.

4.10 Security

Security is both a driver and a theme for research. The high levels of connectivity, enabling information to be transferred to millions of systems very rapidly, provide huge benefits in terms of situation awareness for system users, but also pose significant threats to them in terms of loss of commercially sensitive information, invasions of privacy, national security breaches, etc.; see, for example [41]. The relationship to the global drivers is clear, of course, but it must be noted that the global nature is, of itself, problematic, since ethics are culturally located, but the systems which are vulnerable to security breaches span borders and cultures.

4.11 Human Aspects

The human aspects of all the themes above are clear, but this theme was highlighted specifically because very often the notion of Systems of Systems is thought to be a technical problem, even though many of the problems that must be resolved have solutions very much in the sociological domain. A partner project (ROAD2SOS) that created roadmaps for SoSE derived a set of barriers to implementing the road maps, all of which required significant consideration of the human aspects of the operation of the constituent systems [42]. An important implication of the significance of the human sciences in the SoS domain is the need to involve human scientists in their development as key members of the teams. With reference to the global drivers, this is particularly the case, because the effect on and the effect of human behaviors generally form the most significant impact.

4.12 Energy

Energy emerged as a somewhat surprising theme in the study; not because it is not important, but because it is different in nature to the other themes. However, it is highly relevant because every SoS uses energy and in many cases the energy is used in an unaccounted, but likely inefficient, manner. It is recognized that energy use must be reduced to protect resources and minimize the risks to the natural environment. In terms of the Rebovich problem [9], it is important to incentivize good behaviors at the individual system level, but also to be aware that optimizing the energy consumption of one device may cause another part of the system to have to use more energy.

5 Conclusions

In some ways, it is hard to draw conclusions because, like SoS, the themes discussed in this chapter are ongoing and will not be finally resolved, simply improved as research and methods improve understanding of the behaviors of SoS. It was noted that SoS have always existed, but the ever increasing levels of interconnectivity are leading to more and more complex challenges and impacting the environmental and human landscape in ways that must be mitigated urgently. Global drivers for improving understanding and management of SoS were identified as:

- Population demographics
- Food security
- Energy security
- Resource utilization
- Emissions and global climate
- Community security and safety
- Transportation
- Globalization of economic and social activity

It was observed that for Europe these represented a perfect storm of interrelated challenges that have a daily impact on the lives of citizens.

The problem of individual systems operators optimizing their own systems, potentially at the expense of the performance of the overall SoS as identified by Rebovich [9] has been highlighted. Reasons for this problem include selfishness, lack of appropriate business models for SoS, but also simply unawareness of individual owners of the impact of their actions (or inactions) on the wider SoS. Even though SoS enable much greater distribution of information, nevertheless, appreciating and using huge amounts of information is not straightforward.

The twelve research themes that make up the strategic research agenda in SoSE have been discussed, with particular reference to how research in those themes can impact the global drivers in a positive way. These are:

- Characterization and description of SoS
- Theoretical foundation for SoS

- Emergence
- Multi-level modelling of SoS
- Measurement and metrics for SoS
- Evaluation of SoS
- Definition and evolution of SoS architecture
- Prototyping SoS
- Trade-off in SoS
- Security in SoS
- Human aspects of SoS
- Energy efficient SoS

Systems of Systems have always existed and, naturally, they are here to stay. There is an urgent need to address the challenges that increased connectivity pose. There have been several pioneers of research in SoSE, and the contribution of Mo Jamshidi through his personal research, publication of influential texts, and his corralling of a community of SoSE researchers and practitioners has been a significant contribution to the development of this area of study. These have been the foundation upon which a strategic research agenda for SoSE has been built.

Acknowledgements. The work of the T-AREA-SoS team is gratefully acknowledged: Mo Jamshidi, Vishal Barot, Sharon Henson, Dan DeLaurentis, Huseyin Dogan, Cornelius Ncube, Soo Ling Lim.

The support of the European Commission (Grant number 287593) is acknowledged.

Thank you to all those SoSE experts who contributed time and brain power to the creation of the Strategic Agenda.

References

[1] Barroso, J.M.: President of European Commission (2010)
[2] Henshaw, M.J.D.: T-AREA-SoS Final Report, Loughborough University, Loughborough (2013)
[3] European Commission, Horizon 2020 (September 7, 2014), `http://ec.europa.eu/programmes/horizon2020/en/what-horizon-2020`
[4] Henshaw, M.: The Systems of Systems Engineering Strategic Research Agenda. Loughborough University, Loughborough (2013)
[5] Jamshidi, M.: Introduction in Systems of Systems Engineering - Principles and Applications, p. 1. CRC Press, Boca Raton (2009)
[6] Dahmann, J.: In various presentations in the INCOSE SoS Working Group (2011)
[7] Ackoff, R.L.: Towards a System of Systems Concepts. Management Science 17(11), 661–671 (1971)
[8] Maier, M.: Architecting Principles for Systems-of-Systems. In: Proc. of 6th Int. Symp. INCOSE, Boston, MA (1996)

[9] Rebovich, G.: Enterprise Systems of Systems. In: Systems of Systems Engineering - Principles and Applicaitons, p. 169. CRC Press, Boca Raton (2009)
[10] Dahmann, J., Baldwin, K.: Understanding the Current State of US Defense Systems of Systems and the Implications for Systems Engineering. In: 2nd IEEE Systems Conference, Montreal (2008)
[11] Department of Defense, Systems Engineering Guide for Systems of Systems Essentials, US Dept. of Defense, Washington DC (2010)
[12] Kemp, D., Evans, R., Elphick, J., Camm, D.: Steampunk Systems of Systems Engineering: A Case Study of Successful Systems of Systems Engineering. In: 23rd Int. Symp. INCOSE, Philadelphia, US (2013)
[13] DeLaurentis, D., Crossley, W.A., Mane, M.: Taxonomy to Guide Systems-of-Systems Decision-Making in Air Transportation Problems. AIAA Jour. Aircraft 48(3), 760–770 (2011)
[14] Mangeruca, L., et al.: Specificaiton of the goal contracts specification language. In: DANSE (2013)
[15] Global Europe, Competing in the World: A contribution to the EUs growth and jobs strategy. EC, Brussels (2006)
[16] Tainter, J.: The Collapse of Complex Societies. Cambridge University Press, Cambridge (1988)
[17] Wilbanks, T.J., Lankao, P.R., et al.: Industry, Settlement and Society. In: Climate Change 2007: Working Group II: Impacts, Adaptation and Vulnerability, Ch. 7 (2007)
[18] Eurostat, EU27 population is expected to peak around 2040. Eurostat PRess Office, Brussles (2011)
[19] The Economist. Clean Plates - too much food gets thrown away. The Economist (September 1 2012)
[20] Archive-IT, Ukranian Conflict (2014), `http://archive-it.org/collections/4399`
[21] European Commission, Energy Roadmap 2050, EC, Brussels (2011)
[22] Prior, T., Giurco, D., et al.: Resource Depletion, Peak Minerals and the Implications for Susteainable Resource Management. In: 11th Biennial Conf. Int. Soc. for Ecological Economics, Oldenburg/Bremen, Germany (2011)
[23] Allwood, J.M., Ashby, M.F., Gutowski, T.G., Worrell, E.: Material Efficiency: A White Paper. REsources, Conservation and REcycling 55, 362–381 (2011)
[24] European Commission, Roadmap to a Single European Transport Area - towards a competitive and resource effieicnt transport system, EC, Brussels (2011)
[25] Alberts, D.S.: The Agility Advantage - A survival guide for complex enterprises and endeavours. In: CCRP (2013)
[26] Urwin, E., Gunton, D.J., Reay Atkinson, S.R., Henshaw, M.J.D.: Through-Life NEC Scenario Development. Systems Jour. 5(3), 342–351 (2011)
[27] Neaga, E.I., Henshaw, M.J.D.: A Stakeholder-Based Analysis of the Benefits of Network Enabled Capability. Defense & Security Analysis 27(2), 119–134 (2011)
[28] Barot, V., Henshaw, M., Siemieniuch, C., Sinclair, M., Lim, S.L., Henson, S., Jamshidi, M., DeLaurentis, D.: State of the Art Report, Loughborough Univ., Loughborough (2013)
[29] Henshaw, M., et al.: Assessment of open architectures within defence procurement issue 1: systems of systems approach community forum working group 1 - open systems and architectures. Crown owned copyright, London (2011)
[30] Amin, S., Cox, M., Goldstein, M.: Using data against disasters: Overview and synthesis of lessons learned. The International Bank for Reconstruction, London (2008)

[31] Henson, S.A., Henshaw, M.J.D., Barot, V., Siemieniuch, C.E., Sinclair, M.A., Jamshidi, M., Dogan, H., Lim, S.L., Ncube, C., DeLaurentis, D.: Towards a systems of systems engineering EU strategic research agenda. In: Proc. IEEE 8th Int. Conf. SoSE, Maue, Hawaii (2013)

[32] Dahmann, J.: Keynote address - Systems fo Systems: Perspectives, Pain Points and Prospects. In: IEEE 9th Int. Conf. SoSE, Adelaide, Australia (2014)

[33] Kinder, A., Barot, V., Henshaw, M., Siemieniuch, C.: Systems of Systems: Defining the System of Interest. In: 7th IEEE int. conf. SoSE, Genova (2012)

[34] Kinder, A., Henshaw, M., Siemieniuch, C.: System of systems modelling and simulation - an outlook and open issues. Int. J. System of Systems Engi. 5(2), 150–192 (2014)

[35] IBM, Big Data (September 7, 2014) http://ibm.com/big-data/us/en

[36] National Audit Office, The National Programme for IT in the NHS: an update on the delivery of detailed care records systems, NAO, London (2011)

[37] Zachmann, J.A.: A framework for information systems architecture. IBM Systems Jour. 26(3), 276–292 (1987)

[38] Bruseberg, A.: Human Views for MODAF as a Bridge Between Human Factors Integration and Systems Engineering. J. Cognitive Eng. and Decision Making 2(3), 220–248 (2008)

[39] Kazman, R., Schmid, K., Bellagaard, C., Klein? Understanding, J.: patters for systems of systems integration. In: IEEE 8th Int. Conf. SoSE (2013)

[40] Gezgin, T., Etzein, C., Henkler, S., Rettburg, A.: Towards a Rigorous Modeling Formalism for Systems of Systems. In: 3rd IEEE Workshop on Slef-Organizing Real-Time Systems (2012)

[41] DETICA, The Cost of Cyber Crime - A Detica report in partnership with the Office of Cyber Security and Information Assurance in the Cabinet Office, Cabinet Office, London (2011)

[42] Albrecht, C.: Report on commonalities in the domains and recommendations (Del 5.1 and 5.2). ROAD2SOS (2013)

Design Considerations of Dexterous Telerobotics

Sigal Berman

Dept. of Industrial Engineering and Management
Ben-Gurion University of the Negev
Beer-Sheva, Israel
sigalbe@bgu.ac.il

Abstract. Dexterous telerobotic systems are system in which the operation of a robotic hand is remotely controlled by the operator's hand and finger motion. Such systems have been slower to develop than general telerobotic systems which do not utilize finger motion. This is due to the requirements and challenges placed by dexterous telerobotics on both the remote and the local sites. The current chapter presents the challenges and design considerations of dexterous telerobotic systems and discusses them in the context of different remote control methodologies, namely direct, bilateral, supervisory, and shared control.

Keywords: robotic, telerobotics, dexterous telerobotics.

1 Introduction

Telerobotics refers to remote control of a robotic system where the human is in the loop during run-time operation. Dexterous telerobotics refers to remotely controlling a robotic hand manipulating objects by motion of the operator's fingers. Various dexterous telerobotic systems that facilitate execution of complex manipulation tasks have been developed [1-4] yet with current systems mission execution is slow and requires significant effort from the human operator [5,6]. To-date design of dexterous telerobotic systems is far from being resolved, lagging behind that of remote manipulator and remote mobile robot control systems.

The telerobotics field has made massive strides since its early beginnings in the 1940th during the dawn of the robotic era (thorough reviews can be found in 7-10). There are successful telerobotic systems for a numerous applications. Planetary rovers are remotely controlled by NASA personal from earth and from space, e.g., when flying an orbiting spacecraft[1]. Surgical robots are remotely controlled by the surgeons. Although currently the surgeon and the robot are typically co-located in the same room, the feasibility of a transcontinental remote surgery has been demonstrated [11]. Remotely controlled robotic vehicles take part in underwater exploration expeditions, e.g. the discovery of the titanic[2].

---

[1] http://www.nasa.gov/mission_pages/station/research/news/rover_from_space/

[2] http://www.titanicandco.com/discovery.html

A. El-Osery and J. Prevost (eds.), *Control and Systems Engineering*,
Studies in Systems, Decision and Control 27, DOI: 10.1007/978-3-319-14636-2_10

The basic building blocks of a telerobotic system (Fig. 1) include operator action sensing and sensation generation modules at the local site. These are connected through a communication channel to robotic action generation and sensing modules at the remote site. Internal connections may exist within each site between the sensing and generation modules. Telerobotic systems can be categorized based on their general control methodology and the degree of robot autonomy at the remote site. In direct control systems the motion of the operator is directly mapped to robotic motion and the robot has no autonomy. When direct control is achieved through a physical device, e.g., a joystick, that captures the operator's motion, and when force feedback from the remote system is directly integrated in the control loop of the master device the system is termed bilateral control [9]. At the other end of the control spectrum, in supervisory control systems [10] operator commands are high level commands typically expressed using a symbolic vocabulary of signs, words, or gestures. In such systems the robot possesses a high degree of autonomy since it must translate the high level commands to actuator commands. Shared control represents the middle ground between abstract supervisory control and direct or bilateral control. While the remote system can exert control over some task aspects, the human operator maintains some access to low-level forces and motions [1].

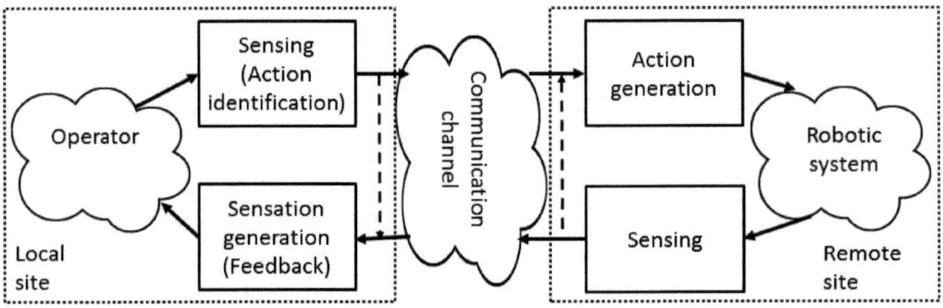

Fig. 1. The basic building blocks of a telerobotic system

Similar in notion to telerobotic systems, dexterous telerobotic systems can also be constructed based on the principles of these control methodologies. Yet, the characteristics of the human and robotic hands impose special requirements and challenges for dexterous telerobotic system design. The current treatise reviews these design challenges in the context of the different control methodologies, namely direct, bilateral, supervisory, and shared control. The remainder of this chapter is organized as follows: design considerations and challenges are presented for the local site in section 2, and for the remote site in section 3. Conclusions are presented in section 4.

2 Local Site Design Considerations and Challenges

2.1 Sensing (Action Identification).

Direct dexterous telerobotic systems require sensors that can track hand and finger motion. Such systems can be implemented without a physical master control device.

Bilateral systems require a master device that can both capture hand and finger motion and exert force on them. In dexterous telerobotic systems implementing supervisory control, symbolic finger gestures are used for controlling the operation of the remote robotic hand thus the sensing apparatus in such systems is in fact a form of a gesture recognition system (GRS) with a symbolic gesture vocabulary. Finally dexterous systems with shared control implement a middle ground between these methods and their sensing requirements depend on the specific methodology implemented.

Gesture Recognition Systems. GRS based on hand gestures are a common form of human-computer interaction (HCI). In symbolic hand gestures, meaning can be conveyed by multiple cues including motion, configuration, and proximity. A gesture may implement a single cue or a combination of cues. Constructing a gesture vocabulary requires balancing multiple objectives as gestures should be adapted to the requirements of both the GRS (facilitate robust recognition) and the user (be intuitive, memorable, not strenuous, and socially acceptable) [12].

Various sensors have been used for GRS [13]. For capturing hand configuration or finger motion the most commonly used sensors are touch, vision, or glove-based sensors. Touch-based systems have become popular with their rising integration in various popular consumer electronic devices. The size of the gesture vocabulary they facilitate is very limited, as only finger contact points, rather than the whole hand configuration, are sensed. Vision-based systems facilitate unencumbered operation and a much larger vocabulary. The most popular cameras used today for vision-based GRS are 3D sensors, e.g., the Kinect camera, since such sensors alleviate difficulties related to background motion and ambient light changes. For gesture vocabularies that encode meaning in finger motion or configuration, high resolution cameras are required, or the distance to the camera must be limited (typically the distance in such systems is about 1 m). Data-gloves offer a robust and precise system for finger motion and hand configuration sensing, facilitating large gesture vocabularies based on fine finger motion. Their processing requirements are typically high and the high quality gloves are expensive.

A GRS algorithm is composed of a pipeline of three sub-systems, which include a tracking algorithm (required for vision-based systems), feature extraction, and a gesture classification algorithm. Classification of finger-based gestures can be complex especially in cases where available resolution limits the feasible features. This difficulty is enhanced for multi-user, large-size vocabularies (more than ten gestures).

A gesture classifier was developed for a symbolic hand gesture vocabulary based on compound gestures (gestures comprised of both configuration and motion cues) from the American Sign Language (ASL) [14]. To make recognition more challenging the gestures were selected such that some of their cues were similar, e.g., two gestures may have a similar arm path but different hand configurations. The gestures were captured using a Kinect sensor. For a vocabulary of 10 gestures, recognition accuracy of 95.5% was attained using a database of 11 subjects. Figure 2 depicts images of misclassified gestures due to common problems in hand configuration recognition. Figure 2A presents an intermediate configuration of the gesture "magic" between a closed fist and an open hand stretched forward. Correct determination of gesture onset and offset is difficult yet imperative for classification. Figure 2B presents an executed hand configuration of the gesture "Chicago" wich includes both arm motion and hand configuration cues. The gesture was executed while the hand was turned sideways with respect to the camera and thus critical features were lost. This can be seen when comparing Figure 2B to the rotated 2D image of the correct hand configuration presented in Figure 2C.

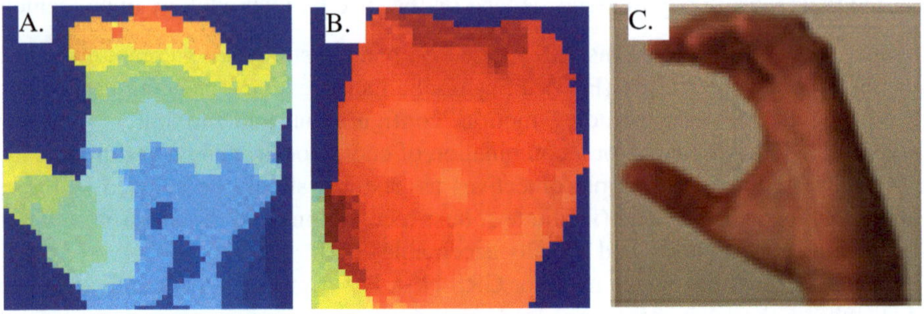

Fig. 2. Abstract. Problematic issues with hand configuration classification. A. and B. 3D hand images where depth is represented by a warm (close-red) to cold (far-blue) color palette. C. A rotated 2D image of the hand configuration presented in B.

Manipulation Motion Tracking. Object manipulation starts with a reach-to-grasp motion. After which the object is grasped and manipulated. Successive object re-grasp and manipulation intervals may follow until the object is released. The reach-to-grasp movement is commonly analyzed as being comprised of two components: the arm transport and the formation of the hand grip [15]. Both components are affected by the task, the object, and the characteristics of the hand and arm. Marteniuk et al. [16] observed that some kinematic features of the arm motion, such as the relative time of peak velocity, are more dependent on the task being performed rather than the object. Friedman and Flash [17] observed similar findings for the hand. These findings are supported by neural recording studies, as some hand-manipulation-related neurons show selectivity for the type of grasp, while others show selectivity for the size, shape, or orientation of the object [18,19].

For tracking arm and hand (finger) motion without a master device commonly used sensors are data-gloves for hand configuration tracking and magnetic or optic marker-based systems for tracking arm motion. When a master device is employed arm and hand motion are captured by the master device. While there are several commercially available master devices suitable for arm motion capture there are very few master devices that can capture finger motion, these are discussed in the next section on sensation generation as they are typically used for motion recording and for generation of haptic sensation.

2.2 Sensation Generation (Feedback)

Telepresence is related to the psychological phenomenon of presence. Presence in a mediated environment is related to the sense of "being there" rather than in the real environment, i.e., the physical space in which the body is located (e.g., the laboratory). It is a mental state that emerges out of sensory stimulus and bodily interaction with the environment. A view shared by several researchers relates presence to action, emphasizing that reality is grounded in action rather than mental filters or sensations. This approach maintains that the reality should be defined relative to functionality rather than appearance [20]. Although there is a widespread agreement regarding the concept of presence, measuring it is a different matter and various methods have been proposed. The most common approach used is subjective accounts usually by means of a questionnaire. Immersion is distinguished by some authors as the technology that gives rise to presence. Or the "overall fidelity in relation to physical reality provided by the display and interaction systems" [21]. In the context of telerobotics, telepresence is closely associated to the sense of distal attribution, i.e., inclusion of remote tools that become extensions of one's own body, even if they are not physically part of it [22]. Accumulating evidence shows that the human brain supports an adaptable body image, and this adaptation can take place within several minutes, given the right kind of sensory stimulus [23]. The rubber hand illusion [24] is a related intermodal phenomena, in which a person who's hand is hidden from view, watches a fake rubber hand stroked in synchrony with strokes applied to his hand. Within several minutes the person feels as if the rubber hand is an actual part of his body. Similar sensations have also been elicited when the rubber hand was replaced by a table [25] showing the illusion is relatively robust to form or location.

Simultaneous activation of different sensory modalities can improve perception and enhance performance yet contradicting, misleading, or lacking sensory inputs impair or even prevent motor learning [26]. Most telemanipulation systems today employ vision systems as the main sensory modality, yet haptics is gaining increasing attention [26,27]. The haptic sense can be broadly divided into two modalities, vibrotactile and kinesthetic, where both are important for dexterous telemanipulation. The vibrotactile modality is concerned with the subcutaneous tissue and is important for indicating that contact has occurred during grasp closure while the kinesthetic modality is concerned with gross body movement (and associated forces) and is important during object manipulation [28]. Thus far, in many cases the addition of haptics in telemanipulation has not elicited any noticeable improvement, e.g., [29],[4]. Although this may be attributed to various causes including the manner in which the sensory stimulus was presented and the delays it was subject to, it may also be caused by the failure of the system to evoke distal attribution.

The role of haptics in attaining presence (the sense of "being there") has been far less researched than that of vision base stimuli [26,27]. Depending on task characteristics, the tactile and kinesthetic systems play roles of varying influence. The kinesthetic stimulus is more important during object manipulation, while the tactile stimulus is critical for grasp closure, e.g., vibrations can indicate that contact has occurred [30]. There are very few systems that convey finger level haptic sensation. Two commercially available systems are depicted in Figure 3. The Phantom 6DOF (sensegraphics) has an additional 7th DOF that facilitates pinch grasps and the CyberGlove (cyberglovesystems) exoskeleton has five actuators capable of exerting normal forces.

Fig. 3. Finger haptic devices. Left: a 7th pinch grasp DOF that compliments the Phantom 6DOF device[3] Right: CyberGrasp[4] exoskeleton device.

3 Remote Site Design Considerations and Challenges

3.1 Action Generation

Traditional robotic end effectors have very limited flexibility in the range of objects they can grasp and manipulate. A much higher degree of flexibility is offered by various multi-fingered robotic hands that have been developed to enhance robotic capabilities. Reviews of such systems can be found in [31,32], and some specific examples in [33,34]. Robotic hands, even anthropomorphic robotic hands, significantly differ from human hands in every dimension (mechanics, shape, size, make, etc.). Thus mapping human hand motion to robotic hand motion is non-trivial.

A common approach for mapping human grasps to robot grasps is to limit the allowable human motion, either physically by an exoskeleton that is kinematically similar to the robot [35] or by verbal instruction [4]. The use of anthropomorphic hands [3] can circumvent the need for complex mapping at the expense of required robotic hand complexity. An additional mapping approach devises a mapping based on configurational resemblance between the operator's hand and that of the robot, either in joint space [36] or enforcing a resemblance of contact points in Cartesian space [37]. Such mappings often rely on the *virtual finger* schema [38,39] that enables concatenating several fingers as a unified "virtual" finger.

Constraining human motion limits the spectrum of possible operations and may greatly degrade human performance. Direct mapping may not be advantageous even in simple cases, due to the structural differences between human and robotic hands [40]. Additionally, as most robotic arms and hands differ considerably from the human arm and hand, limiting telemanipulation to anthropomorphic hands greatly reduces the range of possible applications. Devising a generalized mapping between human and robot grasp that accounts for the differences in their structural and dynamical characteristics still remains an open research question [8].

Example for Shared Control Decoupling Operator and Robot Grasps. An example of a shared control system that facilitate an intermediate-level mapping of operator's

[3] http://www.sensegraphics.com/

[4] http://www.cyberglovesystems.com/

actions is the object-action abstraction methodology [41,42]. This methodology was developed for facilitating intuitive control of robotic hands that considerably differ from the human hand when there is a need for decoupling the elementary manipulation actions of the human operator from those of the robotic system. The central concept of the methodology is that the operator's manipulation actions are identified in the context of the object on which they are being performed. The object context serves to reduce the recognition problem space, limiting the action set to the actions that can be afforded by the object on which the robot is operating, where affordance, introduced by Gibson [43], means the possible actions latent in an object for an individual. The action set of each object depends on the state the object is in. An example of the action set for a jar object is given in Figure 4. For each elementary action an object-action 2-tuple, e.g., cup-lift, is formed. Such object-action 2-tuples provide basic building blocks for translating human motion into movements of a robotic system.

Virtual reality was used for portraying the remote scene to the operator, and for establishing the operator's sense of tele-presence. The action identification was implemented using a linear discriminant classifier based on kinematic data extracted from a data-glove (CyberGlove, Cyberglovesystems). Identification takes place after the system establishes a stable grasp has been formed, based on identification of contact between two or more fingers and an object. During the reach-to-grasp stage the operator's arm motion is tracked and the manipulator follows the presented arm path. During the manipulation stage the motion of the virtual object is tracked and the robot arm moves such that the motion of the physical object follows that of the virtual object. The grasp configuration executed by the robot is based on a grasp library and is decoupled from the configuration used by the operator.

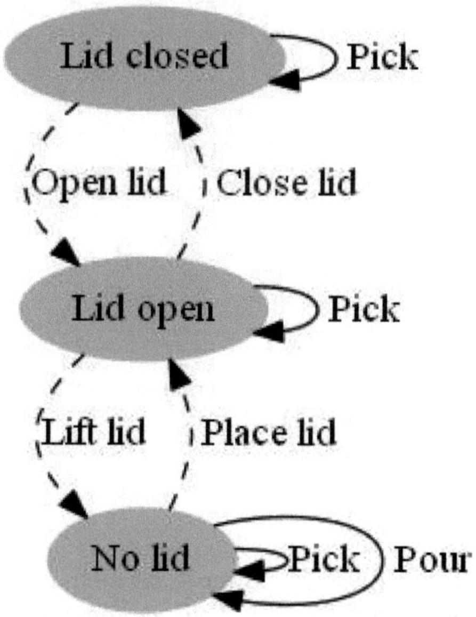

Fig. 4. Elementary manipulation actions for a jar object, that can be in three states

3.2 Sensing

Vision is typically the main sensing device in the remote scene. Camera placement in the remote site in dexterous telerobotic implementations is non-trivial. When using a static camera maintaining the line of site to the robotic hand while not abstracting the manipulators pathway may be impossible. A commonly used alternative camera placement is to use a moving camera placed on the manipulator. With a moving camera the pose (position and orientation) with respect to the environment must be maintained for correctly representing the scene.

Grasp force measurement is an even greater challenge especially for soft fingers. In recent years several human inspired sensors for measuring finger contact forces have been developed [44,45] yet they are still not commonly used due to cost and integration issues. Grasp force is modulated by contact properties which in turn depend on the nature of the bodies in contact, e.g., their material and elastic properties. Exact contact dynamic is difficult to model and in many cases very simplified, yet inaccurate, models are used such as Hook's-law based models. Lumped parameter models offer a manageable compromise for modeling soft-finger contacts against rigid surfaces [46]. An apparatus for determining model parameters of a lumped parameter power-law model for a Barrett hand with printed finger caps and for a human finger is depicted in Figure 5. Concurrent position and force readings were used to establish a force-displacement graph based on which model parameters were fit. The model relates the growth of the circular contact radius to the applied normal force for soft fingers.

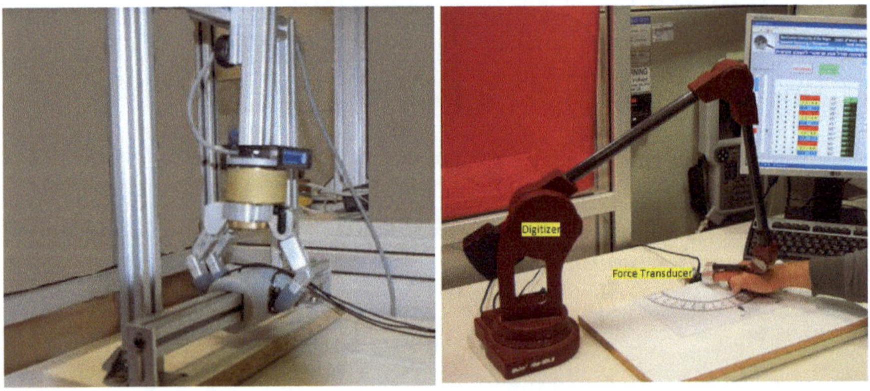

Fig. 5. Grasp force measurement apparatus for determining contact model parameters. Left: Barrett hand (Barrett Technology), Right: human finger.

4 Conclusions

Dexterous telerobotic systems have been slower to develop than telerobotics systems due to the substantial requirements placed in both the remote and the local sites. As rapid strides are made in the development of various system components, e.g., finger force sensors and finger level haptic force generators, dexterous telerobotic systems are expected to close the gap and become more common.

Performance measurement is vital for ensuring system improvement. Common telerobotic system measures can be adapted for assessing performance of dexterous telerobotic systems. The major issues of concern in telerobotic systems are stability and transparency which are generally in conflict. Karniel and Nisky [47] have suggested a multidimensional transparency measurement framework for bilateral teleoperation. The framework is based on taking three measurement views, the human perception of the remote environment, the local transparency of the operators' motion and the remote transparency of the robotic motion.

Acknowledgements. Research supported by the Helmsley Charitable Trust through the Agricultural, Biological and Cognitive Robotics Center of Ben-Gurion University of the Negev.

References

1. Griffin, W.B., Provancher, W.R., Cutkosky, M.R.: Feedback strategies for telemanipulation with shared control of object handling forces. Presence: Teleoperators and Virtual Environments 14(6), 720–731 (2005)
2. Hu, H., Li, J., Xie, Z., Liu, B.W.H., Hirzinger, G.: A robot arm/hand teleoperation system with telepresence and shared control. In: IEEE/ASME International Conference on Advanced Intelligent Mechatronics Monterey, California, USA (2005)
3. Jau, B.M.: Dexterous telemanipulation with four fingered hand system. In: IEEE International Conference on Robotics and Automation, Nagoya, Japan (May 1995)
4. Turner, M.L., Findley, R.P., Griffin, W.B., Cutkosky, M.R., Gomez, D.H.: Development and testing of a telemanipulation system with arm and hand motion. In: ASME IMECE Conference on Haptic Interfaces for Virtual Environments and Teleoperator System Symposium, Orlando, Florida (November 2000)
5. Kemp, C.C., Edsinger, A., Torres-Jara, E.: Challenges for robot manipulation in human environments. IEEE Robotics & Automation Magazine, 20–29 (2007)
6. Burke, J.L., Murphy, R.R., Rogers, E., Lumelsky, V.J., Scholtz, J.: Final report for the DARPA/NSF interdisciplinary study on human–robot interaction. IEEE Transactions Systems, Man, and Cybernetics—Part C: Applications And Reviews 34(2), 103–112 (2004)
7. Niemeyer, G., Preusche, C., Hirzinger, G.: Telerobotics. In: Siciliano, B., Khatib, O. (eds.) Springer Handbook of Robotics, Springer, Berlin (2008)
8. Ferre, M., Buss, M., Aracil, R., Melchiorri, C., Balaguer, C. (eds.): Advances in Telerobotics, Springer Tracts in Advanced Robotics. Springer, Berlin (2007)
9. Hokayem, P.F., Spong, M.W.: Bilateral Teleoperation: An historical survey. Automatica 42, 2035–2057 (2006)
10. Sheridan, T.B.: Telerobotics, Automation, and Human Supervisory Control. MIT Press, Cambridge (1992)
11. Marescaux, J., Leroy, J., Rubino, F., Smith, M., Vix, M., Simone, M., Mutter, D.: Transcontinental Robot-Assisted Remote Telesurgery: Feasibility and Potential Applications. Ann Surg. Apr. 235(4), 487–492 (2002)

12. Stern, H., Wachs, J., Edan, Y.: A Method for Selection of Optimal Hand Gesture Vocabularies. In: Sales Dias, M., Gibet, S., Wanderley, M.M., Bastos, R. (eds.) GW 2007. LNCS (LNAI), vol. 5085, pp. 57–68. Springer, Heidelberg (2009)
13. Berman, S., Stern, H.: Sensors for Gesture Recognition Systems. IEEE Transactions on Systems, Man, and Cybernetics-Part C 42(3), 277–290 (2012)
14. Stern, H., Smilansky, K., Berman, S.: Depth Based Dual Component Dynamic Gesture Recognition. In: The International Conference on Image Processing, Computer Vision, and Pattern Recognition IPCV 2013, Nevada, USA, July 22-25 (2013)
15. Jeannerod, M.: Intersegmental coordination during reaching at natural visual objects. In: Long, J., Baddeley, A. (eds.) Attention and Performance IX, pp. 153–169. Lawrence Erlbaum Associates, Mahwah, Mahwah (1981)
16. Marteniuk, R.G., MacKenzie, C.L., Jeannerod, M., Athenes, S., Dugas, C.: Constraints on human arm movement trajectories. Canadian Journal of Psychology 41(3), 365–378 (1987)
17. Friedman, J., Flash, T.: Task-dependent selection of grasp kinematics and stiffness in human object manipulation. Cortex 43(3), 444–460 (2007)
18. Murata, A., Gallese, V., Luppino, G., Kaseda, M., Sakata, H.: Selectivity for the shape, size, and orientation of objects for grasping in neurons of monkey parietal area AIP. Journal of Neurophysiology 83(5), 2580–2601 (2000)
19. Rizzolatti, G., Camarda, R., Fogassi, L., Gentilucci, M., Luppino, G., Matelli, M.: Functional organization of inferior area 6 in the macaque monkey. II. Area F5 and the control of distal movements. Experimental Brain Research 71(3), 491–507 (1988)
20. Flach, J.M., Holden, J.G.: The Reality of Experience: Gibson's Way. Presence: Teleoperators and Virtual Environments 7(1), 90–95 (1998)
21. Slater, M., Wilbur, S.: A framework for immersive virtual environments (FIVE): Speculations on the role of presence in virtual environments. Presence: Teleoperators and Virtual Environments 6, 603–616 (1997)
22. IJsselsteijn, W.A., Reiner, M.: On the importance of reliable real-time sensorimotor dependencies for establishing telepresence, Presence, Valencia, Spain, October 13-15, 2004, pp. 298–304 (2004)
23. IJsselsteijn, W.A., de Kort, Y.A., Haans, A.: Is this my hand i see before me? The rubber hand illusion in reality, virtual reality, and mixed reality. Presence: Teleoperators and Virtual Environments 15(4), 455–464 (2006)
24. Botvinick, M., Cohen, J.: Rubber hands 'feel' touch that eyes see. Nature 391(6669), 756 (1998)
25. Armel, K.C., Ramachandran, V.S.: Projecting sensations to external objects: Evidence from skin conductance response. Proceedings of the Royal Society of London B 270, 1499–1506 (2003)
26. Reiner, M.: The Role of haptics in immersive telecommunication environments. IEEE Transactions on Circuits & Systems for Video Technology 14(3), 392–401 (2004)
27. Lindeman, R.W., Templeman, J.N.: Vibrotactile feedback for handling virtual contact in immersive virtual environments, Usability Evaluation and Interface Design. In: Smith, M.J., Salvendy, G., Harris, D., Koubek, R.J. (eds.) Cognitive Engineering, Intelligent Agents and Virtual reality, pp. 21–25 (2001)
28. Kontarinis, D.A., Howe, R.D.: Tactile Display of High-Frequency Information in Teleoperation and Virtual Environments. Presence: Teleoperators and Virtual Environments 4(4), 387–402 (1995)
29. Campbell, C.L., Peters, R.A., Bodenheimer, R.E., Bluethmann, W.J., Huber, E., Ambrose, R.O.: Superpositioning of Behaviors Learned Through Teleoperation. IEEE Transactions on Robotics 22(1), 79–91 (2006)

30. Kontarinis, D.A., Howe, R.D.: Tactile display of vibratory information in teleoperation and virtual environments. Presence: Teleoperators and Virtual Environments 4(4), 387–402 (1995)
31. Bicchi, A.: Hands for dexterous manipulation and robust grasping: a difficult road towards simplicity. IEEE Transactions on Robotics and Automation 16(6), 652–662 (2000)
32. Pons, J.L., Ceres, R., Pfeiffer, F.: Multifingered dexterous robotics hand design and control: a review. Robotica 17, 661–674 (1999)
33. Lovchik, C.S., Diftler, M.A.: The Robonaut hand: a dexterous robot hand for space. In: IEEE Conference on Robotics and Automation (ICRA 1999), vol. 2, pp. 907–912 (1999)
34. Wilkinson, D.D., Weghe, M.V., Matsuoka, Y.: An extensor mechanism for an anatomical robotic hand. In: IEEE Conference on Robotics and Automation (ICRA 2003), vol. 1, pp. 238–243 (2003)
35. Howe, R.D.: A force-reflecting teleoperated hand system for the study of tactile sensing in precision manipulation. In: IEEE Conference on Robotics and Automation (ICRA), Nice, France (1992)
36. Kheddar, A.: Teleoperation based on the hidden robot concept. IEEE Trans. SMC – Part A 31(1), 1–13 (2001)
37. Ekvall, S., Kragic, D.: Interactive grasp learning based on human demonstration. In: IEEE Conference on Robotics and Automation (ICRA), New Orleans, LA, USA, April 26-May 1, vol. 4 (2004)
38. Arbib, M.A., Iberall, T., Lyons, D.: Coordinated control programs for control of the hands. Hand function and the neocortex. Experimental Brain Research 10, 111–129 (1985)
39. Iberall, T.: Human prehension and dexterous robot hands. International Journal of Robotics Research 16, 285–299 (1997)
40. Griffin, W.B., Findley, R.P., Turner, M.L., Cutkosky, M.R.: Calibration and mapping of a human hand for dexterous telemanipulation. In: ASME IMECE Haptic Interfaces for Virtual Environments and Teleoperator System Symposium (2000)
41. Berman, S., Friedman, J., Flash, T.: Object-action abstraction for teleportation. In: IEEE Systems Man and Cybernetics (SMC) Conference, Hawaii (October 2005)
42. Berman, S., Friedman, J., Bakır, G., Flash, T.: Action identification for teleoperation based on object - action abstraction. In: IEEE SMC International Conference on Distributed Human-Machine Systems (DHMS), Greece (March 2008)
43. Gibson, J.J.: The Ecological Approach to Visual Perception. Lawrence Erlbaum Inc., New Jersy (1979)
44. Roke, C., Melhuish, C., Pipe, T., Drury, D., Chorley, C.: Lump localisation through a deformation-based tactile feedback system using a biologically inspired finger sensor. Robotics and Autonomous Systems 60(11), 1442–1448 (2012)
45. Wettels, N., Fishel, J.A., Loeb, G.E.: Multimodal Tactile Sensor, the Human Hand as an Inspiration for Robot Hand Development. In: Balasubramanian, R., Santos, V.J. (eds.) Springer Tracts in Advanced Robotics (STAR) series, Springer, Heidelberg (2014)
46. Kravits, M., Shapiro, A., Berman, S.: Parametric Compliant contact model for human finger. In: The 7th Computational Motor Control Workshop (CMCW), Beer-Sheva. Israel (2011)
47. Nisky, I., Mussa-Ivaldi, F.A., Karniel, A.: Analytical Study of Perceptual and Motor Transparency in Bilateral Teleoperation. IEEE Transactions on Human-Machine Systems 43(6), 570–582 (2013)

Real-Time Neural Control of Mobile Robots

Edgar N. Sanchez[1], Alma Y. Alanis[2], Michel Lopez-Franco[1], Nancy Arana-Daniel[2], and Carlos Lopez-Franco[2]

[1] CINVESTAV, Unidad Guadalajara, Jalisco, Mexico
sanchez@gdl.cinvestav.mx
[2] CUCEI, Universidad de Guadalajara, Jalisco, Mexico
almayalanis@gmail.com

Abstract. This chapter presents two application in real-time using neural controls for mobile robots. First, a decentralized inverse optimal neural control is developed for a Shrimp robot, which is a kind of mobile robot with has terrain adaptability. Additionally, a neural control is designed for driving a nonholonomic mobile robot integrating stereo vision feedback. The desired trajectory of the robot is computed during the navigation process using the stereo camera sensor. The proposed neural control approaches are based on discrete-time High Order Neural Networks (RHONN's) trained with an extended Kalman filter (EKF).

Keywords: Neural control, tracking control, differential mobile, identifier, Computer vision, visual control, inverse optimal control.

1 Introduction

Robotics has achieved a great success for industrial manufacturing. Nevertheless, these commercial robots suffer from a fundamental disadvantage: lack of mobility. A fixed manipulator has a limited range of motion which depends on where it is bolted down. In contrast, a mobile robot would be able to travel throughout the manufacturing plant, flexibly applying its talents wherever it is most effective [1]. Fixed manipulators are typically programmed to perform repetitive tasks with perhaps limited use of sensors, whereas mobile robots are less structured in their operation and likely to use more sensors [2].

In applications which are too risky or too demanding for humans, or where a fast response is crucial, multirobot systems can play an important role thanks to their capability to cover the area. Possible applications are planetary exploration, urban search and rescue, monitoring, surveillance, cleaning, maintenance, among others. In order to successfully perform the tasks, robots require a high degree of autonomy and a good level of cooperation [3],[4].

A mobile robot needs locomotion mechanisms which enable it to move unbounded throughout its environment. However there are a large variety of possible ways to move, and so the selection of a robot's approach to locomotion is an important aspect. There are already research robots which can walk, jump, run, slide, skate, swim, fly, and, of course, roll. Most of these locomotion mechanisms have been inspired by their biological counterparts. There is, however,

A. El-Osery and J. Prevost (eds.), *Control and Systems Engineering*,
Studies in Systems, Decision and Control 27, DOI: 10.1007/978-3-319-14636-2_11

one exception: the actively powered wheel is a human invention which achieves extremely high efficiency on flat ground [1].

A extensive class of controllers have been proposed for mobile robots [5], [6], [7], [8], [9], [10], [11]. Most of these references present only simulation results and the controllers are implemented in continuous time. A common problem when applying standard control theory is that the required parameters are often either unknown at time, or are subject to change during operation. For example, the inertia of a robot as seen at the drive motor has many components, which might include the rotational inertia of the motor rotor, the inertia of gears and shafts, rotational inertia of its tires, the robot's empty weight, and its payload. Worse yet, there are elements between these components such as bearings, shafts and belts which may have spring constants and friction loads [12].

In this chapter a decentralized inverse optimal neural control for a Shrimp robot is presented first. Then, a neural control for a nonholonomic mobile robot integrating stereo vision feedback is discussed. A recurrent high order neural network (RHONN) is used to identify the plant model in a decentralized way, under the assumption of all the state is available for measurement. The learning algorithm for the RHONN is implemented using an Extended Kalman Filter (EKF).

The objective of optimal control theory, as applied to nonlinear systems, is to determine the control signals which will force a process to satisfy physical constraints and at the same time to minimize a performance criterion [13]. Unfortunately it is required to solve the associated Hamilton Jacobi Bellman (HJB) equation, which is not an easy task. The target of the inverse optimal control is to avoid the solution of the HJB equation [14]. For the inverse approach, a stabilizing feedback control law is developed and then it is established that this law optimizes a cost functional. The main characteristic of the inverse approach is that the cost function is a posteriori determined from the stabilizing feedback control law [6], [15], [16].

This chapter is organized as follows: Section 2 presents general discrete-time decentralized systems. Section 3 describes a brief review on nonlinear system neural identification; in addition, explain the Extended Kalman Filter training algorithm. Section 4 establishes the inverse optimal control. Section 5 illustrates the applicability of the proposed inverse optimal controller by means of an Shrimp robot. Section 6 presents a controller for mobile robots using visual data; the reference for the controller are computed using visual data, acquired from a camera mounted on the robot. Section 7 presents the experimental results of our proposed discrete-time inverse optimal neural controller with stereo vision feedback.

2 Decentralized Systems

Decentralized systems are basically defined as systems composed of more than two subsystems. A local control station directly observes the state of each subsystem, without known the state of other subsystems. As an advantage of this technique, even though the whole system is complex, the subsystem controllers are simple [17], [18]. Many of today technological and social problems involve very complex and large scale systems. For such systems, is usually necessary to avoid centralized controllers due to the cost of implementation, complexity of on-line computations, complexity of controller design, and reliability. In such a situation, a decentralized or partially decentralized feedback structure may be advantageous [19].

This chapter consider a class of disturbed discrete-time nonlinear and interconnected system, which can be presented in the block strict feedback form (BSFF) [14], consisting of r blocks, as:

$$\begin{aligned}
\chi_{i,k+1}^{1} &= f_i^1(\chi_i^1) + B_i^1(\chi_i^1)\chi_i^2 + \Gamma_{i\ell}^1 \\
\chi_{i,k+1}^{2} &= f_i^2(\chi_i^1, \chi_i^2) + B_i^2(\chi_i^1, \chi_i^2)\chi_i^3 + \Gamma_{i\ell}^2 \\
&\vdots \\
\chi_{i,k+1}^{j-1} &= f_i^{j-1}(\chi_i^1, \chi_i^2, \ldots, \chi_i^{j-1}) \\
&\quad + B_i^{j-1}(\chi_i^1, \chi_i^2, \ldots, \chi_i^{j-1})\chi_i^j + \Gamma_{i\ell}^{j-1} \\
\chi_{i,k+1}^{r} &= f_i^r(\chi_i) + B_i^r(\chi_i)u_i + \Gamma_{i\ell}^r
\end{aligned} \tag{1}$$

where $\chi_i \in \mathbb{R}^{n_i}$, $\chi_i = [\chi_i^{1T}, \chi_i^{2T} \ldots \chi_i^{rT}]^T$ and $\chi_i^j \in \mathbb{R}^{n_{ij}\times 1}$, $\chi_i^j = [\chi_{i1}^j, \chi_{i2}^j \cdots \chi_{il}^j]^T$, $i = 1, \ldots, N$; $j = 1, \ldots, r$; $l = 1, \ldots, n_{ij}$; N is the number of subsystems, $u_i \in \mathbb{R}^{m_i}$ is the input vector, the rank of $B_i^j = n_{ij}$, $\sum_{j=1}^{r} n_{ij} = n_i$, $\forall \chi_i^j \in D_{\chi_i^j} \subset \mathbb{R}^{n_{ij}}$. We assume that f_i^j , B_i^j and Γ_i^j are smooth and bounded functions, $f_i^j(0) = 0$ and $B_i^j(0) = 0$. The integers $n_{i1} \leq n_{i2} \leq \ldots \leq n_{ij} \leq m_i$ define the different subsystem structures. The interconnection terms are given by

$$\begin{aligned}
\Gamma_{i\ell}^1 &= \sum_{\ell=1,\ell\neq i}^{N} \gamma_{i\ell}^1(\chi_\ell^1) \\
\Gamma_{i\ell}^2 &= \sum_{\ell=1,\ell\neq i}^{N} \gamma_{i\ell}^2(\chi_\ell^1, \chi_\ell^2) \\
&\vdots \\
\Gamma_{i\ell}^{j-1} &= \sum_{\ell=1,\ell\neq i}^{N} \gamma_{i\ell}^{j-1}(\chi_\ell^1, \chi_\ell^2, \ldots, \chi_\ell^{j-1}) \\
\Gamma_{i\ell}^r &= \sum_{\ell=1,\ell\neq i}^{N} \gamma_{i\ell}^r(\chi_\ell)
\end{aligned} \tag{2}$$

where χ_ℓ represent the state vector of the ℓ-th subsystem with $1 \leq \ell \leq N$ and $\ell \neq i$. The terms (2) reflect the interaction between the i-th subsystem and the others subsystems.

3 Nonlinear System Identification

3.1 Neural Identification

Changing operating conditions, climatic and other environmental factors can cause either system wear or degradation of its components. It is so extremely difficult to consider all the physical laws involved in the behavior of the system [20]. System identification, which has held the attention of researchers for several years, consists on determining the system mathematical model [21]. Neural networks have grown to be a well-established methodology, which allows solving very difficult problems in engineering, as exemplified by their applications to identification and control of general nonlinear and complex systems. In particular, the use of recurrent neural networks for modeling and control has rapidly increased in recent years ([22] and references there in).

In this section, we consider the problem to identify nonlinear system

$$\chi_{k+1} = F\left(\chi_k, u_k\right) \tag{3}$$

where $\chi_k \in \Re^n$ is the state of the system, $u_k \in \Re^m$ is the control input and $F \in \Re^n \times \Re^m \to \Re^n$ is nonlinear function. To identify system (3), we use a RHONN defined as:

$$x_{j,k+1} = w_j^\top \varphi_j(\chi_k, u_k), \quad j = 1, \cdots, r \tag{4}$$

where x_j is the plant state of the j-th neuron, L_j is the respective number of high order connections, $\{I_1, I_2, \cdots, I_{L_j}\}$ is a collection of non-ordered subsets of $\{1, 2, \cdots, r+m\}$, r is the state dimension, m is the number of external inputs, w_j is the respective on-line adapted weight vector, with $\varphi_j(\chi_k, u_k)$ a vector which contains the high order terms.

Consider the problem to approximate the general discrete-time nonlinear system (3), by the following discrete-time RHONN series-parallel representation [23]:

$$x_{j,k+1} = w_j^{*\top} \varphi_j\left(\chi_k, u_k\right) + \epsilon_{\varphi_j} \tag{5}$$

where ϵ_{φ_j} is a bounded approximation error, which can be reduced by increasing the number of the adjustable weights [23]. Assume that there exists an ideal weights vector w_j^* such that $\left\|\epsilon_{\varphi_j}\right\|$ can be minimized on a compact set $\Omega_{\varphi_j} \subset \Re^{L_j}$ The ideal weight vector w_j^* is an artificial quantity required for analytical purpose [23]. In general, it is assumed that this vector exists and is constant but unknown. Let us define its estimate as w_j and the estimation error as

$$\widetilde{w}_{j,k} = w_{j,k} - w_j^* \tag{6}$$

The RHONN is trained with a Extended Kalman Filter (EKF) algorithm (13). Then, the dynamics of the identification error (15) can be expressed as

$$e_{j,k+1} = \widetilde{w}_{j,k} \varphi_j\left(\chi_k, u_k\right) + \epsilon_{\varphi_j} \tag{7}$$

On the other hand the dynamics of (6) is

$$\widetilde{w}_{j,k+1} = \widetilde{w}_{j,k} - \eta_j K_{j,k} e_k \tag{8}$$

The following decentralized RHONN modified model is proposed to identify (1):

$$\begin{aligned} x^1_{i,k+1} &= w^1_{i,k}\varphi(\chi^1_{i,k}) + w_i^{'1}\chi^2_{i,k} \\ x^2_{i,k+1} &= w^2_{i,k}\varphi(\chi^1_{i,k},\chi^2_{i,k}) + w_i^{'2}\chi^3_{i,k} \\ &\vdots \\ x^{j-1}_{i,k+1} &= w^{j-1}_{i,k}\varphi(\chi^1_{i,k},\chi^2_{i,k},\dots,\chi^{j-1}_{i,k}) + w_i^{'j-1}\chi^j_{i,k} \\ x^r_{i,k+1} &= w^r_{i,k}\varphi(\chi^1_{i,k},\chi^2_{i,k},\dots,\chi^r_{i,k}) + w_i^{'r}u_{i,k} \end{aligned} \tag{9}$$

where $x_i^j = [x_i^1 x_i^2 \dots x_i^r]^T$ is the j-th block neuron state with $i = 1,\dots,N$; $j = 1,\dots,r$; N is the number of subsystems, $w^j_{i,k}$ are the adjustable weights, $w_i^{'j}$ are fixed parameters with $rank(w_i^{'j}) = n_{ij}$, and u_{ik} represents the input vector. It is worth to note that, (9) constitutes a series-parallel identifier [24], [25] and does not consider explicitly the interconnection terms, whose effects are compensated by the neural network weights update.

Proposition 1. *The tracking of a desired trajectory x^j_{id}, defined in terms of the plant state χ_i^j formulated as (1) can be established as the following inequality [26]*

$$\| x^j_{i\delta} - \chi_i^j \| \leq \| x_i^j - \chi_i^j \| + \| x^j_{i\delta} - x_i^j \| \tag{10}$$

where $\| \bullet \|$ stands for the Euclidean norm, $i = 1,\dots,N$, $j = 1,\dots,r$; $x^j_{i\delta} - \chi_i^j$ is the system output tracking error; $x_i^j - \chi_i^j$ is the output identification error; and $x^j_{i\delta} - x_i^j$ is the RHONN output tracking error.

We establish the following requirements for the neural network tracking and control solution:

Requirement 1

$$\lim_{t\to\infty} \| x_i^j - \chi_i^j \| \leq \zeta_i^j \tag{11}$$

with ζ_i^j a small positive constant.

Requirement 2

$$\lim_{t\to\infty} \| x^j_{i\delta} - x_i^j \| = 0. \tag{12}$$

An on-line decentralized neural identifier based on (9) ensures (11), while (12) is guaranteed by a discrete-time decentralized inverse optimal control. It is possible to establish Proposition 1 due to separation principle for discrete-time nonlinear systems [27].

3.2 The EKF Training Algorithm

The best well-known training approach for recurrent neural networks (RNN) is the back propagation through time learning [28]. However, it is a first order gradient descent method and hence its learning speed can be very slow [29]. Recently, Extended Kalman Filter (EKF) based algorithms have been introduced to train neural networks [30], [15], with improved learning convergence [29]. The EKF training of neural networks, both feedforward and recurrent ones, has proven to be reliable and practical for many applications over the past fifteen years [15]. The training goal is to find the optimal weight values which minimize the prediction error. The EKF-based training algorithm is described by [31]:

$$\begin{aligned} K^j_{iq,k} &= P^j_{iq,k} H^j_{iq,k} M^j_{iq,k} \\ w^j_{iq,k+1} &= w^j_{iq,k} + \eta^j_{iq} K^j_{iq,k} e^j_{iq,k} \\ P^j_{iq,k+1} &= P^j_{iq,k} - K^j_{iq,k} H^{j\top}_{iq,k} P^j_{iq,k} + Q^j_{iq,k} \end{aligned} \tag{13}$$

with

$$M^j_{iq,k} = \left[R^j_{iq,k} + H^{j\top}_{iq,k} P^j_{iq,k} H^j_{iq,k} \right]^{-1} \tag{14}$$

$$e^j_{iq,k} = \chi^j_{iq,k} - x^j_{iq,k} \tag{15}$$

where $e^j_{iq,k}$ is the identification error, $P^j_{iq,k+1}$ is the state estimation prediction error covariance matrix, $w^j_{iq,k}$ is the jq-th weight (state) of the i-th subsystem, η^j_{iq} is a design parameter such that $0 \leq \eta^j_{iq} \leq 1$, $\chi^j_{iq,k}$ is the jq-th plant state, $x^j_{iq,k}$ is the jq-th neural network state, q is the number of states, $K^j_{iq,k}$ is the Kalman gain matrix, $Q^j_{iq,k}$ is the measurement noise covariance matrix, $R^j_{iq,k}$ is the state noise covariance matrix, and $H^j_{iq,k}$ is a matrix, in which each entry of $(H^j_{q,k})$ is the derivative of jq-th neural network state $(x^j_{iq,k})$, with respect to all adjustable weights $(w^j_{iq,k})$, as follows

$$H^j_{q,k} = \left[\frac{\partial x^j_{iq,k}}{\partial w^j_{iq,k}} \right]^T_{w^j_{iq,k} = w^j_{iq,k+1}}, \tag{16}$$

$$i = 1, ..., N \; and \; j = 1, ..., r_i \tag{17}$$

Usually P^j_{iq}, Q^j_{iq} and R^j_{iq} are initialized as diagonal matrices, with entries $P^j_{iq}(0)$, $Q^j_{iq}(0)$ and $R^j_{iq}(0)$, respectively [32]. It is important to note that $H^j_{iq,k}$, $K^j_{iq,k}$ and $P^j_{iq,k}$ for the EKF are bounded [33]. Then the dynamics of (15) can be expressed as

$$e^j_{iq,k} = \widetilde{w}^j_{iq,k} \varphi^j_{iq,k}(x_k, u_k) + \epsilon_{\varphi^j_{iq,k}} \tag{18}$$

on the other hand, the dynamics of weight estimation error $\widetilde{w}_{iq,k}^{j}$ is

$$\widetilde{w}_{iq,k+1}^{j}=\widetilde{w}_{iq,k}^{j}-\eta_{iq}^{j}K_{iq,k}^{j}e_{iq,k}^{j} \tag{19}$$

The stability analysis for the i-th subsystem of RHONN (9) identification of the i-th nonlinear plant subsystem (1), is based on the following theorem.

Theorem 1 *[34]: Assume that the solution of i-th subsystem of RHONN (9) is satisfied by the bounds b_1 and b_2, then the RHONN (9) with $i=1,\ldots,N$ trained with the EKF-based algorithm (13) to identify the nonlinear plant (1) in presence of interconnections, ensures that the identification error (15) and the RHONN wights are semiglobally uniformly bounded (SGUUB).*

4 Inverse Optimal Control

This section closely follows [35]. The main goal of this section is to design of an inverse optimal control. First, we give briefly details about optimal control methodology and their limitations. Let consider the nonlinear discrete-time affine system

$$x_{k+1}=f\left(x_k\right)+g(x_k)u_k,\ x_0=x(0) \tag{20}$$

where $x_k\in\mathbb{R}^n$ is the state of the system at time $k\in\mathbb{N}$, $u\in\mathbb{R}^m$ $f:\mathbb{R}^n\rightarrow\mathbb{R}^n$, $g:\mathbb{R}^n\rightarrow\mathbb{R}^{n\times m}$, are smooth and bounded mappings. We assume $f(0)=0$. $\mathbb{N}$ denotes the set of nonnegative integers. The following meaningful cost functional is associated with the trajectory tracking problem for system (20)

$$\mathcal{L}(z_k)=\sum_{n=k}^{\infty}\left(l(z_n)+u_n^TR(z_n)u_n\right) \tag{21}$$

where $z_k=x_k-x_{\delta,k}$ with $x_{\delta,k}$ as the desired trajectory for x_k; $z_k\in\mathbb{R}^n$; $\mathcal{L}(z_k):\mathbb{R}^n\rightarrow\mathbb{R}^+$; $l(z_k):\mathbb{R}^n\rightarrow\mathbb{R}^+$ is a positive semi-definite function and $R(z_k):\mathbb{R}^n\rightarrow\mathbb{R}^m\times m$ is a real symmetric positive definite weighting matrix. The entries of $R(z_k)$ can be fixed or can be functions of the system state in order to vary the weighting on control efforts according to the state value [13]. Considering the state feedback control design problem, we assume that the full state x_k is completely available. Using the optimal value function $\mathcal{L}^*(x_k)$ for (21) as Lyapunov function $V(x_k)$, equation (21) can be rewritten as

$$V(z_k)=l(z_k)+u_k^TR(z_k)u_k+V(z_{k+1})$$

where the following boundary condition $V(0)=0$ is required, so that $V(z_k)$ becomes a Lyapunov function. From Bellman optimality principle [36], [37], it is known that, for the infinite horizon optimization case, the value function $V(z_k)$

becomes time invariant and satisfies the discrete-time (DT) Bellman equation [38], [36], [39].

$$V(z_k) = \min_{u_k} \left\{ l(z_k) + u_k^T R(z_k) u_k + V(z_{k+1}) \right\} \tag{22}$$

where $V(z_{k+1})$ depends on both z_k and u_k by means of z_{k+1} in (20). Note that the DT Bellman equation is solved backward in time [38]. In order to establish the conditions that the optimal control law must satisfy, we define the discrete-time Hamiltonian $H(z_k, u_k)$ as

$$H(z_k, u_k) = l(z_k) + u_k^T R(z_k) u_k + V(z_{k+1}) - V(z_k) \tag{23}$$

A necessary condition which the optimal control law should satisfy is $\frac{\partial H(z_k, u_k)}{\partial u_k} = 0$, then

$$0 = 2R(z_k)u_k + g^T(x_k)\frac{\partial V(z_{k+1})}{\partial u_k}$$

Therefore, the optimal control law to achieve trajectory tracking is formulated as

$$u_k^* = -\frac{1}{2} R^{-1}(z_k) g^T(x_k) \frac{\partial V(z_{k+1})}{\partial z_{k+1}} \tag{24}$$

with the boundary condition $V(0) = 0$. For determining trajectory tracking inverse optimal control, it is necessary to solve the following HJB equation:

$$l(z_k) + V(z_{k+1}) - V(z_k) + \frac{1}{4} \frac{\partial V^T(z_{k+1})}{\partial z_{k+1}} g R^{-1}(z_k) g^T \frac{\partial V(z_{k+1})}{\partial z_{k+1}} = 0 \tag{25}$$

which is a challenging task. To overcome this problem, we propose to solve the inverse optimal control problem.

Definition 1. *Consider the tracking error as $z_k = x_k - x_{\delta,k}$, being $x_{\delta,k}$ the desired trajectory for x_k. Let define the control law*

$$u_k^* = -\frac{1}{2} R^{-1}(z_k) g^T(x_k) \frac{\partial V(z_{k+1})}{\partial z_{k+1}} \tag{26}$$

It will be inverse optimal (globally) stabilizing along the desired trajectory $x_{\delta,k}$ if:

(i) *It achieves (global) asymptotic stability of $x_k = 0$ for system (20) along reference $x_{\delta,k}$;*

(ii) *$V(z_k)$ is (radially unbounded) positive definite function such that inequality*

$$\bar{V} := V(z_{k+1}) - V(z_k) + u_k^{*T} R(z_k) u_k^* \leq 0 \tag{27}$$

is satisfied.

Selecting $l(z_k) := -\bar{V}$, then $V(z_k)$ is a solution for (25) and cost functional (21) is minimized.

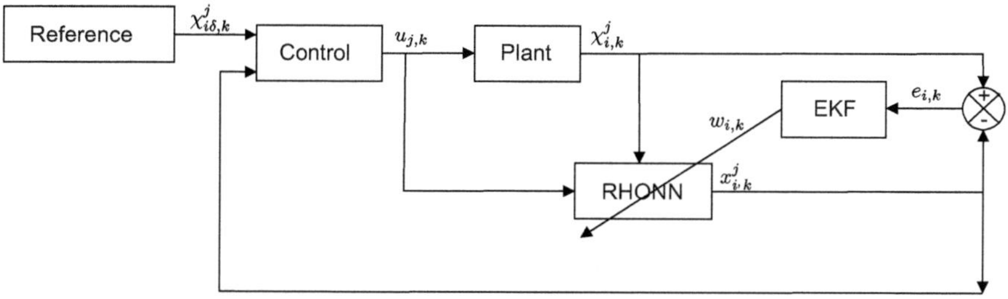

Fig. 1. Control scheme for i-th subsystem

As established in *Definition* 1, the inverse optimal control law for trajectory tracking is based on knowledge of $V(z_k)$. Then, a control Lyapunov function (CLF) $V(z_k)$ is proposed, such that (i) and (ii) are guaranteed. Hence, instead of solving (25) for $V(z_k)$ a quadratic candidate CLF $V(z_k)$ for (26) is proposed as:

$$V(z_k) = \frac{1}{2} z_k^T P z_k \;\; P = P^T > 0 \tag{28}$$

in order to ensure stability of the tracking error z_k, where

$$z_k = x_k - x_{\delta,k} = \begin{bmatrix} x_{1,k} - x_{1\delta,k} \\ \vdots \\ x_{n,k} - x_{n\delta,k} \end{bmatrix} \tag{29}$$

The control law (26) with (28), which is referred to as the inverse optimal control law, optimizes a meaningful cost functional of the form (21). Consequently, by considering $V(z_k)$ as in (28), control law (26) takes the following form:

$$u_k^* = -\frac{1}{2}\left(R(z_k) + \frac{1}{2} g^T(x_k) P_k g(x_k)\right)^{-1} \times g^T(x_k) P_k (f(x_k) - x_{\delta,k+1}) \tag{30}$$

P_k and $R(z_k)$ are positive definite and symmetric matrices; thus, the existence of the inverse in (30) is ensured.

The RHONN system model (9) can be presented in a general form as

$$x_{i,k+1} = f_i(\chi_{i,k}) + g_i u_{i,k} \tag{31}$$

For system (31), we apply the inverse optimal control law (30) as

$$u_{i,k}^* = -\frac{1}{2}\left(R_i(z_{i,k}) + \frac{1}{2} g_i^T(x_{i,k}) P_{i,k} g_i(x_{i,k})\right)^{-1} \times g_i^T(x_{i,k}) P_{i,k} (f_i(x_{i,k}) - x_{i\delta,k+1}) \tag{32}$$

The whole identification and optimal control, for the i-th subsystem, is displayed in (Fig. 1).

5 Shrimp Robot Application

To move on uneven terrain, a special mechanism is required. Mobile robotics in rough terrain has become a growing field for various applications like space exploration, rescue in case of catastrophes, firefighting and remote inspection. Basically, environments which are inaccessible to humans usually demand vehicles, which are extremely maneuverable and capable of dealing with rough terrain. Robotic rover is a kind of mobile robot which has terrain adaptability. In general, rovers can be classified into two types of mechanisms: active locomotion and passive locomotion. Active locomotion requires to have prior knowledge of ground condition. Passive locomotion implies that no active intervention is needed to handle roughness. On other words the robot can adapt to the terrain, i.e. shrimp robots [40], [41]. Applicability of the scheme is illustrated via real time implementation for a Shrimp robot, whose model is considered to be unknown as well as all its parameters and disturbances.

We consider a Shrimp robot with 8 actuated wheels as shown in (Fig. 2). The dynamic of the DC motor actuating each one of the lateral wheels can be expressed as the following state-space model [42]:

Fig. 2. Shrimp Robot

$$\begin{bmatrix} \dot{x}_{1i} \\ \dot{x}_{2i} \end{bmatrix} = \begin{bmatrix} -\frac{b}{J} & \frac{K_t}{J} \\ -\frac{K_b}{L_a} & -\frac{R_a}{L_a} \end{bmatrix} \begin{bmatrix} x_{1i} \\ x_{2i} \end{bmatrix} + \begin{bmatrix} 0 \\ \frac{1}{L_a} \end{bmatrix} u_i \tag{33}$$

(34)

$R_a = 0.6\Omega$	$K_b = 0.8\frac{Vs}{rad}$
$L_a = 0.012H$	$J = 0.0167\frac{kgm^2}{s^2}$
$K_t = 0.8\frac{Nm}{A}$	$b = 0.0167Nms$

where x_{1i} represent the angular velocity in $\frac{rad}{s}$ for each motor respectively with $i = 1 \dots 6$. x_{2i} is the armature current in Amp. The input terminal voltage U_i is taken to be the controlling variable. R_a and L_a are the armature inductance in H and resistance in ohm respectively. K_t is the torque factor constant in $\frac{Nm}{Amp}$. K_b is the back emf constant in $\frac{Vs}{rad}$. J represent the moment of inertia in $\frac{kgm^2}{s^2}$ and b is the coefficient of viscous friction which opposes the direction of motion in Nms.

We apply the neural identifier, developed in Section 3.1, to obtain a discrete-time neural model which is trained with the EKF (13) for the electrically driven Shrimp robot (33), as follows:

$$\begin{aligned} x_{1i,k+1} &= w_{11i,k}S(\chi_{1i,k}) + w_{12i,k}S(\chi_{2i,k}) + w'_{1i}\chi_{2i,k} \\ x_{2i,k+1} &= w_{21i,k}S(\chi_{1i,k}) + w_{22i,k}S(\chi_{2i,k}) + w'_{2i}u_{i,k} \end{aligned} \tag{35}$$

where x_{1i} and x_{2i} identify the angular velocities χ_{1i} and the motor currents χ_{2i}, respectively. The NN training is performed on-line, and all of its weights are initialized, randomly. The RHONN parameters are heuristically selected as:

$$\begin{array}{lll} P_{1i}(0) = 1 \times 10^8 & R_{1i}(0) = 1 \times 10^1 & Q_{1i}(0) = 5 \times 10^9 \\ P_{2i}(0) = 1 \times 10^8 & R_{2i}(0) = 5 \times 10^1 & Q_{2i}(0) = 5 \times 10^9 \\ w'_{1i} = 0.1 & w'_{2i} = 1 & \end{array}$$

It is important to consider that for the EKF-learning algorithm the covariances matrices are used as design parameters [15], [32]. The neural network structure (35) is determined heuristically in order to minimize the state estimation error.

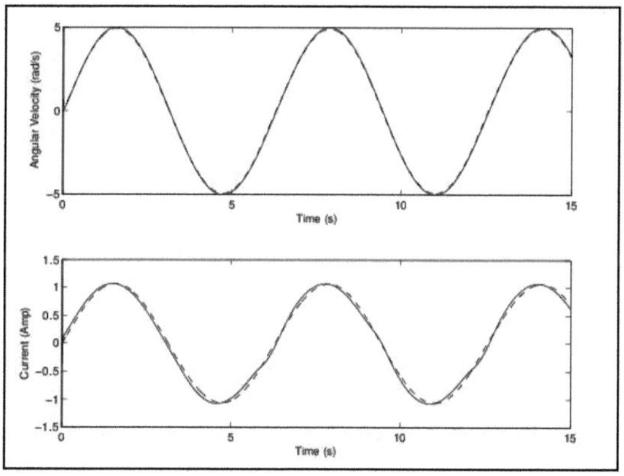

Fig. 3. Tracking performance of $\chi^r_{1,\delta}$(solid line) and χ^r_1(dashed line)

The goal is to force the state $x_{1i,k}$ to track a desired reference signal $\chi_{1i\delta,k}$, which is achieved by a control law as described in section 4. First the tracking error is defined as

$$z_{1i,k} = x_{1i,k} - \chi_{1i\delta,k}. \tag{36}$$

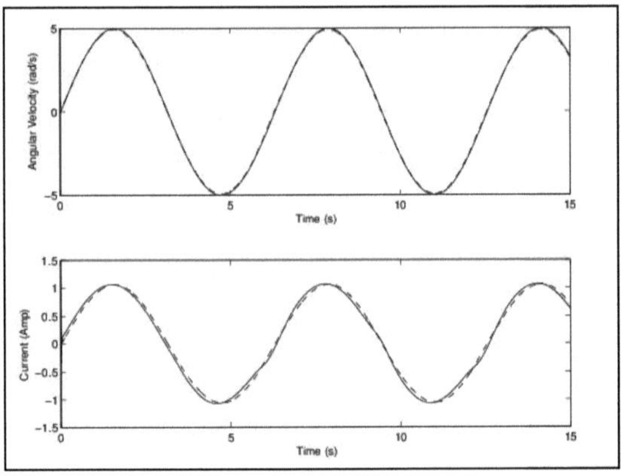

Fig. 4. Tracking performance of $\chi^r_{2,\delta}$(solid line) and χ^r_2(dashed line)

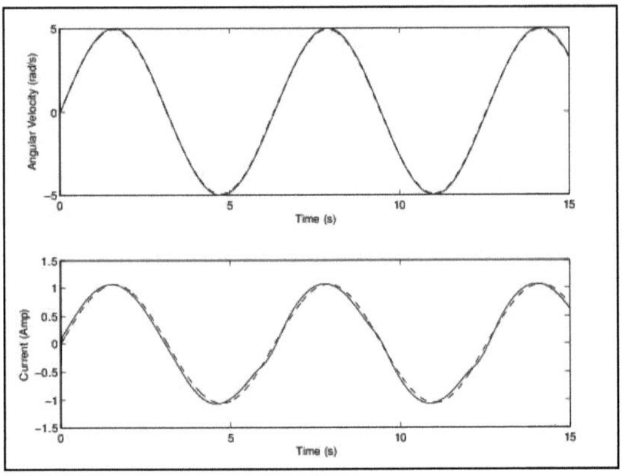

Fig. 5. Tracking performance of $\chi^r_{3,\delta}$(solid line) and χ^r_3(dashed line)

Then, using (35) and introducing the desired dynamics for $z_{1i,k}$, we obtain

$$z_{1i,k+1} = w_{1i,k}\varphi_1(\chi_{1i,k}) + w'_{1i,k}\chi_{2i,k} - \chi_{1i\delta,k+1} = K_{1i}z_{1i,k} \tag{37}$$

where $|K_{1i}| < 1$. The desired value $\chi_{2i\delta,k}$ for the pseudo-control input $\chi_{2i,k}$ is calculated from (37) as

$$\chi_{2i\delta,k} = (w'_{1i,k})^{-1}(-w_{1i,k}\varphi_{1i}(\chi_{1i,k}) + \chi_{1i\delta,k+1} + K_{1i}z_{1i,k}) \tag{38}$$

At the second step, we introduce a new variable as

$$z_{2i_k} = x_{2i,k} - \chi_{2i\delta,k} \tag{39}$$

Taking one step ahead, we have

$$z_{2i,k+1} = w_{2i,k}\varphi_{2i}(\chi_{1i,k}, \chi_{2i,k}) + w'_{2i,k}u_{i,k} - \chi_{2i\delta,k+1} \tag{40}$$

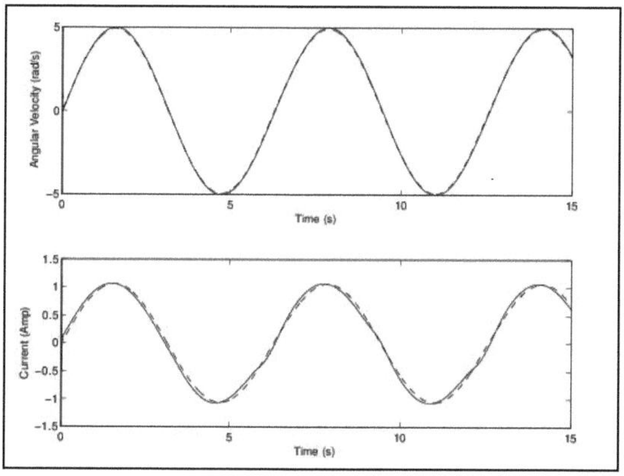

Fig. 6. Tracking performance of $\chi_{4,\delta}^{r}$(solid line) and χ_{4}^{r} (dashed line)

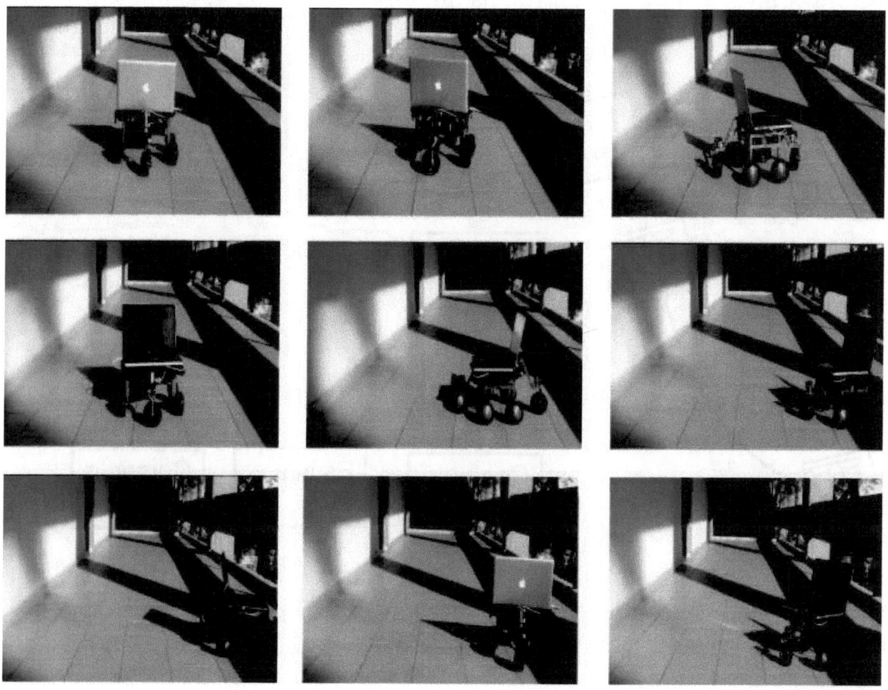

Fig. 7. Navigation result represented as a sequence of images seen from a third person's perspectives

where $u_{i,k}$ is defined as

$$u_{i,k} = -\frac{1}{2}\left(R_i(z_k) + g_i^T(x_{i,k})P_i g_i(z_k)\right)^{-1} \times g_i^T(x_k)P_i(f_i(x_{i,k}) - x_{i\delta,k+1}) \quad (41)$$

The results are displayed as follows: Fig. 3 to Fig. 6 present on the top the trajectory tracking performance for the angular velocities of i-th wheel and on the bottom the trajectory tracking performance for the currents of i-th motor respectively. Fig. 7 represents the navigation result in a sequence of images seen from a third person's perspectives.

6 Vision Systems Application

The objective of this section is to present a controller for mobile robots which includes vision. The reference for the controller are calculated using visual data, acquired from a camera mounted on the robot. Using visual data the controller drives the nonholonomic robot from its current pose toward a desired pose (Fig. 8.a). A recurrent high order neural network (RHONN) as described in section 3, is first used to identify the plant model, under the assumption that all the state is available for measurement; second, based on the real-time identifier, a controller which uses inverse optimal control, as discussed in section 4, is implemented.

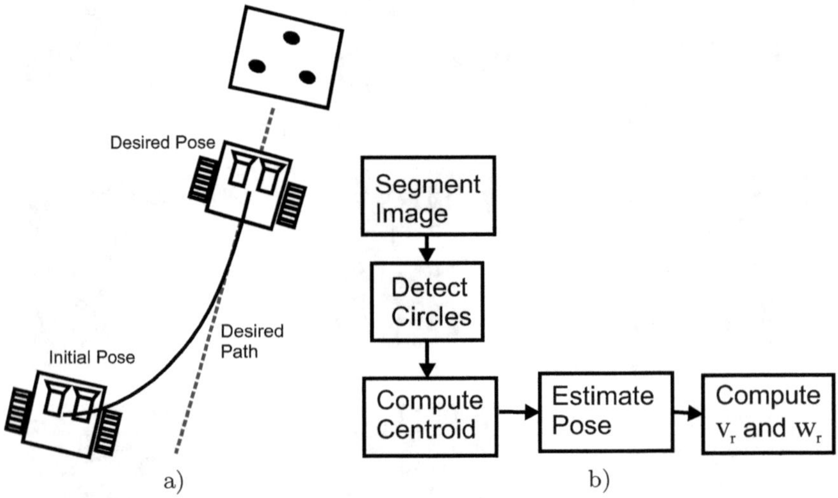

Fig. 8. a) Robot moving from initial pose to the desired pose. b) Translational v_r, ω_r and angular velocities computations from visual data.

The use of visual feedback to control a robot is commonly termed as *visual servoing* or *visual control* [43], [44], [45]. In this section, the visual data is acquired from a stereo vision system which is mounted directly on the mobile robot.

The visual control objective is to minimize the error $e(t)$ defined as [46]

$$e(t) = s(t) - s^* \tag{42}$$

where $s(t)$ denote the features extracted from the current pose, and s^* denote the features extracted from the desired pose.

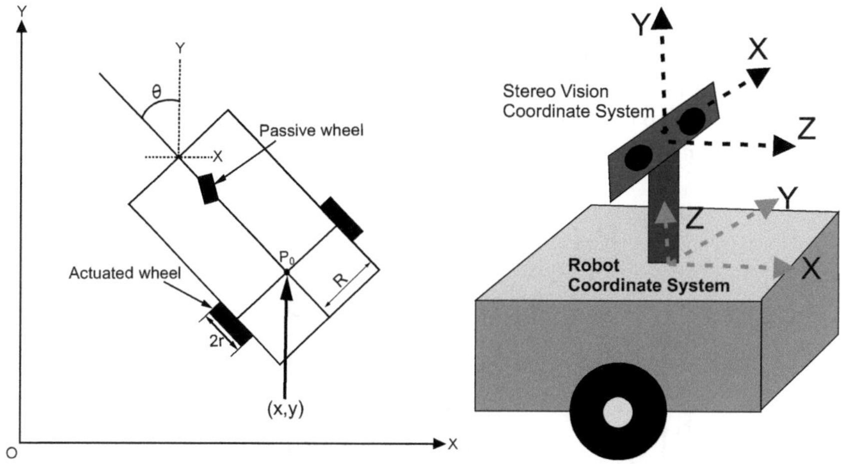

a) b)

Fig. 9. a) Mobile robot with two actuated wheels. b) Coordinates systems of the mobile robot and stereo vision system.

In this secttion we consider a nonholonomic mobile robot moving on a plane as shown in (Figure 9.a). Its pose is defined as $[x\ y\ \theta]^t$. Its kinematics model is

$$\begin{aligned}\dot{x} &= v_r \cos\theta \\ \dot{y} &= v_r \sin\theta \\ \dot{\theta} &= \omega_r\end{aligned} \tag{43}$$

where v_r and ω_r represent the translational and angular velocities, respectively.

In order to estimate v_r and ω_r by using visual data, several steps must be done, (Fig. 8.b.) First, the image is converted to HSV (Hue Saturation Value) colorspace [47]. Using this image we apply a mask, previously computed from a reference image, and then we obtain a segmented image. From the segmented image, we compute the boundaries using the Moore-neightbor tracing algorithm [48]; then for each boundary we compute the following metric $m = 4\pi area/perimeter^2$, which if close to 1, then the boundary is more likely to be a circle.

From the detected circles, we compute their centroid. Later, using the centroids of the circles from the desired image, the current image and the corresponding depths, we estimate the robot's pose. Finally, with the current and desired poses, we compute the velocities v_r, ω_r to drive the robot from the current pose to the desired pose.

6.1 Stereo Vision

The principle of stereo vision with parallel optical axes is displayed in Fig. 10. The 3D point P is projected onto the image plane of the left camera as $p_L =$

$[x_l, y_l]$, similarly $p_R = [x_R, y_R]$ represents the projection of P onto the image plane of the right camera. Since the image plane of the left and right camera are located on the same plane, the y-coordinates in these two images are the same ($y_L = y_R$), and the disparity is equal to the difference between the horizontal coordinates ($x_R - x_L$).

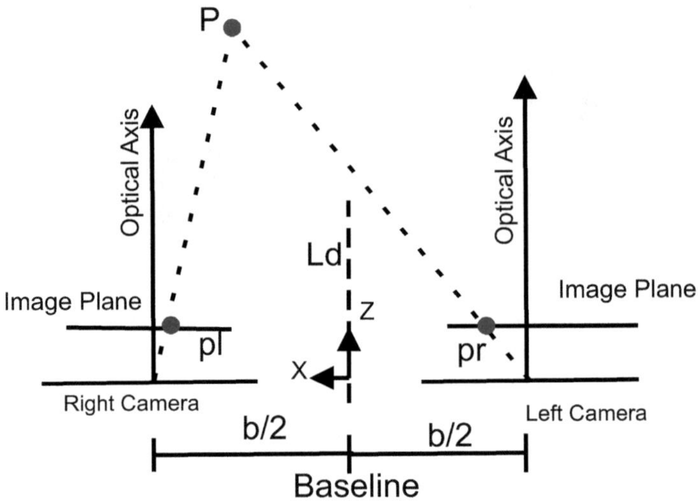

Fig. 10. Image formation of a stereo vision with parallel optical axes

Let $P = (X, Y, Z) \in \mathbb{R}^3$ denote a 3D point in the world. The coordinates of P on the left camera are

$$P_L = [X + b/2, Y, Z]^T \tag{44}$$

Similarly, the point P on the right camera is

$$P_R = [X - b/2, Y, Z]^T \tag{45}$$

Using the standard projective camera projection, we obtain

$$x_L = (X + b/2)f/Z \tag{46}$$
$$x_R = (X - b/2)f/Z \tag{47}$$

Similarly

$$y_L = Yf/Z \tag{48}$$
$$y_R = Yf/Z \tag{49}$$

The depth of the point P can be recovered from the x-coordinate of the images points x_L and x_R, subtracting (47) from (46) we obtain

$$Z = \frac{bf}{x_L - x_R} \tag{50}$$

Similarly, we can also solve for X using (46), (47) and (50) and obtain

$$X = \frac{b(x_L + x_R)}{2(x_L - x_R)} \tag{51}$$

The Y value can be recovered with (48) or (49), since they have the same value, and (50 to get

$$Y = \frac{by}{x_L - x_R} \tag{52}$$

The mobile robot moves on a 2D plane, thus we only need two coordinates to fully determine its pose (x,y,θ). Since the robot can not move in the Y direction (orthogonal to the plane), we can estimate its pose with respect to the planar target using only the Z and X values of the point P.

Let $Q_i^* = (Z_i^*, X_i^*)$ and $Q_i = (Z_i, X_i)$ represent the 2D Euclidean point of the feature point P_i expressed in the frames $\mathcal{F}^*$ and $\mathcal{F}$, respectively. From Euclidean geometry, the relationship between the features is defined as

$$Q_i^* = RQ_i + t \tag{53}$$

where $R \in \mathbb{R}^{2\times 2}$ is the 2D rotation matrix and $t = (t_x, t_y) \in \mathbb{R}^2$ is the translation vector, Fig. 11.

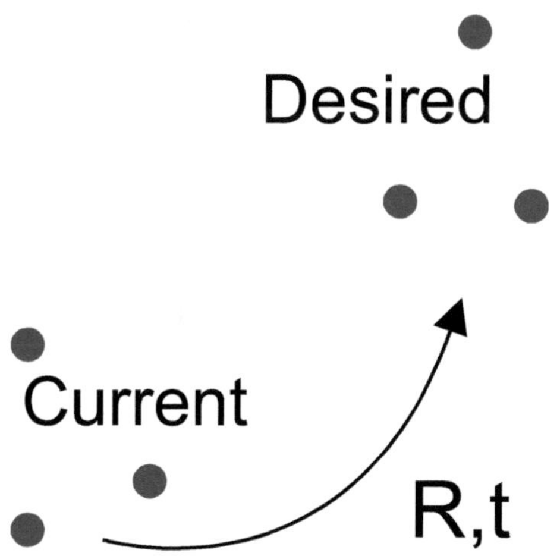

Fig. 11. Pose estimation problem

To estimate the pose of the robot given the points Q_i and Q_i^* from the current and desired pose, we need to solve the following least-squares problem

$$E(\theta, t) = \sum_{i=1}^{n} |R_\theta X_i + t - X_i^*|^2 \tag{54}$$

which can be solved in a closed form [49].

6.2 Kinematic Planer

Once that the pose of the robot has been estimated, the next step is the determined the robot velocities, which minimize the error between the current pose of the robot and the desired one.

The path, to be tracked is defined as the line L_d passing through the center of the stereo rig parallel to the optical axes of the left and right cameras (Fig. 11). A line on the plane can be defined using the general equation of the line ($ax + by + c = 0$), therefore the desired line is defined at the desired pose as

$$L_d = [0\ 1\ 0] \tag{55}$$

The signed distance from the current pose of the robot and principal axis L_d at the desired pose is defined as

$$d = [t_x\ t_y\ 1]^t \cdot L_d \tag{56}$$

The angular velocity of the robot must drive the robot toward the line L_d with

$$\beta_d = -K_d d, \quad K_d > 0 \tag{57}$$

and adjust the orientation of the robot (heading angle) with

$$\beta_o = K_o(\theta^* - \theta), \quad K_o > 0 \tag{58}$$

Then, the combined kinematic control law [50], used to generate the robot angular velocity for path following, is defined as

$$\omega_r = \beta_d + \beta_o \tag{59}$$

v_r is set to a constant value (e.g. 0.2 m/s). However, when the robot is close to the desired pose the velocity is computed with

$$v_r = \kappa_v \sqrt{t_x^2 + t_y^2} \tag{60}$$

7 Results

In this section we present experimental results of our proposed discrete-time inverse optimal neural controller with stereo vision feedback.

Fig. 12. Represent navigation result seen from the third persons perspectives

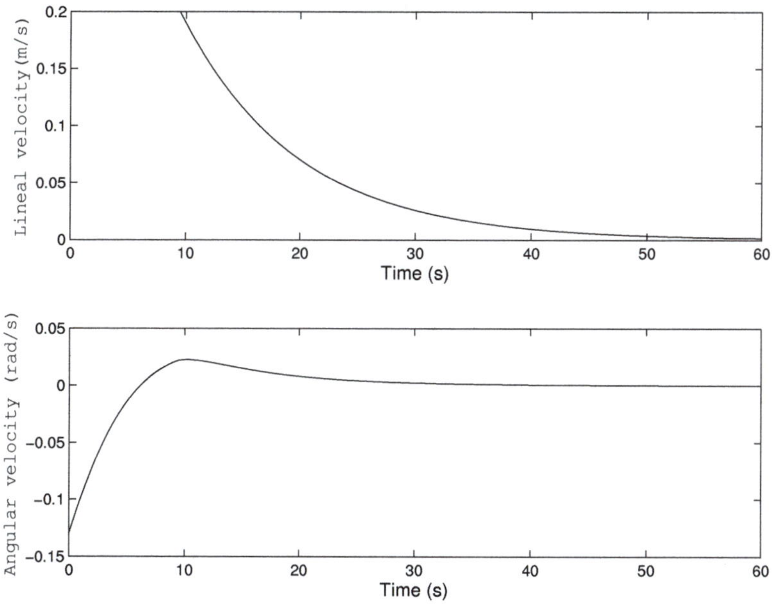

Fig. 13. Linear (top) and angular (bottom) velocities generated by the stereo vision algorithm

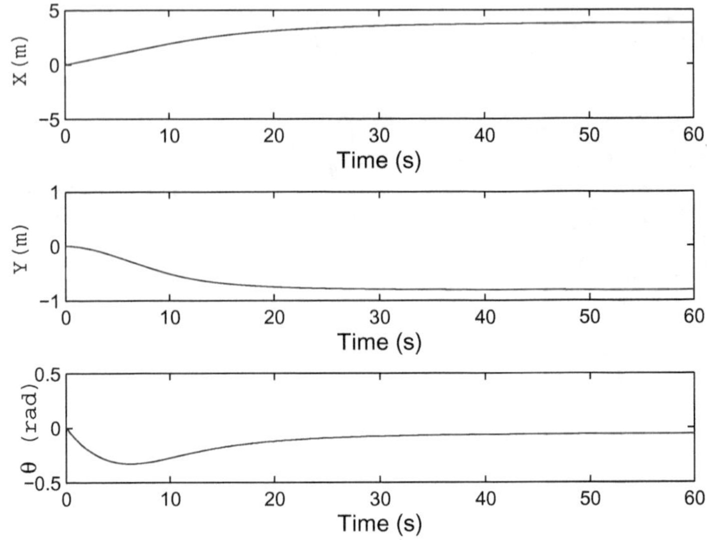

Fig. 14. x-axis identification (top), y-axis identification (middle) and θ angle (bottom), plant signal in solid line and neural signal in dashed line

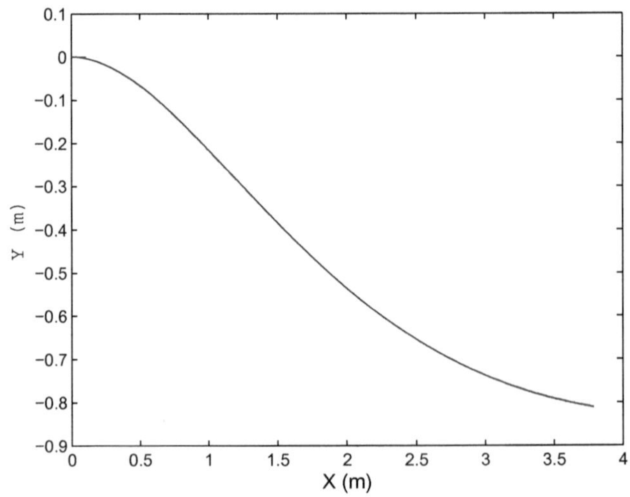

Fig. 15. Trajectory tracking result (plant signal in solid line and neural signal in dashed line)

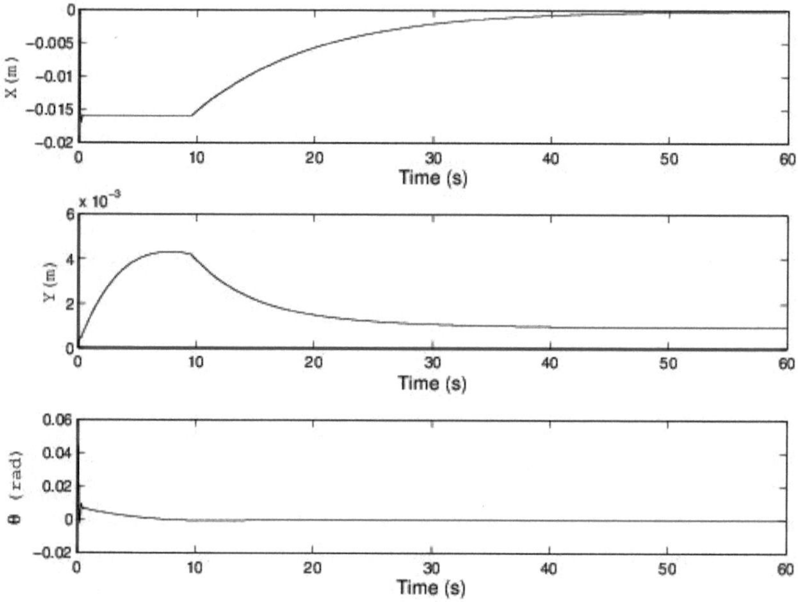

Fig. 16. Tracking errors, x-axis (top), y-axis (middle) and θ angle (bottom)

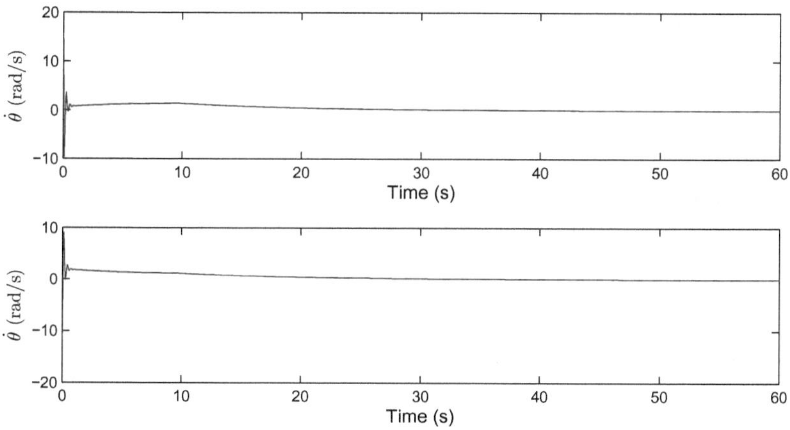

Fig. 17. Angular velocity identification for the left and right wheels, respectively (plant signal in solid line and neural signal in dashed line)

The robot moves under the action of the proposed controller; which uses as references the linear and angular velocities computed from the stereo vision algorithm. The initial pose of the robot is $[0\ 0\ 0]^T$, and the desired pose is $[3.8 - 0.8\ 0]^T$.

Results are presented as follows: Fig. 12 display a sequence images of navigation seen from a third persons perspective. In Fig. 13, we show the linear and angular velocities used as references by the proposed controller; these velocities are computed by the stereo vision algorithm using the current and desired images of the target object. Fig. 14 portrait the identification performance for x-axis, y-axis and θ angle. Fig. 15 shows the trajectory tracking results. In Fig. 16 we present the tracking errors. Finally, Fig. 17 exhibition the angular velocity identification for the left and right wheels.

8 Conclusions

This chapter has presented a real-time discrete-time decentralized inverse optimal neural control, which achieve tracking for non-linear systems and is inverse optimal in the sense that, a posteriori, minimizes a meaningful cost functional. The training of the neural network is performed on-line using an extended Kalman filter. Real-time results confirm the effectiveness of the proposed scheme. In addition, we have presented a discrete-time inverse optimal neural control with visual feedback. The controller is able to accomplish trajectory tracking of a nonlinear system; this controller is inverse optimal in the sense that it minimizes a meaningful cost functional. The mobile robot dynamics at the actuator level as well as its kinematics and dynamics uncertainties are considered in the construction of the controller by means of neural identification. Visual data acquired from a stereo vision sensor is used to estimate the robots pose; with this pose we were able to compute the reference velocities vr, wr that allow the controller to drive the nonholonomic robot from its current pose toward the desired pose. Results are presented to illustrate the effectiveness of the proposed controller.

Acknowledgement. The authors thank the support CONACYT Mexico, through Projects CB-103191, CB-131678, CB-106838, CB-156567 and INFR-229696.

References

1. Siegwart, R., Nourbakhsh, I.R.: Introduction to Autonomous Mobile Robots. Bradford Company, Scituate (2004)
2. Cook, G.: Mobile Robots: Navigation, Control and Remote Sensing. Wiley-IEEE Press, Hoboken (2011)
3. Jevtic, A., Gutierrez, A., Andina, D., Jamshidi, M.: Distributed bees algorithm for task allocation in swarm of robots. IEEE Systems Journal 6(2), 296–304 (2012)

4. Shaneyfelt, T., Joordens, M., Nagothu, K., Prevost, J.: Control and simulation of robotic swarms in heterogeneous environments. In: IEEE International Conference on Systems, Man and Cybernetics, SMC 2008, Suntec, Singapore, pp. 1314–1319 (October 2008)
5. Akhavan, S., Jamshidi, M.: Ann-based sliding mode control for non-holonomic mobile robots. In: Proceedings of the 2000 IEEE International Conference on Control Applications, Anchorage, AK, USA, pp. 664–667 (2000)
6. Do, K.D., Jiang, Z.P., Pan, J.: Simultaneous tracking and stabilization of mobile robots: an adaptive approach. IEEE Transactions on Automatic Control 49(7), 1147–1151 (2004)
7. Fierro, R., Lewis, F.L.: Control of a nonholonomic mobile robot using neural networks. IEEE Transactions on Neural Networks 9, 589–600 (1998)
8. Jiang, Z.-P., Nijmeijer, H.: A recursive technique for tracking control of nonholonomic systems in chained form. IEEE Transactions on Automatic Control 44(2), 265–279 (1999)
9. Kumbla, K.K., Jamshidi, M.: Neural network based identification of robot dynamics used for neuro-fuzzy controller. In: Proceedings of the IEEE International Conference on Robotics and Automation, Albuquerque, NM, USA, vol. 2, pp. 1118–1123 (April 1997)
10. Raghavan, V., Jamshidi, M.: Sensor fusion based autonomous mobile robot navigation. In: IEEE International Conference on System of Systems Engineering, SoSE 2007, San Antonio, TX, USA, pp. 1–6 (April 2007)
11. Yang, J.-M., Kim, J.-H.: Sliding mode control for trajectory tracking of nonholonomic wheeled mobile robots. IEEE Transactions on Robotics and Automation 15(3), 578–587 (1999)
12. Holland, J.: Designing Autonomous Mobile Robots: Inside the Mind of an Intelligent Machine. Newnes, Burlington (2003)
13. Kirk, D.E.: Optimal Control Theory: An Introduction. Dover Publications, Englewood Cliffs (April 2004)
14. Krstic, M., Kokotovic, P.V., Kanellakopoulos, I.: Nonlinear and Adaptive Control Design, 1st edn. John Wiley & Sons, Inc., New York (1995)
15. Feldkamp, L.A., Prokhorov, D.V., Feldkamp, T.M.: Simple and conditioned adaptive behavior from Kalman. Neural Networks 16, 683–689 (2003)
16. Park, B.S., Yoo, S.J., Park, J.B., Choi, Y.H.: A simple adaptive control approach for trajectory tracking of electrically driven nonholonomic mobile robots. IEEE Transactions on Control Systems Technology 18(5), 1199–1206 (2010)
17. Kageyama, T., Ohnishi, K.: An architecture of decentralized control for multi-degrees of freedom parallel manipulator. In: 7th International Workshop on Advanced Motion Control, Maribor, Slovenia, pp. 74–79 (2002)
18. Mahajan, A.: Optimal decentralized control of coupled subsystems with control sharing. In: 2011 50th IEEE Conference on Decision and Control and European Control Conference (CDC-ECC), Orlando, FL, USA, pp. 5726–5731 (December 2011)
19. Iftar, A.: Decentralized optimal control with overlapping decompositions. In: IEEE International Conference on Systems Engineering, Dayton, OH, USA, pp. 299–302 (1991)
20. Viñuela, P.I., Galvan, I.M.: Redes de neuronas Artificiales Un Enfoque Práctico. Pearson, Prentice Hall, Madrid, Spain (2004)
21. Talebi, H.A., Abdollahi, F., Patel, R.V., Khorasani, K.: Neural Network-Based State Estimation of Nonlinear Systems. Springer, N.Y (2010)

22. Sanchez, E.N., Ricalde, L.J.: Trajectory tracking via adaptive recurrent control with input saturation. In: Proceedings of the International Joint Conference on Neural Networks, Portland, USA, vol. 1, pp. 359–364 (2003)
23. Rovithakis, G.A., Chistodoulou, M.A.: Adaptive Control with Recurrent High - Order Neural Networks, London, UK (2000)
24. Felix, R.A.: Variable Structure Neural Control. Ph.D thesis, Cinvestav, Unidad Guadalajara, Guadalajara, Jalisco, Mexico (2003)
25. Ioannou, P.A., Sun, J.: Robust Adaptive Control. Prentice Hall, Inc., New Jersey (1996)
26. Felix, R.A., Sanchez, E.N., Loukianov, A.G.: Avoiding controller singularities in adaptive recurrent neural control. In: Proceedings of the 16th IFAC World Congress, Prague, Czech Republic (2005)
27. Lin, W., Byrnes, C.I.: Design of discrete-time nonlinear control systems via smooth feedback. IEEE Transactions on Automatic Control 39(11), 2340–2346 (1994)
28. Haddad, W.M., Chellaboina, V.-S., Fausz, J.L., Abdallah, C.: Optimal discrete-time control for non-linear cascade systems. Journal of The Franklin Institute 335(5), 827–839 (1998)
29. Chi-Sing, L., Lai-Wan, C.: Dual extended Kalman filtering in recurrent neural networks. Neural Networks 16(2), 223–239 (2003)
30. Alanis, A.Y., Sanchez, E.N., Loukianov, A.G.: Real-time output tracking for induction motors by recurrent high-order neural network control. In: 17th Mediterranean Conference on Control and Automation, MED 2009, Thessaloniki,Grecia, pp. 868–873 (June 2009)
31. Grover, R., Hwang, P.Y.C.: Introduction to Random Signals and Applied Kalman Filtering. John Wiley and Sons, New York (1992)
32. Haykin, S.: Kalman Filtering and Neural Networks. John Wiley and Sons, New York (2001)
33. Song, Y., Grizzle, J.W.: The extended Kalman filter as a local asymptotic observer for discrete-time nonlinear systems. Journal of Mathematical Systems, Estimation and Control 5, 59–78 (1995)
34. Garcia-Hernandez, R.: Control Neuronal Descentralizado Discreto para Manipuladores Robóticos. Ph.D thesis, Cinvestav, Unidad Guadalajara, Guadalajara, Jalisco, Mexico (2005)
35. Sanchez, E.N., Ornelas-Tellez, F.: Discrete-Time Inverse Optimal Control for Nonlinear Systems. CRC Press, Boca Raton (2013)
36. Basar, T., Olsder, G.J.: Dynamic Noncooperative Game Theory. Academic Press, New York (1995)
37. Lewis, F.L., Syrmos, V.L.: Optimal Control. John Wiley & Sons, N.Y (1995)
38. Al-Tamimi, A., Lewis, F.L.: Discrete-time nonlinear hjb solution using approximate dynamic programming: Convergence. IEEE Transactions on Systems 38, 943–949 (2008)
39. Ohsawa, T., Bloch, A.M., Leok, M.: Discrete hamilton-jacobi theory and discrete optimal control. In: Proceddings of the 49th IEEE Conference on Decision and Control (CDC), pp. 5438–5443 (2008)
40. Low, K.H., Loh, W.K.: Motion study of an omni-directional rover for step climbing. In: Proceedings of the 2005 IEEE International Conference on Robotics and Automation ICRA 2005, pp. 1585–1590 (April 2005)
41. Tao, J., Yang, F., Deng, Z., Fang, H.: Kinematic modeling of a six-wheeled robotic rover with a passive/active suspension, Taipei, Taiwan (2008)

42. Ch, U.M., Babu, Y.S.K., Amaresh, K.: Sliding Mode Speed Control of a DC Motor. In: Communication Systems and Network Technologies (CSNT), Katra, Jammu, India (2011)
43. Benavidez, P., Jamshidi, M.: Mobile robot navigation and target tracking system. In: 2011 6th International Conference on System of Systems Engineering (SoSE), Albuquerque, NM, USA, pp. 299–304 (June 2011)
44. Espiau, B., Chaumette, F., Rives, P.: A new approach to visual servoing in robotics. IEEE Transactions on Robotics and Automation 8(3), 313–326 (1992)
45. Hutchinson, S., Hager, G.D., Corke, P.I.: I Corke. A tutorial on visual servo control. IEEE Transactions on Robotics and Automation 12(5), 651–670 (1996)
46. Chaumette, F., Hutchinson, S.: Visual servo control. i. basic approaches. IEEE Robotics Automation Magazine 13(4), 82–90 (2006)
47. Gonzalez, R.C., Woods, R.E.: Digital Image Processing, 3rd edn. Prentice-Hall, Inc., Upper Saddle River (2006)
48. Gonzalez, R.C., Woods, R.E., Eddins, S.L.: Digital Image Processing Using MATLAB. Prentice-Hall, Inc., Upper Saddle River (2003)
49. Feng, L., Milios, E.E.: Robot pose estimation in unknown environments by matching 2d range scans. In: IEEE Computer Society Conference on Computer Vision and Pattern Recognition Proceedings CVPR 1994, pp. 935–938 (June 1994)
50. Canudas de Wit, C., Siciliano, B., Bastian, G.: Theory of Robot Control. Springer, London (1997)

Low-Cost Inertial Navigation

Stephen Bruder[1] and Aly El-Osery[2]

[1] Department of Computer, Electrical, and Software Engineering
Embry-Riddle Aeronautical University, Prescott, Arizona, USA
bruder@erau.edu
[2] Electrical Engineering Department
New Mexico Institute of Mining and Technology, Socorro, New Mexico, USA
elosery@ee.nmt.edu

Abstract. Recent advances in sensor technology have led to opportunities for the development of inertial-centric navigation systems, even for smaller Unmanned Aerial Vehicle (UAV) platforms. In particular, the emergence of micro-machined electromechanical system (MEMS) inertial sensors, which are encroaching on performance levels heretofore the exclusive domain of larger more expensive devices, offers the potential of small, low-cost navigation grade inertial measurement units (IMU). This chapter seeks to present a self-contained introduction to the topic of inertial navigation, with an emphasis on application of MEMS inertial sensors, and a brief foray into the subject of aided inertial navigation.

Keywords: Inertial navigation, aided inertial systems, micro-machined electromechanical systems, global positioning systems.

1 Notation

$\mathbf{x}$	Truth value
$\tilde{\mathbf{x}}$	Measured value
$\hat{\mathbf{x}}$	Estimated or computed value
$\delta\mathbf{x} = \mathbf{x} - \hat{\mathbf{x}}$	Error
$\mathcal{I}_{n\times m}$	$n \times m$ identity matrix
$0_{n\times m}$	$n \times m$ Null matrix
C_a^b	Orientation of the a-frame *wrt* to the b-frame as rotation-matrix
$[\boldsymbol{\omega}\times]$	Skew symmetric matrix
$\mathbf{v}_{b,c}^a$	Vector $\mathbf{v}$ of c with respect to b coordinatized in the a frame
dt	Sampling time

2 Inertial Navigation

This section presents a concise description of the development of an inertial navigation system. First the representation of position and orientation across

A. El-Osery and J. Prevost (eds.), *Control and Systems Engineering*,
Studies in Systems, Decision and Control 27, DOI: 10.1007/978-3-319-14636-2_12

the relevant coordinate systems is presented, the basics of inertial sensors is next discussed (with an emphasis on MEMS), and finally the algorithms necessary to process the sensor data in order to propagate a position, velocity, and attitude (PVA) solution are described.

2.1 Navigation Coordinate Frames

A basis for the discussion of a variety of navigation algorithms first requires a brief introduction of the relevant coordinate frames in which PVA will be represented. Figure 1 presents an illustration of the four principal coordinate frames of interest in the context of inertial-based navigation.

Referring to Figure 1, the Earth-Centered Inertial (ECI) coordinate frame (i-frame) has its origin at the center of mass of the Earth with the z-axis ($\mathbf{z}_i$) pointing along the nominal axis of rotation of the Earth (i.e., points to true North), the x-axis ($\mathbf{x}_i$) along the intersection of the equatorial plane and Earth-Sun orbital plane at the vernal equinox, and the y-axis ($\mathbf{y}_i$) completes a right-hand coordinate system.

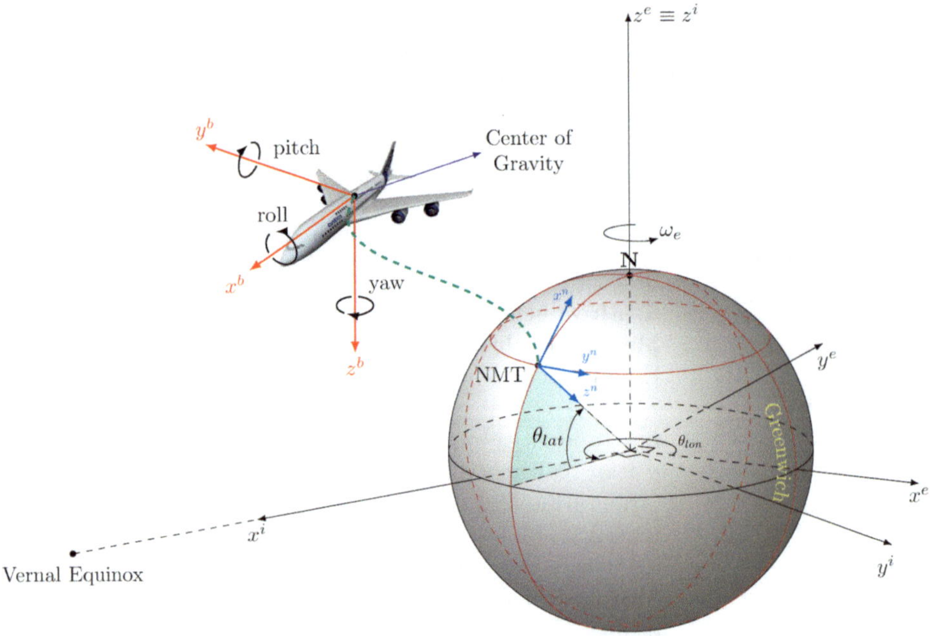

Fig. 1. Illustration of the four principal coordinate frames

The Earth-Centered Earth-Fixed (ECEF) frame (e-frame) shares the same origin as the ECI frame but is fixed with respect to the Earth. The ECI and ECEF frames differ only by a rotation about their common z-axis at a rate of approximately 15°/hr (ω_{ie}). Thus, the relationship between the ECI and ECEF frames is described by the positional vector $\mathbf{r}^i_{ie} = \mathbf{0}$ and rotation matrix $C^i_e = R_{(\mathbf{z},\theta_{ie})}$[1], where $\theta_{ie} = \omega_{ie}t + \theta_{GMST}$ and θ_{GMST} is the Greenwich Mean Sidereal time as an angle.

The navigation frame (n-frame) is a tangential (x/y plane is tangential to the Earth's ellipsoid) or locally level coordinate frame, with its origin defined by geodetic latitude (L_b), longitude (λ_b), and height (h_b) above the ellipsoid. As shown in Figure 1, the North-East-Down (NED) version of the navigation frame has the x-axis ($\mathbf{x}_n$) pointing North and the z-axis ($\mathbf{z}_n$) pointing down. Thus, the orientation of the n-frame with respect to the e-frame is $C^e_n = R_{(\mathbf{z},\lambda_b)}R_{(\mathbf{y}-L_b-90°)}$ and the angular velocity of the n-frame with respect to the e-frame coordinatized in the e-frame becomes

$$\boldsymbol{\omega}^e_{en} = \left[\sin(\lambda_b)\dot{L}_b, -\cos(\lambda_b)\dot{\mathrm{L}}_\mathrm{b}, \dot{\lambda_\mathrm{b}}\right]^T$$

Finally, the body frame (b-frame), attached to the platform of interest, shares a common origin with the navigation frame and has its x-axis ($\mathbf{x}_b$) pointing "forward" and z-axis pointing loosely "down" as illustrated in Figure 1. Thus, the vector describing the origin of the body frame is

$$\mathbf{r}^e_{eb} = \begin{bmatrix} (R_E + h_b)\cos(\mathrm{L_b})\cos(\lambda_\mathrm{b}) \\ (R_E + h_b)\cos(\mathrm{L_b})\sin(\lambda_\mathrm{b}) \\ (R_E(1-e^2) + h_b)\sin(L_b) \end{bmatrix} = \mathbf{r}^e_{en}$$

and the orientation of the b-frame is, $C^n_b = R_{(\mathbf{z},\psi)}R_{(\mathbf{y},\theta)}R_{(\mathbf{x},\phi)}$ (using fixed-axis rotation), in terms of roll (ϕ), pitch (θ), and yaw (ψ) angles.

2.2 Inertial Sensor Technology

Inertial sensors measure motion with respect to an inertial coordinate frame. They benefit from a lack of reliance on external fields (e.g., GPS, magnetometer), however, due to the fact that they typically measure rates, they suffer from inherent drift errors due to integration of said imperfect rates. Accelerometers actually measure specific force, typically, in the body frame as

$$\mathbf{f}^b_{ib} = \mathbf{a}^b_{ib} - \gamma^b_{ib}$$

where $\mathbf{f}^b_{ib}$ is due to motion and $\boldsymbol{\gamma}^b_{ib}$ is due to gravitational mass attraction. Most gyroscopes measure angular speed, typically, in the body frame as $\boldsymbol{\omega}^b_{ib}$. Inertial sensor technology can be decomposed as shown in Figure 2.

The pendulous mass accelerometer is comprised of a suspended mass and a sensing element, as shown in Figure 3 (left), and typically comes in either an open-loop or a closed-loop configuration, wherein the displacement is proportional to the measured acceleration. The vibratory accelerometer measures a shift in frequency as a function of the applied acceleration (see Figure 3 right).

[1] Rotation about the z-axis by an angle of θ_{ie}.

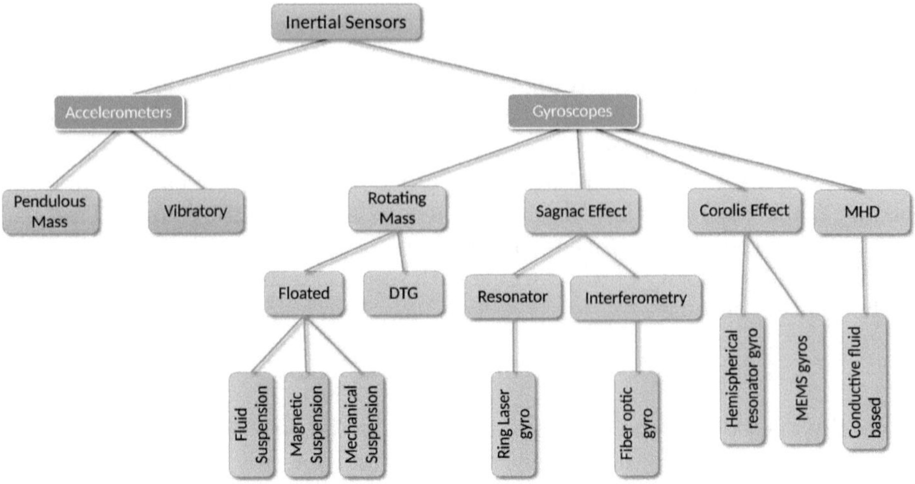

Fig. 2. Inertial Sensor Technology

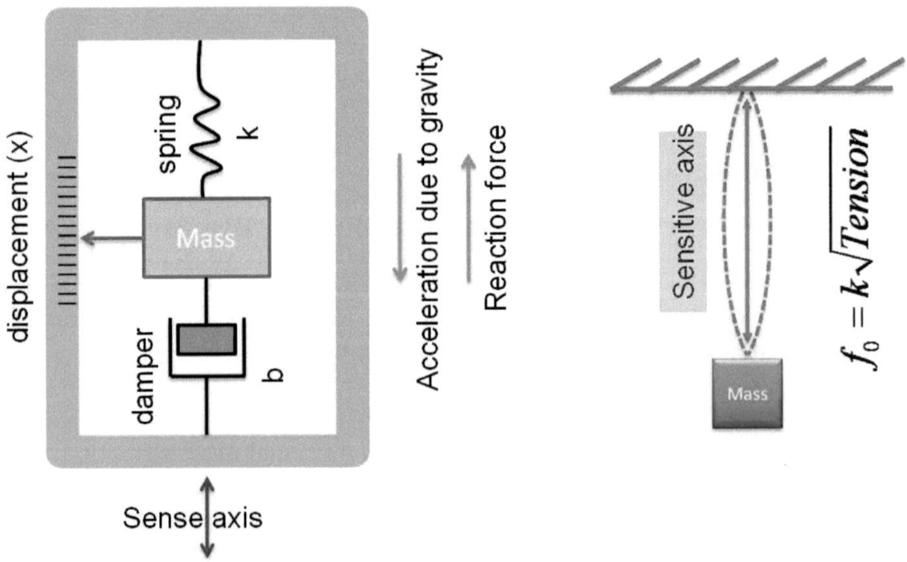

Fig. 3. Pendulous mass (left) and vibratory (right) accels

Rotating mass gyroscopes measure torque due to precession, about the y-axis, as a function of the angular speed about the z-axis, as illustrated in Figure 4 (left). MEMS gyroscopes typically measure displacements due to the Coriolis acceleration ($a_{Coriolis}$) caused by the interaction of an induced linear velocity with an external angular velocity to be measured (see Figure 4 right).

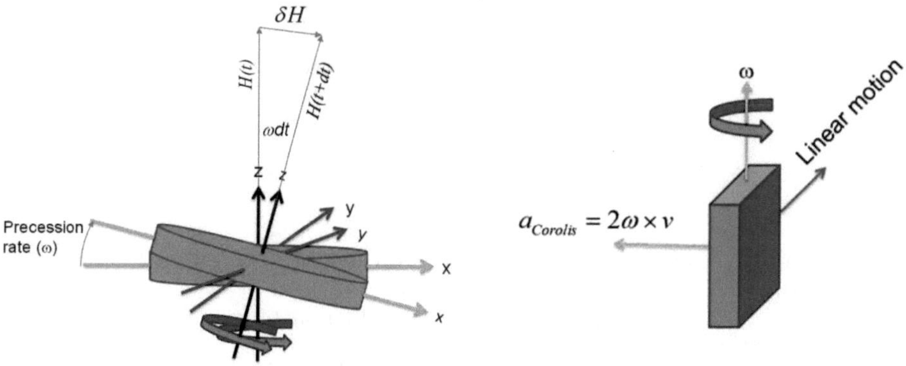

Fig. 4. Rotating mass (left) and Coriolis (right) gyros

A fiber optic gyroscope (FOG) measures the phase shift due to asymmetry in an optical path length that is caused by rotation about the normal to the plane of the optical path. This behavior is referred to as the Sagnac effect. A ring laser gyro (RLG) operates on a similar principle to measure the change in wavelength of asymmetric beams in a resonant cavity.

2.3 Inertial Sensor Errors

Inertial sensors suffer from a variety of error contributors due to physical limitations, thermal effects, quantization, and other imperfections. It is therefore important to be able to characterize these error contributors. Characterizing the sensor errors serve three main purposes:

1. provide a basis for the comparative evaluation of sensor alternatives;
2. support the empirical verification of simulation model parameters; and
3. substantiate the efficacy of INS system level performance predictions.

The dominant error contributors can be categorized as relating to bias, scale factor and alignment, noise sources, or sensor specific (see Figure 5).

Bias Error. The bias error (b) can be decomposed into a deterministic fixed bias term, b_{FB} (often a function of temperature), a bias stability term, b_{BS}, which typically varies from turn-on-to-turn-on as a random constant, and finally

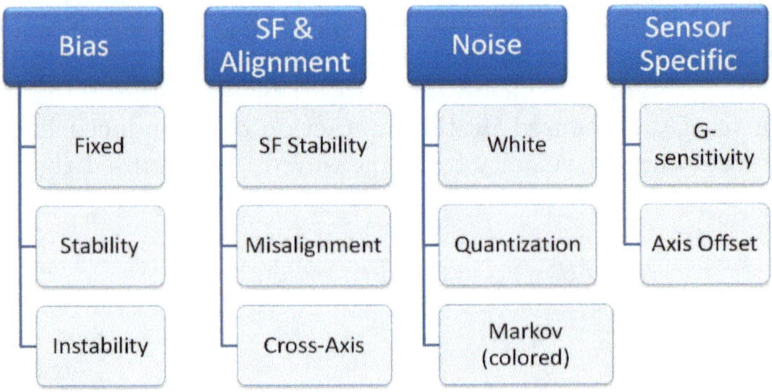

Fig. 5. Inertial sensor error sources

a bias instability component, b_{BI}, which is an in-run drift, often modeled as a random walk as

$$b_{BI}(k) = e^{-\frac{dT}{\tau_{BI}}} b_{BI}(k-1) + w(k) \tag{1}$$

Flicker noise in the electronics can cause the bias of a MEMS gyroscope to wander over time. Flicker noise characterized by the spectrum of $1/F$ is observed mostly at low frequencies. In contrast, at high frequencies, white noise is dominant. Modeling noise characterized by $1/F$ may be achieved by several methods. One approach involves frequency shaping white noise. Another approach is to sum the output of several 1st order Gauss-Markov process.

Accumulating the bias error terms, for the case of accelerometers yields

$$b_a = b_{a,BI} + b_{a,FB} + b_{a,BS}$$

and for the gyroscopes yields

$$b_g = b_{g,BI} + b_{g,FB} + b_{g,BS}$$

Scale Factor and Alignment. The scale factor error can be expressed in terms of a stochastic relative scale factor stability (SSF) and a thermally dependent deterministic term. In particular, the SSF term $s_{a,x}$ refers to the scale factor stability in parts-per-million (ppm) of the x-axis accelerometer.

The alignment error sources are described via sense axis misalignment terms (e.g., m_{zy}) and unrecoverable cross-axis sensitivity. The cross axis error is attributable to a sensor providing a response to a stimulus that is orthogonal to its sense axis (see Figure 6).

Noise Error Terms. Noise sources include white noise sources that are often described in terms of a random walk equivalent. For example, thermo-mechanical noise in a MEMS gyro can cause perturbation in its output resulting in fluctuations at frequencies much greater than the sampling rate of the sensor. Consequently, the samples of the gyroscope experience white noise disturbances. This

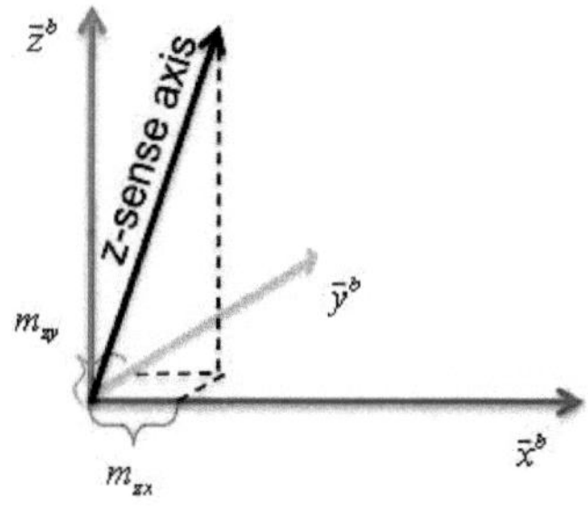

Fig. 6. z-sense axis

noise term typically causes some confusion as to its effect. It is thus worth discussing it in more detail. Assuming the rectangular rule is used for integration, and given a sampling period of T_s and a time span of nT_s

$$\int_0^t \epsilon(\tau)d\tau = T_s \sum_{i=1}^{n} \epsilon(t_i) \tag{2}$$

Since $\mathbb{E}[\epsilon(t_i)] = 0$ and $Cov(\epsilon(t_i), \epsilon(t_j)) = 0$ for all $i \neq j$, $Var[\epsilon(t_i)] = \sigma^2$, then

$$\mathbb{E}\left[\int_0^t \epsilon(\tau)d\tau\right] = T_s n \mathbb{E}[\epsilon(t_i)] = 0, \forall i \tag{3}$$

$$Var\left[\int_0^t \epsilon(\tau)d\tau\right] = T_s^2 n Var[\epsilon(t_i)] = T_s t \sigma^2, \forall i \tag{4}$$

A plot of an ensemble of integrated white rate noise processes is shown in Figure 7 (left). Furthermore, the ensemble standard deviation derived from the analytical expression (ideal) is compared to the computed equivalent (from data sequence) in Figure 7 (right).

The integrated noise resulted in zero-mean random walk with standard deviation that grows with time as

$$\sigma_\theta = \sigma\sqrt{T_s t}$$

Defining the angle random walk, ARW, as

$$ARW = \sigma_\theta(1) \tag{5}$$

in units $(°/\sqrt{\text{hr}})$, suggests a relationship to the power spectral density (double-sided) of the rate noise as

$$ARW(°/\sqrt{h}) = \frac{1}{60}\sqrt{PSD((°/h)^2/Hz)} \tag{6}$$

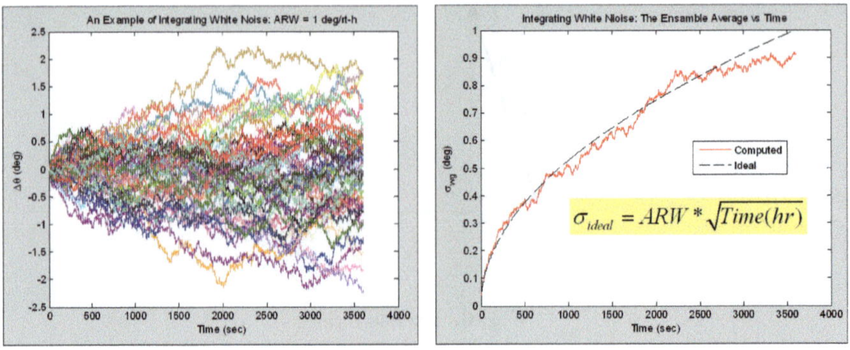

Fig. 7. The relationship between rate white noise and ARW

Quantization Noise. Quantization noise is often a result of the analog to digital conversion at the front end of the sensors. Rate random walk is a random walk type error on the rate output of the sensor and has units of $(°/hr^{3/2})$.

These noise contributors can be identified empirically by developing a PSD plot or Allan deviation graph of the quiescent noise of the sensor (see Figure 8 and Table 1).

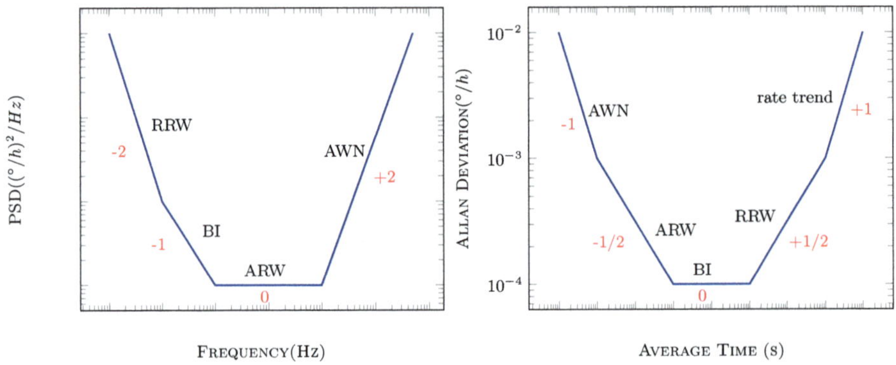

Fig. 8. A typical single-sided PSD (left) and Allan deviation (right) plot

Sensor Specific Errors. Sensor specific errors include effects such as gyroscope g-sensitivity that is primarily due to device mass asymmetry in MEMS gyroscopes and can be modeled as $\delta\boldsymbol{\omega}_{ib}^b = G_g \mathbf{f}_{ib}^b$. Some gyroscopes also exhibit g^2 sensitivities due to anisoelastic effects. Accelerometers often exhibit $\omega^2 r$-type errors due to center of mass offsets. For example, for the x-axis accelerometer

$$\delta f_x = \omega_y^2 \Delta x + \omega_z^2 \Delta x = (\omega_y^2 + \omega_z^2)\Delta x \tag{7}$$

Table 1. Noise parameters

Noise Type	AV $\sigma^2(\tau)$	PSD (2-sided)
Quantization Noise	$3\frac{\alpha^2}{\tau^2}$	$(2\pi f)^2\alpha^2 T_s$
Angle/Velocity Random Walk	$\frac{\alpha^2}{\tau}$	α^2
Flicker Noise	$\frac{2\alpha^2\ln(2)}{\pi}$	$\frac{\alpha^2}{2\pi f}$
Angular Rate/Accel Random Walk	$\frac{\alpha^2\tau}{3}$	$\frac{\alpha^2}{(2\pi f)^2}$
Ramp Noise	$\frac{\alpha^2\tau^2}{2}$	$\frac{\alpha^2}{(2\pi f)^3}$

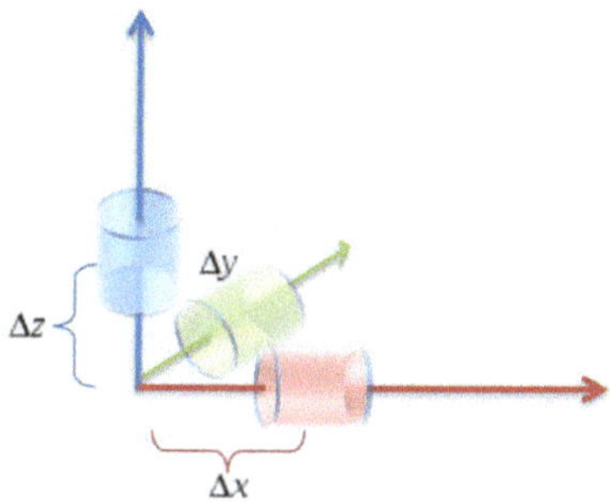

Fig. 9. Sensor Offset

In summary, a medium fidelity accelerometer measurement model might be expressed as

$$\tilde{\mathbf{f}}_{ib}^b = \mathbf{f}_{ib}^b + \Delta\mathbf{f}_{ib}^b = \mathbf{b}_a + (\mathcal{I} + M_a)\mathbf{f}_{ib}^b + \mathbf{w}_a \tag{8}$$

and a gyroscope model as

$$\tilde{\boldsymbol{\omega}}_{ib}^b = \boldsymbol{\omega}_{ib}^b + \Delta\boldsymbol{\omega}_{ib}^b = \mathbf{b}_g + (\mathcal{I} + M_a)\boldsymbol{\omega}_{ib}^b + G_g\mathbf{f}_{ib}^b + \mathbf{w}_g \tag{9}$$

where $\tilde{\mathbf{f}}_{ib}^b$ denotes a measurement of the true quantity $\mathbf{f}_{ib}^b$ corrupted by the measurement error $\Delta\mathbf{f}_{ib}^b$. Similarly, for gyroscopes, $\tilde{\boldsymbol{\omega}}_{ib}^b$ denotes a measurement of the true quantity $\boldsymbol{\omega}_{ib}^b$ corrupted by the measurement error $\Delta\boldsymbol{\omega}_{ib}^b$. The term M captures both scale factor and misalignment errors, $\mathbf{w}$ represents the noise terms, and G_g represents the g-sensitivity effects.

3 Inertial Sensor Performance Comparison

Inertial sensors can be organized into four major performance grades, in ascending order, namely, consumer, tactical, navigation, and strategic. MEMS inertial sensors have typically been relegated to consumer grade, however, some have transitioned to the tactical grade and are beginning to encroach on navigation grade. Figure 10 shows the different classes of gyroscopes and their relationships in terms of their bias instability and angle random walk parameters. Key gyroscopes presented in the figure are: Northrop HRG SIRU [1], EMCORE FOG EMP-1.2k [2], Honeywell HRG GG 1320 [3], Honeywell OPG [4], Boeing DRG [5], Sensonor MEMS SAR500 [6], KVH FOG 1750 IMU [7], Analog Devices MEMS ADIS16137 [8], and VectorNav MEMS [9].

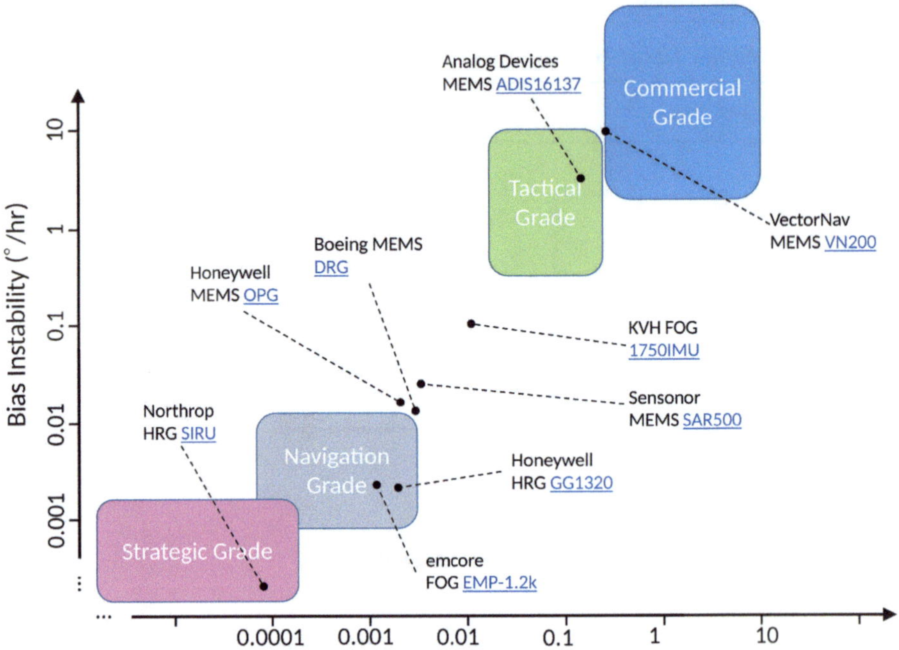

Fig. 10. Gyro performance comparison

3.1 Description of an ISA, IMU, IRU, and INS

Referring to Figure 11, an Inertial Sensor Assembly (ISA) typically contains three gyroscopes, three accelerometers, and basic power and signal conditioning electronics such as analog to digital conversion.

An Inertial Measurement Unit (IMU – see Figure 11) adds correction factors from an *a priori* calibration (e.g., bias and scale factor) and compensation algorithms such as characterization over temperature and g and g^2 sensitivity. The outputs of the IMU are measurements of the specific force ($\tilde{\mathbf{f}}_{ib}^{b}$) of the body with respect to an inertial frame resolved in the body frame and also a measurement of the angular velocity ($\tilde{\boldsymbol{\omega}}_{ib}^{b}$) of the body with respect to an inertial frame resolved in the body frame. Furthermore, the Inertial Navigation System

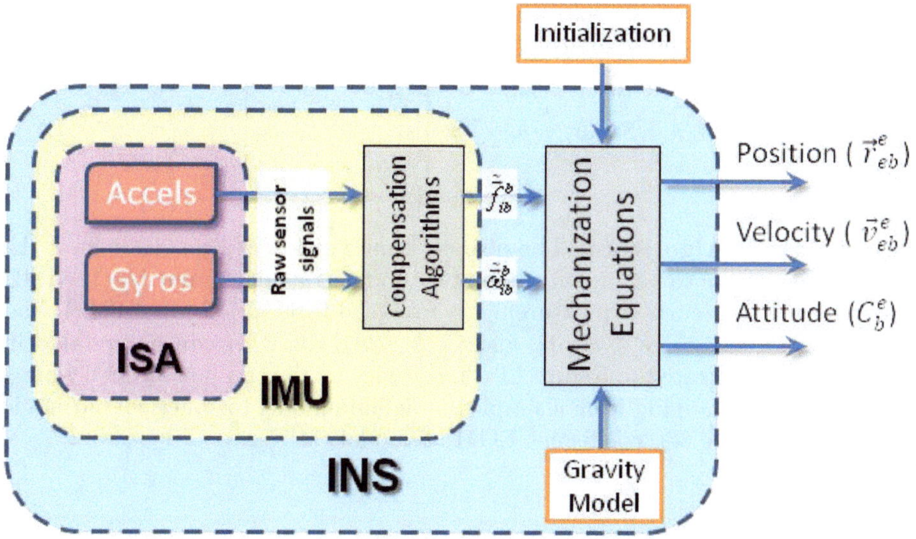

Fig. 11. The relationship between an ISA, IMU, and INS

(INS) (see Figure 11) subsumes the IMU and requires a considerable increase in processing resources to implement more sophisticated algorithms in order to propagate PVA.

3.2 INS-Only Based Navigation

INS (see Figure 11) accepts an external initialization (e.g., from a GPS receiver), inputs from the IMU, includes a gravity/gravitational model, and implements "mechanization equations." These equations propagate the position ($\mathbf{r}^e_{eb}$), velocity ($\mathbf{v}^e_{eb}$), and attitude (C^e_b) (i.e., PVA) of each sample interval as new IMU measurements are obtained. Herein, the term Inertial Reference Unit (IRU) refers to an IMU containing only gyros, and as such, measures only orientational changes.

The PVA of the "body" of interest (b-frame) can be resolved (i.e., coordinatized) in a variety of coordinate frames, to include the Earth Centered Earth Fixed (ECEF) frame, employed herein (e-frame). The state-space version of the differential equations describing the INS mechanization (in the e-frame) can be described by [10]

$$\dot{\boldsymbol{x}}^{\,e}_{INS} = \begin{bmatrix} \dot{\boldsymbol{r}}^{\,e}_{eb} \\ \dot{\boldsymbol{v}}^{\,e}_{eb} \\ \dot{C}^i_b \end{bmatrix} = \begin{bmatrix} \mathbf{v}^e_{eb} \\ C^e_b \mathbf{f}^b_{ib} + \mathbf{g}^e_b - 2\Omega^i_{ie}\mathbf{v}^e_{eb} \\ C^e_b \Omega^b_{eb} \end{bmatrix} \tag{10}$$

wherein, $\mathbf{g}^e_b$ is the acceleration due to gravity experienced by the body and resolved in the e-frame, and Ω^i_{ie} is the skew-symmetric version of the Earth's rotation vector.

Specifically, defining an angular velocity vector as $\boldsymbol{\omega} = [\omega_x, \omega_y, \omega_z]^T$, the skew-symmetric matrix formed from this vector is

$$\Omega = Sk(\boldsymbol{\omega}) = [\boldsymbol{\omega}\times] = \begin{bmatrix} 0 & -\omega_z & \omega_y \\ \omega_z & 0 & -\omega_x \\ -\omega_y & \omega_x & 0 \end{bmatrix}$$

The INS-Only Algorithm. The discrete-time version of Eqn. 10 defines the INS-Only algorithm to be implemented at an update rate determined by the IMU sample rate. Given current measurements from the IMU ($\tilde{\boldsymbol{\omega}}_{ib}^b(k)$ and $\tilde{\mathbf{f}}_{ib}^b(k)$) and prior PVA ($\mathbf{r}_{eb}^e(k-1)$, $\mathbf{v}_{eb}^e(k-1)$, and $C_b^e(k-1)$), the PVA can be updated in three steps as illustrated in Figure 12. The gravity model can be obtained from a variety of sources ranging from a simple Somigliana model to spherical harmonic models, such as the 360 coefficient EGM 96 model [11].

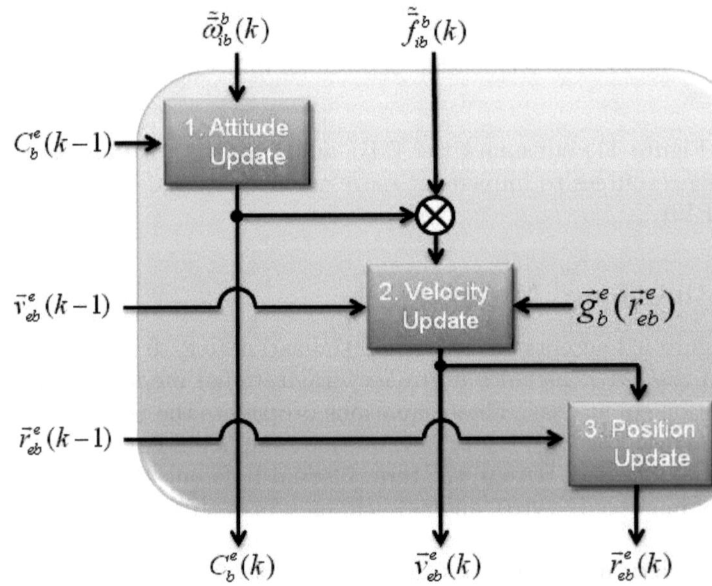

Fig. 12. The discrete INS-Only Algorithm [12]

The discrete versions of the attitude, velocity, and position update equations are given below for the case of mechanization in the ECEF coordinate frame.

Attitude Update. The discrete attitude update equation is given by

$$C_b^e(k) \approx C_b^e(k-1)\left(\mathcal{I} + \Omega_{ib}^b \Delta t\right) - \Omega_{ie}^i C_b^e(k-1)\Delta t \tag{11}$$

A higher fidelity version of the attitude update equation presented in Eqn. 11 can be realized by noting that

$$C_b^e(k) = e^{\Omega_{eb}^e \Delta t} C_b^e(k-1)$$

and thus, forming the instantaneous normalized axis of rotation $\mathbf{k}$, defined by $\boldsymbol{\omega}_{eb}^e dt = \mathbf{k}\Delta\theta$, where $\mathcal{K}$ is the skew symmetric of $\mathbf{k}$ and $\boldsymbol{\omega}_{eb}^e = C_b^e \tilde{\boldsymbol{\omega}}_{ib}^b - \boldsymbol{\omega}_{ie}^e$, yielding:

$$C_b^e(k) = \left[\mathcal{I} + \sin(\Delta\theta)\mathcal{K} + [1 - \cos(\Delta\theta)]\mathcal{K}^2\right] C_b^e(k-1) \tag{12}$$

Update Velocity. The discrete velocity update equation is given by

$$\mathbf{v}_{eb}^e(k) = \mathbf{v}_{eb}^e(k-1) + \mathbf{a}_{eb}^e \Delta t \tag{13}$$

where $\mathbf{a}_{eb}^e = C_b^e(k)\tilde{\mathbf{f}}_{ib}^b + \mathbf{g}_b^e - 2\Omega_{ie}^i \mathbf{v}_{eb}^e(k-1)$

Update Position. The discrete position update equation is given by

$$\mathbf{r}_{eb}^e(k) = \mathbf{r}_{eb}^e(k-1) + \mathbf{v}_{eb}^e(k-1)\Delta t + \mathbf{a}_{eb}^e \frac{\Delta t^2}{2} \tag{14}$$

More sophisticated integration schemes (e.g., forth-order Runge-Kutta) can be employed to realize improved fidelity from the position and velocity update equations.

As the measurements $\tilde{\mathbf{f}}_{ib}^b$, and $\tilde{\boldsymbol{\omega}}_{ib}^b$ suffer from noise as described in Eqn. 8 and 9, respectively, all inertial-only navigation systems will suffer from position and attitude drift as an inherent consequence of integrating error contributors, and hence, the desire to complement these very reliable navigation sensors arises. Furthermore, pursuing higher performance inertial sensors to mitigate this inherent drift error has an exponentially detrimental impact on cost.

3.3 An INS–Only Simulation Example

A simulation example was developed for the case of a ship traveling at approximately 20 nautical miles per hour and experiencing ±2 meter vertical translations in the presence of mild roll and pitch excursions. The motion profile for this three-minute simulation is shown in Figure 13.

This relative (to starting position) motion profile is presented with respect to a locally-level coordinate frame which is oriented North, West, and Up (the navigation frame is actually North, East, and Down (NED)). A more detailed depiction of the position and velocity of this motion profile are given in Figure 14. The estimated quantities are derived from a discrete (at 100 Hz) implementation of the mechanization equations presented in Eqn. 10 and employing the inertial sensor models described in Eqns. 8 and 9 for the accelerometers and gyroscopes, respectively.

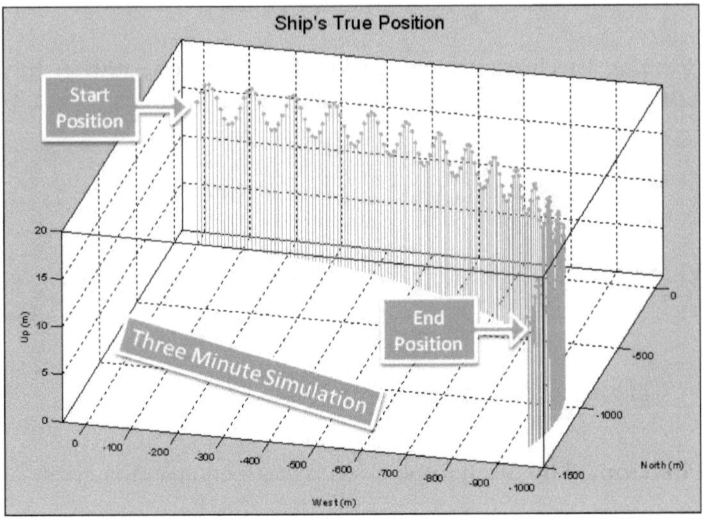

Fig. 13. Basic Motion Profile

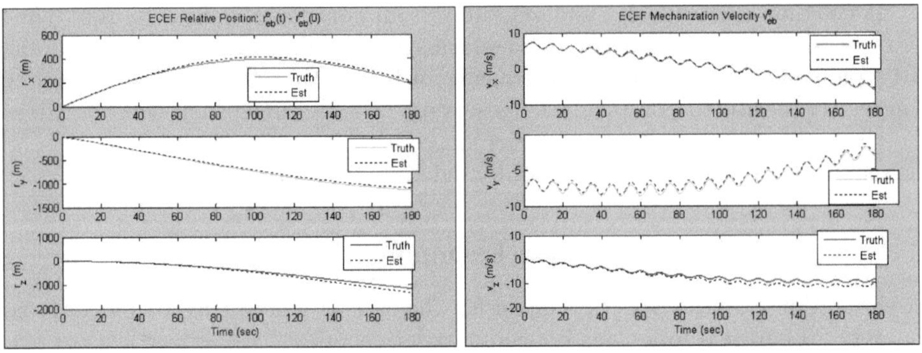

Fig. 14. Relative position (left) and velocity (right)

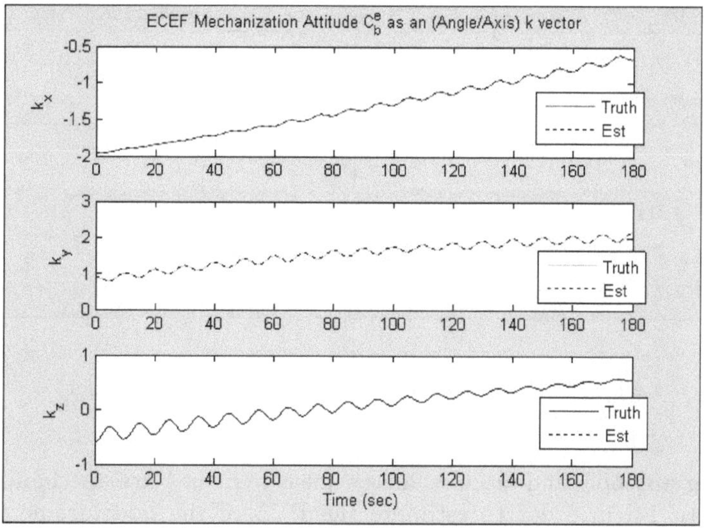

Fig. 15. Attitude of the profile

The attitude profile is expressed as an angle-axis product vector ($C_b^e = e^{[\mathbf{k}\times]}$) corresponding to the rotation matrix that describes the body frame with respect to the ECEF frame (i.e., C_b^e) and is shown in Figure 15.

The corresponding angular velocity ($\tilde{\boldsymbol{\omega}}_{ib}^b$) and specific force ($\tilde{\mathbf{f}}_{ib}^b$) measurements observed in the body frame, as a result of this motion profile, are shown in Figure 16 and contrast the measured values (i.e., a KVH 1750 IMU) against the ground truth (i.e., ideal) velocities and forces.

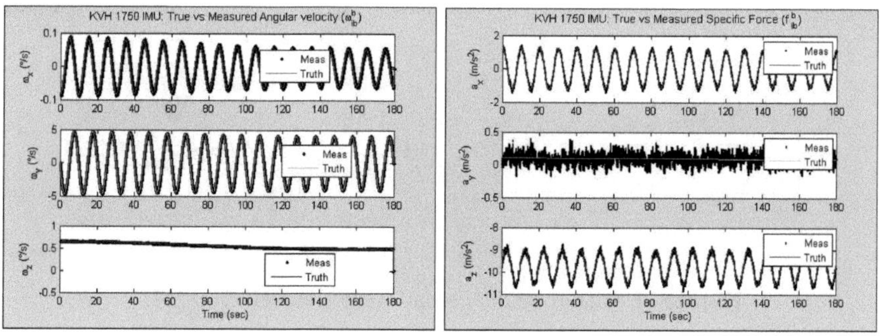

Fig. 16. Angular velocity (left) and specific (right)

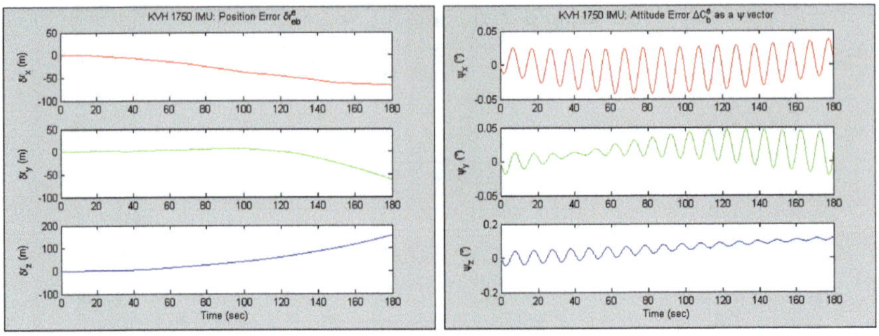

Fig. 17. INS position (left) and attitude errors (right)

These gyroscope and accelerometer measurements are mechanized, using Eqns. 12 to 14, in order to estimate the PVA of the body frame. The associated position and attitude errors are illustrated in Figure 17. The resulting 3D motion profiles are presented in Figure 18 with the KVH 1750 derived position vs truth.

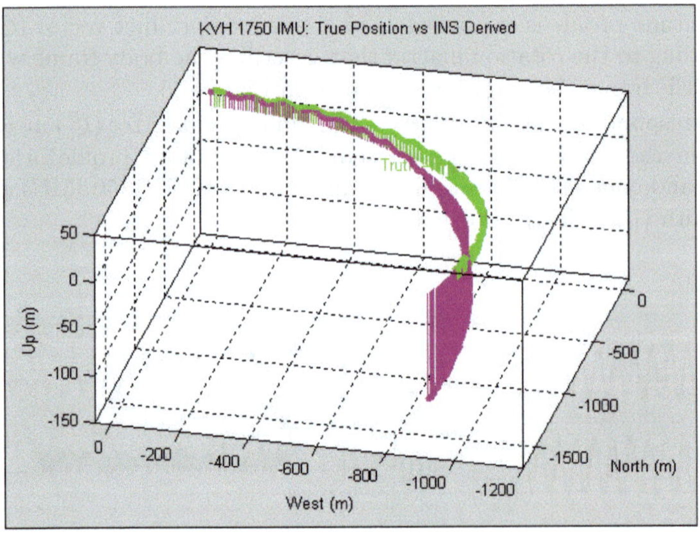

Fig. 18. Position profile: simulating a KVH 1750 IMU

This simple simulation presents the effects of inertial sensor error contributors on PVA estimation. Clearly, however, the need for an aiding sensor(s) is apparent in order to support longer navigation intervals.

3.4 INS Error Modeling

In preparation for GPS/INS integration it is prudent to analyze the error dynamics of the INS mechanization (see Eqn. 10). These errors are primarily a result of the fact that this mechanization is driven by imperfect inertial measurements (see Eqns. 8 and 9) and other error contributors (attitude approximations, IMU alignment, ...). To that end, define the INS PVA-error state vector as

$$\delta\mathbf{x}_{INS}^{e} = \begin{pmatrix} \delta\boldsymbol{\psi}_{eb}^{e} \\ \delta\mathbf{v}_{eb}^{e} \\ \delta\mathbf{r}_{eb}^{e} \end{pmatrix} \tag{15}$$

where the position error is $\delta\mathbf{r}_{eb}^{e} = \mathbf{r}_{eb}^{e} - \hat{\mathbf{r}}_{eb}^{e}$ (i.e., truth minus the estimate), and the velocity error is $\delta\mathbf{v}_{eb}^{e} = \mathbf{v}_{eb}^{e} - \hat{\mathbf{v}}_{eb}^{e}$. The attitude error, $\delta\boldsymbol{\psi}_{eb}^{e}$, is a bit more involved. Consider the attitude error in terms of a rotation matrix defined as (i.e., truth $\times$ estimate)

$$\delta C_b^e = C_b^e \hat{C}_e^b = e^{[\delta\boldsymbol{\psi}_{eb}^{e}\times]} \approx \mathcal{I} + [\delta\boldsymbol{\psi}_{eb}^{e}\times] \tag{16}$$

which can then be related to a "small" angle approximation using the above attitude error vector ($\delta\boldsymbol{\psi}_{eb}^{e}$). This approximation is only reasonable if the attitude estimate ($\hat{C}_b^e = (\mathcal{I} - [\delta\boldsymbol{\psi}_{eb}^{e}\times])C_b^e$ is "close" to the truth ($C_b^e = (\mathcal{I} + [\delta\boldsymbol{\psi}_{eb}^{e}\times])\hat{C}_b^e$).

Looking ahead, the errors in the IMU measurements can be estimated (both off-line and real-time), and hence, used to improve the navigation solution. As such, consider the estimated angular velocity and specific force to be represented by $\hat{\boldsymbol{\omega}}_{ib}^{b} = \tilde{\boldsymbol{\omega}}_{ib}^{b} - \Delta\hat{\boldsymbol{\omega}}_{ib}^{b}$ and $\hat{\mathbf{f}}_{ib}^{b} = \tilde{\mathbf{f}}_{ib}^{b} - \Delta\hat{\mathbf{f}}_{ib}^{b}$, where the estimates of the IMU errors are $\hat{\mathbf{f}}_{ib}^{b}$ and $\hat{\boldsymbol{\omega}}_{ib}^{b}$, respectively. Furthermore, akin to the PVA errors defined in Eqn. 15, the angular velocity estimation error can be defined as (i.e., truth - estimate) $\delta\boldsymbol{\omega}_{ib}^{b} = \boldsymbol{\omega}_{ib}^{b} - \hat{\boldsymbol{\omega}}_{ib}^{b}$ and the specific force estimation error defined as $\delta\mathbf{f}_{ib}^{b} = \mathbf{f}_{ib}^{b} - \hat{\mathbf{f}}_{ib}^{b}$. These equations can be combined into a state-space equation as

$$\begin{pmatrix} \delta\dot{\boldsymbol{\psi}}_{eb}^{e} \\ \delta\dot{\mathbf{v}}_{eb}^{e} \\ \delta\dot{\mathbf{r}}_{eb}^{e} \end{pmatrix} = \begin{bmatrix} -\Omega_{ie}^{e} & 0_{3\times3} & 0_{3\times3} \\ -[\hat{C}_b^e \hat{\mathbf{f}}_{ib}^{b}\times] & -2\Omega_{ie}^{e} & \frac{2g_0(\hat{L}_b)}{r_{eS}^{e}(\hat{L}_b)} \frac{\hat{\mathbf{r}}_{eb}^{e}}{|\hat{\mathbf{r}}_{eb}^{e}|^2} (\hat{\mathbf{r}}_{eb}^{e})^T \\ 0_{3\times3} & \mathcal{I}_{3\times3} & 0_{3\times3} \end{bmatrix} \begin{pmatrix} \delta\boldsymbol{\psi}_{eb}^{e} \\ \delta\mathbf{v}_{eb}^{e} \\ \delta\mathbf{r}_{eb}^{e} \end{pmatrix} + \begin{bmatrix} 0 & \hat{C}_b^e \\ \hat{C}_b^e & 0 \\ 0 & 0 \end{bmatrix} \begin{pmatrix} \delta\mathbf{f}_{ib}^{b} \\ \delta\boldsymbol{\omega}_{ib}^{b} \end{pmatrix} \tag{17}$$

4 GPS-Only Based Navigation

The Global Positioning System (GPS) is one of four currently operational space-based Global Navigation Satellite System (GNSS – see Figure 19). Unlike the

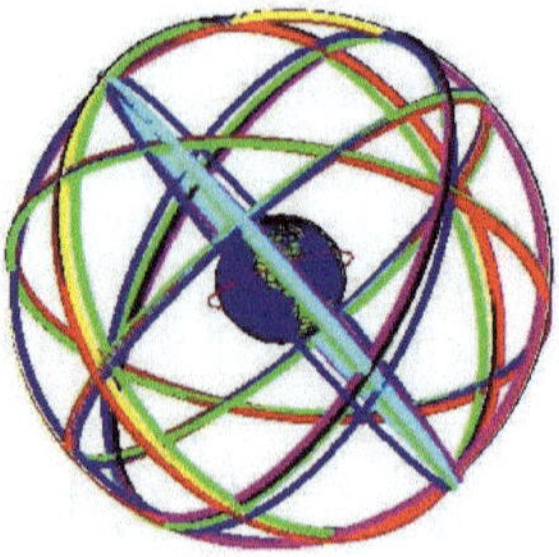

Fig. 19. GPS Constellation

"dead reckoning" (i.e., relative) navigation solution provided by an INS, GPS offers a "position fixing" (i.e., absolute) solution devoid of INS-like drift errors.

By providing the pseudo-range from four (or more) GPS satellites to a receiver, the receiver's ECEF (or geodetic) location can be determined via multilateration. The current GPS system offers two carrier frequencies (L1/L2 = 1,575.42/ 1,227.6 MHz). The civilian (standard positioning service SPS) 10-bit, 1 MHz course acquisition (C/A) codes modulate the L1 carrier and the encrypted 40-bit, 10 MHz code modulates the L2 carrier to provide a precise positioning system (PPS) intended for military use. The pseudorange measurement ($\tilde{\rho}$) contains errors from a variety of contributors.

$$\tilde{\rho} = \rho + \epsilon_{ephemeris} + \epsilon_{iono} + \epsilon_{tropo} + \epsilon_{sat\ clock} + \epsilon_{receiver\ clock} + \epsilon_{multipath} \quad (18)$$

The ephemeris error, $\epsilon_{ephemeris}$, relates to uncertainty in the control segments predicted location of a given GPS satellite. The ionospheric, ϵ_{iono}, and tropospheric, ϵ_{tropo}, errors relate to variation in signal propagation speed through the atmosphere, and as these effects vary with signal frequency, a dual frequency receiver can realize substantial improvement. The satellite and receiver clock errors, $\epsilon_{sat\ clock}$ and $\epsilon_{receiver\ clock}$, respectively, are primarily due to local oscillator noise and can be corrected by differential GPS. Multipath errors, $\epsilon_{multipath}$, relate to the receiver resolving reflected versions of the same signal source, each resulting in a different pseudorange. Differential GPS (DGPS) refers to a configuration wherein a stationary base GPS receiver at a precisely surveyed location is used to provide corrections to a roving receiver resulting in substantial error reduction.

By representing the measured pseudorange ($\tilde{\rho}_i$) from the ith GPS satellite to a GPS receiver as $\sqrt{(x_i - x)^2 + (y_i - y)^2 + (z_i - z)^2}$ and combining the error terms into n_i, except the receiver clock error ($\epsilon_{receiver\ clock} \approx c\Delta T$), Eqn. 18 can be restated as

$$\tilde{\rho}_i = \sqrt{(x_i - x)^2 + (y_i - y)^2 + (z_i - z)^2} + c\Delta T + n_i \quad (19)$$

wherein x, y, and z are the positional coordinates of the receiver to be estimated and x_i, y_i, and z_i are the coordinates of the ith GPS satellite. A Taylor series expansion of Eqn. 19 about an initial estimate of the receiver position and clock delay (i.e., $\hat{\chi} = [\hat{x} \;\; \hat{y} \;\; \hat{z} \;\; \widehat{c\Delta T}]^T$) yields

$$\Delta\hat{\chi} = [J^T \;\; J]^{-1} J^T \left[\begin{pmatrix} \tilde{\rho}_i(\chi) \\ \vdots \\ \tilde{\rho}_n(\chi) \end{pmatrix} - \begin{pmatrix} \hat{\rho}_i(\hat{\chi}) \\ \vdots \\ \hat{\rho}_n(\hat{\chi}) \end{pmatrix} \right] \tag{20}$$

where $J = \frac{\partial \rho}{\partial \chi}\Big|_{\chi=\hat{\chi}}$ is the Jacobian matrix, thus allowing the receivers location to be iteratively estimated in a non-sequential (with respect to the measurements) manner, as

$$\hat{\chi}_{new} = \hat{\chi}_{old} + \Delta\hat{\chi} \tag{21}$$

5 GPS/INS–Based Navigation

5.1 Introduction

All purely inertial navigation systems suffer from drift as an unavoidable consequence that arises from integrating accelerations and angular rates to produce PVA estimates. The predominant contributors include errors relating to misalignment, scale factor variation, bias instability, and various other noise sources (e.g., angle random walk). Current state-of-the-art inertial navigation systems are able to constrain these accumulated errors to less than 500 m/hr drift under ideal conditions, however, even in the presence of ideal sensors (i.e., noise-free), the fundamental IMU mechanization equations themselves give rise to fundamental sources of error. With the advent of the global positioning system (and other GNSS) and ever shrinking GPS receiver technology, this low-frequency non-drifting positional (and velocity) information source has been successfully employed as a capable aiding sensor to the traditional INS. The synergistic nature of the complementary nature between the GPS and INS has led to the growth of a variety of GPS-aided INS systems. The strengths and/or weaknesses of inertial navigation systems (INS) and/or GPS-based systems are highlighted in Table 2.

In developing an aided IMU to obtain Position (P), Velocity (V) and Attitude (A) estimates and having decided to pursue a linear minimum variance estimation approach (i.e., Kalman filter family of algorithms), various integration architectures are possible. The simplest is an uncoupled configuration wherein the GPS and INS have little interaction. At the other end of the spectrum is the deeply coupled architecture that allows both the GPS and INS to operate symbiotically. In between the two extremes lie the loosely and tightly coupled configurations that will be described herein in more detail. Aside from some potential performance gains, the dominant discriminator between the various integration schemes relates to the system's ability to degrade gracefully in increasingly challenging environments.

Table 2. Comparison of INS, GPS, and integrated INS+GPS

	Strength	*Weakness*
INS	– High-bandwidth – Good short-term accuracy (PVA) – Unaffected by RF jamming	– Initialization required – Suffers from drift errors – High-cost – Requires a precise gravity model
GPS	– Good long-term position & velocity accuracy – Not sensitive to gravity – Relatively low-cost	– Low-bandwidth system – Difficult to obtain attitude – Susceptible to satellite signal blockage/jamming
GPS & INS	– High-bandwidth system – Good short & long-term accuracy – Can perform during GPS loss – INS can aid GPS (& vice-versa) – Robust system	– Greater complexity

5.2 Uncoupled GPS/INS Integration

The most straight forward integration approach involves the use of the GPS and IMU sensors in a distributed estimation approach and as such requires a dynamic model of the position and attitude of the host. Such an approach is undesirable as the complexity of the dynamic model would be unnecessarily high and the complete filter would be required to run at the high update rate of the inertial sensors. An alternative approach, commonly referred to as complementary filtering, relies on the complementary nature of the information sources to use one set to estimate errors in the other set of sensors. Typically, this occurs when one set of sensors are low-pass in nature (GPS) and the other is high-pass in nature (IMU).

Figure 20 presents an uncoupled GPS/INS integration architecture wherein the GPS effectively resets or initializes the INS at a regular interval. This approach has the benefit of simplicity, also, if either GPS or INS should fail, due to the lack of interdependency, at least a position and velocity solution remains available.

5.3 Loosely-Coupled GPS/INS Integration

The simplest form of coupled GPS/INS integration is the so-called loosely-coupled configuration. This approach transitions from a "full-state" to an "error-state" estimation configuration. In this architecture, measurements of the error-states

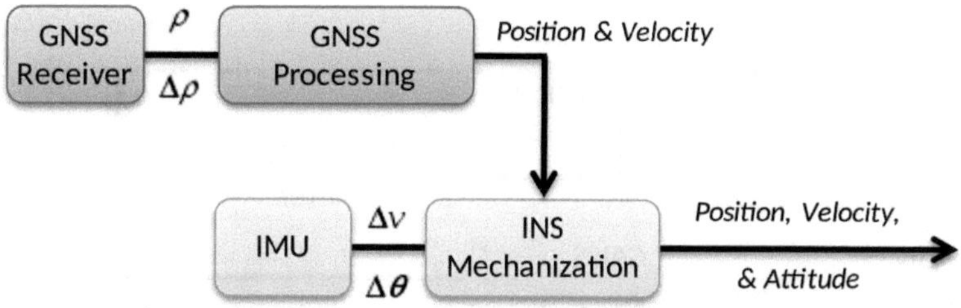

Fig. 20. Uncoupled GPS/INS integration

are derived in the position and velocity domain, which is native to the INS. The error-space filter is typically a Kalman filter, however, more sophisticated estimation schemes such as the Unscented Kalman or particle filters can be employed. Figure 21 presents an open-loop configuration wherein the INS derived PVA estimates are corrected. The principal disadvantage to this configuration relates to the potentially unbounded growth of the error in the INS-derived PVA, which when estimated by the filter can lead to divergence, as local-linearization assumptions may be violated.

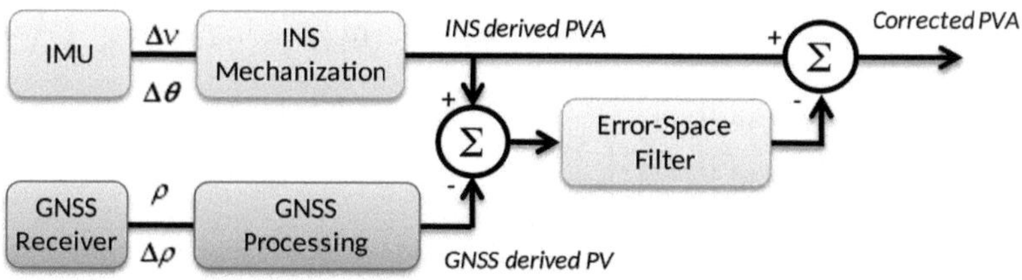

Fig. 21. Loosely-coupled GPS/INS integration (open loop)

A closed-loop version (see Figure 22) of the loosely-coupled configuration addresses the potential divergence problem exhibited by the open-loop configuration, as estimates of the errors in the INS are fed back into the INS mechanization and other IMU error sources such as accelerometer and gyroscope bias can also be estimated in real-time.

A limitation of the loosely coupled integration architecture is that a minimum of four satellites are required to obtain a viable GNSS-derived PV solution and, hence, is not well suited to a GPS degraded environment.

The error-space filter starts with Eqn. 17 and typically augments the PVA state-space error (see [13]) to include attributes, such as, dynamics in the sensor

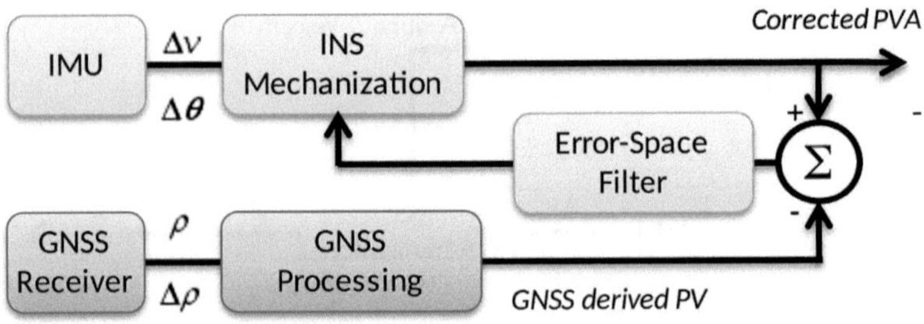

Fig. 22. Loosely-coupled GPS/INS integration (closed loop)

bias (b_a b_g) that can often be modeled as Markovian ($\delta\dot{\mathbf{b}}_b = -1/T_{c_a}\delta\mathbf{b_a} + \mathbf{w}_a$, where T_c is the serial correlation time constant) in nature to give,

$$\begin{pmatrix} \delta\dot{\psi}^e_{eb} \\ \delta\dot{\mathbf{v}}^e_{eb} \\ \delta\dot{\mathbf{r}}^e_{eb} \\ \delta\dot{\mathbf{x}}_a \\ \delta\dot{\mathbf{x}}_g \end{pmatrix} = \begin{bmatrix} -\Omega^e_{ie} & 0_{3\times3} & 0_{3\times3} & 0_{3\times3} & -\hat{C}^e_b \\ -[\hat{C}^e_b\hat{\mathbf{f}}^b_{ib}\times] & -2\Omega^e_{ie} & \frac{2g_0(\hat{L}_b)}{r^e_{eS}(\hat{L}_b)}\frac{\hat{\mathbf{r}}^e_{eb}}{|\hat{\mathbf{r}}^e_{eb}|^2}(\hat{\mathbf{r}}^e_{eb})^T & -\hat{C}^e_b & 0_{3\times3} \\ 0_{3\times3} & \mathcal{I}_{3\times3} & 0_{3\times3} & 0_{3\times3} & 0_{3\times3} \\ 0_{3\times3} & 0_{3\times3} & 0_{3\times3} & -\mathcal{I}_{3\times3}/T_{ca} & 0_{3\times3} \\ 0_{3\times3} & 0_{3\times3} & 0_{3\times3} & 0_{3\times3} & -\mathcal{I}_{3\times3}/T_{c_g} \end{bmatrix} \begin{pmatrix} \delta\psi^e_{eb} \\ \delta\mathbf{v}^e_{eb} \\ \delta\mathbf{r}^e_{eb} \\ \delta\mathbf{x}_a \\ \delta\mathbf{x}_g \end{pmatrix} + \begin{bmatrix} -\hat{C}^e_b & 0_{3\times3} & 0_{3\times3} & 0_{3\times3} \\ 0_{3\times3} & -\hat{C}^e_b & 0_{3\times3} & 0_{3\times3} \\ 0_{3\times3} & 0_{3\times3} & 0_{3\times3} & 0_{3\times3} \\ 0_{3\times3} & 0_{3\times3} & \mathcal{I}_{3\times3} & 0_{3\times3} \\ 0_{3\times3} & 0_{3\times3} & 0_{3\times3} & \mathcal{I}_{3\times3} \end{bmatrix} \begin{pmatrix} \varsigma_g \\ \varsigma_a \\ \mathbf{w}_a \\ \mathbf{w}_g \end{pmatrix} \tag{22}$$

In more compact form, Eqn. 22 can be re-written as

$$\dot{\mathbf{x}} = F\mathbf{x} + G\mathbf{w} \tag{23}$$

Furthermore, the measurements driving the error-space filter are simply the difference in PV between the GNSS and INS as

$$\begin{aligned} \mathbf{z}^e &= \begin{pmatrix} \tilde{\mathbf{r}}_{GPS} - \hat{\mathbf{r}}_{INS} \\ \tilde{\mathbf{v}}_{GPS} - \hat{\mathbf{v}}_{INS} \end{pmatrix} = \begin{pmatrix} 0_{3\times3} & 0_{3\times3} & \mathbf{I}_{3\times3} & 0_{3\times3} & 0_{3\times3} \\ 0_{3\times3} & \mathbf{I}_{3\times3} & 0_{3\times3} & 0_{3\times3} & 0_{3\times3} \end{pmatrix} \begin{pmatrix} \delta\psi^e_{eb} \\ \delta\mathbf{v}^e_{eb} \\ \delta\mathbf{r}^e_{eb} \\ \delta\mathbf{x}_a \\ \delta\mathbf{x}_g \end{pmatrix} + \mathbf{v} \\ &= H\mathbf{x} + \mathbf{v} \end{aligned} \tag{24}$$

Discretized versions of Eqns. 23 and 24 are used to "drive" the Kalman filter equations to provide aided PVA estimates [13].

Figure 23 presents the positional errors (left) resulting from a loosely-coupled (with feedback) integration scheme. Clearly, a significant improvement over the INS-only case (see Figure 17) has been realized, as the errors are now bounded across time. Similar gains have been realized in the velocity errors, as shown by Figure 23 (right).

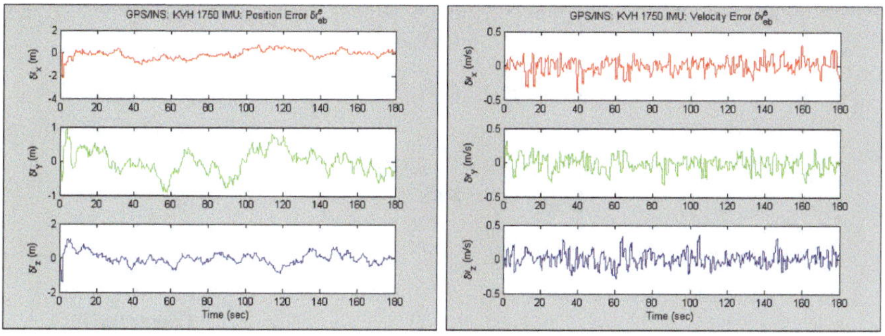

Fig. 23. GPS/INS position (left) and Velocity (right) errors: KVH 1750

The 3D positional profile shown in Figure 24 demonstrates the benefits of the GPS to significantly compensate for the deficiencies of the INS, as the two IMU configurations do not differ by the prior margins (see Figure 16). If the GPS signal "drops out," however, the impact of the INS once again becomes significant, as illustrated in Figure 24 (right), wherein the GPS is lost for the last 80 seconds of the simulation.

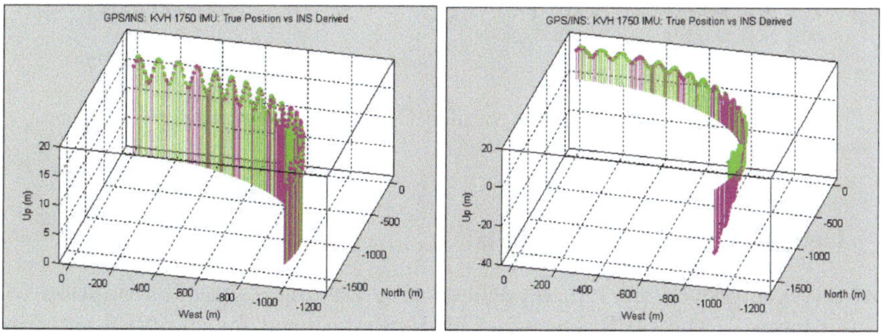

Fig. 24. GPS/INS position (left) and with "drop-outs" (right)

The obvious benefits of aiding are the bounding of the PV errors that are a direct result of the complementary nature of the GPS and INS.

5.4 Tightly-Coupled GPS/INS Integration

The second general integration category is a tightly-coupled configuration, shown in Figure 25 , which unlike the loosely-coupled configuration, derives estimates in the pseudo range (and pseudo range rate) error domain and may provide aiding to the GPS carrier tracking loops. This configuration can operate in the presence of less than four satellites and degrades more gracefully than the loosely-coupled configuration. The principal change is within the measurement model that now is

$$\mathbf{z} = \begin{bmatrix} \rho_{GPS_1} - \rho_{INS_1} \\ \rho_{GPS_2} - \rho_{INS_2} \\ \vdots \\ \rho_{GPS_n} - \rho_{INS_n} \end{bmatrix} = H\mathbf{x} + \mathbf{v} \tag{25}$$

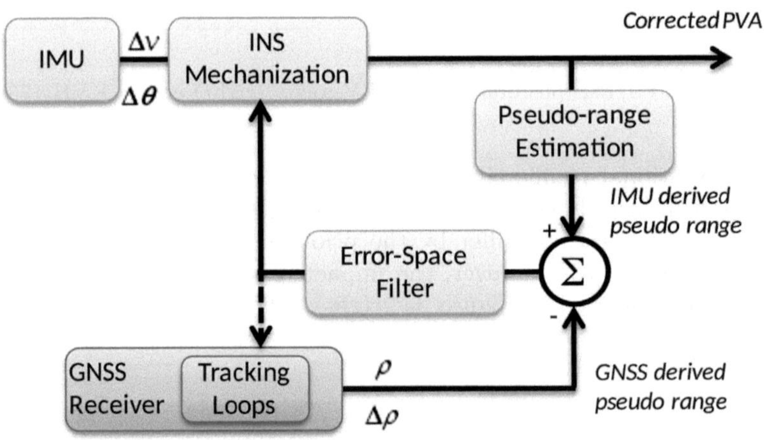

Fig. 25. Tightly-coupled GPS/INS Integration

5.5 Deeply-Coupled GPS/INS

Substantial improvement can be achieved by feeding aiding information to the tracking loops of the GPS. By using an INS, the need for the GPS to estimate the dynamics of the receiver may be eliminated, hence, the required bandwidth can be reduced. Bandwidth reduction leads to significant noise reduction. In addition, such coupling offers higher performance in jamming environments. An ultra-tightly coupled integration, also known as deep integration of the GPS and INS, represents a fundamental coupling between the two. Unfortunately, there isn't a strict definition of what constitutes ultra-tightly coupled and there are various implementations thereof. In general there are two main categories for implementing an ultra-tightly coupled GPS/INS. One uses a single master filter to estimate all the code and carrier tracking errors as well as the navigation errors. In this approach, the data needs to be processed at the GPS rate, which may reach 50Hz, consequently, the computational overhead is very high. The second category, the focus of this section, employs a federated filtering, as shown in Figure 26. A federated ultra-tightly coupled architecture employs a master filter and multiple smaller filters also called pre-filters.

The front-end block is responsible for down converting the received signal to intermediate frequency (IF) for ease of sampling. The IF signal is of the form

$$s_{IF}(t) = A_0 C(t) D(t) \cos\left[2\pi(\mathrm{f_{IF}} + \Delta\mathrm{f_c})\mathrm{t} + \phi_{\mathrm{IF}}\right] + \mathrm{n_{IF}(t)}$$

where A_0 is the signal amplitude, $C(t)$ is the C/A PRN code modulation, , $D(t)$ is the navigation data, f_{IF} is the IF frequency, Δf_c is the Doppler shift, and n_{IF} is the band-limited noise.

The next stage is the Doppler removal, which is accomplished by multiplying the input with, in-phase and quadrature of the receiver generated carrier. The carrier is generated by a numerically controlled oscillator (NCO) that attempts to match the true carrier frequency, including the Doppler shift. The in-phase (I_0) may be expressed as

$$I_0(t) = \frac{A_0}{2} C(t-\tau) D(t-\tau) \cos\left[2\pi(\Delta\mathrm{f_c} + \Delta\tilde{\mathrm{f}}_\mathrm{c})\mathrm{t} + (\phi_\mathrm{c} - \tilde{\phi}_\mathrm{c}) + \delta\phi_{\mathrm{IF}}\right] + \mathrm{n_I(t)}$$

Similarly, the quadrature component may be expressed as

$$Q_0(t) = \frac{A_0}{2} C(t-\tau) D(t-\tau) \sin\left[2\pi(\Delta f_c + \Delta\tilde{f}_c)t + (\phi_c - \tilde{\phi}_c) + \delta\phi_{IF}\right] + n_Q(t)$$

where τ is the code phase delay, $\Delta\tilde{f}_c$ is the ranging processor's measurement of the Doppler shift, ϕ_c is the phase offset, $\tilde{\phi}_c$ is the measurement of the phase offset, $\delta\phi_{IF}$ is a common phase due to the translation of the received signal to IF frequency, and finally, $n_I(t)$ and $n_Q(t)$ are noise on the in-phase and quadrature components, respectively.

The in-phase and quadrature components are then passed through a correlator block followed by an integrate and dump stage. The code correlation is accomplished by multiplying the precorrelation signals with early, prompt and late reference codes given, respectively, by $C_E(t) = C(\tilde{t}_{st} + d/2)$, $C_P(t) = C(\tilde{t}_{st})$, and $C_L(t) = C(\tilde{t}_{st} - d/2)$ where $C(t)$ is the pseudo random code, $\tilde{t}_{st}$ is the range processors measurement of the signal transmission time, and d is the code-phase offset between the early and late reference signals.

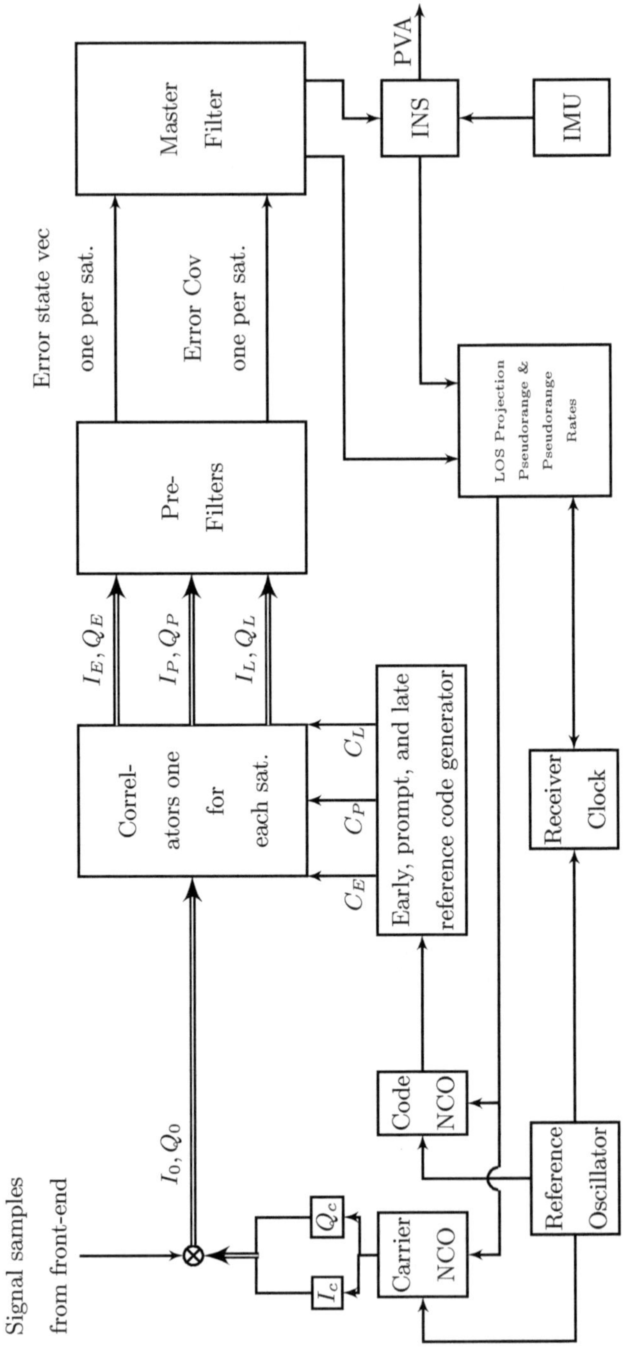

Fig. 26. Deeply-coupled GPS/INS federated architecture [14]

If the code is properly tracked, the early and late signals mirror each other about the prompt one. The integrate and dump operation offers further reduction to noise generating the following six signals

$$I_E(t_k) = A_n R\left[\epsilon - \frac{d}{2}\right] M_T D(t_k)\text{sinc}\left[\frac{\delta\dot{\phi}_\text{c}\text{T}}{2}\right]\cos\left[\delta\phi_\text{c} + \frac{\delta\dot{\phi}_\text{c}\text{T}}{2}\right] + \text{n}_\text{IE}$$

$$Q_E(t_k) = A_n R\left[\epsilon - \frac{d}{2}\right] M_T D(t_k)\text{sinc}\left[\frac{\delta\dot{\phi}_\text{c}\text{T}}{2}\right]\sin\left[\delta\phi_\text{c} + \frac{\delta\dot{\phi}_\text{c}\text{T}}{2}\right] + \text{n}_\text{QE}$$

$$I_P(t_k) = A_n R(\epsilon) M_T D(t_k)\text{sinc}\left[\frac{\delta\dot{\phi}_\text{c}\text{T}}{2}\right]\cos\left[\delta\phi_\text{c} + \frac{\delta\dot{\phi}_\text{c}\text{T}}{2}\right] + \text{n}_\text{IP}$$

$$Q_P(t_k) = A_n R(\epsilon) M_T D(t_k)\text{sinc}\left[\frac{\delta\dot{\phi}_\text{c}\text{T}}{2}\right]\sin\left[\delta\phi_\text{c} + \frac{\delta\dot{\phi}_\text{c}\text{T}}{2}\right] + \text{n}_\text{QP}$$

$$I_L(t_k) = A_n R\left[\epsilon + \frac{d}{2}\right] M_T D(t_k)\text{sinc}\left[\frac{\delta\dot{\phi}_\text{c}\text{T}}{2}\right]\cos\left[\delta\phi_\text{c} + \frac{\delta\dot{\phi}_\text{c}\text{T}}{2}\right] + \text{n}_\text{IL}$$

$$Q_L(t_k) = A_n R\left[\epsilon + \frac{d}{2}\right] M_T D(t_k)\text{sinc}\left[\frac{\delta\dot{\phi}_\text{c}\text{T}}{2}\right]\sin\left[\delta\phi_\text{c} + \frac{\delta\dot{\phi}_\text{c}\text{T}}{2}\right] + \text{n}_\text{QL}$$

where A_n is the normalized amplitude, T is the accumulation time of the correlators, M_T is the number of samples during the accumulation period, ϵ is the code tracking error in chips, $\delta\phi_c$ is the carrier phase tracking error, and $\delta\dot{\phi}_c$ is the carrier frequency tracking error. σ_{IQ} is the noise standard deviation. The terms n_{IE}, n_{QE}, n_{IP}, n_{QP}, n_{IL}, and n_{QL} are the normalized noise terms. Finally, the term R is the autocorrelation function.

The carrier frequency error, carrier phase error, and code error may be derived, respectively, to be $\delta\dot{\phi}_c = 2\pi(\Delta f_c - \Delta\tilde{f}_c)$, $\delta\phi_c = \phi_c - \tilde{\phi}_c + \delta\phi_{IF}$, and $\epsilon = (t_{sa} - \tilde{t}_{sa})f_{co}$, where f_{co} is the chip rate. These errors directly relate to pseuderange and pseudorange rates. A discriminator is then used to measure the misalignment of the code/carrier phase. Coherent or noncoherent delay lock loops (DLL) are the two types of discriminators typically used. Coherent DLL requires phase. Therefore, noncoherent is more desired, as it does not require phase lock.

Each of the pre-filters, comprised typically of an extended Kalman filter, estimates tracking errors represented by the state vector: carrier phase error, $\delta\phi_k$; carrier frequency error, $\delta\dot{\phi}_k$; carrier phase acceleration, $\delta\ddot{\phi}_k$; code phase error, ϵ_k; and GPS signal amplitude/noise sigma, A. The measurement vector is comprised of I_E, Q_E, I_P, Q_P, I_L, and Q_L, which is related to the state vector via a non-linear equation.

The output of the pre-filter block feeds into a master integration filter that estimates the INS and GPS receiver clock errors. This master integration filter processes measurements from N pre-filters. Each pre-filter generates three measurements: pseudorange, $\delta\rho$, delta pseudorange, $\Delta\delta\rho$, and delta pseudorange rates, $\Delta\delta\dot{\rho}$, residuals. The measurements are computed as [14]

$$z_{MF} = \begin{bmatrix} \delta\rho \\ \Delta\delta\rho \\ \Delta\delta\dot{\rho} \end{bmatrix} = \begin{bmatrix} k_\epsilon \epsilon \\ k_\phi(\delta\phi_k - \delta\phi_{k-1}) \\ k_\phi(\delta\dot{\phi}_k - \delta\dot{\phi}_{k-1}) \end{bmatrix}$$

where k_ϵ and k_ϕ are conversion factors of chips to meters and radians to meters, respectively. These measurements are used at the rate of 1Hz. Once these measurements are fed back to correct the master filter states, the pre-filter states are zeroed. The relationship between the measurement equation and the state vector of the master filter, $\begin{bmatrix}\delta\boldsymbol{\psi}^e_{eb} \; \delta\mathbf{v}^e_{eb} \; \delta\mathbf{r}^e_{eb} \; \mathbf{b}_a \; \mathbf{b}_g \; \delta c_b \; \delta c_d\end{bmatrix}^T$, is given by

$$H_{MF} = \begin{bmatrix} 0_{1\times3} & 0_{1\times3} & -\mathbf{u}_k^T & 0_{1\times3} & 0_{1\times3} & 1 & 0 \\ \frac{1}{2}T^2\mathbf{u}_{k-1}^T\Omega^e_{ie} & -T\mathbf{u}_{k-1}^T & -\Delta\mathbf{u}^T & \frac{1}{2}T^2\mathbf{u}_{k-1}^T C^e_b & 0_{1\times3} & 0 & T \\ -T\mathbf{u}_{k-1}^T\Omega^e_{ie} & -\Delta\mathbf{u}_{k-1}^T & 0_{1\times3} & -T\mathbf{u}_{k-1}^T C^e_b & 0_{1\times3} & 0 & 0 \end{bmatrix}$$

where $\mathbf{u}$ is the line-of-sight vector from the receiver to the satellite, and $\Delta\mathbf{u}$ is defined as

$$\Delta\mathbf{u} = \mathbf{u}_k - \mathbf{u}_{k-1} \tag{26}$$

6 Conclusion

Inertial navigation is essential in a variety of applications ranging from consumer electronics to highly sophisticated commercial and military applications. As MEMS technology advances, more devices will integrate inertial-based navigation systems. A key advantage of these systems is the ability to provide a navigation solution at a high data rate without the reliance on external signals that may be unavailable or prone to external manipulation (e.g., jamming). Inertial sensors, however, suffer from unbounded drift due to inherent error/noise sources, unless these sensors are periodically aided. The rate of drift is directly related to the grade of the inertial sensors.

In this chapter, an introduction to inertial navigation is provided. This includes key components, such as coordinate frames, navigation equations, sensor technologies and noise characteristics. Furthermore, different types of integration, such as loosely-coupled, both open and closed loops, tightly-coupled and ultra-tightly coupled are discussed. Finally, simulation examples are presented to highlight the advantages of using aiding (GPS is used here) over INS-only systems.

References

1. Northrop, http://www.northropgrumman.com/Capabilities/SIRU/Pages/default.aspx

2. Emcore, `http://www.emcore.com/emp-1-2k-lithium-niobate-fiber-optic-gyroscope-fog/`
3. Honeywell, `http://www51.honeywell.com/aero/common/documents/myaerospacecatalog-documents/Missiles-Munitions/GG1320AN_Digital_Laser_Gyro.pdf`
4. Johnson, B.R., Cabuz, E., French, H.B., Supino, R.: Development of a mems gyroscope for northfinding applications. In: IEEE/ION Position Location and Navigation Symposium (PLANS), Indian Wells, CA, USA, pp. 168–170 (May 2010)
5. Challoner, A.D., Ge, H.H., Liu, J.Y.: Boeing disc resonator gyroscope. In: IEEE/ION Position Location and Navigation Symposium (PLANS), Monterey, CA, pp. 504–514 (May 2014)
6. Lapadatu, D., Blixhavn, B., Holm, R., Kvisteroy, T.: Sar500 - a high-precision high-stability butterfly gyroscope with north seeking capability. In: IEEE/ION Position Location and Navigation Symposium (PLANS), pp. 6–13 (May 2010)
7. Kvh, `http://www.kvh.com/Military-and-Government/Gyros-and-Inertial-Systems-and-Compasses/Gyros-and-IMUs-and-INS/IMUs/1750-IMU.aspx`
8. Analog devices, `http://www.analog.com/static/imported-files/data_sheets/ADIS16137.pdf`
9. Vectornav, `http://www.vectornav.com/products/vn200-rugged/`
10. Farrell, J.A.: Aided Navigation Systems: GPS and High Rate Sensors. McGraw-Hill, New York (2008)
11. Malys, S., et al.: Refinements to the world geodetic system 1984. In: ION GPS 1997, Kansas, MO, pp. 915–920 (September 1997)
12. Groves, P.: Principles of GNSS, Inertial, and Multisensor Integrated Navigation Systems, Artech House (2008)
13. Aly El-Osery and Stephen Bruder. Ee 570: Advanced topics in electrical engineering: Location and navigation: Theory & practice (2014), `http://ee.nmt.edu/~elosery/spring_2014/ee570/`
14. Ohlmeyer, E.J.: Analysis of an ultra-tightly coupled GPS/INS system in jamming. In: IEEE/ION Position Location and Navigation Symposium (PLANS), San Diego, CA, pp. 44–53 (2006)

Hardware Implementation of Fuzzy Logic Controller -Designability, Stability and Robustness-

Takeshi Yamakawa

Fuzzy Logic Systems Institute, 650-41 Kawazu, Iizuka, Fukuoka 820-0067, Japan
yamakawa@flsi.or.jp

Abstract. The concept of Fuzzy Logic was presented in order to translate the knowhow of a human expert to formalities acceptable for a digital computer. Fuzzy information processing by employing a digital computer takes a time, and thus ths paper describes the fuzzy intrinsic electronic circuits and also designing of fuzzy logic controller. In order to prove the usability of the controller, it is applied to an inerted pendulum of 5mm in diameter, 5cm in length, 3g in weight, which is usually employed to prove control systems. The fuzzy logic controller is applied to the wine glass stabilization, the center of gravity of which is changed by the external disturbance. This experimental results implies the robustness of the fuzzy logic controller. The controller is also applied to the "mouse stabilization", where a mouse is moving arround on the table attached on the pole, and the primary control target is accomplished at the sacrifice of accuracy of positioning. The commercial fuzzy chips developed by the author are also described as well as the application to robotics.

1 Introduction

The concept of Fuzzy logic was presented in order to formalize the knowhow of a human expert to formalities acceptable for a digital computer. Since it takes a time to accomplish fuzzy inference in a digital computer, it is necessary to develop a new type of hardware system which is peculiar to fuzzy inference . It should be able to accomplish a fuzzy inference within a drastically high speed. Thus the circuit of continuous mode is considered.

Although an analog computer, one of continuous mode circuits, was used to solve a differential equation, it has vanished away because of its low accuracy and limitted dynamic range. However the range of value is 0 though 1, and the accuracy of several % is acceptable in Fuzzy Logic. Thus the problem of a traditional analog circuit is resolved in Fuzzy Logic. Rather, the problem of fuzzy intrinsic electronic circuits is the response time and the complexity of the circuit. High speed performance and simple circuit configuration (small number of devices) are targetted on for desgning fuzzy hardware systems.

Hardarwe implementation of fuzzy logic systems includes current mode circuits and voltage mode circuits. The former represents a grade of membership by a current. For instance, grade of 0 through 1 is represented by 0A though 10μA. On the other

A. El-Osery and J. Prevost (eds.), *Control and Systems Engineering*,
Studies in Systems, Decision and Control 27, DOI: 10.1007/978-3-319-14636-2_13

hand, the latter represents a grade of 0 through 1 by 0V through 5V. Both of modes are employed and discussed in this paper.

In order to prove the usability of the fuzzy logic hardware systems, the fuzzy logic controller is designed and fabricated. Forthermore, it is applied to the inverted pendulum, the wine glass stabilization and mouse stabilazation and its stability and robustness are discussed. This paper touches upon the commercial chips developed by the author.

2 Practical Hardware Systems

Practical hardware includes current mode and voltage mode, and also the electronic circuits on the board and the integrated circuits.

2.1 Current Mode Integrated Circuits[1,2,3,4,5,6,7,8,9,10,11,12,13,14]

A signal or a grade of membership is represented by current, e.g. 0 through 1by 0A through 10μA. The distinctive features of a current mode circuit are (1) algebraic summation can be easily implemented by connection of signal lines which represent membership grades to be summed up, (2) a lower power supply voltage than the voltage mode can drive the circuit. A diode can be easily implemented by so-called diode connection of bipolar transistors or MOS transistors.

All the fuzzy logic functions were proved to be represented by simple fuzzy logic functions, Bounded Difference.

Fig.1 shows a microphotograph of a basic fuzzy logic circuit, Bounded Difference, which is implemented with a current mirror (two transistors) and a diode connected transistor, and integrated by p-MOS semiconductor process.

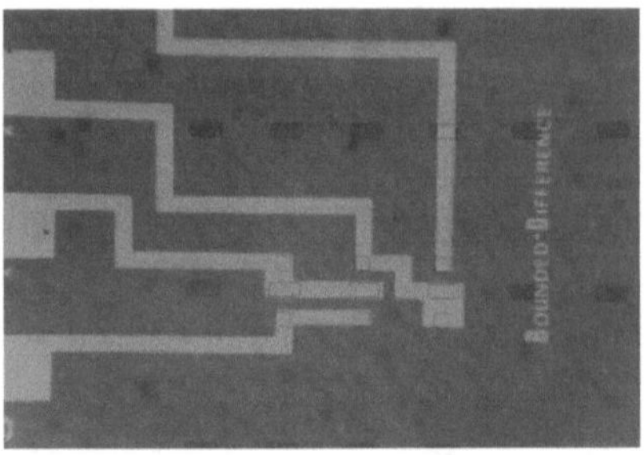

Fig. 1. Bounded Difference circuit in current mode implemented in pMOS integrated circuit. A current mirror and a diode are employed for this circuit.

Nine basic fuzzy logic functions are implemented with Bounded Difference circuits and integrated into one chip as shown in Fig.2.

Any complicated fuzzy hardware systems made up of nine basic fuzzy logic functions can be realized with only one kind of master slice including the Bounded-Difference (basic logic cell) and multiple-output n-MOS current mirror array. CMOS fuzzy logic semi-custom IC presented here exhibits the following distinctive features. (1) The effect of the variation in V_{TH} and g_m on the electrical characteristics of the hardware systems can be cut off by adjusting the supply voltage to the appropriate value. Thus the expensive ion implanter is not needed. (2) The basic logic cell exhibits good linearity which cannot be easily realized in voltage mode. (3) Since it does not need resistors nor isolation, it is suitable for a large scale fuzzy hardware system. (4) Circuits presented here can be appropriated to a binary logic and even to a 10-valued logic. (5) It exhibits the advantage over fuzzy information processing by using binary digital circuits, which arises from the ability to provide much more "functions per unit area". (6) It presents a low cost and a short term of design and fabrication.

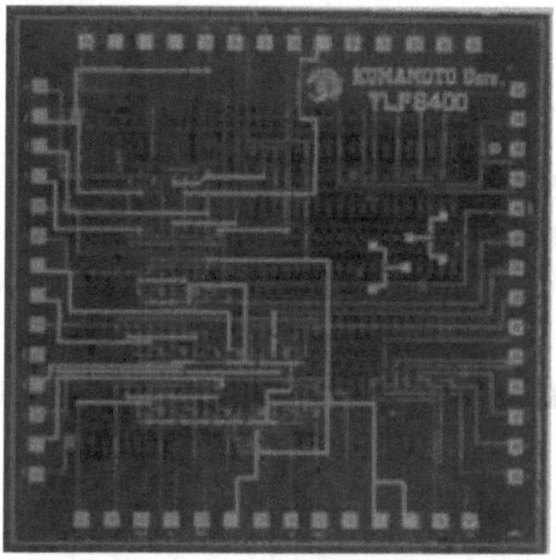

Fig. 2. CMOS fuzzy logic integrated circuits

2.2 Membership Function Generator in Voltage Mode[15]

Fuzzy inference is a very significant algorithm in fuzzy logic control which facilitates intelligent control with the knowledge of human experts. The fuzzy inference engine is typically constructed with a fuzzifier (membership function circuit), Membership function circuits, Minimum circuits taking fuzzy logic AND and a membership function generator (MFG) the label of which is assigned by an external binary digital signal. The membership function generator is indispensable for a fuzzy computer and a fuzzy logic controller.

On the other hand, the membership function generator is also indispensable for a fuzzy memory device storing the membership functions and labels in the knowhow of human experts. The fuzzy memory device in voltage mode consists of a binary RAM, a register and an MFG. Labels of membership functions are represented with only 3-bit binary codes and stored in 21 binary RAM. 'The 3-bit binary signals are transformed into distributed analog voltages corresponding to grades of membership function through an MFG. The MFG shown in Fig.3 was fabricated by a normal c-MOS semiconductor process. The fuzzy memory device presented here exhibits the following distinctive features. (1) The shape of membership function can be

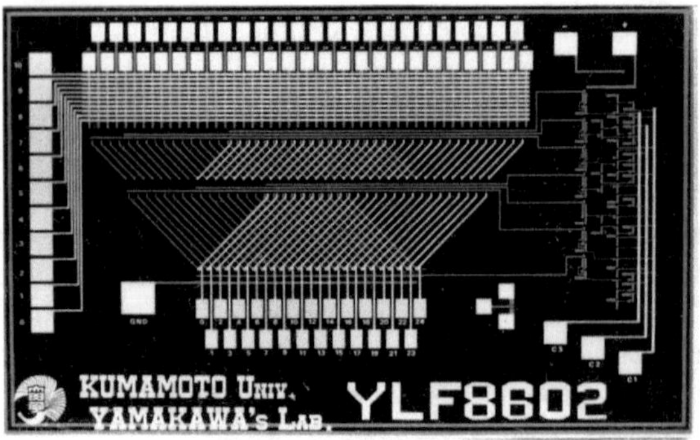

Fig. 3. The membership function generator necessary for constructing a fuzzy memory device (CMOS integrated circuit)

programmed by users. (2) The response time of MFG exhibits high operation speed (5ns). (3) The fuzzy memory device is as robust as a binary RAM against the noise. This fuzzy memory is indispensable for a fuzzy computer and a fuzzy logic controller.

2.3 Fuzzy Inference Engine Board of Voltage Mode Circuit[16]

The linguistic information naturally possesses fuzziness, vagueness, uncertainty, incompleteness cu" impreciseness and ii; characterized by a membership function. The grade of membership is represented by a number ranging from 0.0 to 1.0 and changes gradually over the universe of discourse. When linguistic information is handled in a digital computer, grades of membership representing it is transformed to many sets of binary codes. The binary coded grades are stored, transferred and operated in electronic binary circuits over and over again in accordance with the stored-program. Therefore, it takes a long period to process fuzzy information by a digital system. Moreover, the binary coded grades require incredibly large number of storing and operating elements. Although a digital computer is an universal machine, it is not most suitable for processing fuzzy information in real time. We have to grope for another type of machine able to process fuzzy information effectively and at high speed.

The inference engine is designed employing MIN and MAX operations. Intrinsic fuzzy building blocks necessary for the inference engine are described in detail. The fuzzy inference engine presented here has the distinctive features such as high speed (100nsec) and robustness against the supply voltage fluctuation, ambient temperature change, external noise and variance of transistors. These features suggest us that the inference engine in this paper has extensibility to a large scale fuzzy computer which we are dreaming of.

Fig.4 shows a fuzzy inference engine constructed on a board with 228 discrete bipolar transistors, 3 capacitors and 2 resistors. It shows 10nsec of operating speed.

(a)

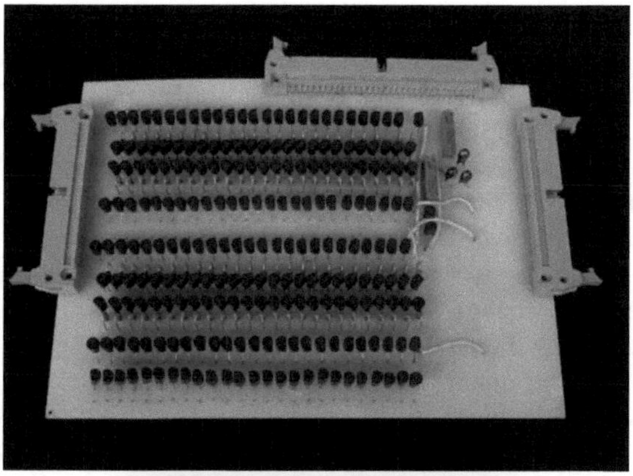

(b)

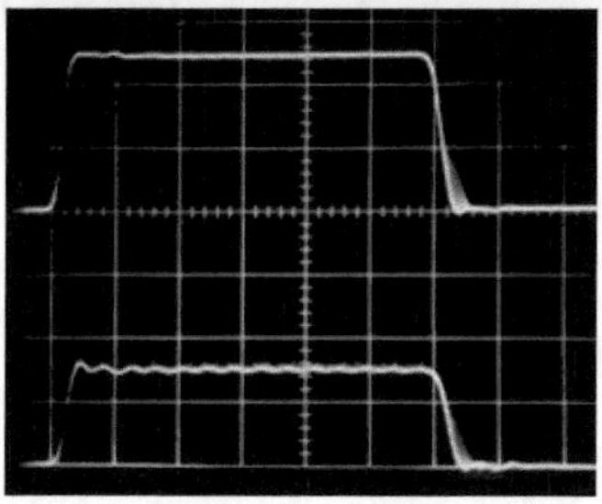

Fig. 4. (a) Fuzzy Inference Engine Board constructed with 288 discrete bipolar transistors, 3 capacitors and two variable resistors. (b) Operating speed of the fuzzy inference engine board. It shows the operating time of about 10 nsec. Upper: Input signal. Lower: Output signal.

2.4 Commercial Hardware for Fuzzy Logic Control in Voltage Mode[17,18,19]

A high-speed fuzzy controller hardware system (Fig.5) employing Min-Max operations was developed by the collaboration of the author and OMRON Corporation in 1987 and it was demonstrated in the 2nd IFSA Congress held at Gakushuin University, Japan. It has fifteen control rule boards and one defuzzifier board. Grades of membership are represented with 0V~5 V corresponding to 0.0~ 1.0. This fuzzy controller facilitates approximate reasoning at 1 000 000 FIPS (fuzzy inferences per second) and is able to be used for various purposes by programming on each control rule board (Fig.5 (c)). It handles fuzzy linguistic information of the form NL (negative large), NM (negative medium), NS (negative small), ZR (approximately zero), PS (positive small), PM (positive medium), PL (positive large) and NG (negation) as shown in Fig.6 (a). The fuzziness is assigned by the slop of the membership function which is controlled by the variable resistor on the Rule Board (Fig.6 (b)). The user can select the shape of membership function to be triangular or trapezoidal by the switch on the Rule Board (Fig.6 (c)). In case of this control of an inverted pendulum, the author employed two input variables and one output variable as shown in Table 1. One input variable is the emergency on angle which is represented by a linear summation of angle and angular velocity. The other input is the emergency on position which is represented by a linear summation of position and velocity. The output variable is the velocity of the vehicle. This fuzzy controller hardware system makes an inverted pendulum stand on a vehicle by using eleven control rules in the feedback environment shown in Fig.7. Eleven fuzzy IF-THEN rules (Table 1) are assigned on the Rule Boards. The parenthetical two rules in Table 1 are necessary, when the severe disturbance is expected. It is much easier to produce control rules than to solve the nonlinear differential equations representing the dynamics of an inverted pendulum. Two pendulums (5mm in diameter, 15cm in length, 3.5g in weight; respectively 10 mm, 50 cm, and 50 g) can be stabilized by the same rule set. This means the fuzzy logic controller exhibits a significant robustness against the changes of mass, center of gravity, length of the object to be controlled and also against the disturbance. Fig.8 shows the phase diagram of the fuzzy logic control system with the inverted pendulum. This figure shows the response of the controller to the disturbance in position. It proves the good stability of the fuzzy logic control. This fuzzy controller hardware system can be applied to attitude control of a space booster rocket and satellite, an automatic aircraft landing system, pattern recognition, stabilization of nuclear fuel rods in a reactor, an intelligent sensor, and many other applications which need swift approximate reasoning.

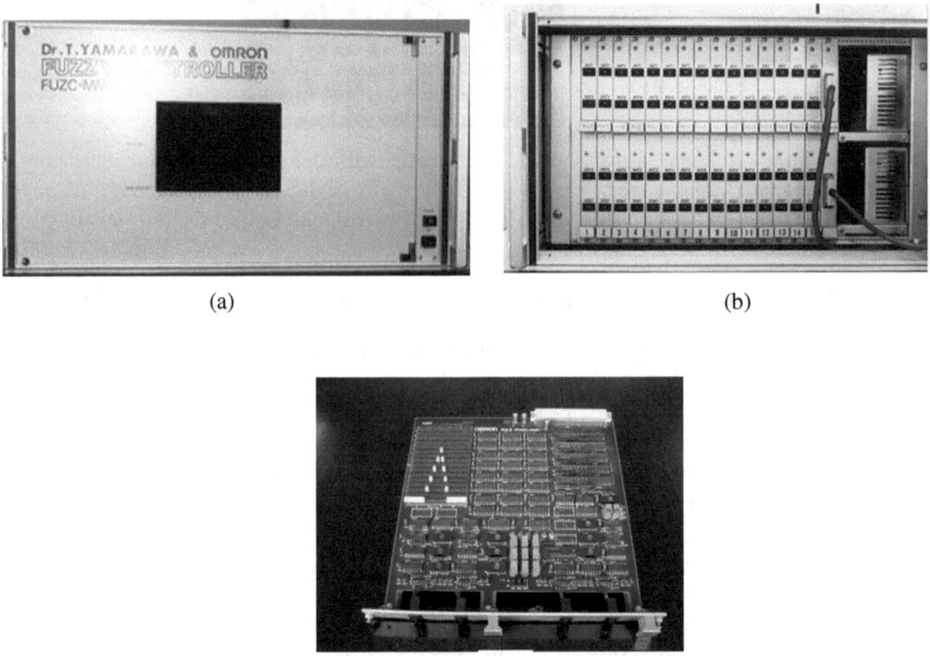

Fig. 5. The commercial fuzzy logic controller developed by the author and OMRON. (a) A front panel of the fuzzy logic controller, on which the fuzzy inference result is displayed in the form of membership function and the defuzzified value (center of gravity) is also indicated. (b) The front panel is opened. 14 boards of fuzzy inference engine (Rule Boards) and one defuzzifier board are mounted. (c) Rule Board.

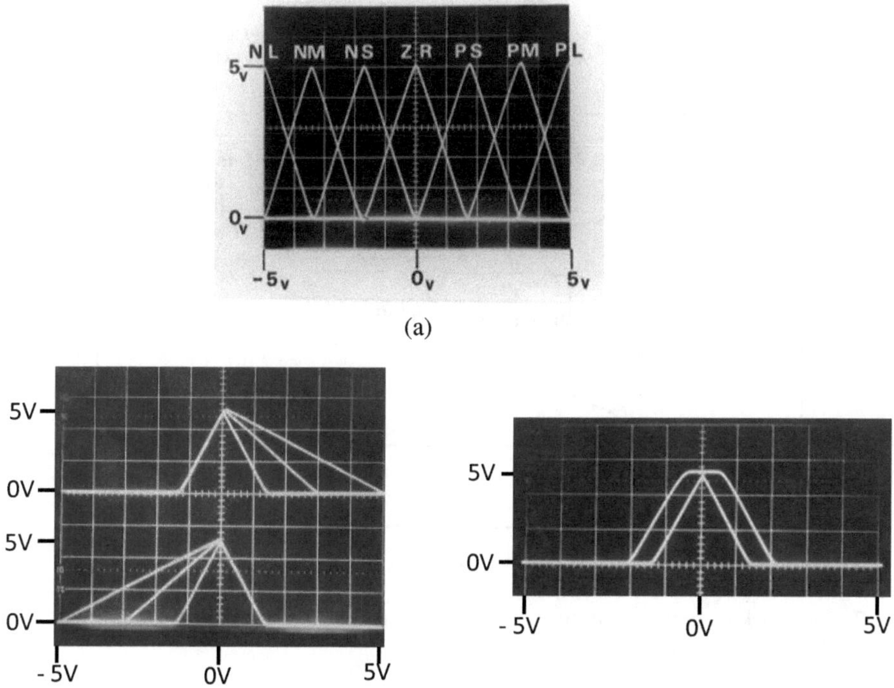

Fig. 6. Responses of the membership function circuit in the input stage of the fuzzy inference engine. (a) Seven labels can be assigned NL (negative large) through PL (positive large). (b) Fuzziness of the linguistic term is assigned by the slope of the membership function which is controlled by the variable resistor on the Rule Board. (c) The user can select the shape of membership function to be triangular or trapezoidal by the switch on the Rule Board.

Table 1. A set of fuzzy IF-THEN rules for describing the input-output characteristics of a fuzzy logic controller

X=A ·ϑ+B ·(dϑ/dt)=NS
Urgency on Tilting

Y=C ·x+D ·(dx/dt)=NS Urgency on Positioning	NM	NS	ZR	PS	PM
PM	NS		PS		(PL)
PS		NS		PM	
ZR	NM		ZR		PM
NS		NM		PS	
NM	(NL)		NS		PS

PL:Positively Large
PM:Positively Medium
PS:Positively Small
ZR:Approximately Zero
NS:Negatively Small
NM:Negatively Medium
NL:Negatively Large

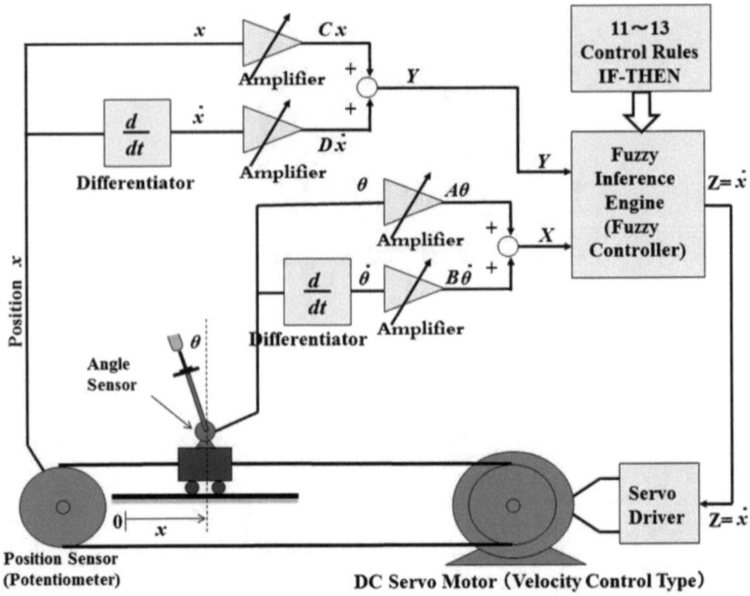

Fig. 7. The block diagram of the fuzzy logic controller for an inverted pendulum, wine glass stabilization and mouse stabilization

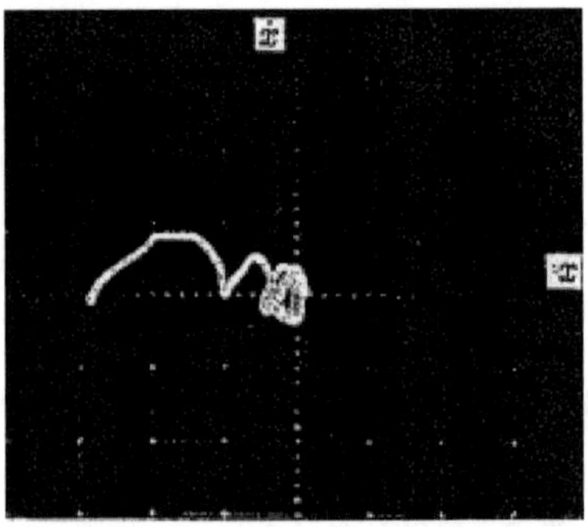

Fig. 8. Phase diagram of this fuzzy logic control system (wine glass stabilization). H: position (rightward position of the vehicle is positive x). V: velocity. This figure shows the response of the pendulum to the negative disturbance in position. After the disturbance is removed, the vehicle approaches to the equilibrium point in a vibrating manner.

2.5 Wine Glass/Mouse Stabilization by Employing the Fuzzy Logic Controller[20]

An inverted pendulum is a rigid body and thus the mathematical modeling of this system is not so difficult, even though it exhibits a nonlinear system. Next, let's consider a glass with wine is put on the plate which is attached to the top of the pendulum. In this case, the mathematical equation of dynamics of the simplified model is very difficult to derive, if not impossible. The external disturbance to the glass causes the swinging of the surface of wine as well as the pendulum with glass and wine. The center of gravity is also made swinging by the disturbance. Therefore, the mathematical approach offers little hope. On the other hand, linguistic approach with fuzzy IF-THEN rules is rather easy and hopeful, if the designer imagines the behavior of the system. The fuzzy IF-THEN rules employed in wine glass stabilization are the same to the previous inverted pendulum. Fig.9 shows the glass with wine stabilized by the fuzzy logic controller.

In the same rules, a glass with wine is replaced with a mouse. The fuzzy logic controller accomplishes the stabilization, even though the mouse is going around on the plate. This behavior of the mouse cannot be described in a mathematical way. However, the linguistic strategy can be done. In this case we have two control targets; one is the priority target, i.e. stability of the wine glass, the other is the subsidiary target, i.e. positioning accuracy. The interesting point of this fuzzy control system with sophisticated fuzzy IF-THEN rules is that the stability of the wine glass is accomplished at the sacrifice of the positioning accuracy as shown in Fig.10. When the mouse moves to the left, the vehicle is driven to the left by the controller in order to stabilize the mouse. As a result, the positioning error is produced. It is reasonable for a controller.

Fig. 9. Wine glass stabilization

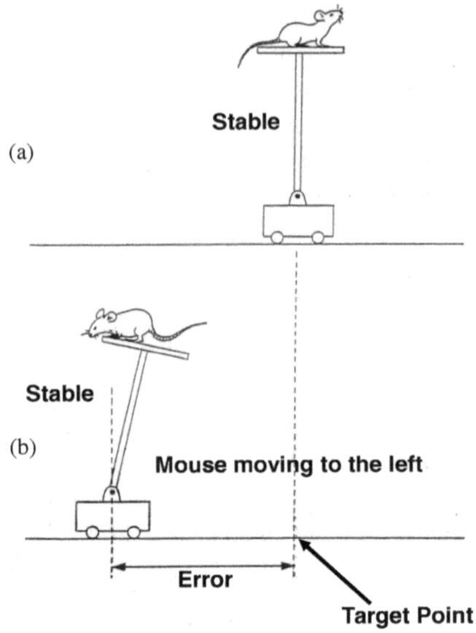

Fig. 10. The most distinctive feature of this fuzzy logic controller. The priority target, i.e. stability of the wine glass, is accomplished by this controller at the sacrifice of the subsidiary target, i.e. positioning accuracy.

2.6 Fuzzy Chips –Rule Chip and Defuzzifier Chip–

A fuzzy inference engine on a board and a defuzzifier on a board were redesigned and fabricated to a Bi-CMOS integrated circuit (Rule Chip) and to a hybrid integrated circuit (Defuzzifier Chip), respectively, which are shown in Fig.11. The former one achieves one fuzzy inference and the latter one achieves the defuzzification in the center-of-gravity method. The rule chip was designed to have three input variables in order that it can be used in the fuzzy PID control. Thus in order to establish a fuzzy logic controller, we need the rule chips the number of which is the same to necessary fuzzy IF-THEN rules and the defuzzifier chips the number of which is the same to the output. These chips appeared in the market from OMRON Corporation in 1988.

(a) (b)

Fig. 11. (a) Rule Chip and (b) Defuzzifier Chip

3 Applications

The distinctive features of fuzzy logic control can be summarized as follows;

1. How it should work is described with fuzzy natural languages and the structure of the knowledge is very clear. Therefore, the control strategy, know-how, knowledge, or data are very easy to represent, very easy to understand, very easy to remember, very easy to debug.
2. Since all the rules are independent of each other, it is easy to update the rule to follow the change of the system under control. Thus a fuzzy system is suitable for a learning system, a self-organizing system, an adaptive system, etc.
3. It can accept the exceptional data (knowledge) and the contradictory data, so that the fuzzy logic controller is suitable for nonlinear and time-variant complicated systems.
4. Compound sensory signals such as outputs from a chemical sensor, an odor sensor and other contaminated cheap sensor can be accepted.
5. Since a fuzzy inference is one kind of interpolation with very few data, drastic reduction of data and software/hardware systems can be achieved.

Taking advantage of these features, the fuzzy logic control is employed in designing the grasping robot which can grasp hard or soft (damageable), and heavy or light materials. Fig.12 shows the Tofu- grasping robot. Tofu (bean curd) is soft, damageable and rather heavy, so that human beings grasp a Tofu very carefully not to drop down and not crush it. The grasping robot developed by the author and OMRON Corporation is shown in Fig.12. It has sensors of force produced on two flat fingers, deformation of flat fingers (i.e. hardness of the material). The input signals for the fuzzy logic controller are the force applied to the material, deformation of flat fingers and the weight of the material, and the output signal of the controller is the force applied to the material.

This robot can also grasp a small bottle, can, and “clear cellophane wrapping” of a cigarette packet. There is no need to change the fuzzy IF-THEN rules. The behavior was demonstrated in IIZUKA 88 (International workshop on Fuzzy System Applications, Iizuka, Japan, August 22-24, 1988).

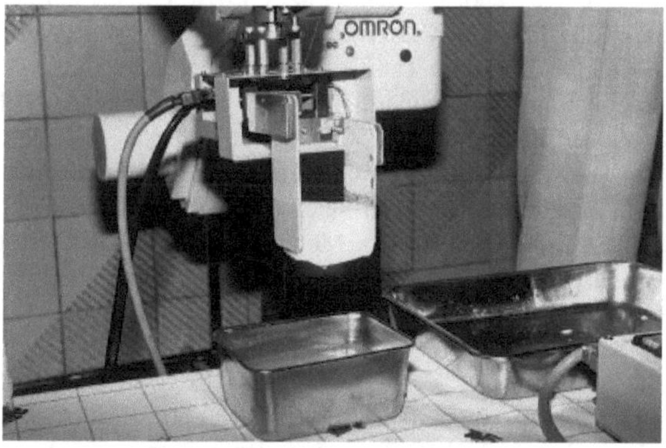

Fig. 12. Tofu-grasping Robot

4 Conclusions

Fuzzy sets and theory is the Practical Science leading the researchers to practical applications. In the practical application, hardware design and fabrication is very important to realize the concept of fuzziness to the real life. The demonstration of the inverted pendulum by employing the fuzzy logic control in the Second IFSA Congress in 1987stimulated not only researchers but also the lay people, especially newspeople, to excite a feeling of hope of breakthrough in the conventional artificial intelligence. Exposure of the essence of fuzzy concepts is achieved by employing the hardware demonstration easy to understand. This paper described the history of hardware implementation of fuzzy logic in its early days.

Acknowledgement. The author would like to thank Prof. Mo Jamshidi for his warm support and good suggestions through more than a quarter-century.

References

1. Yamakawa, T., Inoue, T., Shirai, Y., Ueno, F.: Implementation of Fuzzy Logic Hardware Systems -Three Fundamental Arithmetic Circuits-. Trans. IECE 63-C, 720–721 (1980)
2. Yamakawa, T., Inoue, T., Shirai, Y., Ueno, F.: Implementation of Fuzzy Logic (Complement, Bounded-Difference, Bounded-Sum and Absolute-Difference) by Current Mode Circuits. Trans. IECE 63-C, 722–723 (1980)
3. Yamakawa, T., Inoue, T., Shirai, Y., Ueno, F.: Implementation of Fuzzy Logic Circuits and a Programmable Multifunction Circuit with Current Mode. Trans. IECE 64-C, 147–148 (1981)
4. Yamakawa, T., Miki, T., Ueno, F.: The Fuzzy Logic Semi-Custom IC Fabricated by Using the Standard p-MOS Process. Trans. IECE 64-D, 600–601 (1984)
5. Shirai, Y., Yamakawa, T., Ueno, F.: A CAD Oriented Synthesis of Fuzzy Logic Circuits. Trans. IECE 67-D, 708–714 (1984)
6. Yamakawa, T., Miki, T., Ueno, F.: The Fuzzy Logic Semi-Custom IC Fabricated by Using the Standard p-MOS Process. Trans. IECE 67-C, 600–601 (1984)
7. Yamakawa, T., Miki, T., Ueno, F.: Basic Fuzzy Logic Circuit Formed by Using p-MOS Current Mirror Circuits. Trans. IECE 67-C, 1022–1029 (1984)
8. Yamakawa, T., Miki, T., Ueno, F.: The Design and Fabrication of the Current Mode Fuzzy Logic Semi-Custom IC in the Standard Cmos IC Technology. In: IEEE Proc. 15th Int'l Symp. Multiple-Valued Logic, May 28-30, pp. 76–82 (1985)
9. Yamakawa, T., Miki, T.: The Current Mode Fuzzy Logic Integrated Circuits Fabricated by the Standard CMOS Process. IEEE Trans. Computers C-35, 161–167 (1986)
10. Yamakawa, T., Kabuo, H.: Synthesis of Membership Function Circuits in Current Mode and its Implementation in the p-MOS Technology. In: Proc. 2nd Fuzzy System Symp., June 16-18, 1986, pp. 115–121 (1986)
11. Yamakawa, T., Kabuo, H., Miki, T.: Implementation of Programmable Membership Function Circuits in the p-MOS Technology. Trans. IECE. J69-C, 1472–1475 (1986)
12. Yamakawa, T.: Fuzzy Logic Circuits in Current Mode. In: Bezdek, J.C. (ed.) Mathematics and Logic, vol. I, pp. 241–262. CRC Press, Inc., Boca Raton (1987)

13. Yamakawa, T.: Fuzzy Hardware Systems of Tomorrow. In: Sanchez, E., Zadeh, L.A. (eds.) Approximate Reasoning in Intelligent Systems, Decision and Control, pp. 1–20. Pergamon Press (1987)
14. Yamakawa, T., Kabuo, H.: A_Programmable Fuzzifier Integrated Circuit — Syntheis, Design, and Fabrication. Information Sciences 45, 75–112 (1988)
15. Yamakawa, T.: Fuzzy Memory Device, July 20-25. Second IFSA Congress, pp. 551–555 (1987)
16. Yamakawa, T.: A Simple Fuzzy Computer Hardware System Employing Min & Max Operations -A Challenge to 6th Generation Computer, July 20-25. Second IFSA Congress, pp. 827–830 (1987)
17. Yamakawa, T.: Fuzzy Controller Hardware System, July 20-25. Second IFSA Congress (1987)
18. Yamakawa, T.: High-Speed Fuzzy Controller Hardware System: The Mega-FIPS Machine. Information Sciences 45, 113–128 (1988)
19. Yamakawa, T.: Stabilization of an Inverted Pendulum by a High-Speed Fuzzy Logic Controller Hardware System. Fuzzy Sets and Systems 32, 161–180 (1989)
20. Yamakawa, T.: A Fuzzy Inference Engine in Nonlinear Analog Mode and Its Application to a Fuzzy Logic Control. IEEE Transactions on Neural Networks 4, 496–522 (1993)

Decision Making under Z-Information

Rafik Aliev

Department of Business Administration, Azerbaijan State Oil Academy,
20 Azadlig Ave., AZ1010 Baku, Azerbaijan
raliev@asoa.edu.az

Abstract. In decision analysis we often use natural language to represent real-world information due to imperfect nature of the latter. On the one hand, such information is often characterized by fuzziness. This implies that we impose soft constraints on values of variables of interest. The other essential property of information is partial reliability. Indeed, any estimation of values of interest, are subject to the confidence in sources of information we deal with – knowledge, assumptions, intuition, envision, experience – which cannot completely cover the whole complexity of real-world phenomena. Thus, fuzziness and partial reliability are strongly associated to each other. In order to take into account this fact, L.A. Zadeh suggested the concept of a Z-number as a more adequate formal construct for description of real-world information. In this chapter we present new decision making theory based on Z-information.

Keywords: decision making, Z-information, Z-number, combined states, Z-restriction, Z-valued joint probability.

1 Critical Analysis of the Existing Decision Theories

The main drawback of the existing theories of decisions is the use of numerical, precise techniques whereas real-world information is imprecise and constraints are soft, therefore the obtained results are not trustful. Precise non-additive measure used is in CEU to model choices under uncertainty whereas human behavior cannot be precisely modeled. ambiguity aversion in α -Maximin Expected Utility(MMEU)[10] is described by a precise value α , whereas real information about ambiguity attitude is imprecise.

Gain-loss and risk attitudes are described in Cumulative Prospect theory (CPT)[9] by precise values whereas relevant information is imprecise. A DM is considered as risk averse for gains and as risk seeking for losses whereas in reality such complete information is not available. Ambiguity and risk attitudes are described by means of precise numbers in the Smooth ambiguity model (SAM)[11]. Modeling mental and social behavioral aspects as *altruism*, *fairness*, *trust*, *reciprocity,* etc [6,7,8] ignores the fact that these concepts are rather qualitative and the relevant information is intrinsically imperfect. In the existing mental-level models outcomes are considered deterministic, utilities are precise.

A. El-Osery and J. Prevost (eds.), *Control and Systems Engineering,*
Studies in Systems, Decision and Control 27, DOI: 10.1007/978-3-319-14636-2_14

Existing theories require the use of mutually exclusive and exhaustive states of nature whereas our knowledge about future and its possible forms is quite dim. Unfortunately, in the existing decision theories a small attention paid to the essence and structure of states of nature, consideration of them is very abstract (formal) and is unclear from perception point of view. The existing utility theories are based on Savage's formulation of states of nature as "a space of mutually exclusive and exhaustive states". In real-life a DM often cannot exhaustively determine each objective condition that may be faced and cannot precisely differentiate them. Each state of nature is, essentially, some area under consideration which collects in some sense similar objective conditions, that is some set of "elementary" states or quantities. The strong accent is put on the use of precise probability measures even in advanced models, whereas a long time ago J. Keynes mentioned that real probabilities are imprecise. In the SAM DM's beliefs are modeled by a second-order subjective probability distribution over priors. This is counterintuitive as humans can not follow so strict and 'inanimate' model due to highly constrained computational abilities and influence of psychological factors. In the existing mental-level models[12,13] information both on states of nature and states of mind is described by precise probabilities. Lack of data, statistics, vagueness of data rule out determination of actual precise probabilities.

In the existing theories, the fact that a human being reasons with linguistic description of information is not taken into account. Human being often reasons with perceptions described in NL. A lot of important aspects of decision making are qualitative like non-monetary issues including health, time, reputation, quality, worker satisfaction which are usually described linguistically.

The existing theories use mainly binary logic-based preference relations. Real preferences are often vague: humans find difficulty to make choices due to incomplete information, uncertainty of future, multiple criteria of choice etc. We compare alternatives linguistically using evaluations such as "much better", "much worse", "a little better", "almost equivalent"[4,5]. Comparing alternatives, a DM may not have clear preference, but may consider that each alternative is to some measure better than the others. In the existing theories they consider mainly parametrical modeling of behavioral determinants without taking into account their interaction as a driving force of decision making. Additive combination of ambiguity aversion and seeking in are used α -MMEU. Parametrical modeling of gain-loss and risk attitudes is used in CPT. Parametrical modeling of ambiguity attitudes in the SAM. Interaction of risk, ambiguity, gain-loss attitudes is missed in MMEU, SAM, CPT. Parametrical model is complicated and not transparent.

We have to mention also missing partial reliability of information in main existing theories. As a rule, in the existing theories decision-relevant information on all the main elements of a decison problem – states of nature, outcomes, probabilities, behavioral determinants etc is considered as completely reliable.

2 Isights for Development of New Theory of Decisions

It is needed to develop such a theory of decisions that would be free of the mentioned restrictions, every existing decision making theory would be a particular case, a unified model is used for decision making with imperfect information in the realm of decision making under uncertainty, multicriteria decision making, group decision making.

The following steps will help to create such a theory:

1. modeling interaction of the behavioral determinants such as risk, ambiguity and losses attitudes, fairness, reciprocity and others. Indeed, a behavior of a DM is some 'fusion' of intensities of behavioral determinants
2. formalization of imperfect decision relevant information which is, as a rule, supported by perceptions. Often, perception-based information has linguistic representation. From the other side, in science any problem is initially described verbally, and then a formal model is constructed to reflect observed regularities.
3. Joint consideration of subjective conditions (behavior of a DM) and objective conditions at the same fundamental level. This will help to create a model which is more close to real decisions and is more transparent for decision analysis. In the existing theories the basis of decision analysis is the space of states of nature only, whereas behavioral determinants are just introduced parametrically.

3 The Evolution of Modeling Decision-Relevant Information

Formally, imperfect information can be considered as a restriction R(x) on a value of a variable under consideration, X. The concept of a restriction has greater generality than the concept of interval, set, fuzzy set and probability distribution. The principal types of restrictions are: possibilistic restrictions, probabilistic restrictions and **Z-restrictions. Z-restrictions** is the highest level of hierarchy of restrictions is. Assume that X is a real-valued random variable. A Z-restriction is expressed as

R(X): X is Z,

where Z is a combination of possibilistic and probabilistic restrictions defined as

Z: Prob(X is A) is B,

where A and B are fuzzy numbers. Usually, A and B are labels drawn from a natural language. The ordered pair, (A,B), is referred to as a Z-number[1]. The first component, A, is a possibilistic restriction on X. The second component, B, is a possibilistic restriction on the certainty (probability) that X is A.

The hierarchy of the restriction is formed along the evolution of modeling imperfect information:

- Numerical information (ground level),
- Interval-valued information (first level),

- Fuzzy information (second-level),
- Second-order imprecise hierarchical model (third level),
- Z-information and visual information (fourth level).

Z-information is encoded in **Z-valuations**. A Z-valuation is an ordered triple of the form (X,A,B), and (A,B) is a Z-number. Equivalently, a Z-valuation, (X,A,B), is a Z-restriction on X:

$$(X,A,B) \rightarrow X \text{ is } (A,B).$$

4 Basic Principles of the Suggested New Theory of Decisions

The suggested new theory of decisions is based on the following principles:

1. The use of Z-restriction which is able to reflect imprecision, vagueness, partial truth, and what is very important, partial reliability of real decision relevant information
2. Description of a human behavior by a space of states of a DM. The states of a DM represent his/her possible principal mental conditions inspired by various intensities of behavioral determinants

The space of states of a DM and the space of states of nature compose the space of CSs (CSs) as a basis of decision analysis.

4.1 Z-Restriction

In the existing theories almost no attention is paid to such important aspect as reliability of information. It is often assumed, especially in the most famous theories, that the relevant information is trustful. However, real-world information is, as a rule, partially reliable. The reasons for partial reliability of information may be partial reliability of the source of information, misperceptions, psychological biases, incompetence. Therefore, the obtained results of decision analysis will naturally be partially reliable.

What is important is that Z-information is not only the most general representation of information but also has the highest descriptive power from human perception point of view. The use of Z-information is very important to describe both decision environment and behavioral determinants. The reason is that information on states of nature, future outcomes and behavioral determinants is naturally not completely known, and, therefore, taking into account reliability of estimated values is necessary.

4.2 Combined States

Objectives and subjective conditions should be linked together as a human behavior naturally depends on objective conditions. A combined state[3] is a pair of a state of nature and a state of a DM may reflect whether the considered human **subjective**

condition can take place at the considered **objective condition** and to what **degree of likelihood**. CSs is more fundamental and a transparent basis of behavioral decision analysis. A state of a DM should in general be considered as a vector whose components are behavioral factors. For example, a state of an investor may be characterized by risk attitude, ambiguity attitude, reciprocity, social responsibility etc.

The same applies to a state of nature, for which the following indexes may be considered as inflation rate, GDP, unemployment level etc. Consideration of a CS, therefore, becomes a promising task.

The proposed basics of the new theory of decisions open a door to improvement and progress of such important directions as **group decision making** and **multicriterial decision making**.

For **group decision making**, the use of a formalism of states of a DM may help to discover at a more fundamental level an expertise and professionalism of members of a group. This, in turn would lead to more adequate and clear ranking of them and, therefore, a better determination of an optimal group decision.

For **multicriterial decision making**, the use of states of a DM would help to overcome incorrect, improper choices of a human being which can be conditioned by the fact that emotions, biases etc may hinder constructive consideration of criteria and their importance.

Thus, the third requirement for a general theory of decisions is satisfied: basics of a unified model based on space of CS with Z-valued information is suggested.

4.3 Short Description of a New Theory

The suggested theory must possess a unified decision model which could unite the existing models into a single whole. As a result, the existing decision theories would be particular cases of the new general theory of decisions, because they are developed for modeling particular evidences. The basic principles of the suggested new general theory are outlined below:

1. Z-valued granulation of space of behavioral determinants into a "states of a DM" with each state describing one principle behavior.
2. states of nature and states of a DM constitute a single space of CSs. A CS represents possible combination of objective conditions and a DM's behavior.
3. Likelihood of a CS is described by Z-valued joint probability determined on base of available information about dependence of DM's behavior on objective conditions.
4. Z-values of outcomes are distributed over CSs reflecting various evaluation of the outcomes by a DM in various states.

Development of a new theory requires to formulate arithmetic operations over Z-numbers information.

5 Statement of the Problem in the Suggested Theory of Decisions

A set of ***Z-valued states of nature*:**

$$\mathcal{S}=\{\hat{s}_1,\hat{s}_2,...,\hat{s}_n\}\text{, where } \hat{s}_i=(\tilde{A}_{s_i},\tilde{B}_{s_i}),\ i=1,...,n.$$

A set of ***Z-valued states of a DM*:**

$$\mathcal{H}=\{h_1,h_2,...,h_m\}\text{, where } h_j=(\tilde{A}_{h_j},\tilde{B}_{h_j}),\ j=1,...,m.$$

A set of ***Z-valued outcomes***:

$$\mathcal{X}=\{x_1,...,x_l\}\text{, where } x_k=(\tilde{A}_{x_k},\tilde{B}_{x_k}).$$

A set of ***Z-valued actions*:**

$$\mathcal{A}=\{f\,|\,f:\mathcal{S}\times\mathcal{H}\to\mathcal{X}\}.$$

We denote $\Omega=\mathcal{S}\times\mathcal{H}$ a space "nature-DM", elements of which are CSs $\hat{w}=(\hat{s},\hat{h})$.

Only some partial information is available in form of Z-valued marginal probability distributions over states of nature and DM.

Nature:

$$\hat{P}_{\mathcal{S}}=\hat{p}(\hat{s}_1)/\hat{s}_1+\hat{p}(\hat{s}_2)/\hat{s}_2+...+\hat{p}(\hat{s}_n)/\hat{s}_n,\quad \hat{p}(\hat{s}_i)=(\tilde{A}_{p(s_i)},\tilde{B}_{p(s_i)})$$

DM:

$$\hat{P}_{\mathcal{H}}=\hat{p}(\hat{h}_1)/\hat{h}_1+\hat{p}(\hat{h}_2)/\hat{h}_2+...+\hat{p}(\hat{h}_n)/\hat{h}_n,\quad \hat{p}(\hat{h}_j)=(\tilde{A}_{p(h_j)},\tilde{B}_{p(h_j)})$$

That is, $\hat{p}()$ is a Z-valued probability. $\hat{p}(\hat{h}_i)/\hat{h}_i$ can be formulated as

"*a probability that a state of a DM is* $\hat{h}_i$ *is equal to* $\hat{p}(\hat{h}_i)$".

Value of an outcome $\hat{x}=\hat{f}(\hat{s},\hat{h})$ in various states of a DM is formalized as a value $\hat{u}(\hat{f}(\hat{s},\hat{h}))$ of a Z-valued function $\hat{u}$. We can claim that the value function of Kahneman and Tversky $v=v(f(s))$ appears then as a special case.

An overall utility $\hat{U}(\hat{f})$ of an action $\hat{f}$ is to be determined as a Z-number-valued aggregation of $\hat{u}(\hat{f}(\hat{w}))$ over Ω. Then the problem of behavioral decision making under partially reliable imperfect information consists in determination of an optimal action:

$$\text{Find } \hat{f}^* \in \mathcal{A} \text{ with} \hat{U}(\hat{f}^*) = \max_{\hat{f} \in \mathcal{A}} \int_{\Omega} \hat{U}(\hat{f}(\hat{w}))d\hat{\eta} .$$

Thus, the first requirement for a new general theory of decisions is satisfied. In the suggested statement of problem the mentioned above restrictions of the existing decision theories are weakened:

1. Membership functions and possibility distributions are used instead of precise estimates
2. Fuzzy granulation of objective conditions into overlapping states of nature is used instead of mutually exclusive states of nature
3. Z-valued imprecise probabilities are used instead of precise probability measure
4. Fuzzy logic-based preference relations are used instead of binary logic-based preference relations to represent natural vagueness of preferences
5. Fuzzy set theory is used to formalize linguistic description of information usually used by humans
6. A state of a DM representing ensemble of behavioral determinants is used instead of parametrical modeling
7. Z-numbers are used to model partially reliable imprecise information

The axiomatization of our utility model is an extension of axiomatization of a bi-capacity formulated by Labreuche and Grabisch for mutliattribute decision making (MADM)[2]. Bi-capacity is a generalization of capacities which is able to describe interaction between attractive and repulsive values of attributes. The suggested general theory of decisions extends bi-capacity axiomatization on the base of formal parallelism between MADM and decision making under uncertainty as follows:

1. space of CSs Ω is used instead of a set of attributes $\mathcal{I}$;
2. Actions $\mathcal{A} = \{\hat{f} \,|\, \hat{f} : \mathcal{S} \times \mathcal{H} \to \mathcal{X}\}$ are used instead of alternatives $\mathcal{A} = \{f \,|\, f : \mathcal{I} \to \mathcal{X}\}$ in MADM
3. Linguistic preference relation (LPR) is used instead of classical preference relations

According to 1 and 2 we use:

- **intra-combined state information** to determine utilities of outcomes $\hat{f}(\hat{w}_i) = x_i$, $\hat{w}_i = (\hat{s}_{i_1}, \hat{h}_{i_2})$, with understanding that these are utilities at state of nature $\hat{s}_{i_1}$ and are conditioned by a state $\tilde{h}_{i_2}$ of a DM
- **inter-combined states information** to determine fuzzy-valued bi-capacity representing dependence between CSs $\hat{w}_i = (\hat{s}_{i_1}, \hat{h}_{i_2})$.

According to 3, we use LPR $\succsim_l$ that assigns to any two actions $\tilde{f}$ and $\tilde{g}$ degrees $Deg(\hat{f} \succsim_l \hat{g}), Deg(\hat{g} \succsim_l \hat{f}) \in [0,1]$ as extents to which one action is better than another one. The degrees are used to model the fact that imperfect decision-relevant information rules out to definite determination of preferences. For special case, when $Deg(\hat{g} \succsim_l \hat{f}) = 0$ and $Deg(\hat{f} \succsim_l \hat{g}) \neq 0$ we have a classical preference, i.e. we say that $\hat{f}$ is preferred to $\hat{g}$.

6 Model

Let us look at a general representation of space Ω (table 1):

Table 1. General representations of a CSs space

	$\hat{s}_1$	$\cdots$	$\hat{s}_i$	$\cdots$	$\hat{s}_n$
$\hat{h}_1$	$(\hat{s}_1, \hat{h}_1)$	$\cdots$	$(\hat{s}_i, \hat{h}_1)$	$\cdots$	$(\hat{s}_n, \hat{h}_1)$
$\vdots$	$\vdots$	$\vdots$	$\vdots$	$\vdots$	$\vdots$
$\hat{h}_j$	$(\hat{s}_1, \hat{h}_j)$	$\cdots$	$(\hat{s}_i, \hat{h}_j)$	$\cdots$	$(\hat{s}_n, \hat{h}_j)$
$\vdots$	$\vdots$	$\vdots$	$\vdots$	$\vdots$	$\vdots$
$\hat{h}_m$	$(\hat{s}_1, \hat{h}_m)$	$\cdots$	$(\hat{s}_i, \hat{h}_m)$	$\cdots$	$(\hat{s}_n, \hat{h}_m)$

A fuzzy overall utility $\hat{U}(\hat{f})$ of any action $\hat{f} \in \mathcal{A}$ is fuzzy-valued Choquet-like aggregation of $\hat{u}(\hat{f}(\hat{s}, \hat{h}))$ over space Ω w.r.t. fuzzy number-valued bi-capacity:

$$\hat{U}(\hat{f}) = \sum_{l=1}^{n} (\hat{u}(\hat{f}(\hat{w}_{(l)})) - \hat{u}(\hat{f}(\hat{w}_{(l+1)})))\hat{\eta}(C,D) =$$

$$= \sum_{l=1}^{n} ((\tilde{A}_{u(f(w_{(l)}))}, \tilde{B}_{u(f(w_{(l)}))}) - (\tilde{A}_{u(f(w_{(l+1)}))}, \tilde{B}_{u(f(w_{(l+1)}))}))(\tilde{A}_{\eta(C,D)}, \tilde{B}_{\eta(C,D)}) \quad (1)$$

provided $\hat{u}(\hat{f}(\hat{w}_{(l)})) \geq \hat{u}(\hat{f}(\hat{w}_{(l+1)}))$; $C = \{\hat{w}_{(1)}, ..., \hat{w}_{(l)}\} \cap N^+$, $D = \{\hat{w}_{(1)}, ..., \hat{w}_{(l)}\} \cap N^-$) $N^+ = \{\hat{w} \in \Omega : \hat{u}(\hat{f}(\hat{w})) \geq 0\}$, $N^- = \Omega \setminus N^+$. $\hat{\eta} : \Omega \times \Omega \to \mathcal{Z}^1_{[-1,1]}$ is a Z-number-valued bi-capacity, $\mathcal{Z}^1_{[-1,1]}$ is a set of Z-numbers defined over [-1,1] interval.

An optimal $\hat{f}^* \in \mathcal{A}$, that is $\hat{f}^* \in \mathcal{A}$ for which $\hat{U}(\hat{f}^*) = \max_{\hat{f}\in\mathcal{A}} \left\{ \int_\Omega \hat{u}(\hat{f}(\hat{s},\hat{h}))d\hat{\eta} \right\}$, is determined by using the method of ranking of Z-numbers.

Z-number-valued bi-capacity $\hat{\eta}(\cdot,\cdot)$ is to be constructed from a Z-valued joint distribution $\hat{P} = \hat{p}(\hat{w}_1)/\hat{w}_1 + \hat{p}(\hat{w}_2)/\hat{w}_2 + \ldots + \hat{p}(\hat{w}_t)/\hat{w}_t$ on Ω.

$\hat{P}_\Omega$ is to be determined from marginal distributions $\hat{P}_{\mathcal{S}}$ and $\hat{P}_{\mathcal{H}}$ and information on dependence between $\hat{s} \in \mathcal{S}$ and $\hat{h} \in \mathcal{H}$.

Given $\hat{P}_\Omega$, a Z-number-valued bi-capacity $\hat{\eta}(\cdot,\cdot)$ can be constructed as the difference of two Z-number-valued capacities:

$$\hat{\eta}(\mathcal{C},\mathcal{D}) = \hat{\eta}_1(\mathcal{C}) - \hat{\eta}_2(\mathcal{D}),\ \mathcal{C},\mathcal{D} \subset \Omega,$$

where $\hat{\eta}_1(\mathcal{C}),\hat{\eta}_2(\mathcal{D})$ can be chosen as fuzzy number-valued lower or upper previsions.

Thus, $\hat{U}(\hat{f})$ depends on likelihood of states of a DM.

Let us show that the suggested theory is a generalization of the existing decision theories. First, let us suppose that all the decision relevant information is completely reliable, that is, for any Z-number $Z = (\tilde{A},\tilde{B})$ the part $\tilde{B}$ is taken as a singletone $\tilde{B} = 1$. In this case we will have the framework of behavioral decision making under Z-valued information (patially reliable information) will reduce to behavioral decision making under fuzzy information and the utility model will reduce to fuzzy-valued aggregation with respect to fuzzy-valued bi-capacity:

$$\hat{U}(\hat{f}) = \sum_{l=1}^{n} (\hat{u}(\hat{f}(\hat{w}_{(l)})) - \hat{u}(\hat{f}(\hat{w}_{(l+1)})))\hat{\eta}(\mathcal{C},\mathcal{D}) =$$

$$= \sum_{l=1}^{n} ((\tilde{A}_{u(f(w_{(l)}))}, \tilde{B}_{u(f(w_{(l)}))}) - (\tilde{A}_{u(f(w_{(l+1)}))}, \tilde{B}_{u(f(w_{(l+1)}))}))(\tilde{A}_{\eta(\mathcal{C},\mathcal{D})}, \tilde{B}_{\eta(\mathcal{C},\mathcal{D})}) =$$

$$= \sum_{l=1}^{n} ((\tilde{A}_{u(f(w_{(l)}))},1) - (\tilde{A}_{u(f(w_{(l+1)}))},1))(\tilde{A}_{\eta(\mathcal{C},\mathcal{D})},1)$$

$$= \sum_{l=1}^{n} (\tilde{A}_{u(f(w_{(l)}))} - \tilde{A}_{u(f(w_{(l+1)}))})\tilde{A}_{\eta(\mathcal{C},\mathcal{D})}$$

$$= \sum_{l=1}^{n} (\tilde{u}(\tilde{f}(\tilde{w}_{(l)})) - \tilde{u}(\tilde{f}(\tilde{w}_{(l+1)})))\tilde{\eta}(\mathcal{C},\mathcal{D})$$

$$= \sum_{l=1}^{n} (\tilde{u}(\tilde{f}(\tilde{w}_{(l)})) - \tilde{u}(\tilde{f}(\tilde{w}_{(l+1)})))\tilde{\eta}(\mathcal{C},\mathcal{D})$$

Thus, a fuzzy-valued utility functional is obtained:

$$\tilde{U}(\tilde{f})=\sum_{l=1}^{n}(\tilde{u}(\tilde{f}(\tilde{w}_{(l)}))-\tilde{u}(\tilde{f}(\tilde{w}_{(l+1)})))\tilde{\eta}(\mathcal{C},\mathcal{D}).$$

Let us now simplify the framework of behavioral decision making under fuzzy information and fuzzy decision model to its non-fuzzy variant. This will result in the following model:

$$U(f)=\sum_{l=1}^{n}(u(f(w_{(l)}))-u(f(w_{(l+1)})))\eta(\mathcal{C},\mathcal{D})$$

This is a real-valued utility functional as an aggregation of real-valued utilities with respect to a real-valued bi-capacity. Consider its relations with the Expected Utility (EU) and CEU as the basic existing utility models applied on space S.

EU Criterion. Let η in (1) be a probability measure on Ω. Then we have (1) as:

$$U(f)=\sum_{k=1}^{mn}u(f(w_k))p(w_k)=\sum_{j=1}^{m}\sum_{i=1}^{n}u(f(s_i,h_j))p(s_i,h_j) \tag{2}$$

In traditional EU (EU applied on a space S only) they consider that a DM exhibits the same behavior in any state of nature. In our terminology this means that only one state of a DM exists. Then, to model a traditional EU within (1) we exclude all h_j except one, say h_k. This means $p(s_i,h_j)=0,\forall j\neq k$ (as all $h_j, j\neq k$ don't exist) and we have

$$U(f)=\sum_{i=1}^{n}u(f(s_i,h_k))p(s_i,h_k)$$

As a DM is always at h_k whatever s_i takes place, we have $p(s_i,h_k)=p(s_i)$. Furthermore, in traditional EU a DM is considered as either risk averse or risk seeking or risk neutral. So, h_k can represent one of these behaviors. For example, if h_k represents risk aversion then $u(f(\cdot,h_k))$ will be concave, for risk seeking $u(f(\cdot,h_k))$ will be convex etc. If we use notation $u'(f(\cdot))$ for $u(f(\cdot,h_k))$ then we have (1) as

$$U(f)=\sum_{i=1}^{n}u'(f(s_i))p(s_i)$$

which is nothing but a traditional EU.

The suggested theory as opposed to EU allows to take into account DM's various risk attitudes at various states of nature.

CEU Criterion. Let η in (1) be a capacity. Then we have (1) as:

$$U(f)=\sum_{l=1}^{n}(u(f(w_{(l)}))-u(f(w_{(l+1)})))\eta(\{w_{(1)},...,w_{(l)}\}) \quad (3)$$

$w_{(l)}=(s_j,h_k),u(f(w_{(l)}))\geq u(f(w_{(l+1)}))$. Assuming that only some h_k exists, we have $\forall w\in\Omega, w=(s_i,h_k)$, that is $\Omega=S\times\{h_k\}$. Then $u(f(w_{(l)}))-u(f(w_{(l+1)}))=0$ for $w_{(l)}=w_{(l+1)}=(s_i,h_k)$. Only differences $u(f(w_{(l)}))-u(f(w_{(l+1)}))=0$ for which $w_{(l)}=(s_i,h_k),w_{(l+1)}=(s_j,h_k),i\neq j$ may not be equal to zero.

As a result, we have:

$$U(f)=\sum_{i=1}^{n}(u(f(s_{(i)},h_k))-\tilde{u}(f((s_{(i+1)},h_k))))\eta(\{(s_{(1)},h_k),...,(s_{(i)},h_k)\})$$

Now, using notations $u'(f(\cdot))=u(f(\cdot,h_k))$ and $\eta'(\{s_{(1)},...,s_{(i)}\})= =\eta(\{(s_{(1)},h_k),...,(s_{(i)},h_k)\})$ we can write

$$U(f)=\sum_{i=1}^{n}(u'(f(s_{(i)}))-u'(f(s_{(i+1)})))\eta'(\{s_{(1)},...,s_{(i)}\})$$

This is CEU representation. CEU is often used to represent uncertainty attitude. So, if h_k represents ambiguity aversion (ambiguity seeking) then $\eta(\{(s_{(1)},h_k),...,(s_{(i)},h_k)\})$ can be chosen as lower prevision (upper prevision). It can be shown that the CPT representation is also a special case of the suggested theory as this representation is a sum of two Choquet integrals.

Thus, the famous existing decision theories are particular case of the suggested decision model based on a Z-valued Choquet integration over space of CS.

A few words about Z-valued joint probability. To measure a joint probability of two events we need information on marginal probabilities and information on a type of dependence referred to as a sign of dependence. There exist three types of dependence:

- **positive dependence:** events have tendency to occur together
- **negative dependence:** events do not commonly occur together, e.g. one precludes occurrence of another.
- **independence:** occurrence of one event does not depend on occurrence of another one

The problems of construction of fuzzy joint probabilities are being discovered and some approaches are suggested. It is needed to develop arithmetic operations necessary for construction of Z-valued joint probabilities over the space of CSs.

7 Solving Methodology for a Suggested Theory of Decisions

The problem of behavioral decision making under partially reliable imperfect information is the following:

$$\text{Find } \hat{f}^* \in \mathcal{A} \text{ with } \hat{U}(\hat{f}^*) = \max_{\hat{f}\in\mathcal{A}} \left\{ \int_{\Omega} \hat{u}(\hat{f}(\hat{s},\hat{h}))d\hat{\eta} \right\}.$$

The problem is solved as follows.

At the **first** stage we need to assign fuzzy utilities $\hat{u}(\hat{f}(\hat{s},\hat{h}))$ to every action $\hat{f} \in \mathcal{A}$ taken at a state of nature $\hat{s}_i \in \mathcal{S}$ when a DM's state is $\hat{h}_j \in \mathcal{H}$.

The second **stage** consists in construction of a Z-valued joint probability $\hat{P}_\Omega$ over Ω proceeding from partial information on marginal distributions over $\mathcal{S}$ and $\mathcal{H}$ represented by Z-valued probabilities for all states except one. This requires constructing unknown Z-valued probability for each space. Given marginal distribution of Z-valued probabilities for all the states, it is needed to verify consistency, completeness and redundancy of this distribution.

Finally, on the base of fuzzy marginal distributions and information on dependence between states of nature $\hat{s} \in \mathcal{S}$ and a DM's states $\hat{h} \in \mathcal{H}$ it is needed to construct Z-valued joint probability distribution $\hat{P}_\Omega$ over Ω.

At the **third stage** it is necessary to construct a Z-valued bi-capacity $\hat{\eta}(\cdot,\cdot)$ based on $\hat{P}_\Omega$ over Ω. For simplicity one can determine a $\hat{\eta}(\cdot,\cdot)$ as the difference of two Z-valued capacities.

At the **fourth stage** an overall utility $\hat{U}(\hat{f})$ for every action $\hat{f} \in \mathcal{A}$ is calculated by using (1).

Finally, an optimal action $\hat{f}^* \in \mathcal{A}$ as the action with the maximal Z-valued utility $\hat{U}(\hat{f}^*) = \max_{\hat{f}\in\mathcal{A}} \left\{ \int_{\Omega} \hat{u}(\hat{f}(\hat{s},\hat{h}))d\hat{\eta} \right\}$ is determined by comparing Z-overall utilities $\hat{U}(\hat{f})$ for all $\hat{f} \in \mathcal{A}$.

8 Conclusion

The suggested new theory of decisions is developed to a considerable extent and a series of results of a great generality has been provided.

References

1. Zadeh, L.A.: A Note on Z-numbers. Inform. Sciences 181, 2923–2932 (2010)
2. Labreuche, C., Grabisch, M.: Generalized Choquet-like Aggregation Functions for Handling Bipolar Scales. Eur. J. of Oper. Res. 172, 931–955 (2006)
3. Aliev, R.A., Pedrycz, W., Huseynov, O.H.: Behavioral Decision Making with Combined States Under Imperfect Information. Int. J. Inf. Tech. Decis. 12(3), 619–645 (2013)
4. Aliev, R.A., Pedrycz, W., Fazlollahi, B., Huseynov, O.H., Alizadeh, A.V., Guirimov, B.G.: Fuzzy Logic-based Generalized Decision Theory with Imperfect Information. Information Sciences 189, 18–42 (2011)
5. Aliev, R.A., Pedrycz, W., Huseynov, O.H.: Decision Theory with Imprecise Probabilities. International Journal of Information Technology and Decision Making 11(02), 271–306 (2012)
6. Cox, J.C., Friedman, D., Gjerstad, S.: A Tractable Model of Reciprocity and Fairness. Games and Economic Behavior 59, 17–45 (2007)
7. Cox, J.C., Sadiraj, K., Sadiraj, V.: Implications of Trust, Fear, and Reciprocity for Modeling Economic Behavior. Experimental Economics 11, 1–24 (2008)
8. Cox, J.C., Ostrom, E., Walker, J.M., Castillo, A.J., Coleman, E., Holahan, R., Schoon, M., Steed, B.: Trust in Private and Common Property Experiments. Southern Economic Journal 75(4), 957–975 (2009)
9. Tversky, A., Kahneman, D.: Advances in Prospect Theory: Cumulative Representation of Uncertainty. Journal of Risk and Uncertainty 5(4), 297–323 (1992)
10. Ghirardato, P., Maccheroni, F., Marinacci, M.: Differentiating Ambiguity and Ambiguity Attitude. Journal of Economic Theory 118, 133–173 (2004)
11. Klibanoff, P., Marinacci, M., Mukerji, S.: A Smooth Model of Decision Making Under Ambiguity. Econometrica 73(6), 1849–1892 (2005)
12. Shoham, Y., Cousins, S.B.: Logics of Mental Attitudes in AI: A Very Preliminary Survey. In: Lakemeyer, G., Nebel, B. (eds.) ECAI-WS 1992. LNCS, vol. 810, pp. 296–309. Springer, Heidelberg (1994)
13. Compte, O., Postlewaite, A.: Mental Processes and Decision Making, Working paper. Yale University, New Haven (2009), http://econ.as.nyu.edu/docs/IO/19118/Postlewaite_2011Mar22.pdf

Agencies of Intelligence: From the Macro to the Nano

Mohammad-R. Akbarzadeh-T.

Departments of Electrical Engineering and Computer Engineering,
Center of Excellence on Soft Computing and Intelligent Information Processing,
Ferdowsi University of Mashhad, Mashhad, Iran
Akbazar@um.ac.ir

Abstract. 'Homo sapiens' (Latin for 'Wise Man') is what we call our species. Wisdom and Intelligence stands at the center of how we define ourselves. But what is intelligence, and how does it influence an agent's ability to face the world? In this chapter, we review an array of perspectives from the outer behavioral aspects of intelligence such as generalization, optimization and learning to its inner compositional that emphasizes intelligence in terms of networking and connectivity. Our journey will walk us through the concept of omnipotency where an entity has it all, knows it all, and does it all; to the nanopotency where the entity has little, knows little, and does little. From the original manifestation of the human dream to create the omnipotent being, we now come to its recent realization that perhaps less can be more. We will illustrate by sharing a few of our findings on traditionally hard problems such as robotics, urban traffic, fault detection and isolation, portfolio selection and judicial/medical decision making to the more evasive and humanly profound problems such as the atherosclerosis and cancer.

Keywords: Omnipotent, Intelligence, Learning, Generalization, Optimization, Complex Systems, Agent.

1 Introduction

Man defines its kind in various ways; the animal that talks, that manipulates, and that is intelligent. Intelligence is, perhaps, at the forefront of how we see ourselves, one that could bind all of these differing perspectives. And yet, this is most mysterious of all. What are its outer attributes, what is it composed of and how is it generated from within an entity? Is it an attribute of an agent's interactions with its environments or can it be defined independent of the environment? Is it an attribute of 'the mind and the soul' as the philosophers of the East proclaim, or of the 'brain and body' as those in the West perceive? From whichever camp that we belong, one thing that we all may concur is that 'intelligence' remains the most glorious of all wonders in this universe. Yet, knowing intelligence and its effect on an agent can greatly impact our ability to solve the problems of the modern world.

Intelligence is indeed difficult to define, as this process is closely aligned with how man defines itself. Realizing this difficulty, Turing proposed a test based on the outer

A. El-Osery and J. Prevost (eds.), *Control and Systems Engineering*,
Studies in Systems, Decision and Control 27, DOI: 10.1007/978-3-319-14636-2_15

features of intelligence, proclaiming that while a clear definition of intelligence may remain an ever evading problem, we can always recognize intelligence once we encounter it. The Turing test is one such occasion where intelligence or a lack of it is determined by communicating with the entity from behind a wall. Turing proclaimed that the entity behind the wall is intelligent if it is mistaken for human.

But within the concept of omnipotency lies a different approach to the definition of intelligence. The omnipotent here is a singular concept that could do everything, generalize, learn/adapt, and optimize, without needing to socialize and seek assistance from others. Realizing this threesome has been a dream of artificial intelligence (AI) since its inception.

Generalization, perhaps, is the first pillar of such kingdom of omnipotency. Among the frameworks of intelligence that address generalization, one can mention neural networks and fuzzy logic. Incidentally, these paradigms are based on two opposing perspectives to human intelligence, first being inspired by the study of the human brain and the second being based on the human mind. Many biologists may proclaim that there is no such thing as human mind, and that there are only biological/electrical/chemical processes. And yet, the impact and utility of studies that stem from human mind and outer behavior such as psychology, sociology and linguistics is undeniable.

One such useful paradigm that stem from psychology is reinforcement learning (RL). RL enables the learning behavior of an intelligent entity, realizing the second pillar of intelligence with least supervision, i.e. with only coarse feedback from the environment. Hence, in contrast to supervised and adaptive/learning techniques, RL agents can function in unstructured ill-defined environments where the level of 'reward' is not clear and where the environment can change in time with other 'intelligent' entities interacting with it.

Optimization is the third pillar of intelligence of our omnipotency. In this domain, examples of nature inspired algorithms are abundant, such as those that are rooted in biology, ethology and physics. One can name a few such as evolutionary and genetic algorithms that stem from biology; particle swarm, leaping frogs, ant colonies and cuckoo's nests optimization that stem from ethology as well as those that imitate the laws of physics such as magnetism and gravity. This list is endless.

Admittedly, the above approaches, as well as many others that could not be mentioned in this brief, have done well to address many of the aspects of our 'intelligent' omnipotency. And yet, the complexity and size of the future problems are big and our desires are many. The question is whether the above omnipotent perspective is a sufficient framework to address the problems of future, that are highly distributed, with a great amount of (big) data with high heterogeneity, ubiquity, time variation as well as high uncertainty. As we proclaim here, a different perspective of nanopotency to intelligence can perhaps address it more profoundly. We believe that this new perspective can complement the existing approaches and form a synergism with them in reaching higher levels of intelligence.

This chapter is a report of the author's quest in search of the 'artificial' intelligence over a course of 15 years that commences with his own PhD dissertation and follows up with the works of some of his graduates. Due to large span of the topic, the discussion on each illustrated example is inevitably brief, and the interested reader is encouraged to refer to the cited references for more details. In Section 2, we will begin from a macro perspective to intelligence and explore its various attributes. A micro view to intelligence is then provided in Section 3. The concept of 'social intelligence' is then discussed in Section 4, including the aspect of intelligence from the view of human-machine interaction and shared autonomy. Finally, we introduce the concept of 'nanopotency' within the concept of swarm intelligence or the 'intelligence of the many' in Section 5. Conclusions are made in Section 6.

2 What Is Intelligence? A Macro Perspective

There are three main attributes to intelligence here: generalization, optimization and learning. It should be mentioned that we include knowledge integration and fusion as a part of generalization. In the following subsections, we will review some of the works that address these aspects.

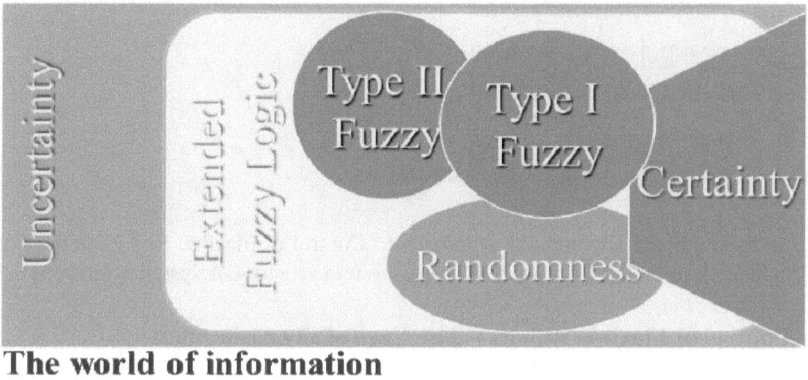

Fig. 1. The Type I fuzzy, Type II fuzzy, and finally the extended fuzzy logic aim to progressively handle more aspects of the uncertainty in the world of information

2.1 Generalization

For the past fifty years, we have witnessed a growing appeal for fuzzy logic and its various partners that form a 'soft' perspective to computing and problem solving. This is due to the realization that there is much more to the world of information that is uncertain as compared to being certain and that the paradigm of probability is inept in handling this vast domain by itself. This begins with Zadeh's seminal paper on fuzzy sets [1] that aim to provide a precisiated human-like reasoning for machine and follows up to his later works that addresses the uncertainty in sets, and to his more recent works on extended fuzzy logic that aim to provide an all-inclusive unifying framework for uncertainty handling [2], as illustrated in Fig. 1.

Fig. 2. The Experimental Testbed for Flexible Robot Arm (left) [3], The fly fisherman (right)

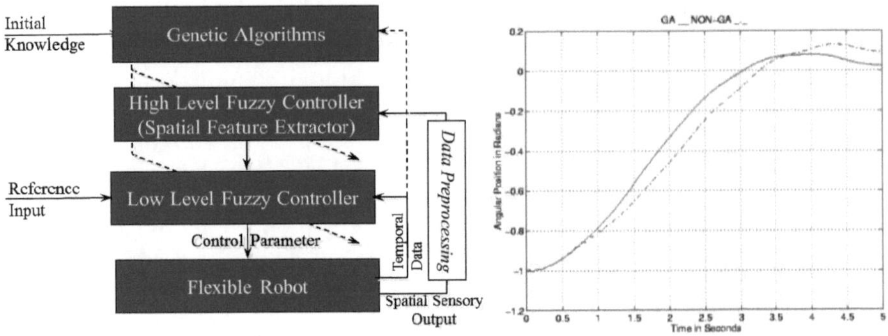

Fig. 3. An Evolutionary Hierarchical Approach to Control of Flexible Robots (left), Comparison of Angular Position for GA-optimized fuzzy system vs. non-GA system (right) [3]

Fuzzy Control of Flexible Robots – The Case of Fly Fisherman

My PhD research began with the application of fuzzy control for a flexible link. Flexible robotics was chosen as a testbed due to its complex and distributed nature and its exacting requirements. At the time, it was generally accepted that (Type I) fuzzy control could be useful if there was reliable rule-based human expertise. But the question was if there is indeed such expert knowledge for the case of flexible robots, particularly for the space applications that were the target of this research. Fortunately, there was an example in fly fishing and the exceptional details of how the fishing line is manipulated to smoothly land the fly on the water surface, as illustrated in Fig. 2 (right). Inspired by this example, the resulting hierarchical controller could then reduce the number of rules from the possible 3^9 rules, i.e. partitioning the universe by only three membership functions for nine inputs, to only 52 rules. This was done by separating variables that had more spatial significance (five strain gauges) from those that had a more temporal significance (two angular positions and velocities of robot base and tip). Genetic algorithms (GA) was then used to optimize this rule-based controller, as illustrated in Fig. 3 (right). The key idea in this process was to gainfully utilize the expert knowledge that is often available in case of fuzzy control design to improve the evolutionary process. This aspect of evolutionary fuzzy optimization may separate it from other applications of evolutionary algorithms that begin with a purely random initial population.

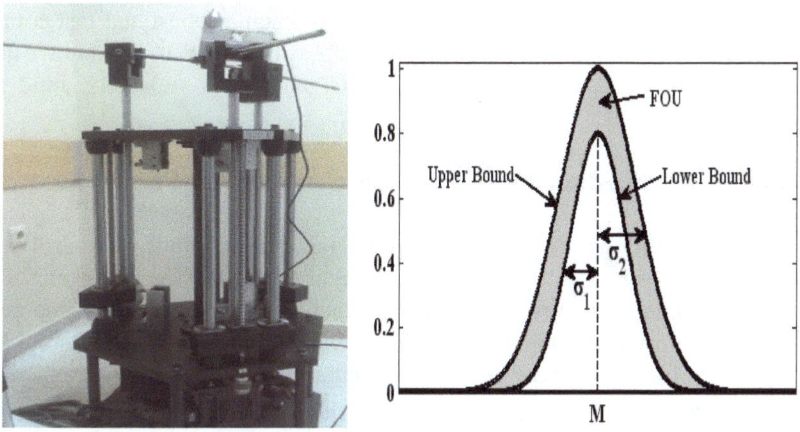

Fig. 4. The experimental testbed for the parallel 3PSP robotic arm (Left), The Interval Fuzzy Sets showing lower and upper bounds of uncertainty (the foot print of uncertainty – FOU) in a fuzzy set (Right) [4]

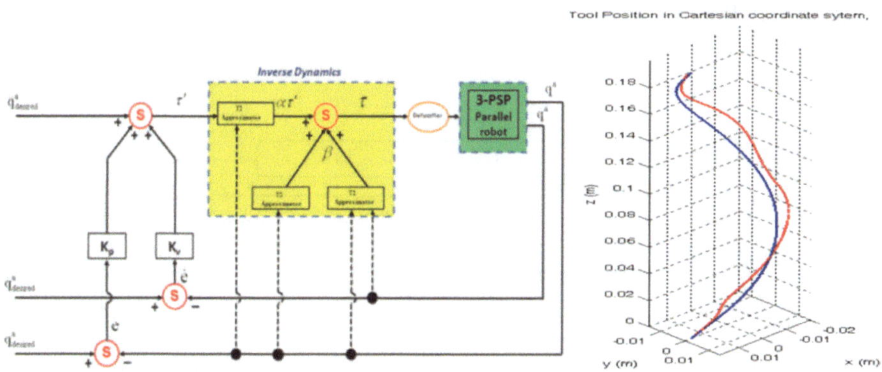

Fig. 5. The model based computed torque approach to control of 3PSP parallel robot (left), The time response of the tool position in following a Helix trajectory (right) [4]

Interval Valued Fuzzy Logic for Computed Torque Control of a 3PSP Robot
The interval valued fuzzy sets aim to handle the uncertainty in defining fuzzy sets as intervals, creating what is known as a footprint of uncertainty (FOU), as illustrated in Fig. 4 (right). This is a step towards Type II fuzzy sets, but with the advantage of significantly less computations. Here, we use the interval valued fuzzy logic framework for modeling the nonlinear dynamics of the parallel 3PSP robot as shown in Fig. 4 (left). This estimated dynamics is then used in the computed torque control as

illustrated in Fig. 5. While the computed torque control is considered as a highly effective linearization technique and renders high quality theoretical results, its application in real life are often hampered by the inaccuracies of the model. They also impose high computational cost, particularly for parallel robots as the case here. In contrast, the interval valued fuzzy approach aims to reach a robust and fast computational framework for this model based approach. As illustrated in this study, the robot's computed torque approach follows the assigned Helix trajectory amid noisy environmental conditions.

Stable Adaptive Sliding-Based Interval Fuzzy Control of the Parallel 3PSP Robotic Arm

The concept of stability has consistently been an important attribute of control design. To the novice, fuzzy systems, by the chosen word of fuzzy, may be mistaken for 'ambiguous' and 'imprecise' control; and yet it is exactly this ambiguity and imprecision that fuzzy logic aims to remove by *clarifying* and *precisiating*. The interval fuzzy approach aims to further address imprecision and ambiguity by *managing* them. The approach proposed here utilizes the basic concepts of fuzzy basis functions, sliding and Lyapunov-based controller design, and interval fuzzy systems to reach a stable adaptive controller, as illustrated in Fig 6. As an example, the proposed approach is applied to control a first order nonlinear system with bounded disturbances. The comparison with a conventional controller as well as a Type I fuzzy controller illustrates the better handling of uncertainty by the interval approach, as illustrated in Fig. 7. More details can be found in [5].

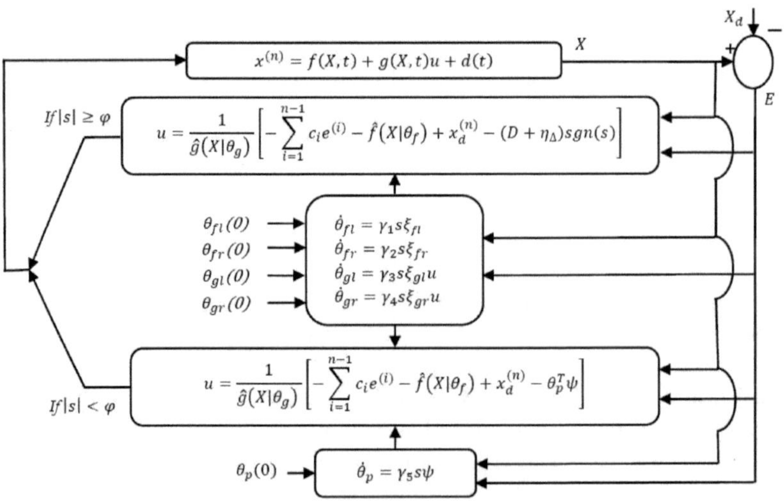

Fig. 6. The Adaptive Interval Fuzzy Control of the parallel 3PSP robotic arm [5]

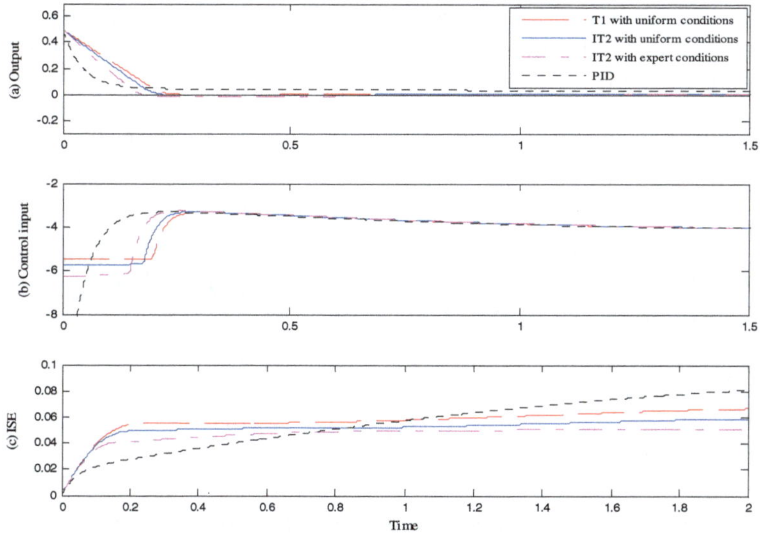

Fig. 7. (Top) System output; (Middle) control input signal; (Bottom) Integral squared Error (ISE) of the sliding based approach in [5]

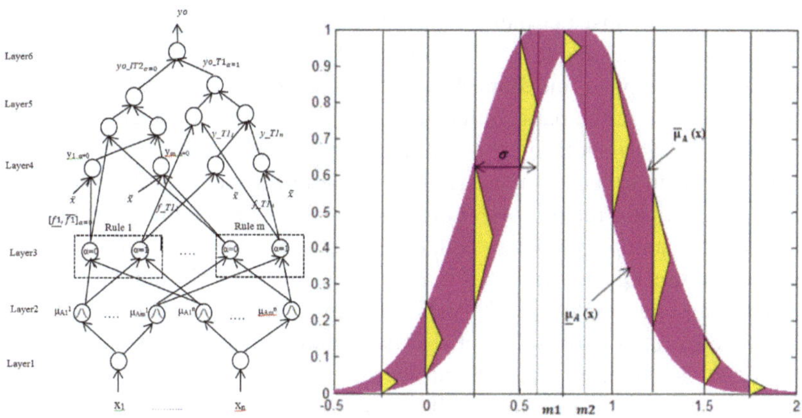

Fig. 8. The Dynamic Growing General Type II Fuzzy Neural Network (left), The General Type-II Fuzzy Sets (right) [6]

Dynamic Growing General Type II Fuzzy Neural Structure (DGGT2FNS) for Control of a Parallel 3PSP Robotic Arm

Here, a fuzzy neuro-based self-organizing approach is considered for the controller design. But the fuzzy neurons here are using general type II fuzzy sets. Hence, it is important to keep the number of parameters to a possible minimum, thereby reducing computational cost [6]. Figure 8 illustrates the proposed dynamic growing structure as well as a general type II fuzzy structure that addresses the ambiguity in defining fuzzy sets. Table 1 displays the simulation results of applying the proposed approach to a nonlinear control system with uncertain dynamics. As is shown, the DGGT2FNS delivers a better or competitive error performance with significantly fewer parameters and computation time.

Table 1. Simulation Results of Dynamic Growing General Type II Fuzzy Network System (DGGT2FNS) vs. The non-self-organizing approach for a nonlinear control system [6]

Approaches	# of rules	# of Parm eters	RMSE	RMSE [60, 85)	Training Time	Testing Time
GT2FNS	4	131	0.6235	0.3339	0.0826	0.0401
Proposed DGGT2FNS	4	61	0.6248	0.2105	0.0178	0.0039

Extended Fuzzy Logic for Judicial and Medical Decision Making

As illustrated by the earlier discussions, fuzzy logic and its various forms hope to address imprecision by its fuzzy set structure and manage uncertainty by its extension to general type II fuzzy sets. The remaining question is whether this structure can now be used to solve the most pertinent of problems such as those that are open and ill-defined. Extended fuzzy logic aims to address the large class of problems where the information is incomplete and imperfect such as in the case of judicial and medical decision making where the data sources are either highly noisy or untrustworthy. Being inspired by the various forms that ambiguity is played out and processed in human languages, Zadeh introduced the concept of generalized constraints where variables and their valuations are related by *f*-constraints such as *f-p* for fuzzy probabilistic, *f-v* for fuzzy veristic, and *f-u* for fuzzy usuality. The basic concept of *f*-transformation as applies to definition of parallel lines is illustrated in Fig. 4. In the case study of the present work, a general concept of *f*-validity is introduced [7]. This validity measure is then applied to fuzzy probability of risk factors in coronary heart disease [8] and to conviction analysis in an actual judicial decision making process [9]. In particular, a controversial case was chosen in which the Supreme Court had overturned the local court's decision. In this case, the proposed approach decided in consistence with that of the Supreme Court.

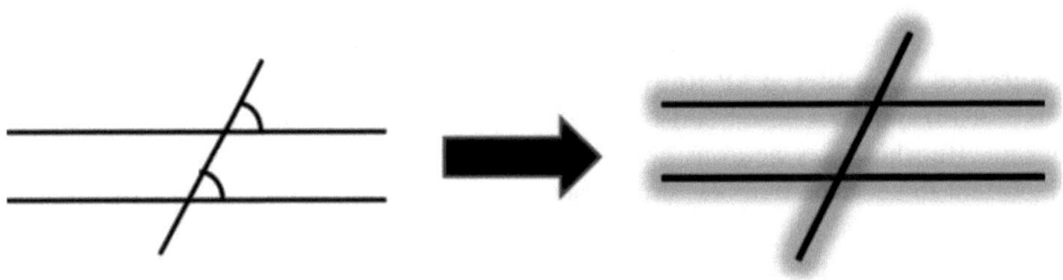

Fig. 9. Parallel lines and their *f*-transforms [7]

2.2 Optimization by Nature

Inspirations for optimization are abundant in nature. These are often seen in various studies such as biology, ethology as well as physics. To name a few that are based on biology, one can mention artificial immune algorithms as well as genetic algorithms, genetic programs, evolutionary strategies, and evolutionary programming that are inspired by various interpretations of the Darwainian process of natural evolution. Leaping frogs and their fly hunting process, migrating birds and their exchange of leadership role during long distance flights, honey-bees with their communication dance for their colonies about sweet flowers, spiders with their web-based communication, Cukoos' clever use of other birds' nests to host their eggs, ants with their pheromone and stigmergic communication, and bacteria with their foraging behavior form the bases for some of the more successful algorithms that come from ethology. Physics has also been a great source of inspiration such as in spin glasses and their temperature-based magnetic behavior, as well as gravitational forces, quantum-inspired variable coding and operations, and water drops and their path formation towards rivers.

Exploration and Exploitation are two main functions in almost any optimization process. These are addressed by the above algorithms by encouraging cooperation or competition among solutions, decomposing solutions into several sub-solutions, random motion for better exploration, and recombining genetic codes for better exploitation. Some of these involve parametric operators and some are structure alternating behaviors. Here, a Magnetic Optimization Algorithm (MOA) is chosen to illustrate this process.

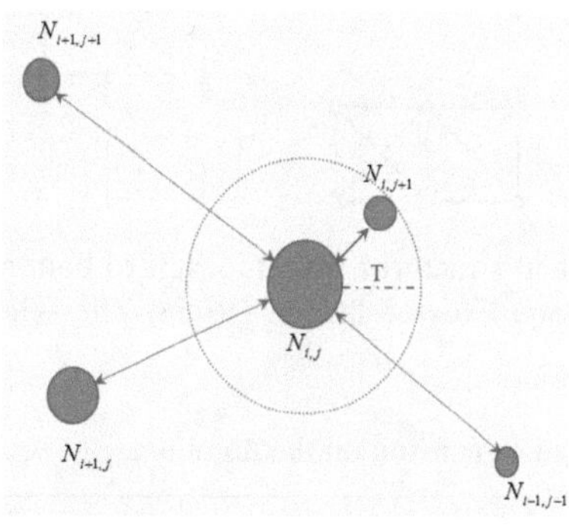

Fig. 10. Illustration of short-range repulsion force [10]. If the distance between two particles is more than T_{SRR}, the particles attract each other, else repulse each other.

The Magnetic Optimization Algorithms: Operators and Structures
MOA is inspired by the principles of magnetic field theory [10]. Here, the possible solutions are magnetic particles scattered in the search space. Each magnetic particle has a measure of mass and magnetic field according to its fitness. The fitter magnetic particles are those with higher magnetic field and higher mass. To complement the force of attraction, several operators are used including a repulsive short-range force

between the particles, an explosion operator, a combination of short-range force and explosion operator, and a crossover interaction between the neighboring particles, as shown in Fig. 10. To define neighborhood of interaction, various forms of structured populations can be used. Here, ten general lattice structures are investigated including the cellular, grid, ring, star, binary tree, ladder, crossed ladder, cluster, Km,m and Random. The cellular lattice-like population, as shown in Fig. 11 (right) is then chosen for its best performance. In order to test the proposed algorithm and the proposed operators, the proposed algorithm is compared with several competing algorithms including GA, PSO, QEA and a Saw-tooth GA on 14 numerical benchmark functions. It is shown that the proposed cellular structure allows a better exploitation of local neighborhoods before they move towards the global best, hence it increases population diversity. Additionally, the hybrid explosion-repulsion operator provides best consistent results. Also, as Table 2 illustrates, when compared with more traditional benchmark algorithms, experimental results show that the MOA and its variants outperforms PSO, and dominates GA, QEA and Saw-tooth GA on most benchmark functions, particularly as problem dimension grows.

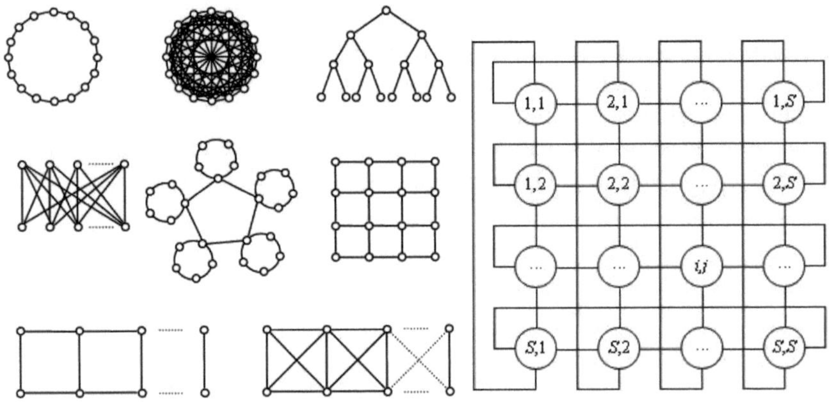

Fig. 11. (Left) The investigated structures from top, left to bottom right are Ring, Star, Btree, $K_{m,m}$, Cluster, Grid, Ladder and Crossed-ladder. (Right) The selected cellular lattice with the size of S [10].

Table 2. The effect of dimension on the dominance of several algorithms [10]

M	MOA	EXP-MOA	SRR-MOA	HER-MOA	GA	PSO	QEA	SAW
5	4	0	1	1	0	0	3	5
10	2	0	4	1	4	0	3	0
25	2	1	2	3	4	0	2	0
100	0	2	4	5	2	0	1	0
250	0	7	1	4	1	0	1	0
500	0	6	2	4	1	0	1	0
1000	0	5	4	3	2	0	0	0
1500	1	5	2	4	2	0	0	0
2000	1	5	2	4	2	0	0	0

2.3 Learning from Psychology

By many, learning is the main attribute of intelligence. There are various mechanisms of learning such as in supervised where the agent learns by feeding back the error, unsupervised where it searches for patterns and clusters in the data, and semi-supervised where it receives only minimal feedback from the environment and has to estimate the utility of its future actions and states. While there are many conventional approaches for supervised and unsupervised learning, there are relatively few approaches that can learn in a semi-supervised way. Psychology has been an inspiration for such approaches, specifically reinforcement learning [11] and learning automata [12].

One of the applications has been the electric deregulation that has been on going in the power industry. The electric power industry is confronted with a major restructuring in which the operation scheduling is decided based on competitive market. In this new arrangement, bidding strategy becomes an important issue. Participants of the deregulated energy marketplace may be able to compete better by choosing a suitable bidding strategy for trading the electricity. In such deregulated market, coalition formation with other participants may also change the diffusion of profit. Therefore, the problem changes to selection of the best participant for coalition and joint bidding strategies, as shown in Fig. 12, [11].

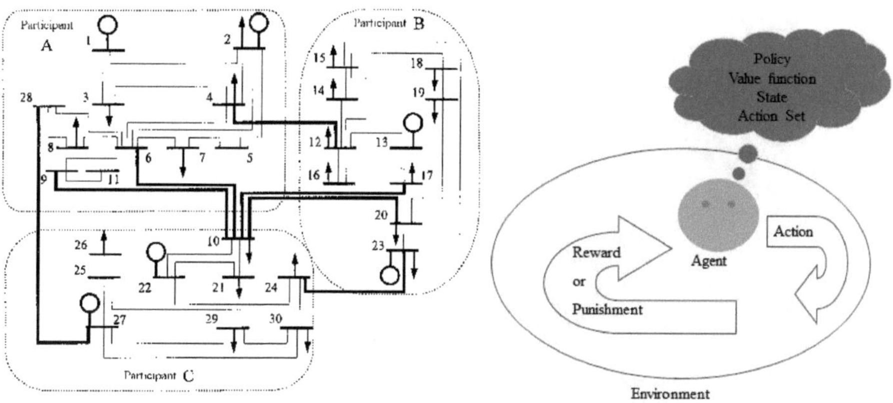

Fig. 12. (left) Three participants in a deregulated energy marketplace, (right) Reinforcement by estimated reward or punishment [11]

Different classic methods for decision making in uncertain environment of the market can be applied to select a suitable strategy. Most of these methods, such as game theory that insures reaching the optimal solution for all market participants, require a lot of information about other market players and the market. However, in real marketplace only little information such as the spot price is available for all participants. In this example, a modified reinforcement learning approach based on Q-adjusted annealing has been applied to determine the optimal strategy for a power supplier in the electricity market with multiple sources. A modified IEEE 30 bus system has been considered and the simulation results are shown to be the same as the standard game theory. The main advantage of the proposed method is that no information about other participants is required. Furthermore, investigation shows that if all participants use this method, they will stay in Nash equilibrium. The deregulated energy marketplace may be able to compete better by choosing a suitable bidding strategy for trading the electricity. In such deregulated market, coalition formation with other participants may also change the diffusion of profit. Therefore, the problem changes to selection of the best participant for coalition and joint bidding strategies [11].

In the above analysis, the learning agents compete against each other for higher profits. In other scenarios, the agents may gain more by cooperating. Then the question is the level of cooperation and the type of message that is communicated. In [12], the two learning agents learn to move a large item which cannot be moved by one agent alone, as illustrated in Fig. 8. Here, a central station provides reward/punishment for the robot agents.

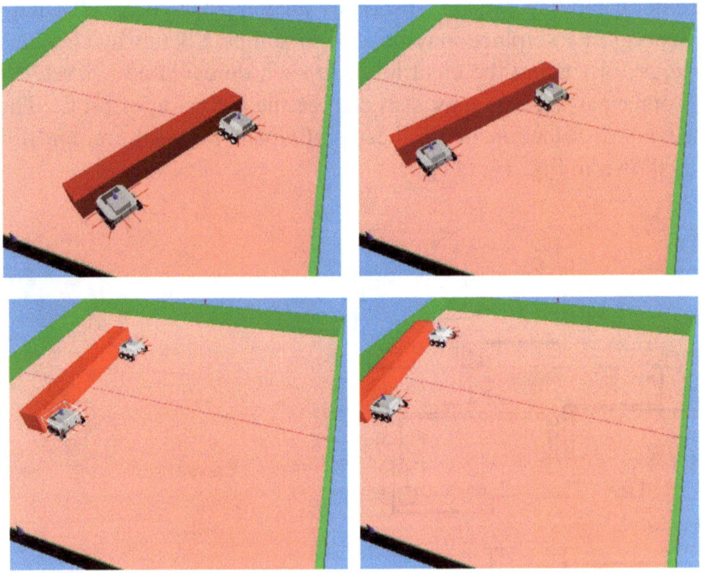

Fig. 13. Snapshots of two robots learning to cooperatively move a large box in a noisy environment without direct communication (from left-top to right-bottom) [12]

3 What Is Intelligence? A Micro Perspective

In contrast to the above, many may proclaim that intelligence is a physical process that occurs in the brain. It is a behavioral accumulation of the contributions of many small neurons that can be probed and analyzed. In this realm, artificial neural networks have made great strides in terms of learning nonlinear mappings and modeling and control of complex systems. Multilayer perceptron with their hidden layers, self-organizing networks with their ability to grow and shrink, as well as the more recent deep neural networks have demonstrated considerable advances in solving difficult problems. However, there are estimated to be 10^{11} Neurons and 10^{14} connections in the living brain. Even when we eventually manage to build such overwhelming memory system, the remaining issues are on how learning is done and where *thoughts* are held and processed. What we know of the brain's learning process is at best primitive and limited by our limitations of experimentation such as the fact that brain is a dynamic living process which alters by the simple act of measurement. Among the few studies that address this aspect, Nielsen [13] suggested that brain works by grouping symbols, which are each represented by a collection of neurons, into modules. It is then the Hebbian interaction among symbols of one module with those of the other modules that creates associations, called cogencies, that we form among various concepts and thoughts.

3.1 Confabulation-Inspired Association Rule Mining (CARM)

The above basic idea is extended to data mining in [14]. Pairwise conditional probabilities are used to reach a measure of 'interestingness' that is inspired by cogency. The algorithm mines association rules by only one pass though the file, hence lower computational cost. The proposed algorithm is also more efficient for dealing with infrequent items. The problem of associative classification is used here for evaluating the proposed algorithm, as illustrated in Figure 14. We evaluate CARM over both synthetic and real benchmark datasets obtained from the UCI repository. Experimental results in Figure 15 show that the proposed algorithm is consistently faster due to its one time file access and consumes less memory space than the competing CFPgrowth algorithm. Also, statistical analysis reveals the superiority of the approach for classifying minority classes in unbalanced datasets.

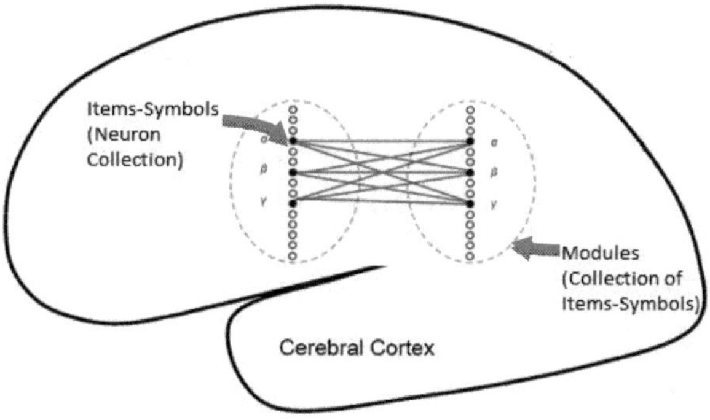

Fig. 14. Links strengthened with transaction "$\alpha\beta\gamma$."

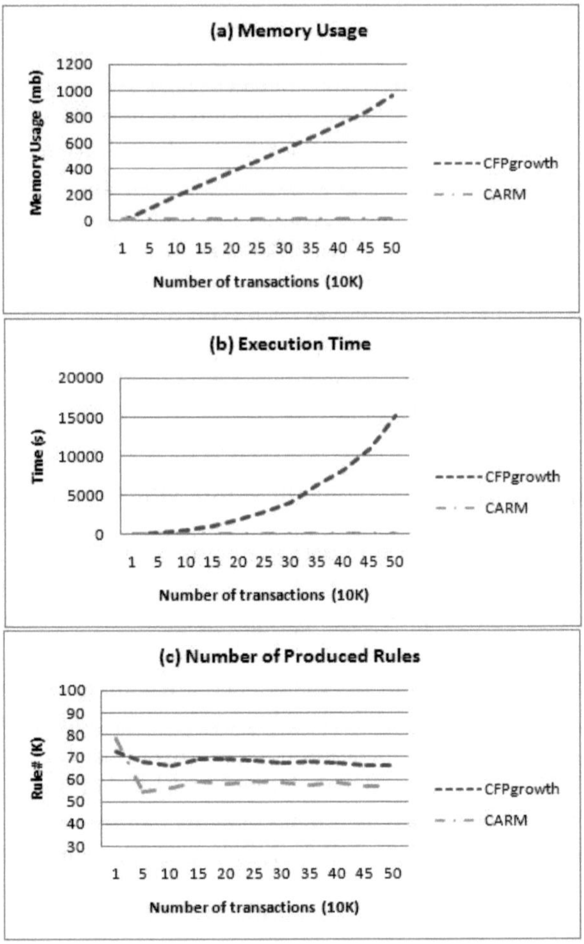

Fig. 15. Memory usage, runtime, and the number of rules on datasets with 1K items [14]

4 Social Intelligence? Intelligence Is as a Result of Interaction (Cooperation/Competition) of Several Beings

A different perspective to intelligence arises from the synergistic interaction among natural entities. This can be in the form of homogeneous interactions such as in human societies or heterogeneous interactions such as in human body and microorganisms support it. It can also be in the form of cooperating such as in group hunting for a prey or competing for limited resources. Here, we would like to discuss three engineering examples, first is an automatic fault detection and isolation system such as for a CNC machine where local experts compete to better represent their fault type, second is a traffic light coordinator that must cooperate with others in order to reach green wave, and third is shared autonomy in a human-machine interactive paradigm.

4.1 Fault Detection and Isolation by a Hybrid Fuzzy and Neuro Approach

A fuzzy and neuro hybrid approach is shown in Figure 16 for fault detection/isolation and fault size estimation of industrial plants, in particular a highly nonlinear CNC X-axis drive system's component fault detection. To simplify the task, we divide it into the number of fault types, m, and train the m local experts using fuzzy clustering, one for each fault type. A moderator is given the task of moderating among the local experts when two or more local experts disagree. The size estimation is then performed by RBF neural networks of the corresponding winning local expert. Application to diagnosis of 12 abrupt fault types of a CNC X-axis drive system shows high success amid significant noise levels [15].

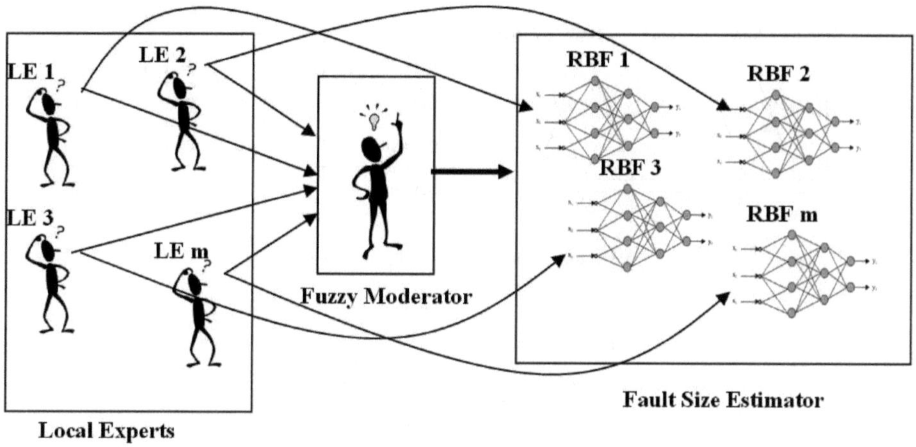

Fig. 16. Hierarchy of local expert decision making, fuzzy moderator, and RBF networks [15]

4.2 Distributed Urban Traffic Control and Modeling

Green wave is defined as the traffic state where vehicles reache subsequent green lights after stopping for a red light, as illustrated in Fig. 17. This is a highly desirable state for urban traffic. However, the time varying nature of intersections, the perpendicular green waves with possibly conflicting objectives, involvement of human behavior in the system and distributed nature of the problem at large makes the urban traffic control a challenging problem. Here, a fuzzy supervisory controller is proposed using a discrete event system that is done via state flow formalism. Specifically, because of this distributed nature, it is important for the decision making paradigm to consider historical values of the linguistic variables as well as their current values. Here, we establish two novel fuzzy systems which designate cycle time and split time based on temporal fuzzy rules, as illustrated in Fig. 18. The temporal quantifiers and rules are designed to resolve inconsistency in temporal information.

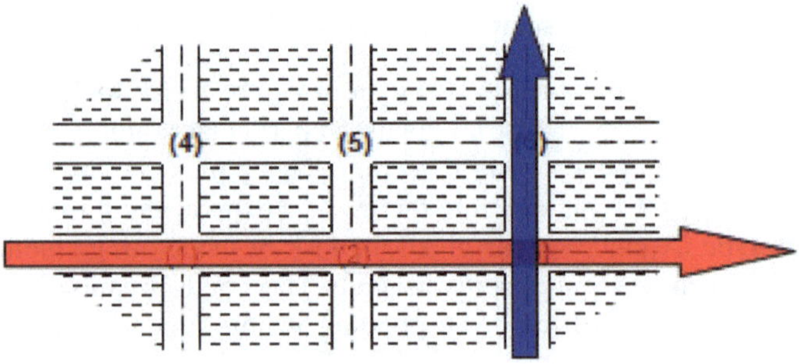

Fig. 17. The Conflicting Strategy for Multiple Green Waves

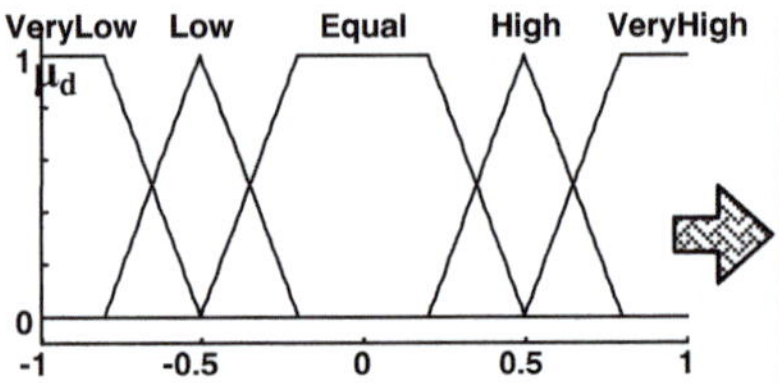

TVF	VL	L	E	H	VH
Often	VL	L	E	H	VH
Occasionally	~L	L	E	H	~H
Seldom	~VL	~L	~E	~H	~VH

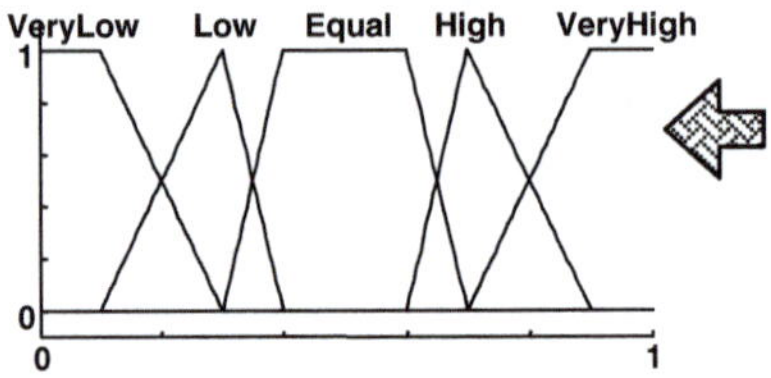

Fig. 18. Temporal fuzzy rule base and its relevant membership functions in split time determination [16]

4.3 What about Shared Autonomy? Social Intelligence with the Human in the Loop

With the increasing ubiquity and pervasiveness of computing and automated machines in everyday lives, the rising question is how one balances the autonomy of the human with that of the machine. There are a number of disciplines that are related to this aspect of intelligence such as affective computing and cognitive sciences. One example that can help illustrate the importance of this aspect is in urban transportation.

In urban traffic, the trend has been in increasing number of vehicles on the roads and increasing vehicle comfort at higher possible speeds. Yet, two things will reluctantly not change in significant terms. First are the roads, and second are the capabilities of the human drivers. In this midst of these constrained resources, the convergence of key technologies from multiple domains may eventually produce the emergence of a "swarm" concept of operations. The swarm, a collection of vehicles traveling at high speeds and in close proximity, will require management techniques to ensure safe, efficient, and reliable vehicle interactions, as depicted in

Fig. 19. We propose a shared-autonomy approach in which the strengths of both human drivers and machines are employed in concert for this management. Inspired by prior research, a fuzzy logic-based control implementation is developed and described in this paper for the purpose of exploring the shared-autonomy concept [17].

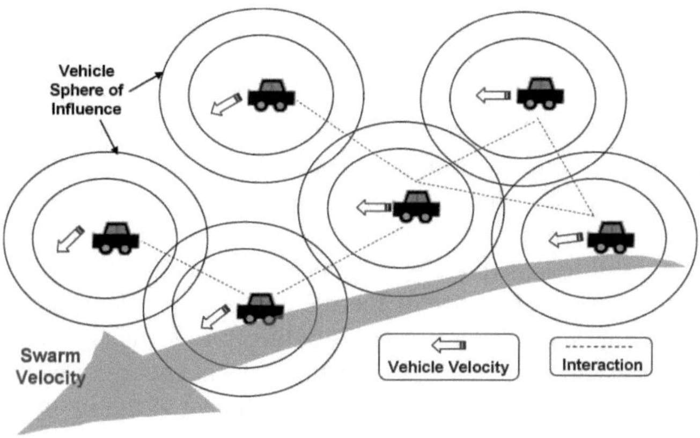

Fig. 19. Notional depiction of interacting vehicles in a swarm [17]

5 What Is Intelligence? The Intelligence of the Many

Swarm intelligence, has in recent years, been a popular strategy for solving many complex problems. It is based on the simple concept that intelligence can result from the interaction of many entities that are of little or no intelligence. Particle swarm optimization is one such instance where optimization of a large and difficult problem is performed by particles with very simple computational behavior. Ant colonies are another form of such swarm intelligence where the ants have a simply behave reactively in a stigmergic environment. Similar to particles and ants, an entity with nanopotency has little or no intelligence, but also has little or insignificant physical/computational capabilities on its own, to a point that it can represent only a piece of a complete solution. This is in contrast to the original concept of omnipotency, where the entity would know it all and could do it all, all on its own. Here, two such physical concepts are explained based on a network of *spin glasses* for solving the classic portfolio selection problem and *swarm control* for solving atherosclerosis and cancer by nanomedicine.

5.1 Spin Glasses for Optimization and Portfolio Selection

Spin glass optimization is a distributed technique inspired by the interactions in natural spin glasses, as illustrated in Fig. 20, whose magnetic behavior is affected by its temperature. Spin glasses are the lattices of spins where each spin is only a part of the entire solution, in contrast to genetic algorithms where each chromosome represents a complete solution. The interaction between spins creates special optimal patterns given appropriate temperature. This optimization paradigm is promising in complex multi-objective optimization tasks because it allows high problem decomposition and high computational parallelism among its member spins. In other words, since the overall network of spins represents only one solution, there is a great promise in computational efficiency when compared with other population based/stochastic approaches such as Genetic Algorithms (GA) and Simulated Annealing (SA). The nature of this method is also entirely different from other distributed frameworks such as Hopfield Neural Network since spins' paradigm of interaction does not have to be fully connected, i.e. the neighborhoods of interactions can expand or collapse, hence less computation and better convergence speed. Here, we apply a heuristic method based on spin glass model that uses migration and elitism operators in addition to temperature control in order to trace out an efficient frontier in the optimization landscape. The proposed methodology is then applied to the problem of portfolio selection. Portfolio selection is one of the NP complete problems where each asset's behavior is similar to spin's behavior and it is therefore suitable as a case study. We show that, in proper circumstances, decrementing local energy of each spin can decrement global energy of the glass; and correspondingly, if the optimization problem can be suitably mapped to the glass, the expected cost function decreases. These results are shown in Table 3.

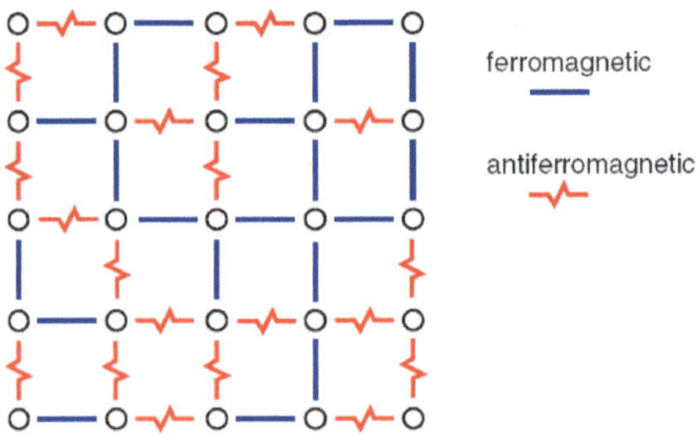

Fig. 20. A two-dimensional spin glass with bond disorder. Spins are placed on the sites of a regular grid. They interact with their neighbors, the interaction is random, either ferromagnetic (straight lines) or anti-ferromagnetic (jagged lines) [18].

5.2 Swarm Control for Atherosclerosis and Cancer

Atherosclerosis, or hardening of the arteries, is one of the major causes of death in humans. High accumulation of Low-Density Lipoprotein (LDL) macromolecules within the arterial wall plays a critical role in initiation and development of atherosclerotic plaques. Here, a proportional drug-encapsulated nanoparticle (PDENP) is discussed that utilizes a simple piecewise-proportional controller to realize collective feedback control of LDL concentration in the interior of the arterial wall [20]. In contrast to the competing strategies on nanorobotics, PDENPs carry simpler hardware architecture in order to be more reasonably realized technologically as well as to

Table 3. The comparison between spin glass behaviors for various neighborhood sizes for different stock market [19]

	Actual Cost Function	Global behavior with $m=N$		Local behavior with $m=N$		Global behavior with $m=8$		Local behavior with $m=8$	
		Cost Function	Time (ms)	Cost Function	Time (ms)	Cost Function	Time (ms)	Cost Function	Time (ms)
Hang Seng (N=31)	-0.0034	-0.0034	4626	-0.0034	3470	-0.0033	3210	-0.0029	2180
DAX 100 (N=85)	-0.0041	-0.0041	65268	-0.0041	51847	-0.0039	20735	-0.0038	17659
FTSE (N=89)	-0.0034	-0.0034	75631	-0.0034	61329	-0.0032	41003	-0.0029	31981
S&P (N=98)	-0.0036	-0.0036	81231	-0.0036	62431	-0.0036	46164	-0.0031	32039
Nikkei (N=225)	-0.0014	-0.0014	418140	-0.0014	291200	-0.0013	121320	-0.0012	81321

penetrate the interior arterial wall. Furthermore, in contrast to the existing targeted DENPs that usually target the surface proteins of atherosclerotic plaque, the PDENPs here directly sense the LDL level in the arterial walls. Hence, they can diagnose abnormal LDL accumulation before plaque formation, prevent critical growth of atherosclerotic plaques, while considerably reducing the unwanted drug side effects in healthy tissue. Simulation results on a well-known mathematical model of the arterial wall demonstrate that the proposed approach can successfully reduce the LDL level to a desired value in the arterial wall of a patient with very high LDL level. Also, the mass of the released drug by PDENPs in a healthy wall is 11 times less than its corresponding value in an unhealthy wall.

Cancer is the another leading cause of death in humans, particularly in the modern society where the less trivial causes of death have been better handled and where the life style changes such as in the use of fast food and less physical activity promote cancer growth. Among various approaches to cancer treatment, Targeted Drug Delivery System (TDDS) is a strong possible alternative to the more traditional approaches. TDDS localizes and targets the drug to the diseased tissue by a swarm of nanodrug carriers, hence hoping for a more effective cancer treatment. Here, considering a biomimicry approach, a novel nanonetwork is proposed for targeted cancer drug delivery [21]. In the proposed system, a swarm of bioinspired nanomachines utilize the blood distribution network and chemotaxis to carry drug through vascular system to the cancer site that is recognized by a high concentration of Vascular Endothelial Growth Factor (VEGF). Our approach is multi-scale and includes processes that occur both within cells and with their neighbors, as illustrated in Fig. 23. For a more appropriate evaluation, we compare our work with a conventional chemotherapy approach using a mathematical model of cancer under actual experimental parameter settings. Simulation results show the merits of the proposed method in target cancer therapy of cancer by improving four relevant cancer cells density and VEGF concentration while following a more organic and natural processes.

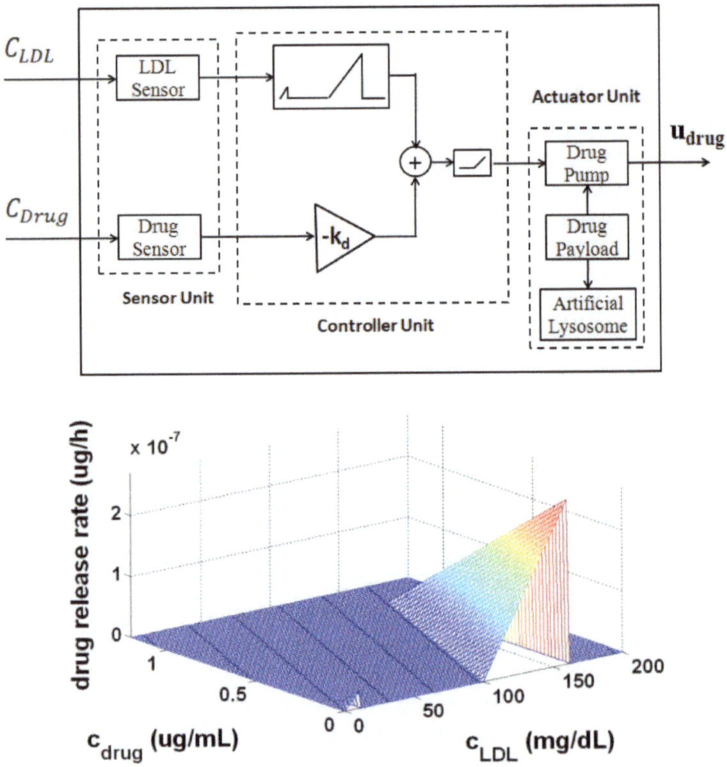

Fig. 21. The Structure of a PDENP (top), The control surface of Each PDENP (bottom) [20]

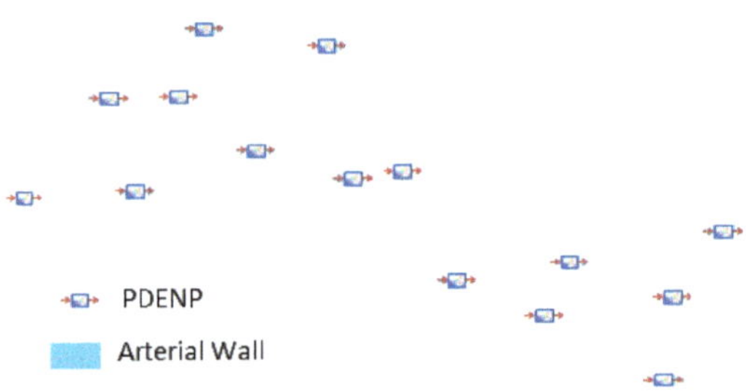

Fig. 22. Notional depiction of PDENPS penetrating the arterial wall for better drug delivery

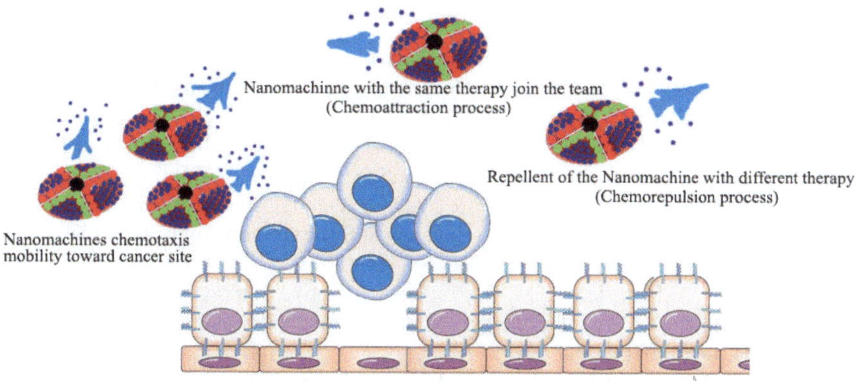

Fig. 23. Chemotaxis processes of different nanomachines [21]

6 Conclusion

One can view intelligence from multiple angles, either from the inside (micro) or the outside (macro), either from the eastern philosophy of mind or from that of the materialistic west, either as an attribute of one or the collective trait of many. I suppose, intelligence will always remain a mystery as its definition is intertwined with how man defines itself. Nevertheless, a clear definition of intelligence can be clear step to solving many of the problems in our modern society. Here, we introduce a new concept of nanopotency that integrates little intelligence with little actuation capability. In contrast to particles of swarm intelligence, a particle with nanoptency is only a small piece of the solution, but learn to perform difficult tasks by their sheer number. While all various aspects of intelligence can be rewarding in their own merits, in author's opinion, it is this nano-scale potency that can be most rewarding in terms of solving humanity's most difficult and evading problems.

In this chapter, we could only shed some light on intelligence from one person's perspective, one that is formed by the author's research and collaboration with his colleagues, students, as well as his own advisor. Hence, this is only the author's personal and subjective opinion and not everyone may agree. As so, I will welcome comments and suggestions.

Acknowledgements. The author would like to sincerely thank his PhD advisor, Prof. Mo Jamshidi, for the many years of support and advisement that began in the Fall of 1989 with a course on Modern Control Systems and has continued to date. The author would also like to thank the many colleagues and students with whom he has had the pleasure of collaborating and some of whose work are presented here.

References

1. Zadeh, L.A.: Fuzzy Sets. Information and Control 8, 338–353 (1965)
2. Zadeh, L.A.: Generalized Theory of Uncertainty (GTU) – Principal Concepts and Ideas. Computational Statistics & Data Analysis 51, 15–46 (2006)
3. Akbarzadeh-T, M.-R., Jamshidi, M.: Evolutionary Fuzzy Control of a Flexible Link. Journal of Intelligent Automation and Soft Computing 9(3), 181–214 (1997)
4. Hassanzadeh, H., Akbarzadeh-T, M.-R., Akbarzadeh-T, A.-R., Rezaei, A.: An Interval-valued Fuzzy Controller for Complex Dynamical Systems with Application to a 3PSP Parallel Robot. Fuzzy Sets and Systems 235, 83–100 (2014)
5. Ghaemi., M., Akbarzadeh-T, M.-R.: Indirect Adaptive Interval Type-2 Fuzzy Sliding Mode Control for a Class of Uncertain Nonlinear Systems. Iranian Journal of Fuzzy Systems 11(5), 1–21 (2014)
6. Fartashtoloue, S.: Designing a Dynamic Growing General Type-2 Fuzzy Neural Controller for a Class of Nonlinear Systems with Experimental Implementation on a 3PSP Parallel Robot, M.S.Thesis, Ferdowsi University of Mashhad (2013)
7. Sabahi, F., Akbarzadeh-T, M.-R.: A qualified description of extended fuzzy logic. Information Sciences 244, 60–74 (2013)
8. Sabahi, F., Akbarzadeh-T, M.-R.: Comparative Evaluation of Risk Factors in Coronary Heart Disease Based on Fuzzy Probability-Validity Modeling. Journal of Zanjan University of Medical Sciences and Health Services 22(91), 73–83 (2014)
9. Sabahi&, F., Akbarzadeh-T, M.-R.: Introducing Validity in Fuzzy Probabilities for Judicial Decision-Making. International Journal of Approximate Reasoning 55(6), 1383–1403 (2014)
10. Tayarani, M.H., Akbarzadeh-T, M.-R.: Magnetic Inspired Optimization Algorithms: Operators and Structures. Swarm and Evolutionary Computation (accepted, 2014)
11. Naghibi-S., M.-B., Akbarzadeh-T., M.-R., Javidi, H., Rajabi-M, H.: Application of Q-learning with Temperature Variation for Bidding Strategies in Market Based Power Systems. Journal of Energy Conversion and Management 47(11/12), 1529–1538 (2006)
12. Aly El-Osery, J., Burge, M., Jamshidi, A., Saba, M., Fathi, M., Akbarzadeh-T, M.-R.: V-Lab: A Virtual Laboratory for Autonomous Agents – SLA-Based Learning Controllers. IEEE Transactions on Systems, Man and Cybernetics – Part B: Cybernetics 32(6), 791–803 (2002)
13. Hecht-Nielsen, R.: Cogent confabulation. Neural Networks 18(2), 111–115 (2005)
14. Soltani, A., Akbarzadeh-T, M.-R.: Confabulation-inspired Association Rule Mining for Rare and Frequent Itemsets. IEEE Transactions on Neural Networks 25(11), 2053–2064 (2014)
15. Sotudeh-Chafi, M., Akbarzadeh-T., M.-R., Moavenian, M.: A Novel Soft Computing Approach to Component Fault Detection and Isolation of CNC X-axis Drive System. International Journal of Intelligent Automation and Soft Computing 16(2), 177–191 (2010)
16. Akramizadeh, A., Akbarzadeh-T, M.R., Khademi, M.: Fuzzy Discrete Event System Modeling and Temporal Fuzzy Reasoning in Urban Traffic Control. In: Proceedings of the 2004 World Automation Congress and Fifth International Symposium on Intelligent Automation and Control, Seville, Spain, June 28-July 1 (2004)
17. Sengstacken, A.J., DeLaurentis, D.A., Akbarzadeh-T, M.R.: Optimization of Shared Autonomy Vehicle Control Architectures for Swarm Operations. IEEE Transactions on Systems, Man, and Cybernetics 40(4), 145–1157 (2010)
18. Hartmann, A.K., Weigt, M.: Phase Transitions in Combinatorial Optimization Problems, Basis, Algorithms and Statistical Mechanics. Wiley-VCH Verlag Co. (2005)

19. Vafaei Jahan, M., Akbarzadeh-T, M.R.: From Local Search to Global Conclusions: Migrating Spin Glass-based Distributed Portfolio Selection. IEEE Transactions on Evolutionary Computations (2009)
20. Rowhanimanesh, A., Akbarzadeh-T, M.-R.: Control of Low-Density Lipoprotein Concentration in the Arterial Wall by Proportional Drug-Encapsulated Nanoparticles. IEEE Transactions on Nanobioscience (2012)
21. Radyraz, N.: Bio-inspired Nanonetworks with Local Fuzzy Vision for Targeted Cancer Drug Delivery. M.S. Thesis, Islamic Azad University, Mashhad Branch (2014)

Erratum: Control and Systems Engineering

Aly El-Osery and Jeff Prevost

Electrical Engineering Department,
New Mexico Institute of Mining and Technology,
Socorro New Mexico, USA

Electrical and Computer Engineering Department,
The University of Texas at San Antonio
San Antonio Texas, USA

A. El-Osery and J. Prevost (eds.), *Control and Systems Engineering,*
Studies in Systems, Decision and Control 18, DOI: 10.1007/978-3-319-14636-2

DOI 10.1007/978-3-319-14636-2_16

The volume number of the original version is incorrect; it should be read as SSDC 27.

The original online version for this chapter can be found at
http://dx.doi.org/10.1007/978-3-319-14636-2

Personal Notes

Mansour Eslami

To all students, friends and colleagues of Professor Mohammad Jamshidi

Greetings from Chicago!

It is indeed an honor and especial privilege to write a few words on this great occasion of celebrating the 70th birthday and, of course, through an accompanying conference for highlighting the lifetime academic achievements of Professor Mohammad Jamshidi.

I met Professor Jamshidi in the Fall of 1975 when as a graduate student, I presented my first paper at the Allerton Conference in Urbana, Illinois; and we have remained in touch ever since, and now I consider him as my dear friend and esteemed colleague and mentor.

This has indeed been a remarkable story of lasting friendship based upon no expectation, certainly none from his end, and only because of his generosity and caring attitude towards those whom he can help. Often, we meet in various conferences and/or airports, and he had accepted my invitation to give a lecture at my school and therefore helping me without any expectation.

In all these years that I have been interacting with him, I have never heard an ill word from him towards anyone, and his general approach in dealing with others, or all matters of interests in this world, is very simple: If he cannot help contributing to a cause constructively, he would not create any harm to anyone. Seldom we meet anyone exhibiting such a consistently distinctive behavior. He is truly an optimist with great sense of humanitarian towards all creeds.

On the academic front, Professor Jamshidi has published seminal research papers and books that are being used all over the world; and yet he has continuously remained active in research, teaching and guiding students towards their higher education and advanced degrees. Indeed, he has continuously dedicated his entire life for educating and mentoring generations of students from all over the world, and certainly in various universities in the United States with large minority students concentrations.

Equally, he has also helped in multiple-academic as well as non-academic directions many professionals for the advancement of their careers; and he has done this noble act because of his genuinely giving humanitarian style.

I salute him for all his academic accomplishments and his lifetime dedication to the education of our youth, and thus the welfare of our global community.

May God bless him and his abundantly.

Mansour Eslami
Center For Agile Research
Chicago, Illinois 60690-1939

Madan Gupta

Some Reflections over the past four decades

Dear Mo.

It was in mid-seventies, almost four decades ago, when I received a letter from you requesting a research position in my Intelligent Systems Research Laboratory at the University of Saskatchewan. Your CV was very attractive, and by the time I managed to find some research funds for you, you got a teaching/research position in USA. Though we missed you but this was your wise decision as I can see today. Mo, it is almost four decades since I first met you and learned many things in the field of soft-computing and fuzzy logic from you. I remember when you invited me to the first ISRAM Conference held in Maui in 1994, and then again in 1996 and 1998. I also enjoyed the thought provoking conferences organized by you in San Antonio in 2009 and 2013. I noticed a large support that you have received in the organization of these conferences from your charming and most dedicated wife Jila.

Thank you Jila for your highly dedicated work and supporting my friend Mo!!

Your seminal work on fuzzy logic and soft-computing has created a worldwide impact as can be seen from the tremendous amount of your research publications in books, technical journals and conference proceedings. Indeed you have created a deep impact on your research colleagues and friends working in the field of soft computing.

In summary, Mo, we highly appreciate your intellectual contributions to our fuzzy society, and it is gives me a great pleasure and honour in decorating you on your seventieth birthday with a new title:

Mohammad (Mo) Jamshidi: "Son of Fuzzy Logic and Soft-Computing"

Dr. Madan M. Gupta
Ph.D.; D.Sc.
PROFESSOR (Emeritus) & DIRECTOR
Distinguished Research Chair
Fellows: LF-IEEE; F-SPIE; & F- IFSA

Gordon Lee

A Tribute to Mo Jamshidi

It is with great pleasure that I write this note in tribute to the vast contributions that Dr. Mo Jamshidi has provided to the systems engineering community over the years. Mo is a visionary who understands not only what should be done but also has the energy and passion to make things happen. There are many examples of the results of his passion; many years ago, he had the idea that the systems community needed a forum where scientists and engineers from various disciplines could meet to present new concepts, discuss applications of these new methods, or discover what others in the technical community are doing. This forum became the World Automation Congress. He also realized a long time ago, while at the University of New Mexico, of the need to develop software for control engineers and worked with his students to generate control modules for design and analysis. He applied his expertise in many fields including space systems, energy, and robotics. Mo worked to advance the field of fuzzy systems and later helped to develop the concept of system-of-systems analysis and design, which are used in many applications today.

Besides all of his technical accomplishments, I must also note that Mo is a people- oriented person who provides many of us the opportunity to grow, to succeed, and to network with others in our field. Always optimistic and cheerful whenever I meet him, Mo seems to get younger each year! It is with deep gratitude that we honor Mo as truly a gentleman and a scholar.

Gordon Lee
Dept. of Electrical & Computer Engineering
San Diego State University

Ali Hessami

Mo Jamshidi, A Distinguished Educator

An IEEE Distinguished Lecture was held at the University of East London on 10th July 2009. The "Control of System of Systems" lecture was delivered by the Lutcher Brown Endowed Chaired Professor, Prof Mo Jamshidi from The University of Texas, St. Antonio.

A the Chair of the IEEE SMC Chapter in the UK & the Republic of Ireland Section, I had arranged this event with Prof Jamshidi who had gone out of his way to come to London and deliver this keynote to a group of academics, Masters and PhD students and systems engineers. Prof Jamshidi explained the challenges in SoS a "super system," or an integration of complex systems coordinated together in such a way to achieve a wider goal with possible higher significance. He explained that applications of SoS are quite extensive and range from future combat mission, Global Warming, Mars missions, Air Traffic System, Global Earth Observation System, Electric Power Grid System, Energy systems, etc. The main challenge being in the design of a controller for SoS's in view of the difficulty or impossibility of developing a comprehensive SoS model, either analytically or through simulation so by and large, SoS control remains an open problem. He illustrated UTSA ACE Center's SoS technology and demonstrated some animated and experimental implementations.

On another occasion and after a long trip to Azerbaijan, we again managed to attract Prof Jamshidi to interrupt his return flight to the US and come to London City University and deliver a System of Systems Engineering lecture on 11 Dec. 2012. This was titled "Intelligent Management Decisions for a System of Conventional and Sustainable Energy Systems with Air Pollution Constraints". Mo presented how Fuzzy Logic-based decision-making framework is implemented for energy management in microgrid systems in order to meet targets such as providing local consumers with required energy demand and making good revenue for the microgrid. Five different scenarios were considered for local load and microgrid assembly operation. Additionally a review of gaps in system of systems engineering advances as well as cooperative efforts between European Union and USA researchers and agencies was presented, based on T-AREA-SoS (Trans-Atlantic Agenda on Research and Education in SoS) an EC-funded research programme.

The two occasions highlighted above are our experience of Mo's dedication to the cause of education, global collaboration and leading edge research on one of the most promising yet challenging areas of technology, the System of Systems paradigm. In spreading this knowledge, Mo has demonstrated dedication, generosity, vision and empathy with the researchers and those attending his landmark lectures. These are no doubt an insignificant sample of Mo's vast work, association and global recognition but they go to show a degree of generosity and humility that sets him apart from many others in his discipline.

Prof. Ali Hessami
London-UK, VC-IEEE UK & Ireland Section

Michael S. Packianather

Professor Mo Jamshidi is a warm and friendly person. He is very unique in many ways. I had the privilege of meeting him in 2008 in Hawaii at the WAC conference which he chaired. The following year he came to Cardiff University as a Distinguished Visiting Fellow of the Royal Academy of Engineering. It was during this visit that I got to know him and his dear wife, Jila. They were a joy to be with. One thing which he shared with me has stayed with me till now and always makes me think. This is what he said, "we must always make sure that what we are doing, work or otherwise, makes an impact on others around us". How true this is and if only we all are willing to adopt this as our motto then, not only we will be following the good example of Mo but, the world will be a better place. While in Cardiff he delivered the keynote lecture at INDIN 2009 conference chaired by Professor D. T. Pham. We had a great conference with our colleagues from University of Polytechnic Madrid headed by Professor Diego Andina. While he was in Cardiff University he visited University of Reading, University of East London and University of Ulster giving seminars on "System of Systems" to which he has dedicated his career. We all enjoyed his lectures packed with real examples showing how this theory can actually help improve the quality of life for everyone. I was impressed of his memory, how he remembered all his PhD students past and present and his concern for their welfare. While he was in Cardiff University he spent a lot of time with our own PhD students and staff exchanging research ideas. He left us with copies of his papers and books disseminating and transferring knowledge, the sign of a born "guru".

I would like to dedicate this to Prof. Mo Jamshidi as he celebrates his happy 70th birthday.

M – Man of good qualities
O – Originator of many scientific ideas

J – Joyful in all situations
A – Author of countless number of publications
M – Master of research
S – System of Systems is his baby
H – Humorous in many ways
I – Impacting on others around him is his motto
D – Devoted to his family
I – Innovative throughout his academic career

Michael, Leena,
Jila and Mo

Michael, Leena,
Jila and Mo

Mo, Jila, Michael,
Pham and Diego

Michael S. Packianather
Cardiff University, UK

Mehdi Shadaram

It was about sixteen years ago when I had the good fortune to meet Professor Mo Jamshidi. Since then, I have had the pleasure of having him as a friend and colleague. Currently, Professor Jamshidi serves as the Lutcher Brown Endowed Distinguished Professor in the Department of Electrical and Computer Engineering at The University of Texas at San Antonio (UTSA). After decades of technical research and development in the field of control and systems engineering, Professor Jamshidi is considered a leader in this field. He has contributed globally to the field of large-scale complex systems and system of systems and their applications and integration with various control and modeling paradigms, including intermixing control theory and intelligent paradigms of artificial intelligence, also known as autonomous control. In addition to his many contributions to his field, Professor Jamshidi has served his profession in many capacities. He has served as the editor-in-chief of several successful and well-known scientific journals and magazines, has organized numerous national and international conferences, and has served on numerous panels. Professor Jamshidi is an excellent and caring teacher. He has advised and mentored more than one hundred masters and doctoral level students. Professor Jamshidi is also an avid advocate of research experience for undergraduates. Throughout his professional career, he has mentored numerous undergraduates in his laboratory. Professor Jamshidi continues to mentor and help his students even after his formal role as an advisor ends, a quality greatly appreciated and admired by his students. I wish him a happy 70th birthday and best of luck in his future endeavors.

Mehdi Shadaram, PhD, PE
Interim Dean of Engineering
Briscoe Distinguished Professor
David and Jennifer Spencer Distinguished Dean's Chair in Engineering
The University of Texas at San Antonio
San Antonio, TX 78258

Bahram Shafai and Mehrdad Saif

It is with sincere enthusiasm that we submit this note on the occasion of Mo Jamshidi's 70th birthday celebration. Mo Jamshidi's contributions to system and control research, teaching, professional and community service is well known and is well documented, therefore, we shall not belabor it here. Briefly however, over the years, he has established himself as a prolific researcher who has published widely and in several areas including but not limited to large-scale systems; hierarchical control; intelligent and fuzzy systems/control; and system-of-systems. Mo is also a dedicated and inspiring teacher who has contributed to training of many students, post-doctoral fellows, and visiting researchers, many of whom have gone to have successful academic or industrial careers themselves. Mo has also given a great deal to the profession through his dedicated and tireless contributions to various professional societies such as IEEE, IFAC, ASME, and many others. As mentioned above Mo's contributions to electrical engineering, and system and control discipline in particular is widely known. As such, we wish to focus the remainder of this note towards our own personal experiences. We have had the good fortune of knowing Mo for over three decades, and from very start of our academic careers. Like an older brother, Mo has been a mentor, a constant source of support, and encouragement to us throughout our career. We know of many other colleagues who feel the same way as we do, because Mo is always there to give a helping hand to those who need it. In our eyes, Mo is a great leader, as he possess many characteristics of a leaderthere is no doubt that he is a great visionary as he has demonstrated over the years (e.g. Mo was the Founding Editor of the IEEE Control System Magazine); he has a great deal of passion and never ending supply of energy when he takes on a task; he delegates (we can attest to that as tasks have been delegated to us knowingly and unknowingly!); he is a team and friend builder as is evidenced in the conferences he organizes; he is a man of good character; and he is a quick decision maker. We feel fortunate to have known Mo for all these years. He has been a role model and a mentor to us, but most important of all, he has been a terrific friend. Thanks for all the memories Mo.

As a token of appreciation and our experience gained from your work, particularly your first book on large-scale systems, we motivated to contribute a chapter entitled "Proportional-Integral Observer in Robust Control, Fault Detection, and Decentralized Control of Dynamic Systems" for this book. We are very proud of your many years of dedication to control system community and are indebted for your support and encouragement throughout the past years. We wish you all the best and looking forward to interact with you for many years to come.

Bahram Shafai and Mehrdad Saif

Bijan & Saied Tadayon

Dear friends and colleagues,

Prof. Jamshidi is one of our best friends and mentors. He and his family have been so kind to us. He is a great example of a researcher and academic pioneer. We have been very fortunate to know him.

He is very helpful to all his friends, colleagues, and students. He considers his students as his own kids. He guides them through their research, as well as their lives. In particular, for the foreign students that come to US for the first time, his help and guidance is extremely useful.

He has published so many papers in various technologies with his students. He has pioneered control systems, robotics, vision, image processing, fuzzy logic, AI, and system-of-systems. He has had so many successful graduate students, coming from many different countries, cultures, and backgrounds. They represent the "United Nations." He is always proud that his students cover so many backgrounds and ethnicities.

He has also contributed to many conferences and journal publications, as well as so many books, many of which he has started from scratch. His students, as well as his wife and family, are involved in all aspects of running his conferences. His wife is a big supporter of him. He could not accomplish as much without her constant help and advice.

His conferences are so unique, because it is as much fun-packed as the academic value. He organizes the conferences so that all people have a great memory and feel at home. It is the best place to do the networking, make new friends, and exchange ideas.

All of his friends, colleagues, professors, students, and researchers have a very high opinion of him, in many aspects. He is a great human, lecturer, teacher, professor, mentor, father, husband, brother, friend, and researcher. We value his friendship, as well as that of his family and students (former and current). His students have also followed his path in academy and humanity. So, we are very proud of all of his group members and students.

Bijan Tadayon, Ph.D., J.D., CEO,
Saied Tadayon, Ph.D., J.D., CTO,
Z Advanced Computing, Inc.
Potomac, MD, 20854, USA
www.ZAdvancedComputing.com

Appendix: Mo Jamshidi Publication List

1 Publication Activities at a Glance

Total Number of published works = 706 - CD ROMS = 13, Books = 68 (Textbooks = 12), book chapters = 32, Reviewed Conference papers = 352, Reviewed Journal papers =190, Special issues of journals =33 and Technical reports = 18 Known languages in which Jamshidi's books have been translated: Mandarin (4), Russian (2), French (1). The publication is grouped into seven categories of CD ROMS, books and book chapters, technical reports, conference papers, journal papers, and special issues of journals.

CD ROMS

1. M. JAMSHIDI, R. Lumia, E. Tunstel, Jr., B. White, J. Malone, and P. Sakimoto, *Proceedings NASA URC Conference on Education, Earth, Environment and Space*, Vol. 1, ACE Center Series, Albuquerque, NM: ACE Center, 1997.
2. M. JAMSHIDI, C. W. de Silva, F. Pierrot, M. Fathi, and M. Kamel, *Proceedings World 3rd Automation Congress* – Anchorage, AK, TSI Press, Albuquerque, NM, May 1998.
3. M. JAMSHIDI, P. Borne, A. A. Macijewski, M. Fathi, S. Nahavandi, R. Lumia, and T. Furuhashi, *Proceedings 4th World Automation Congress* - Maui, HI, USA, TSI Press, Albuquerque, NM, June 2000.
4. M. JAMSHIDI, Y. Hata, F. Proctor, J. Fedemma, B. Shafai, and A. Homaifar, *Proceedings 5th World Automation Congress* – Orlando, FL, USA, TSI Press, Albuquerque, NM, June 2002.
5. M. JAMSHIDI, A. Ollero, L. Foulloy, M. Reuter, A. Kamrani, and Y. Hata, *Proceedings 5th World Automation Congress* – Seville, Spain, TSI Press, Albuquerque, NM, USA, June 2004.
6. M. JAMSHIDI, Y. Hata, M. Reuter, D. Cox, S. Nahavandi, and J. S. Jamshidi, *Proceedings 6th World Automation Congress* – Budapest, Hungary, TSI Press, San Antonio, TX, USA, July 2006.
7. M. JAMSHIDI, W. Pedrycz, K. W. Bonfig, R. Aliev, and R. Lewerenz, *Proceedings 7th International Conference on Application of Fuzzy Systems and Soft Computing* – Siegen, Germany, September 2006.
8. R. A. Aliev, K. W. Bonfig, M. JAMSHIDI, W. Pedrycz, and I. B. Turksen (Eds.), *Proceedings 8th International Conference on Applications of Fuzzy Systems and Soft Computing*, b-Quadrat Verlag, ISBN393 3609-26-7, Helsinki, Finland, September 1-3, 2008.
9. M. JAMSHIDI, Y. Hata, G. Parker, M. Reuter, M. Sadat, and D. Cox, *Proceedings 7th World Automation Congress* – Waikoloa, HI, TSI Press, San Antonio, TX, USA, September 2008.
10. R. A. Aliev, K. W. Bonfig, M. JAMSHIDI, W. Pedrycz, and I. B. Turksen (Eds.), *Proceedings 9th International Conference on Applications of Fuzzy Systems and Soft Computing*, b-Quadrat Verlag, ISBN393 3609-26-7, Prague, Czech Republic, August 26-27, 2010.
11. M. JAMSHIDI, Y. Hata, S. Kobashi, G. Parker, D. Andina, and B. Ane, , *Proceedings 8th World Automation Congress* – Kobe, Japan, TSI Press, Albuquerque, NM, USA, September 2010.

12. M. JAMSHIDI, S. Kobashi, G. Muscato, N. Sepehri, D. Andina, and R. Valerdi,, *Proceedings 8th World Automation Congress* – Puerto Vallarta, Mexico, TSI Press, Albuquerque, NM, USA, September 2012.
13. R. A. Aliev, K. W. Bonfig, M. JAMSHIDI, and I. B. Tukesen, Proc. 7th International Conference on Soft Computing, Computing with Words and Perceptions in System Analysis, Decision and Control, Izmir, Turkey, September 2-3, 2013.

2 Books

(see site http://wacong.org/freepublicationsbymojamshidi/ for free selected books and of the author)

1. J. M. Davis and M. JAMSHIDI, *Solution Manual of Engineering of Dynamic Systems* (W. R. Perkins and J. B. Cruz, Jr.) John Wiley and Sons, Inc. New York, NY, 1969.
2. M. JAMSHIDI, *Analog Simulation of Dynamic Processes*, University of Illinois, School of Engineering Publications Office, Urbana, IL., 1971.
3. M. JAMSHIDI and M. H. Nehrir, *Proceedings 4th Iranian Conference* (1st Congress) on Electrical Engineering, (eds.) Volume 1, Shiraz, Iran, Shiraz University Press, May 1974.
4. M. JAMSHIDI and M. H. Nehrir, *Proceedings 4th Iranian Conference* (1st Congress) on Electrical Engineering, (eds.) Volume 2, Shiraz, Iran, Shiraz University Press, May 1974.
5. M. JAMSHIDI, *Large-Scale Systems - Modeling and Control*, Elsevier North- Holland, New York, NY, 1983.
6. M. JAMSHIDI, *Large-Scale Systems - Modeling and Control*, Elsevier North- Holland, New York, NY, 1983. Russian Edition, see http://urss.ru/cgi-bin/db.pl?lang=en&blang=en&page= Book&id=107181.
7. M. JAMSHIDI, *Large Scale Systems - Modeling and Control*, (Chinese Edition) Science Publishers, Beijing, P. R. China, 1986.
8. M. JAMSHIDI and C. J. Herget, *Computer-Aided Control System Engineering*, eds., North-Holland, Amsterdam, 1985.
9. M. JAMSHIDI and C. J. Herget, *Computer-Aided Control Systems Engineering*, Majinostroni Publishing House, (Russian Edition) Moscow, USSR, 1989, this translation sold 20,000 copies in USSR.
10. M. JAMSHIDI and M. Malek-Zavarei, *Linear Control Systems - A Computer- Aided Approach*, Pergamon Press, Oxford, England, 1986.
11. M. JAMSHIDI and M. Malek-Zavarei, *Linear Control Systems - A Computer Aided Approach*, NAI Press, (Chinese Edition) Beijing, P. R. China, 1989.
12. M. Malek-Zavarei and M. JAMSHIDI, *Time Delay Systems: Analysis, Optimization and Applications*, North-Holland Amsterdam, The Netherlands, 1987.
13. M. JAMSHIDI, J. Y. S. Luh, and M. Shahinpoor, *ADVANCES IN ROBOTICS: Modeling, Control and Education*, eds., Elsevier Publishing Co., New York, 1986.
14. M. JAMSHIDI, J. Y. S. Luh, H. Seraji, and G. P. Starr, *ROBOTICS AND MANUFACTURING: Recent Trends in Research, Education, and Applications*, eds., vol. 2, ASME Press, New York, NY, 1988.
15. M. JAMSHIDI, M. Tarokh, and B. Shafai, *Computer-Aided Analysis and Design of Linear Control Systems*, Prentice Hall, Englewood Cliffs, NJ, 1992.
16. M. JAMSHIDI and M. Saif, *ROBOTICS AND MANUFACTURING: Recent Trends in Research, Education, and Applications*, (eds.), vol. 3, ASME Press, New York, NY, 1990.

17. M. JAMSHIDI, M. Ahmadi, and M. Nahvi, (eds.) *Circuits, Systems, and Information*, TSI Press, Albuquerque, NM, 1991.
18. M. JAMSHIDI and P. J. Eicker, (eds.) *Robotics and Remote Systems*, U.S. Government Printing Office, Washington, DC., Feb 1991.
19. M. JAMSHIDI, M. Shahinpoor, and J. H. Mullins, (eds.) *Environmentally Conscious Manufacturing*, ECM Press, Vol. 1, Albuquerque, NM, 1992.
20. M. JAMSHIDI and C. J. Herget, *Advances in Computer-Aided Control System Analysis and Design*, Eds. North-Holland, Amsterdam, The Netherlands 1992.
21. M. JAMSHIDI and P. J. Eicker, (eds.) *Robotics and Remote Systems in Unstructured Environments*, Englewood Cliffs, NJ: Prentice Hall, 1993.
22. M. JAMSHIDI, N. Vadiee, and T. Ross, (eds.) *Fuzzy Logic and Control: Software and Hardware Applications*, Englewood Cliffs, NJ: Prentice Hall, 1993.
23. M. JAMSHIDI, R. Lumia, J. Mullins, and M. Shahinpoor, *Robotics and Manufacturing: Recent Trends in Research, Education, and Applications*, (eds.), Vol. 4, ASME Press, New York, NY, 1992.
24. M. JAMSHIDI, M. Mansour, B. D. O. Anderson, and N. K. Bose (eds.) *Fundamentals of Discrete-Time Systems*, Albuquerque, NM: TSI Press, 1993.
25. M. JAMSHIDI and H. Parsaei, (eds.) *Design and Implementation of Intelligent Manufacturing Systems*, Englewood Cliffs, NJ: Prentice Hall, 1993.
26. M. Shahinpoor, M. JAMSHIDI, B. Granhoff, and C. Berger, (eds.) *Environmentally Conscious Manufacturing*, ECM Press, Vol. 2, Albuquerque, NM, 1994.
27. F. Aminzadeh and M. JAMSHIDI, (eds.) *Soft Computing*, Englewood Cliffs, NJ: Prentice Hall, 1994.
28. M. JAMSHIDI, C. C. Nguyen, R. Lumia, and J. Yuh (eds.) *Robotics and Manufacturing: Robotics Research and Applications*, Vol. 5, New York: ASME Press, 1994.
29. M. JAMSHIDI, C. C. Nguyen, R. Lumia, and J. Yuh (eds.) *Intelligent Automation and Soft Computing*, Vol. 1, Albuquerque, NM: TSI Press, 1994.
30. M. JAMSHIDI, C. C. Nguyen, R. Lumia, and J. Yuh (eds.) *Intelligent Automation and Soft Computing*, Vol. 2, Albuquerque, NM, 1994.
31. M. JAMSHIDI, *Large-Scale Systems: Modeling, Control, and Fuzzy Logic*, Englewood Cliffs, NJ: Prentice Hall, 1997.
32. M. JAMSHIDI, F. Pin, and P. Dauchez, *Robotics and Manufacturing: Robotics Research and Applications*, Vol. 6, New York: ASME Press, 1996.
33. M. JAMSHIDI, F. Pin, and F. Pierrot, *Robotics and Manufacturing: Robotics Research and Applications*, Vol. 3, Albuquerque, NM: TSI Press, 1996.
34. M. JAMSHIDI, J. Yuh, and F. Pierrot (eds.) *Intelligent Automation and Control*, Vol. 4, Albuquerque, NM: TSI Press, 1996.
35. M. JAMSHIDI, M. Fathi, and P. Dauchez (eds.) *Soft Computing for Industry*, Vol. 5, Albuquerque, NM: TSI Press, 1996.
36. M. JAMSHIDI, N. Vadiee, and T. Ross, (eds.) *Fuzzy Logic and Control: Software and Hardware Applications*, (Persian Edition), by M. Makrechi and A. Katebi, Shiraz, Iran, 1996.
37. M. JAMSHIDI, A. Titli, L. A. Zadeh, and S. Boverie, *Applications of Fuzzy Logic - Towards High Machine Intelligence Quotient (MI?) Systems*, Upper Saddle River, NJ: Prentice Hall, 1997.
38. M. JAMSHIDI, R. Lumia, E. Tunstel, Jr., B. White, J. Malone, and P. Sakimoto, *Proceedings NASA URC Conference on Education, Earth, Environment and Space*, Vol. 1, ACE Center Series, Albuquerque, NM: ACE Center, 1997.

39. M. JAMSHIDI, *Large-Scale Systems: Modeling, Control, and Fuzzy Logic*, (French Edition) Paris, France, 1999.
40. M. JAMSHIDI and C. W. de Silva (eds.) *Intelligent Automation and Control*, Vol. 6, Albuquerque, NM: TSI Press, 1998.
41. M. JAMSHIDI, F. Pierrot, and M. Kamel (eds.) *Robotics and Manufacturing*, Vol. 7, Albuquerque, NM: TSI Press, 1998.
42. M. JAMSHIDI, Z. Bien, and M. Fathi (eds.) *Soft Computing, Multimedia and Image Processing: Research and Applications*, Vol. 8, Albuquerque, NM: TSI Press, 1998.
43. M. JAMSHIDI, D. Kauffman, and N. Vadiee *Proceedings of ACE-PURSUE Student Conference, ACE-Center Series*, Vol.2, Albuquerque, NM, April 19-20, 1999.
44. M. JAMSHIDI, P. Borne, and J. S. Jamshidi (eds.) *Intelligent Automation and Control* (Proc. ISIAC 2000), Vol. 9 (Intelligent Automation and Soft Computing Series), Albuquerque, NM: TSI Press, 2000.
45. M. JAMSHIDI, A. A. Maciejewski, S. Nahavandi, and R. Lumia (eds.) *Robotic and Manufacturing Systems: Recent Results in Research, Development and Applications*, (Proc. ISORA and ISOMA 2000), Vol. 10 (Intelligent Automation and Soft Computing Series), Albuquerque, NM: TSI Press, 2000.
46. M. JAMSHIDI, M. Fathi, and T. Furuhashi (eds.) *Soft Computing, Multimedia and Image Processing: Trends, Principles and Application*s (Proc. IFMIP 2000 and ISSCI 2000), Vol. 11 (Intelligent Automation and Soft Computing Series), Albuquerque, NM: TSI Press, 2000.
47. M. JAMSHIDI, B. Turksen, A. Aliev, G. Bonfig, and D. Aliew, (eds.), *Proceedings of International Conference on Soft Computing and Computing with Words*, Antalya, Turkey, June 12-14, 2001.
48. A. Zilouchian and M. JAMSHIDI, (eds.), *Intelligent Control Systems With Soft Computing Methodologies*, CRC Publishers, Boca Raton, FL, 2001.
49. M. JAMSHIDI, R. A. Krohling, L. dos S. Coelho, and P. Fleming, *Robust Control Design Using Genetic Algorithms,* CRC Publishers, Boca Raton, FL, 2003.
50. M. JAMSHIDI, F. Proctor, J. Feddema, and B. Shafai, (eds.) *Robotic, Manufacturing and Control Systems: Recent Results in Research, Development and Applications*, (Proc. ISORA, ISOMA and ISIAC 2002), Vol. 14 (Intelligent Automation and Soft Computing Series), Albuquerque, NM: TSI Press, 2002.
51. M. JAMSHIDI, Y. Hata, and A. Homaifar, (eds.) *Soft Computing, Multimedia and Image Processing: Trends, Principles and Application*s (Proc. IFMIP 2002 and ISSCI 2002), Vol. 15 (Intelligent Automation and Soft Computing Series), Albuquerque, NM: TSI Press, 2002.
52. M. JAMSHIDI, A. Ollero, J. R. Martinez-de Dios, and J. S. Jamshidi, (eds.) *Robotics: Recent Results in Research, Development and Applications*, (Proc. ISORA 2004), Vol. 15 (Intelligent Automation and Soft Computing Series), Albuquerque, NM: TSI Press, 2004.
53. M. JAMSHIDI, L. Foulloy, A. Elkamel, and J. S. Jamshidi (eds.) *Automation and Control: Recent Results in Research, Development and Applications*, (Proc. ISIAC 2004), Vol. 16 (Intelligent Automation and Soft Computing Series), Albuquerque, NM: TSI Press, 2004.
54. M. JAMSHIDI, M. Reuter, D. Andina, and J. S. Jamshidi, (eds.) *Soft Computing: Trends, Principles and Application*s (Proc. ISSCI 2004), Vol. 17 (Intelligent Automation and Soft Computing Series), Albuquerque, NM: TSI Press, 2004.
55. M. JAMSHIDI,Y. Hata, A. Kamrani, and J. S. Jamshidi (eds.) *Multimedia, Image Processing, Biomedicine and Manufacturing* (Proc. IFMIP 2004 and ISOMA 2004), Vol. 18 (Intelligent Automation and Soft Computing Series), Albuquerque, NM: TSI Press, 2004.

56. M. JAMSHIDI (eds.) *Systems and Control*, Albuquerque, NM: TSI Press, 2006.
57. M. JAMSHIDI, W. Pedrycz, K. W. Bonfig, R. Aliev, and R. Lewerenz, *Proceedings 7th International Conference on Application of Fuzzy Systems and Soft Computings* – Siegen, Germany, September 2006.
58. M. JAMSHIDI, D. Cox, S. Nahavandi, and J. S. Jamshidi (eds.) *Robotics, Manufacturing and Automation – Trends, Principles and Applications* (Proc. ISORA 2006, ISOMA 2006 and ISIAC 2006), Vol. 19 (Intelligent Automation and Soft Computing Series), San Antonio, TX, USA: TSI Press, 2006.
59. M. JAMSHIDI, Y. Hata, M. Reuter, and J. S. Jamshidi (eds.) *Image Processing, Biomedicine, and Soft Computing – Trends, Principles and Applications* (Proc. IFMIP 2006 and ISOMA 2006), Vol. 20 (Intelligent Automation and Soft Computing Series), San Antonio, TX, USA: TSI Press, 2006.
60. M. JAMSHIDI (ed.), *Systems of Systems Engineering – Principles and Applications*, CRC – Taylor & Francis Publishers, London, UK, 2008.
61. R. A. Aliev, K. W. Bonfig, M. JAMSHIDI, W. Pedrycz and I. B. Turksen (Eds.), *Proceedings 8th International Conference on Applications of Fuzzy Systems and Soft Computing*, b-Quadrat Verlag, ISBN393 3609-26-7, Helsinki, Finland, September 1-3, 2008.
62. M. JAMSHIDI (ed.), *System of Systems Engineering – Innovations for the 21st Century*, John Wiley & Sons, Publishers, New York, NY, 2009.
63. T. Nanayakkara, F. Sahin, and M. JAMSHIDI, *Intelligent Control Systems with an introduction to System of Systems*, Volume 2, CRC-Taylor Francis Series on System of Systems Engineering (M. JAMSHIDI, Ed.), CRC – Taylor & Francis Publishers, London, UK, 2010.
64. R. A. Aliev, K. W. Bonfig, M. JAMSHIDI, W. Pedrycz, and I. B. Turksen (Eds.), *Proceedings 9th International Conference on Applications of Fuzzy Systems and Soft Computing*, b-Quadrat Verlag, ISBN393 3609-26-7, Prague, Czech Republic, August 26-27, 2010.
65. R. A. Aliev, K. W. Bonfig, M. JAMSHIDI, W. Pedrycz, and I. B. Turksen (Eds.), *Proceedings 10th International Conference on Applications of Fuzzy Systems and Soft Computing*, b-Quadrat Verlag, ISBN393 3609-26-7, Lisbon, Portugal, August 29-30, 2012.
66. R. A. Aliev, K. W. Bonfig, M. JAMSHIDI, and I. B. Turksen, Proceeding 7th International Conference on Soft Computing, Computing with Words and Perceptions in System Analysis, Decision and Control b-Quadrat Verlag, ISBN393 3609-32-1, Seigen, Germany, 2013.
67. M. JAMSHIDI (ed.), *Systems of Systems Engineering – Principles and Applications*, Mandarin Version, China Machine Press, Beijing, China, 2013.
68. M. JAMSHIDI, V. Kreinovitch, and J. Kacprzyk (Eds.), "Advance Trends in Soft Computing," Springer-Verlag, Germany, 2013.

3 Book Chapters

1. M. JAMSHIDI, "Introduction to Large-Scale Systems," in *Systems Modeling and Computer Simulation*, N. Kheir (ed.), Marcel Dekker, Inc., New York, NY, 1987.
2. M. JAMSHIDI, "Large-Scale Systems - An Expository Look," *Encyclopedia of Physical Sciences and Technology*, Academic Press, Inc., San Diego, CA, 1986 and 1992.

3. M. JAMSHIDI, "Large-Scale Systems," McGraw Hill, New York, *Encyclopedia of Science and Technology*, New York, 6th Edition, 1989.
4. C. Abdallah, D. Dawson, P. Dorato, and M. JAMSHIDI, "*Techniques for the robust control of rigid robots*," in Control and Dynamic Systems, C. T. Leondes (ed.), vol. 53, Academic Press, 1992, pp. 387-426..
5. R. L. Colbaugh and M. JAMSHIDI "Adaptive impedance control of solid waste handling robots, " in *Robotics and Remote Systems in Unstructured Environments*, M. Jamshidi and P. Eicker (Eds.), Prentice Hall, Englewood Cliffs, NJ, 1993.
6. M. JAMSHIDI, "Introduction to Fuzzy Logic and Control, in "*Fuzzy Logic and Control: Software and Hardware Applications*, M. Jamshidi, et al. (eds.), Englewood Cliffs, NJ: Prentice Hall, 1993.
7. M. JAMSHIDI, R. Marchbank, K. Bissett, R. Kelsey, D. Barak, and S. Baugh., "Computer-Aided Design of Fuzzy Control Systems, " in *Advances in Computer -Aided Control Systems*, M. Jamshidi and C. J. Herget (eds.), North Holland, Amsterdam, 1 993, pp. 81-126.
8. M. JAMSHIDI, "Fuzzy Logic Software and Hardware", in *Fuzzy Logic and Control: Software and Hardware Applications*, M. Jamshidi, et al. (eds.), Englewood Cliffs, NJ: Prentice Hall, 1993.
9. M. JAMSHIDI, "CAD Techniques in Control Systems, in *Advances in Digital Control Systems*," vol. 79, C. T. Leondes (ed.), Academic Press, 1996, pp. 247-346.
10. M. JAMSHIDI, On Software and Hardware Applications of Fuzzy Logic, in "*Fuzzy Logic Fundamentals*, R. Yager and L. A. Zadeh (eds.), New York, NY: Plenum Publishers, 1993.
11. M. JAMSHIDI," On Fuzzy Control of Non-Chlorofuorcarbon Air Conditioning Systems," *Intelligent Control Systems*, J. Yen, R. Langari and L. A. Zadeh (eds.) IEEE Press, 1994.
12. M. JAMSHIDI, "Large-Scale Systems - an Introduction," in *Systems Modeling and Computer Simulation,* N. Kheir (ed.), 2nd edition, Marcel Dekker, Inc., New York, NY, 1993.
13. M. JAMSHIDI, R. Marchbank, R. Kelsey, K. Kumbla, E. Krijstansson, and D. Barak, " Hardware Applications of Fuzzy Logic Control, in "*Soft Computing,* F. Aminzadeh and M. Jamshidi (eds.), Englewood Cliffs, NJ, Prentice Hall, 1994.
14. C. Schmidt and M. JAMSHIDI, "CADACS for System Analysis, Synthesis, and Real-Time Control", in Chapter 51 of *The Handbook of Software for Engineers and Scientists*, P. Ross (ed.), CRC Press, Inc., Boca Raton, FL, 1996.
15. E. Tunstel and M. JAMSHIDI, "Intelligent Control and Evolution of Mobile Robot Behavior", Chapter 6 in Jamshidi, Titli, Zadeh, & Boverie (Eds.) *Applications of Fuzzy Logic: Towards High Machine Intelligence Quotient Systems,* Vol. 9, Prentice-Hall Series on Environmental & Intelligent Manufacturing Systems, 1997.
16. R. D. Colbaugh, J. Baca, and M. JAMSHIDI., "Adaptive Compliant Motion Control for Robotic Waste management Applications," *Waste Management: From Risk to Remediation*, Vol. 1, Bhada (ed.), Albuquerque, NM: ECM Press, 1994.
17. M.-R. Akbarzadeh-T. and M. JAMSHIDI, chapter in *Proceedings of the Applied Computing Symposium on Applied Computing*, Nashville, Tennessee, 1995.
18. A. Jadbabaie, C. Abdallah, A. Titli, and M. JAMSHIDI, "Observer-Based Controller Synthesis for Model-Based Fuzzy Systems via Linear Matrix Inequalities" in *Intelligent Control*, R. Langari (ed.), IEEE Press. New York,1999
19. M. JAMSHIDI, "Autonomous Complex Systems", McGraw Hill, New York in *Encyclopedia of Science and Technology*, 7th Edition, 1998.

20. E. Tunstel and M. JAMSHIDI, "Intelligent Control and Evolution of Mobile Robot Behavior", Chapter 1 in JAMSHIDI, Titli, Zadeh and Boverie (Eds.), *Applications of Fuzzy Logic: Towards High Machine Intelligence Quotient Systems*, Vol. 9, Prentice Hall Series on Environmental and Intelligent Manufacturing Systems, M. JAMSHIDI, Series (ed.), 1997.
21. M. JAMSHIDI, "Fuzzy Sets and Logic" in *Encyclopedia of Arts and Sciences*, London, UK, 1998.
22. A. El-Osery and M. JAMSHIDI "Color image enhancement: Statistical versus fuzzy expert Systems Approaches," A Chapter in Ross, Booker and Parkinson (eds.) *Fuzzy Logic and Probability Applications.* SIAM, 2002., Chapter 7, pp. 127-144.
23. M. JAMSHIDI, "Soft Computing Control of Complex Systems," Chapter 6, *Intelligent Machines*, CRC Publishers, Boca Raton, FL, pp. 193-214, C. W. de Silva, Ed., 2000.
24. M. JAMSHIDI, "Autonomous Control of Complex Systems by Soft Computing, " in *Intelligent Systems - Myths and Realities*, Chapter 6, C. W. de Silva (Ed.), Boca Raton, FL: CRC Press, 2000.
25. M. JAMSHIDI and A. El-Osery, "Introduction to Fuzzy Sets: Basic Definitions and Relations," *Intelligent Control Systems based on Soft Computing Methodologies*, A. Zilouchian and M. JAMSHIDI (eds.), Chapter 8, CRC Publishers, Boca Raton, FL, 2001.
26. M. JAMSHIDI, A. El-Osery, and T. J. Ross, "Introduction to Fuzzy Logic," in *Intelligent Control Systems based on Soft Computing Methodologies*, A. Zilouchian and M. JAMSHIDI (eds.), Chapter 9, CRC Publishers, Boca Raton, FL, 2001.
27. M. JAMSHIDI and A. El-Osery, "Fuzzy Control Systems and their Stability," in *Intelligent Control Systems based on Soft Computing Methodologies*, A. Zilouchian and M. JAMSHIDI (eds.), Chapter 10, CRC Publishers, Boca Raton, FL, 2001
28. A. El-Osery and M. JAMSHIDI, "Image enhancement using a fuzzy logic approach," in *Intelligent Control Systems based on Soft Computing Methodologies*, T. Ross, W. J. Parkinson and J. (eds.), Chapter 12, SIAM Publishers,, FL, 2003.
29. A. Jevtic, D. Andina, and M. JAMSHIDI, " Distributed Task Allocation in Swarms of Robots," Chapter in "*Swarm Intelligence for Electric and Electronic Engineering*," G. Fornarelli and L. Mescia (eds.), IGI Global, Hershey, Pennsylvania Publishers, USA. (Web: www.igi-global.com), 2012.
30. M. JAMSHIDI, "Introduction to System of systems engineering," Chapter 1, in *System of Systems Engineering – Principles and Applications*, (M. Jamshidi, ed.), China Machine Press, Beijing, China, pp. 1-28, 2013 (In Mandarin).
31. F. Sahin, M. JAMSHIDI, and P. Sridhar, "A System of Systems simulation framework and its applications," Chapter 4, in *System of Systems Engineering – Principles and Applications*, (M. Jamshidi, ed.), China Machine Press, Beijing, China, pp. 77-98, 2013 (In Mandarin).
32. F. Sahin, B. Horan, S. Nahavandi, V. Raghavan, and M. JAMSHIDI, "System of autonomous rovers and their applications," Chapter 14, in *System of Systems Engineering – Principles and Applications*, (M. Jamshidi, ed.), China Machine Press, Beijing, China, pp. 301-315, 2013 (In Mandarin).

4 Technical Reports

1. M. JAMSHIDI, "A New Design Technique of Sub-Optimal Automatic Control Systems Design Using Sensitivity Functions," Coordinated Science Laboratory *CSL Report R-405*, University of Illinois. Urbana, IL., January, 1969.

2. M. JAMSHIDI and P. V. Kokotovic, "An Approximation of Force-Torque Equations in Rolling Mills," *CSL Report R-452,* University of IL., Urbana, IL., January, 1970.
3. M. JAMSHIDI and P. V. Kokotovic, "Dynamic Model of a Cold Rolling Mill," *CSL Report R-459*, University of Illinois, Urbana, IL., February, 1970.
4. M. JAMSHIDI, "An Integrated Near-Optimum Design of Cold Rolling Mills," *CSL Report R-499*, University of Illinois, Urbana, IL., January, 1971.
5. M. JAMSHIDI, "Optimization of Water Resources Systems with Statistical Inflows," *IBM Research Report RC-5720*, Yorktown, NY, November, 1975.
6. M. JAMSHIDI, "Some Computational Algorithms for the Solution of the Matrix Riccati Equations," *IBM Research Report RC-5974*, Yorktown Heights, NY, May, 1976.
7. M. JAMSHIDI, "An Imbedded Initialization of Newton's Algorithm for Matrix Riccati Equation," *IBM Research Report RC-5974*, Yorktown Heights, NY, May 1976.
8. M. JAMSHIDI, "An Input-Output Energy Model for Denmark," *Technical Report,* Electric Power Engineering Dept., Technical University of Denmark, Lyngby, Denmark, July, 1977.
9. M. JAMSHIDI and P. Cueno, "TIMDOM/PC-User's Guide - Software for Multivariable Control Systems," *Technical Report No. LCAD - 84 - 01*, Laboratory for Computer-Aided Design of Systems and Networks, EECE Department, The University of New Mexico, Albuquerque, NM April, 1984.
10. M. JAMSHIDI, "NONLIN-CTR... A Computer - Aided Design Package for Non-Linear Systems," *Report, General Motors Research Laboratories*, Electronics Department, Warren, MI, August, 1984.
11. M. JAMSHIDI and R. Banning, "LSSPAK/PC - User's Guide Software for Large - Scale Control Systems," *Technical Report No. LCAD - 85 - 01*, CAD Laboratory for Systems / Networks, EECE Department, University of New Mexico, Albuquerque, NM January, 1985.
12. J. L. Schotik, M. JAMSHIDI, and T. C. Yenn, "*CONTROL.Lab User's Guide A CAD Language for Multivariable Control and Kalman Filtering*," Technical Report No. LCAD-85-02, CAD Laboratory for Systems / Robotics, EECE Department, University of New Mexico, Albuquerque, NM, March, 1985.
13. M. JAMSHIDI, "TIMDOM/PC User's Guide - A CAD Package for Multivariable Control," *Technical Report No. LCAD-85-04*, CAD Laboratory for Systems/ Robotics, EECE Department, University of New Mexico, Albuquerque, NM, July, 1985.
14. M. JAMSHIDI, "FREDOM/PC A Preliminary Guide to Operation," *Technical Report LCAD 86-01*, CAD Laboratory for Systems / Robotics, UNM, Albuquerque, NM, 1986.
15. F. Kia and M. JAMSHIDI, "POLPAC/PC User's Guide - A CAD Package for Pole Placement," Tec*hnical Report No. LCAD-86-04,* May, 1986.
16. M. JAMSHIDI and G. L. Schotik, "*CONTROL.Lab - A CAD Language for Systems Engineering," Technical Report No. 87-01*, CAD Lab. Systems / Robotics, EECE Dept., UNM, Albuquerque, NM, January, 1987.
17. M. JAMSHIDI, J. A. Meinhardt, R. A. Carreras, and M. G. Baciak, "Model Reference Adaptive Control of a Phased-Array Telescope," *CAD Lab. Report LCAD-88-03*, UNM, Department of EECE, Albuquerque, NM, December, 1988.
18. W. Horne, K. Smith, B. McClung, and M. JAMSHIDI, "A Connection Network for Robotic Gripper Control," *Technical Report LCAD-88-02*, CAD Lab. Systems / Robotics, EECE Dept., University of New Mexico, Albuquerque, NM, August, 1988

5 Special Issues of Technical Journals

1. M. JAMSHIDI and C. J. Herget, (eds.), Special Issue on Computer-Aided Control Systems Engineering, *IEEE Control Systems Magazine*, Vol. 2, No. 4, November, 1982.
2. M. JAMSHIDI, G. S. Axelby, O. I. Franksen, and N.B. Nichols, (eds.), Special IEEE Centennial Issue, *IEEE Control Systems Magazine*, Vol. 4, No. 4, November, 1984.
3. M. JAMSHIDI, (ed.) Special Issue on "Multi-Arm Robotics," Robotics and Autonomous Systems, Guest Editor, vol. 5, no. 4, 1989.
4. M. JAMSHIDI, (ed.) Special Issue on "Robots in Unstructured Environments," *J. of Robotics and Automation*, Guest Editor, 1990.
5. M. JAMSHIDI, (ed.) Special Issue on "Fixed and flexible robots," *Journal. Robotic Systems*, Vol. 6, no. 4, 1989.
6. M. JAMSHIDI, (ed.) Special Issue on "Robots in Manufacturing," *Robotics and Computer Integrated Manufacturing*, Guest Editor, Vol. 6, No. 4, 1989.
7. M. JAMSHIDI, (ed.) Special Issue on "Intelligence in Robotics and Manufacturing," Guest Editor, *J. of Intelligent and Robotic Systems*, Vol. 3, No. 1, 1990.
8. M. JAMSHIDI and V. Salminen (eds.) Special Issue on "Mechatronics," *Int. J. Computers & Electrical Engineering*, Vol. 18, No. 3., 1992.
9. R. Lumia and M. JAMSHIDI, (eds.) Special Issue on "Robotics and Manufacturing Research at National Institute for Standards and Technology," *J. Intelligent and Robotics Systems*, Vol. 4, 1992.
10. M. JAMSHIDI and M. Bayoumi, (eds.) Special Issue on "Robot Control - Position, Force, and Impedance," *J. Intelligent and Robotics Systems*, Vol. 5, 1993.
11. M. JAMSHIDI and C. Gosselin, (eds.) Special Issue on "Trends in Robot Kinematics, Dynamics, and Control," *J. Robotics and Autonomous Systems*, 1992.
12. M. JAMSHIDI (ed.) Special Issue on "Robots in Radioactive and Biochemical Waste Management," *J. Robotics and Autonomous Systems*, 1992.
13. M. JAMSHIDI, D. Cherchas, and R. D. Colbaugh (eds.) Special Issue on "Flexible and Redundant Robots, " *J. Robotics and Computer-Integrated Manufacturing*, Vol. 9, No. 2, 1992.
14. M. JAMSHIDI (ed.) Special Issue on "Robotics and CIM Education and Research," *J. Robotics and Computer-Integrated Manufacturing,* 1992.
15. M. JAMSHIDI and R. D. Colbaugh (eds.) Special Issue on "Redundant Robots," *J. Robotic Systems*, 1993.
16. M. JAMSHIDI and S. Hayati (eds.) Special Issue on "Robots in Unstructured Environments," *J. Robotic Systems*, 1992.
17. M. JAMSHIDI (ed.) Special Issue on "Intelligent Robotic and Manufacturing Systems, "*J. Expert Systems and Applications*, 1992.
18. M. JAMSHIDI (ed.) Special Issue on "Automation in Manufacturing Systems," *Int. J. Systems Automation Research and Applications,* 1992.
19. M. JAMSHIDI and P. Dauchez, (eds.) Special Issue on "Parallel and Multiple Manipulators," *Int. J. Robotics and Automation,* 1993.
20. White, R. Brittain, R. Kisner, and M. JAMSHIDI, (eds.) Special Issue on "Advance Control Architectures for Nuclear Reactor," *Control Theory and Advanced Technologies,* Vol. 8, September, 1992.
21. M. JAMSHIDI and C. C. Nguyen, (eds.) Special Issue on "Parallel Robots and Mechanisms, "*Journal of Robotic Systems*, 1993.
22. M. JAMSHIDI, (ed.) Special Issue on "Environmental and Intelligent Manufacturing Systems" *Journal of Intelligent Manufacturing*, 1994.

23. M. JAMSHIDI Special Section on "Advances in Laboratory Robotic Automation," *Journal of Laboratory Robotic Automation*, Vol... 6, 1994.
24. M. JAMSHIDI (ed.) Special Issue on "Intelligent Robotics and Manufacturing," *J. Robotics and Computer-Integrated Manufacturing*, 1994.
25. M. JAMSHIDI, F.-Y. Wang, and Z. Geng, (eds.) Special Issue on "Intelligent Systems Architecture," *Journal of Intelligent and Fuzzy Systems*, Vol. 2, 1994.
26. M. JAMSHIDI (ed.) Special Issue on "Parallel and Multi-Arm Robotic Systems," *J. Robotics and Autonomous Systems*, 1994.
27. M. JAMSHIDI (ed.) Special Issue on "Mobile, Redundant, and Telerobots Control, "*Journal of Robotica*, 1995.
28. E. Tunstel and M. JAMSHIDI, Guest Editors, Special Issue on Autonomous control Engineering at the NASA ACE Center, *Intelligent Automation and Soft Computing - AutoSoft*, Vol. 3, No. 1, 1997.
29. M. Kamel and M. JAMSHIDI, (eds.) Special Issue on "Distributed Intelligent Systems," *Journal of Intelligent Automation and Soft Computing*, Vol. 6, No. 3, 2000, pp. 171-172.
30. M. Kamel and M. JAMSHIDI, (eds.) "*Intelligent Agents," Intelligent Manufacturing*, 2000.
31. M. JAMSHIDI and J. Johnson III, "Science and Technology Research at NASA University Research Centers," *Int. Journal of Computers and Electrical Engineering*, Vol. 26, No. 1, 2000.
32. M. JAMSHIDI (Ed.), "Special Issue on 6-decades of Contributions of Professor Lotfi A. Zadeh to Science, Emerging Technology and Academic Leadership," *Iranica Scientia*, Vol. 18, No. 3-D1, 2011.
33. M. JAMSHIDI, H. Berenji, and S. Shahbazova (eds.), "Special Issue on Fuzzy Set Theory and Applications- dedicated to Lotfi Zadeh," Applied and Computational mathematics, Vol. 10, 2011, Vol. 10, 2011.

6 Reviewed Conference Papers

1. M. JAMSHIDI, "A Near-Optimal Controller for Nonlinear Systems," *Proc. 7th Allerton Conf.,* Monticello, IL., October 1969.
2. C. Hadlock, M. JAMSHIDI, and P. Kokotovic, "A Near-Optimum Design of Three Time-Scale Systems," Proc. 4th Princeton Conf., Princeton, NJ, March, 1970.
3. P. V. Kokotovic and M. JAMSHIDI, "Optimal Tension Regulation of a Strip Winding Process," *Proc. 1970 JACC*, Atlanta, GA, June, 1970, pp. 1-6.
4. G. D'Ans, Y. Hontoir and M. JAMSHIDI, "Manifold-Imbedding Solution of Optimal Control Problems," *Proc. 8th Allerton Conf.,* Monticello, IL., October, 1970.
5. P. V. Kokotovic and M. JAMSHIDI, "Optimal Control of Cold Rolling Mills," *Proc. 3rd IFAC/IFIP Conf. on Digital Computer Control of Industrial Processes*, Helsinki, Finland, June, 1972.
6. R. Herbrik and M. JAMSHIDI, "Design of an Optimal Regulator for a Once-through Boiler," *Proc. 5th World IFAC Congress*, Paris, France, June, 1972.
7. M. JAMSHIDI, "Sub-Optimal Control of Coupled Time-Delay Systems," *Proc. 10th Allerton Conf.*, Monticello, IL., October, 1972.
8. M. JAMSHIDI, "On the Imbedding Solution of a CSTR Process by Hybrid Computer," *Proc. 1st Iranian Congress on Chemical Engineering*, Shiraz, Iran, May, 1973 (Elsevier Publishers, Amsterdam, the Netherlands)

9. M. JAMSHIDI, "On the Optimal Control of Nonlinear Power Systems," *Proc. IFAC Symp. Control in Power Electronics and Electrical Derives,* Deusseldorf, FRG., October, 1974.
10. M. JAMSHIDI, "A Systematic Approach to Near Optimum Design of Nonlinear Time-Delay Systems," *Proc. 12th Allerton Conf., Monticello*, IL., October, 1974.
11. M. JAMSHIDI, "On the Analog/Hybrid Simulation of Prey-Predator Systems," *Proc. AICA Symposium on Hybrid Computation and Dynamic Systems Design*, Rome, Italy, November, 1974.
12. M. JAMSHIDI, "On the Analog Simulation of Mathematical Programming Problems," *Proc. Simulation 75*, Zurich, Switzerland, June, 1975.
13. M. JAMSHIDI, "Optimal Control of Some Dynamic Operations Research Processes by Analog Computer" *Proc. Simulation 75*, Zurich, Switzerland, June, 1975.
14. M. JAMSHIDI and M. Mohseni, "On the Optimization of Water Resources Systems with Statistical Inputs," *Proc. 2nd IFIP Working Conf., on Biosystems Simulation in Water Resources and Waste Water Pro*blems, Gent, Belgium, September, 1975.
15. M. JAMSHIDI and M. Heidari, "Modeling and Optimization of Khuzestan Water Resources Systems," *Proc. 6th World IFAC Congress*, Boston, MA, August, 1975.
16. M. JAMSHIDI and M. H. Nehrir, *Proceedings 4th Iranian Conference (1st Congress) on Electrical Engineering,* (Eds.) Shiraz, Iran, May, 1974.
17. M. JAMSHIDI and M. Razzaghi, "On the Imbedded Solution of Linear Singular Control Problems with Delay" *Proc. 13th Allerton Conf.,* Monticello, IL, October, 1975.
18. M. JAMSHIDI and F. Boettiger, "On the Imbedded Solutions of the Algebraic Matrix Riccati Equations," *Proc. 1976 JACC*, W. Lafayette, IN, July, 1976.
19. J. Peters, K. C. Chu, and M. JAMSHIDI, "Optimization of a Water Resources System by Stochastic Programming with Resource and Linear Rules," *Proc. IIX Int. Sym. on Mathematical Programming*, Budapest, Hungary, August, 1976.
20. U. Ozguner and M. JAMSHIDI, "Multi-Time Scale Analysis and Control in Microeconomics," *Proc. IFAC/IFORS/IASA Conf. on Dynamic Modeling and Control of National Economies*, Vienna, Austria, January, 1977.
21. M. JAMSHIDI and U. Ozguner, "An Optimal Adaptive Policy for Long-Term Economic Stabilization Problem," *Proc. IFAC/IFORS/IASA Conf. On Dynamic Modeling and Control of National Economies*, Vienna, Austria, January, 1977.
22. M. JAMSHIDI and I. Vakilzadeh, "On some of the Problems with Control Education in Developing Countries," *Proc. IFAC Sym. on Trends in Automatic Control Education*, Barcelona, Spain, March, 1977.
23. K. C. Chu, M. JAMSHIDI, and R. E. Levitan, "Real-Time Urban Power Dispatch with Ambient Air Equality Constraints," *Proc. IFAC Sym. On Environmental Systems Planning, Design and Control*, August, 1977, Kyoto, Japan.
24. M. JAMSHIDI, "A Large Scale Stochastic Model for Optimal Management of Khuzestan Water Resources System," *Proc. IFAC Conf. On Systems Approaches for Development*, Cairo, Egypt, November, 1977.
25. J. Peters, K. C. Chu, and M. JAMSHIDI, "Optimal Operation of a Water Resources System Stochastic Programming," *Proc. VIIth IFAC World Congress,* Helsinki, Finland, June, 1978.
26. M. JAMSHIDI and J. R. Hansen, "a Dynamic Input-Output Energy Model for Denmark," *Proc. Simulation,. Modeling & Decision in Energy System*, Montreal, Canada, June, 1978.

27. M. JAMSHIDI and R. J. Heggen, "A multi-level stochastic management model for optimal conjunctive use of ground and surface water," Proc. IFAC Symposium Water & Land Related Resources Systems, May 1980, Cleveland, OH.
28. M. JAMSHIDI and S. Karni, "Resource sensitivity and stochastic input/output modeling of energy systems," (invited paper), 1980 Optimization Days, Montreal, Canada, May 1980.
29. M. JAMSHIDI and R. J. Heggen, "Optimal management of water resources systems: A survey," (invited paper), 1980 Optimization Days, Montreal, Canada, May 1980.
30. M. JAMSHIDI and M. Malek-Zavarei, "A Hierarchical optimization method of large-scale time-delay systems," Proc. 20th IEEE CDC, San Diego, CA, Dec. 16-18, 1981.
31. M. JAMSHIDI, "An overview on the aggregation of large-scale systems," Proc. VIII IFAC Congress, Kyoto, Japan, Aug. 24-28, 1981, paper no. 44.1.
32. M. JAMSHIDI, "On the reduction of large-scale systems in the frequency domain," Proc. 24th Midwest Symposium, Jun. 29-28, 1981, Albuquerque, NM, paper no. TA4.1.
33. R. J. Heggen and M. JAMSHIDI, "A Hierarchical control algorithm for water resources systems," Proc. IASTED Conf., Geneva, Switzerland, Feb. 1981.
34. M. JAMSHIDI and J. E. Merryman, "On the hierarchical optimization of retarded systems via costate prediction," Proc. ACC, vol. 3, no. 3, pp. 899-904, Arlington, VA, Jun. 1982.
35. M. JAMSHIDI and C. M. Wang, "A computational algorithm for large-scale nonlinear time-delay systems," Proc. 1st IEEE Symposium on Large-Scale Systems, Virginia Beach, VA, Oct. 1, 1982.
36. M. JAMSHIDI and F. Asamoah, "On exponential stabilization of a class of large-scale bilinear systems," Proc. 1st Symposium on Large-Scale Systems, Virginia Beach, VA, Oct. 1982.
37. T. Portas and M. JAMSHIDI, "A computer-based hierarchical model for a large-scale energy system," Proc. 3rd IFAC/IFORS, Warsaw, Poland, Jul. 1983.
38. R. E. Salters and M. JAMSHIDI, "Two interactive programming packages for control systems," Proc IEEE MIT (CAD) Symposium, Cambridge, MA, 1983.
39. Morel and M. JAMSHIDI, "Two interactive programming packages for control systems," Proc. IEEE / MIT (CAD) Symposium, Cambridge, MA, 1983.
40. M. JAMSHIDI, R. E. Owen, and J. E. Cunningham, Jr., "Modeling and simulation of photoconductor discharge voltage control system," Proceedings IBM Workshop on Electrophotographic Printhead Systems, Jul. 28-29, 1983, Boulder, Co.
41. M. Santiago, J. W. Lange, and M. JAMSHIDI, "Trends in modeling and control of large space structures," Workshop on Identification and Control of Flexible Space Structures, San Diego, CA, Jun. 4-5, 1984.
42. M. JAMSHIDI and C. M. Wang, "Optimal operation of large-scale water resources systems," Proc. 9th IFAC Congress, Budapest, Hungary, Jul. 2-6, 1984.
43. M. JAMSHIDI and C. M. Wang, "Hierarchical optimization of large-scale water resources systems," Proc. First European Workshop on the Real Time Control of Large-Scale Systems, Patras, Greece, Jul. 10-12, 1984.
44. M. Santiago, J. W. Lange, Jr., and M. JAMSHIDI, "An overview of modeling and control of large flexible space-structures," Proc. 9th IFAC Congress, pp. 299-304, Budapest, Hungary, Jul. 2-6, 1984.
45. M. JAMSHIDI and M. Etezadi, "On the decentralized control of large-scale power systems," Proc. 1984 ACC, Jun. 6-8, 1984, pp. 1156-1161.
46. M. JAMSHIDI, "Problems of technical development in developing countries," Proceedings IFAC World Congress, Budapest, Hungary, 1984.

47. M. JAMSHIDI and S. Karni, "A hierarchical optimization of large-scale energy resources systems," Proceedings IEEE Midwest Symposium, Morgantown, WV, 1984.
48. M. JAMSHIDI, "Hierarchical optimization of large-scale time-delay systems - A brief overview," Proc. 1984 ACC, San Diego, CA, Jun. 1984, pp. 129-134.
49. H. Seraji, M. Shahinpoor, and M. JAMSHIDI, "Perfect tracking with application to robotics," Proc. Int. Conf. Robotics, Luzanne, Switzerland, June, 1986.
50. J. M. Santiago and M. JAMSHIDI, "Simple extensions of the open-loop balanced approach for model reduction," Proc. 1985 ACC, Boston, MA, Jun. 1985.
51. M. JAMSHIDI, G. L. Schotik, and T. C. Yenn, "CONTROL.lab - a CAD language for multivariable control systems," Proc. IFAC Conference on Systems Approach for Development, Beijing, China, Aug. 1985.
52. M. JAMSHIDI, S. Karni, Y. C. Yenn, and G. L. Schotik, "Two CAD languages for control systems and electrical networks," Proc. IFAC Conference on Computer-Aided Design, Lyngby, Denmark, Jul. 1985.
53. H. Seraji, M. Shahinpoor, and M. JAMSHIDI, "Perfect output control with applications to robotics," Proc. IEEE Conf. Systems Man and Cybernetics, Atlanta, GA, 1986.
54. M. JAMSHIDI, H. Seraji, M. Shahinpoor, and Y. T. Kim, "Regulation of two-link robot manipulators," Proc. IEEE Conference on Systems, Man and Cybernetics, Tucson, AZ, Nov. 1985.
55. M. JAMSHIDI, H. Seraji, and Y. T. Kim, "On decentralized control of nonlinear three-link robots," Proc. 9th Asilomar Conf., Nov. 1985, Monterey, CA.
56. M. Tarokh and M. JAMSHIDI, "Fixed mode elimination with minimum information exchange among controllers," Proc. IEEE Conf. Dec. Contr., Athens, Greece, Dec. 1986.
57. M. Shahinpoor, H. Kalhor, and M. JAMSHIDI, "On magnetically activated robotic tensor arms," Proc. Int. Symp. Robot Manipulators: Modeling, Control, and Education, M. Jamshidi, et al. (eds.), Albuquerque, NM, pp. 517-512, Nov. 1986.
58. M. Shahinpoor and M. JAMSHIDI, "Proc. Int. Symp. on Robot Manipulators: Modeling, Control, and Education," Albuquerque, NM, pp. 351-353, Nov. 1986
59. M. JAMSHIDI, Y. T. Kim, and M. Shahinpoor, "A near-optimum control of robot manipulator," Proc. 10th World IFAC Congress, Munich, FRG, July 26-Aug. 1.
60. M. Tarokh, B. J. Oh, and M. JAMSHIDI, "An iterative method for pole assignment using output feedback," Proc. 1987 MTNS, Tempe, AZ, June 1987.
61. K. W. Smith, W. Horne, B. McClung, A. Young, and M. JAMSHIDI, "Robotic control in a connection neural like environment," Proc. IEEE Int. Conf. Intelligent Control, Philadelphia, PA, Jan. 1987.
62. W. Horne, K. W. Smith, B. McClung, and M. JAMSHIDI, "Application-oriented connection systems for pattern acquisition and recognition," Proc. IEEE Int. Conf. on Neural Networks, San Diego, CA, Jun. 1987.
63. J. Oh, H. Seraji, and M. JAMSHIDI, "Decentralized adaptive control of robotic manipulators," Proc. 1987 MTNS, Tempe, AZ, Jun. 1987.
64. M. Tarokh and M. JAMSHIDI, "On pole assignment by decentralized output feedback," Proc. IEEE SMC Conference, Alexandria, VA, Oct. 1987.
65. J. Oh, H. Seraji, and M. JAMSHIDI, "Decentralized adaptive control of robotic manipulators," Proc. 1987 MTNS, Phoenix, AZ, 1987.
66. J. Oh and M. JAMSHIDI, "A Decentralized adaptive control of robot manipulators with feedforward loop," Proc. IEEE SMC Conference, Alexandria, VA, Oct. 1987.
67. C. S. Tseng and M. JAMSHIDI, "On robust decentralized control of a five-axis robot," Proc. 1987 MTNS, Tempe, AZ, Jun. 1987.

68. J. Oh, M. JAMSHIDI, and H. Seraji, "Decentralized adaptive control," *Proc. IEEE Conf. Robotics and Automation,* Philadelphia, PA, Apr. 1988.
69. M. JAMSHIDI and C. S. Tseng, "On the computer-aided modeling and control of robot manipulators-ROBOTLAB," *Proc. 4th IFAC Symp. On Computer Aided Design in Control Systems*, Beijing, P. R. China, Aug. 1988.
70. J. Oh and M. JAMSHIDI, "A decentralized adaptive control of robot manipulation," *Proc. IEEE SMC Conference*, Alexandria, VA, Oct. 1987.
71. R. Lee and M. JAMSHIDI, "On Lyapunov stability and instability of large-scale nonlinear systems with time delay," *Proc. Asilomar Conf.,* Oct., 1988, Monterey, CA.
72. S. O'Neill and M. JAMSHIDI, "Robot_S: An interactive design and simulation language for robot manipulators," *Proc. 4th IFAC Symposium on CAD*, Aug. 1988, Beijing, P. R. China.
73. N. Vadiee and M. JAMSHIDI, "A design philosophy for multilayer artificial neural networks with applications to robot control," *Proc. 2nd NASA Conf. Telerobotics*, Pasadena, CA, Jan. 1989.
74. W. Horne and M. JAMSHIDI, "A connection network for robotic gripper control," *Proc. 27th IEEE Conf. Dec. and Control*, Austin, TX, Dec.1988, pp. 1070-1075.
75. M. JAMSHIDI and C. S. Tseng, "On the computer-aided modeling and control of robot manipulators - ROBOTLAB," *Proc. 4th IFAC Symp. On Computer Aided Design in Control Systems,* Beijing, P. R. China, Aug. 1988.
76. Z. Geng and M. JAMSHIDI, "An expert self-learning controller for robot systems," *Proc. 27th IEEE Conf. Dec. and Control*, Austin, TX, December, 1988, pp. 1090-1095.
77. Z. Geng and M. JAMSHIDI, "Design of self-learning controllers using expert system techniques," *Proc. 3rd IEEE Symposium on Intelligent Control*, Arlington, VA, Aug. 1988.
78. M. Aldeen and M. JAMSHIDI, "A new decentralized control algorithm via model reduction," *Proc. IFAC Symposium on Large-Scale Systems*, Berlin, GDR, Aug. 1988.
79. M. Aldeen and M. JAMSHIDI, "Decentralized control via static and dynamic compensation," *Proc. 1989 ACC*, Pittsburgh, PA.
80. J. Oh and M. JAMSHIDI, "Decentralized adaptive feedforward/feedback robot manipulator control," *Proc. IFAC/IFORS/IMACS Symposium on Large-Scale Systems - Theory and Applications*, Berlin, GDR, Aug. 1989.
81. M. JAMSHIDI, J. A. Meinhardt, R. Carreras, and M. Baciak, "Adaptive control of a phased array telescope," *Proc. SPIE Int. Society of Optical Engineering Technical Symposium on Aerospace Sensing*, Orlando, FL, Mar., 1989.
82. A. Meinhardt, C. DeHainaut, M. JAMSHIDI, and D. Marker, "Active pupil geometry control in a phased array telescope," *Proc. SPIE-The Int. Optical Society*, Mar., 1989.
83. H. Bolandi, R. L. Carroll, and M. JAMSHIDI, "On the discrete-time model reference adaptive control of a two-link robot," in *Robotics and Manufacturing.* M. Jamshidi, et al, (eds.), Vol. 2, ASME Press, 1988, pp. 203-210.
84. J. N. Liou, M. JAMSHIDI, and G. P. Starr, "On the adaptive edge-following force control of a PUMA 560 robot," in *Robotics and Manufacturing*, M. Jamshidi, et al, (eds.), vol. 2, ASME Press, 1988, pp. 227-236.
85. W. Horne and M. JAMSHIDI, "A connection network for robotic gripper control," in *Robotics and Manufacturing*, M. Jamshidi, et al, (eds.), Vol. 2, ASME Press, 1988, pp. 257-263.
86. Z. Geng and M. JAMSHIDI, "Two-dimensional system models for learning control systems," in Robotics and Manufacturing, M. Jamshidi, et al (eds.), ASME Press, vol. 2, 1988, pp. 273-280.

87. J. Benitez-Read, M. JAMSHIDI, and Z. Geng, "On the adaptive control of a nuclear reactor," Proc. 7th Power Plant Dynamics, Control and Testing Symposium, Knoxville, TN, May, 1989.
88. Z. Huang, M. JAMSHIDI, and H. Xiong, "On linear aggregation in input-output analysis," Proc. National Systems Conference, Beijing, China, 1987 (in Chinese).
89. M. Jacobus, M. JAMSHIDI, P. Dorato, C. Abdallah, and D. S. Bernstein, "Suboptimal strong stabilization using fixed-order dynamic compensation," Proc. American Control Conference, San Diego, CA, May 1990.
90. C. Abdallah, P. Dorato, and M. JAMSHIDI, "Survey of robust control of robots," Proc. American Control Conf., San Diego, CA, May 1990.
91. Z. Geng, R. L. Carroll, and M. JAMSHIDI, "On the learning control of nuclear reactors by a 2-D system theory," Proc. American Control Conference, San Diego, May 1990.
92. T. Tipton, J. A. Meinhardt, and M. JAMSHIDI, "Control system description and performance of a phased array telescope," Proc. SPIE Conf. Active Telescope Control, Tucson, AZ, Feb., 1990.
93. T. Sarkodie-Gyan and M. JAMSHIDI, "Development of a prototype intelligent robot task planning algorithm using sensor fusion," Robotics and Manufacturing, M. Jamshidi and M. Saif, (eds.) vol. 3, ASME press, 1990.
94. J. N. Liou and M. JAMSHIDI, "On the adaptive control of a contour-following system," Robotics and Manufacturing, M. Jamshidi and M. Saif, (eds.) vol. 3, ASME Press, 1990.
95. M. JAMSHIDI and J. H. Mullins, "The New Mexico program for manufacturing engineering education," Proc. UPCADEM Symposium, Georgia Inst. Technology, Atlanta, GA, June 1988.
96. Z. Geng, R. Carroll, M. JAMSHIDI, and J. Xie, "Two dimensional model and algorithm for a class of iterative learning control system," Proc. ACC, 1990.
97. T. Sarkodie-Gyan and M. JAMSHIDI, "An intelligent robot task planning algorithm using sensory fusion", Proc. German National Conf. On Instrumentation, November, 1991.
98. M. JAMSHIDI, W. Horne, and N. Vadiee, "A neural network-based controller for two link robots," Proc. IEEE CDC, Honolulu, HI, December, 1990.
99. M. Jacobus, M. JAMSHIDI, C. Abdallah P. Dorato, and D. S. Bernstein, "Design of strictly positive real, fixed-order dynamic compensators," Proc. IEEE CDC, Honolulu, HI, December, 1990.
100. M. JAMSHIDI and Z. Geng , "A two-dimensional learning controller for robot manipulators," Proc. 4th ANS Topical Meeting on Robotics and Remote Systems, M. Jamshidi and P. Eicker (eds.), US Government Printing Office, Washington, DC., Feb., 1991.
101. M. JAMSHIDI and Z. Geng, "An Expert Learning Approach for Robot Manipulator Control," Proc. 4th ANS Topical Meeting on Robotics and Remote Systems, M. Jamshidi and P. Eicker (eds.), US Government Printing Office, Washington, DC., Feb., 1991, pp. 41-60.
102. J. Benitez-Read, M. JAMSHIDI, and R. Kisner, "On the advanced control techniques for a nuclear power system," Proc. ACC, Boston, MA, June, 1991.
103. Z. Geng, R. L. Carroll, M. JAMSHIDI, and R. A. Kisner, "A learning control scheme with gain estimator," Proc. IEEE Symposium on Intelligent Control, Arlington, VA, August, 1991.
104. W. Honey and M. JAMSHIDI, "ROBO_SIM - A robotic simulation environment on a personal computer" Proc. IFAC Conference on Robots, Vienna, Austria, September, 1991.

105. Z. Geng, R. L. Carroll, and M. JAMSHIDI, "Learning control of 2 D systems," _Proc. IEEE CDC_, Brighton, England, December, 1991.
106. M. JAMSHIDI, S. Baugh, D. Barak, and N. Vadiee, "A Comparison of an Expert and an Adaptive Control Approach," _Proc. IEEE CDC_, Brighton, England, December, 1991.
107. R. D. Colbaugh, K. Glass, and M. JAMSHIDI, "Adaptive Compliant Control of Robots," _Proc. IEEE CDC_, Brighton, England, December, 1991.
108. M. JAMSHIDI, D. Barak, N. Vaidee, and S. Baugh, "A Simulation Environment for Adaptive Fuzzy Control Systems," _Proc. Society of Computer Simulation Conference_, New Port Beach, CA, January 20, 1992.
109. M. JAMSHIDI, T. Ross, D. Barak, S. Baugh, and N. Vadiee, "Fuzzy Control Systems: Laboratory Experiments for an Engineering Curriculum," _Proc. Intelligent and Fuzzy Control Systems Conference_, Louisville, KY, March 16-18, 1992.
110. M. JAMSHIDI, D. Barak, S. Baugh, and D. O'Geary, "Operations and Control of a Reliable Non-Chlorofuorocarbon Based Air Conditioning System via Fuzzy Logic," _Proc. IFAC Workshop on Intelligent Manufacturing_, Dearborn, MI, USA, October, 1992.
111. D. Barak, S. Baugh, and M. JAMSHIDI, "A Fuzzy Control of Thermal Systems," Robotics and Manufacturing, M. Jamshidi, et al. (eds.), ASME Press, NY., Vol. 4, 1992, pp. 649-654.
112. R. Marchbank and M. JAMSHIDI, "Fuzzy Tracking Control of a Laser Beam System, Robotics and Manufacturing, M. Jamshidi, et al. (eds.), ASME Press, NY., Vol. 4, 1992, pp. 681-686.
113. K. Kumbla, J. Moya, R. Baird, S. Rajagopalan, and M. JAMSHIDI, "Fuzzy Control of Three Links of a Robotic Manipulator, " Robotics and Manufacturing, M. Jamshidi, et al. (eds.), ASME Press, NY., Vol. 4, 1992, pp. 687-694.
114. M. JAMSHIDI and R. Marchbank, "On the Fuzzy Control of Optical Systems," Proc. Int. Symposium on Fundamentals of Discrete-Time Systems, Chicago, IL, June, 1992.
115. M. Akbarzadeh-T., M.-R., M. JAMSHIDI, and N. Vadiee, "A Hierarchical Fuzzy Controller Using Line-Curvature Feature Extraction for a Single Flexible Arm," Proceedings of the Third IEEE Conference on Fuzzy Systems, Orlando, Florida, 1993.
116. M. JAMSHIDI, E. Krijstansson, "On Fuzzy Control of Electric Power generation," Proc. IFAC World Congress, Sydney, Australia, July, 1993.
117. M. JAMSHIDI, R. Kelsey, and K. Bisset, "Fuzzy Traffic Control," Proc. IFAC World Congress, Sydney, Australia, July, 1993.
118. E. Krijstansson and M. JAMSHIDI, "A Comparative Study of Fuzzy and Crisp Control of Power Systems, " _Proc. ICEE '93_, Tehran, Iran, May, 18-21, 1993.
119. M. JAMSHIDI, R. Kelsey, and K. Bisset, "Fuzzy Logic Based Traffic Control - Software and Hardware Experiments, _Proc. ICEE '93_, Tehran, Iran, May 18-21, 1993.
120. E. Tunstel and M. JAMSHIDI, "Fuzzy Control of a Hovercraft," _Proc. 3rd IFIS_, Houston, TX, December 1-3, 1993.
121. D. Peterson and M. JAMSHIDI, "Fuzzy Pattern Recognition of Geometric Objects," _Proc. 5th IFSA_, Seoul, Korea, July 4-9, 1993, pp. 135-138.
122. K. Kumbla and M. JAMSHIDI, "Fuzzy Control of A Robot manipulator," Proc. 5th IFSA, Seoul, Korea, July 4-9, 1993, pp. 1410-1413.
123. M. JAMSHIDI, R. Kelsey and K. Bisset, "Fuzzy Traffic Control - Software and Hardware Implementations," Proc. 5th IFSA, Seoul, Korea, July 4-9, 1993, pp. 907-910.
124. M. Akbarzadeh-Totoonchi, H. Xue, and M. JAMSHIDI, " Model Reduction of a Flexible Arm Using Pade'-Routh Approximation, "Proc. ICEE '93, Tehran, Iran, May 18-20, 1993.

125. K. K. Kumbla and M. JAMSHIDI, "Implementation of Fuzzy Logic and Neural Networks Control Algorithm Using a TMS320 DSP Chip," *First Industry/Academy Symposium on Research for Future Supersonic and Hypersonic Vehicles*, A. Homaifar and J. C. Kelly (eds.), Albuquerque, NM: TSI Press, 1994, pp. 124-129.
126. E. Tunstel and M. JAMSHIDI, "Fuzzy Relational Representation of Uncertain Spatial Maps for Autonomous Vehicles," *First Industry/Academy Symposium on Research for Future Supersonic and Hypersonic Vehicles*, A. Homaifar and J. C. Kelly (eds.), Albuquerque, NM: TSI Press, 1994, pp. 1262-167.
127. E. Tunstel and M. JAMSHIDI, "Embedded Fuzzy Logic Based Wall Following Behavior for Mobile Robot Navigation", NAFIPS/IFIS/NASA '94, San Antonio, TX, Dec. 18-20, 1994, pp. 329-330.
128. A. Asgharzadeh and M. JAMSHIDI, "Fuzzy Control of a Video Printer" Proc. SPIE Boston, MA, September, 1993.
129. M. -R. Akbarzadeh-T, M. JAMSHIDI, and R. D. Colbaugh, "Intelligent Control of Flexible and Redundant Manipulators," Waste-Management: From Risk to Remediation, ECM Press Publishers, Albuquerque, New Mexico, 1994.
130. K. Kumbla and M. JAMSHIDI, "Control of Robotic Manipulator Using Fuzzy Logic", *Proc. Third IEEE Conference on Fuzzy Systems*, Orlando, Florida., 1994.
131. M. Akbarzadeh, K. Kumbla, and M. JAMSHIDI, "Intelligent Control of Flexible and Redundant Robot", Waste-management Education and Research Consortium (WERC) Conference, Las Cruses, April 1994.
132. M. JAMSHIDI, C. C. Nguyen, R. Lumia, and J. Yuh (eds.), *Intelligent Automation and Soft Computing,* Preface, Vol. 1, Albuquerque, NM: TSI Press, 1994, pp. vii.
133. M. JAMSHIDI, C. C. Nguyen, R. Lumia, and J. Yuh (eds.), *Intelligent Automation and Soft Computing*, Vol. 2, Albuquerque, NM, 1994, pp. vii.
134. K. Kumbla and M. JAMSHIDI, "Implementation of Fuzzy Logic and Neural Networks Control Algorithm Using a TMS320 DSP Chip", First University/Industry Symposium on High Speed Civil Transport Vehicles, December, 1994., pp. 124-129.
135. K. Kumbla and M. JAMSHIDI, "Implementation of Fuzzy Logic and Neural Networks Control Algorithm Using DSP Chip" Symposium on Applied Computing, Nashville, February 1995.
136. K. Kumbla, M. Akbarzadeh, and M. JAMSHIDI, "TMS320 DSP Chip Based Neuro-Fuzzy controller" IEEE Conference on Man, System and Cybernetics, Vancouver, pp 4015-4020, October 1995.
137. M. Akbarzadeh, K. Kumbla, and M. JAMSHIDI, "Genetic Algorithms in Learning Fuzzy Hierarchical Control of Distributed Parameter 4032, October 1995.
138. K. Kumbla, M. JAMSHIDI, and S. Rodriguez, "Adaptive Neuro-Fuzzy Controller, *First National Student Conference,* NASA, March 1996.
139. K. Kumbla and M. JAMSHIDI, "Intelligent Control of Fault-Tolerant Robots," *Proc. IIZUKA '94*, Iizuka, Japan, August, 1994.
140. M. Akbarzadeh and M. JAMSHIDI, "Fuzzy Control of Flexible Robots," *Proc. IIZUKA '94*, Iizuka, Japan, August, 1994.
141. E. Tunstel and M. JAMSHIDI, "Fuzzy-Neural Control of Mobile Robots," *Proc. IIZUKA '94,* Iizuka, Japan, August, 1994.
142. E. Tunstel and M. JAMSHIDI, "On Embedded Fuzzy Controllers," *Proc. ISRAM '94,* Maui, Hawaii, August, 1994, pp. 619-624.
143. A. Martinez, E. Tunstel, and M. JAMSHIDI, "Collision Avoidance of Mobile Robots Using Fuzzy Logic," *Proc. ISRAM '94*, Maui, Hawaii, August, 1994.

144. N. Vadiee and M. JAMSHIDI, "Advanced Fuzzy Ruled Based Expert Systems, " Proc. ICEE '93, Tehran, Iran, May 18-22, 1993.
145. A. Martinez and M. JAMSHIDI, "A Fuzzy Controller for Engine Idle Speed," Proc. 36th IEEE Midwest Symposium, Detroit, Michigan, August, 1993.
146. H. Hwang, M. Saif, and M. JAMSHIDI, "Fault Detection and Diagnosis of a Nuclear Power Plant Using Artificial Neural Networks" Proc. World IFAC Congress, Sydney, Australia, July, 1993.
147. E. Tunstel and M. JAMSHIDI, "Fuzzy Logic Based Collision Avoidance for a Mobile Robot," Proc. IFIS '93 Conference, Houston, TX, December, 1993.
148. E. Tunstel, S. Hockemeier, and M. JAMSHIDI, "Fuzzy Control for Stabilization of a Hovercraft Platform", Proceedings, Engineering & Architecture Symposium, Workshop, and Exhibition, Prairie View A&M University, Prairie View, Texas, March 21-22, 1994, pp. 294-299.
149. E. Tunstel, A. Martinez, and M. JAMSHIDI, "Mobile Robot Collision Avoidance Using a Fuzzy Logic Controller", Engineering & Architecture Symposium, Workshop, and Exhibition, Prairie View A&M University, Prairie View, TX, March 21-22, 1994, pp. 304-309.
150. E. Tunstel, and M. JAMSHIDI, "Fuzzy Logic and Behavior Control Strategy for Autonomous Mobile Robot Mapping", Proceedings, 3rd IEEE International Conference on Fuzzy Systems, Orlando, FL, June 26-July 2, 1994, pp. 514-517.
151. M. JAMSHIDI, F. Pin, and F. Pierrot, Robotics and Manufacturing: Robotics Research and Applications, Vol. 3, Albuquerque, NM: TSI Press, 1996, pp. viii-ix.
152. M. JAMSHIDI, J. Yuh, and F. Pierrot (eds.) Intelligent Automation and Control, Vol. 4, Albuquerque, NM: TSI Press, 1996, pp. xii-xiii.
153. M. JAMSHIDI, M. Fathi, and P. Dauchez (eds.) Soft Computing for Industry, Vol. 5, Albuquerque, NM: TSI Press, 1996, pp. ix-xi.
154. E. Tunstel, A. Asgharzadeh, and M. JAMSHIDI, "Towards Embedded Fuzzy Control of Mobile Robots", Proceedings, 3rd International Conference on Fuzzy Logic, Neural Nets and Soft Computing, Iizuka '94, Iizuka, Fukuoka, Japan, August 1-7, 1994.
155. A. Asgharzadeh, E. Tunstel, and M. JAMSHIDI, "An Embedded Fuzzy Controller for a Video Printer", Proceedings 3rd International Conference on Fuzzy Logic, Neural Nets and Soft Computing, Iizuka '94, Iizuka, Fukuoka, Japan, August 1-7, 1994.
156. A. Martinez, E. Tunstel , and M. JAMSHIDI, "Collision Avoidance of Mobile Robots Using Fuzzy Logic", Proceedings, 5th Int'l Symposium on Robotics and Manufacturing ISRAM '94, Maui, Hawaii, August 14-17, 1994, pp. 193-198.
157. E. Tunstel, M. JAMSHIDI, "On Embedded Fuzzy Controllers", Proceedings, 1st World Automation Congress (WAC '94), Maui, Hawaii, August 14-17, 1994, pp. 619-624.
158. E. Tunstel and M. JAMSHIDI, "Fuzzy Relational Representation of Uncertain Spatial Maps for Autonomous Vehicles", Proceedings, 1st Industry/Univ. Symp. on High Speed Civil Transport Vehicles (HSCTV), NC A&T State Univ., Greensboro, NC, Dec. 4-6, 1994, pp. 162-167.
159. E. Tunstel and M. JAMSHIDI, "Embedded Fuzzy Logic-Based Wall Following Behavior for Mobile Robot Navigation", Proceedings NAFIPS/IFIS/NASA '94, San Antonio, TX, De. 18-20, 1994.
160. E. Tunstel, M. R. Akbarzadeh-T., K. Kumbla, and M. JAMSHIDI, "Hybrid Fuzzy Control Schemes for Robotic Systems", Proceedings, 10th IEEE Int'l. Symp. on Intelligent Control, Monterey, CA, August 27-29, 1995, pp. 171-176.

161. M. R. Akbarzadeh-T., K. Kumbla, and M. JAMSHIDI, "Genetic Algorithms in Learning Fuzzy Hierarchical Control of Distributed Parameter Systems," *Proceedings, IEEE Conference of Systems, Man and Cybernetics*, Vancouver, Canada, 1995.
162. K. Kumbla, M. R. Akbarzadeh-T., and M. JAMSHIDI, "TMS320 DSP Based Neuro-Fuzzy Controller," *Proceedings, IEEE Conference on Systems, Man and Cybernetics*, Vancouver, Canada, 1995.
163. M. JAMSHIDI and A. Titli, "Stability of Fuzzy Control Systems via Interval Matrix Method", *Proc. IEEE Mediterranean Conference*, Cyprus, July, 1995.
164. M. R. Akbarzadeh-T., E. Medina, and M. JAMSHIDI, "DSP Implementation of Evolutionary Fuzzy Control," *Proceedings of the First National Student Conference, National Alliance of NASA Univ. Research Centers*, NC A&T, Greensboro, NC, March 1996.
165. E. Tunstel, T. Lippincott, and M. JAMSHIDI, "Introduction to Fuzzy Logic Control with Application to Mobile Robotics", *1st National Student Conference, National Alliance of NASA Univ. Research Centers*, NC A&T, Greensboro, NC, March 1996, pp. 174-177.
166. M. JAMSHIDI and C. W. de Silva (eds.), *Intelligent Automation and Control*, Vol. 6, Preface, Albuquerque, NM: TSI Press, 1998, pp. vii-viii.
167. M. JAMSHIDI, F. Pierrot, and M. Kamel (eds.), *Robotics and Manufacturing*, Vol. 7, Preface, Albuquerque, NM: TSI Press, 1998, pp. viii-x.
168. M. JAMSHIDI, Z. Bien, and M. Fathi (eds.), *Soft Computing, Multimedia and Image Processing: Research and Applications*, Preface, Vol. 8, Albuquerque, NM: TSI Press, 1998, pp. viii-x.
169. E. Tunstel, D. E. Gonzales, and M. JAMSHIDI, "Embedded Mobile Robot Control Using Fuzzy Logic Integrated Circuits", *Proceedings, 1st National Student Conference, National Alliance of NASA Univ. Research Centers*, NC A&T, Greensboro, NC, March 1996, pp. 178-181.
170. E. Tunstel, M. R. Akbarzadeh-T, K. Kumbla, and M. JAMSHIDI, "Soft Computing Paradigms for Learning Fuzzy Controllers with Applications to Robotics," *Proceedings of the Biennial Conference of the North American Fuzzy Information Processing Society*, 1996, pp. 355-359.
171. E. Tunstel and M. JAMSHIDI, "Hierarchical Fuzzy Control Approach to Adaptive Behavior Synthesis in Mobile Robots," *Proceedings, From Animals to Animals: 4th Int'l Conference on Simulation of Adaptive Behavior*, Cape Cod, MA, September 1996.
172. A. S. Heger, M. JAMSHIDI, and N. K. Alang-Rashid, "Self-Tuning Fuzzy Logic Nuclear Reactor Controller," FLINS-96, *Second International Workshop on Intelligent Systems and Soft Computing for Nuclear Science and Industry*, Mol, Belgium, (September 25-27, 1996)
173. M. R. Akbarzadeh-T., M. JAMSHIDI, and P. Dorato, "Fuzzy Hierarchical Control of Distributed Parameter Systems, A Case Study on a Heating Slab," *Proceedings of the Applied Computing Symposium on Applied Computing*, Nashville, Tennessee, 1995.
174. M. JAMSHIDI, M-R. Akbarzadeh, and K. Kumbla, "Design and Implementation of Fuzzy Controllers for Complex Systems - Case Study: A Water Desalination Plant," *Proceedings IFIS 96*, Cancun, Mexico, November 12-15, 1996.
175. M. R. Akbarzadeh-T., and M. JAMSHIDI, "Evolutionary Fuzzy Control of a Flixible Link," nominated for Best Paper Award, *Proc. IEEE International Conference on Robotics and Automation*, Albuquerque, New Mexico, April 1997.
176. M. R. Akbarzadeh-T., E. Tunstel, and M. JAMSHIDI, "Genetic Algorithms and Genetic Programming: Combining Strength in One Evolutionary Strategy," *Proc. Joint Conference on Environment*, Albuquerque, New Mexico, April, 1997.

177. E. T. Tunstel, H. Danny, T. Lippincott, and M. JAMSHIDI, "Adaptive Fuzzy-Behavior Hierarchy for Autonomous Navigation," Proceedings of IEEE Int. Conf. On Robotics and Automation, Albuquerque, NM, April 20-25, 1997.
178. M. R. Akbarzadeh-T. and M. JAMSHIDI, "Incorporating A-Priori Expert Knowledge in Genetic Algorithms," nominated for Best Paper Award, Proc. IEEE Conference on Computational Intelligence in Robotics and Automation, Monterey, California, July 1997.
179. M. R. Akbarzadeh-T., K. Kumbla, M. JAMSHIDI, and D. M. K. Al-Gobaisi, "GA-optimization of PID Fuzzy Control of Desalination Plants" Proc. IDA World Congress on Desalination and Water Reuse, Madrid Spain, October 6-9, 1997.
180. P. S. Sarkar, M. R. Akbarzadeh-T., M. JAMSHIDI, and D. M. K. Al-Gobaisi, "Design and Modeling of a Laboratory Scale Single Stage Flash Desalination Plant," Proc. IDA World Congress on Desalination and Water Reuse, Madrid Spain, October 6-9, 1997.
181. K. Kumbla and M. JAMSHIDI, "Neural Network Based Identification of Robot Dynamics Used for Neuro-Fuzzy Controller," Proceedings of IEEE Int. Conf. On Robotics and Automation, Albuquerque, NM, April 20-25, 1997.
182. A. Jadbabaie, A. Titli, and M. JAMSHIDI, "Separation Property of Observer/Controller for Continuous-Time Fuzzy Systems," Proc. 35th Allerton National Conference on Communication & Control, September, 1997, Allerton, IL.
183. K. Kumbla and M. JAMSHIDI, "Real-Time Neuro-Fuzzy Control of a Direct Drive Motor," Proc. IEEE Conference on System, Man and Cybernetics, October 1997, Orlando, Florida.
184. E. Tunstel, H. Danny, T. Lippincott, and M. JAMSHIDI, "Fuzzy Behavior-Based Navigation for Planetary Microrovers," Proc. of NASA University Research Centers Technical Conference, Albuquerque, NM, 1997, pp. 735-740.
185. E. Tunstel and M. JAMSHIDI, "On Decision-Making Among Multiple Rule-Bases in Fuzzy Control Systems, "Proc. of NASA University Research Centers Technical Conference, Albuquerque, NM, 1997, pp. 729-734.
186. M. R. Akbarzadeh, E. Tunstel, and M. JAMSHIDI, "Genetic Algorithms and Genetic Programming: Combining Strengths in One Evolutionary Strategy," Proc. of 1997 Waste Education and Research a Consortium/Hazardous Substance Research Center (WERC/HSRC) Joint Conference on Environment, Albuquerque, NM, 1997, pp. 729-734. (Best Student Paper Award).
187. E. Tunstel, H. Danny, T. Lippincott, and M. JAMSHIDI, "Adaptive Fuzzy-Behavior Hierarchy for Autonomous Navigation, "Proc. IEEE Int. Symposium on Computational Intelligence in Robotics and Automation, Albuquerque, NM, 1997, pp. 829-834.
188. E. Tunstel, H. Danny, T. Lippincott, and M. JAMSHIDI, "Autonomous Navigation Using an Adaptive Hierarchy of Multiple Fuzzy-Behaviors", Proc. IEEE Int. Symposium on Computational Intelligence in Robotics and Automation, Monterey, CA, July, 1997, pp. 276-281.
189. A. Jadbabaie, A. Titli, and M. JAMSHIDI, "Fuzzy Observer-Based Control on Nonlinear Systems" Proc. IEEE Conference on Decision and Control (CDC), December, 1997, San Diego, CA.
190. A. Jadbabaie, A. Titli and M. JAMSHIDI, "Guaranteed-Cost Design of Continuous-Time Takagi-Sugeno Fuzzy Systems via Linear Matrix Inequalities" Proc. IEEE International Conference on Fuzzy Systems (Fuzz IEEE 98), Anchorage, AK.
191. M. R. Akbarzadeh-T., S. Quintano, and M. JAMSHIDI, "Fuzzy Control of a Flexible Link: A Review" Proc. NASA University Research Centers-Technical Conference, Huntsville, AL, February, 1998, pp. 252-256.

192. M. R. Akbarzadeh-T., S. Shaikh, R. Hubbell, and M. JAMSHIDI, "SoftLab®: A Soft Computing Software for Experimental Research with Commercialization Aspects", *Proc. NASA University Research Centers-Technical Conference*, Huntsville, AL, February, 1998, pp. 182-187.
193. M. R. Akbarzadeh-T., M. Johnson, and M. JAMSHIDI, "Control of Flexible-Link Manipulators in Hazardous Environment", *Proc. 1998 Conference on Waste Management*, Tucson, Arizona, March, 1998.
194. M. R. Akbarzadeh-T., K. K. Kumbla, S. Shaikh, and M. JAMSHIDI, "SoftLab: A Soft Computing Software Laboratory for Research and Development", *Proc. 1998 Conference on Waste Management*, Tucson, Arizona, March, 1998.
195. M.-R. Akbarzadeh-T., S. Shaikh, and M. JAMSHIDI, "Hierarchical Fuzzy Control of Flexible Robotic Systems", *Proc. IFAC Symposium on Large Scale Systems*, Patras, Greece, July 1998.
196. A. Jadbabaie, M.J. Jadbabaie, and M. JAMSHIDI, "Guaranteed-Cost Design of Takagi-Sugeno Fuzzy Controllers: Discrete-Time Case," *Proc. 3rd World Automation Congress*, May, 1998. Anchorage, Alaska.
197. S. Sarkar, M-R. Akbarzadeh-T., M. JAMSHIDI, "Modeling for Simulation and Control of a Single Stage Flash Desalination Pilot-Plant," *Proc. 1998 World Automation Congress* (May 1998). Anchorage, Alaska.
198. P. Sarkar, M-R. Akbarzadeh-T., and M. JAMSHIDI, "Decentralized Hierarchical Fuzzy Control Architecture for a Single Stage Flash Desalination Process Plant," *Proc. 3rd World Automation Congress*, May, 1998. Anchorage, Alaska.
199. M. -R. Akbarzadeh-T., E. Tunstel, K. Kumbla, and M. JAMSHIDI, "Soft Computing Paradigms for Hybrid Fuzzy Controllers: Experiments and Applications," *Proc. IEEE International Conference on Fuzzy Systems*, May 1998. Anchorage, Alaska.
200. M. JAMSHIDI, P. Borne, and J. S. Jamshidi (eds.), *Intelligent Automation and Control* (Proc. ISIAC 2000), Vol. 9, Preface, Albuquerque, NM: TSI Press, 2000, pp. viii-x..
201. M. JAMSHIDI, A. A. Maciejewski, S. Nahavandi, and R. Lumia (eds.), *Robotic and Manufacturing Systems: Recent Results in Research, Development and Applications*, (Proc. ISORA and ISOMA 2000), Vol. 10, Preface, Albuquerque, NM: TSI Press, 2000, pp. vii-ix.
202. M. JAMSHIDI, M. Fathi, and T. Furuhashi (eds.), *Soft Computing, Multimedia and Image Processing: Trends, Principles and Applications* (Proc. IFMIP 2000 and ISSCI 2000), Vol. 11, Preface, Albuquerque, NM: TSI Press, 2000, pp. vii-ix.
203. C. Guo, P. Sarkar, and M. JAMSHIDI, "Soft Computing Based Intelligent Controller and its Application to a Nonlinear System," Proc. World Automation Congress, Maui, HI, USA, June, 2000, pp. 340-345.
204. M. JAMSHIDI, "Autonomous Control of Complex Systems: Applications in Robotics, " *Proc. Richard Bellman Memorial Conference*, May, 1999. Santa Fe, NM.
205. S. Berman, M. de Oleivera, and M. JAMSHIDI, "Fuzzy-Behavior Control of Mobile Robots," *Proc. IEEE Mediterranean Conference*, Haifa, Israel, June, 1999.
206. J. Parkinson, M. JAMSHIDI, and P. Wantuck, "Fuzzy Control of a Nonlinear Three-Tank System," *Proc. 4th WAC*, Maui, HI, USA, June, 2000., pp.452-460.
207. M. Johnson, M. JAMSHIDI, K. Firoosbaknsh, and M. Moneim, "Flexor-Tendon-Repair Technique Determination: Decision via soft computing versus Taguchi method," *Proc. 4th WAC*, Maui, HI, USA, June 2000, pp. 124-130.
208. A. El-Osery, R. Lecointe, M. JAMSHIDI, M. H. Nehrir, and J. Lu, "A Takagi-Sugeno Type Fuzzy modeling, control and stability analysis of power systems," *Proc. 4th WAC*, Maui, HI, USA, June 2000, pp. 534-539.

209. P. Sarkar, M. Akbarzadeh-T., and M. JAMSHIDI, "Decentralized hierarchical fuzzy control architecture for a single stage flash desalination process plant," Proc. 4th WAC, Maui, HI, USA, June 2000.
210. S. A. Taheri, M. JAMSHIDI, H. Sarjoughian, and B. P. Zeigler, " DEVS Simulation and Modeling of Multi-robot System", Proceedings of AI, Simulation and Modeling Conference, Tucson Arizona, 6-8 March, 2000.
211. S. A. Taheri and M. JAMSHIDI, "ANN-Based Sliding Mode Control for Nonholonomic Mobile Robots," Proceedings of IEEE Conference on Control Applications, Anchorage, Alaska, 25-27 Sep. 2000.
212. M. JAMSHIDI, "Autonomous Control of Space-Bound Systems", Proc. 7th Conference on Computers, Control and Systems," Cairo, Egypt, Elsevier Publishers, February, 2000, pp. 1-6.. (Keynote Address)
213. M. JAMSHIDI, "Autonomous Control of Complex Systems", Proc. Iizuka 2000, Iizuka, Japan, September, 2000
214. M. A. A. de Oliveira, S. Berman, E. Tunstel, and M. JAMSHIDI, "Remote Surface Exploration With Soft-Computing Based Cooperative Rovers," Proc.4th World Automation Congress, June, 2000, Vol. 10, Maui, HI, USA, pp. 163-168.
215. M. Johnson, M. JAMSHIDI, K. Firoozbakhsh, and M. Moneim, "Flexor-Tendon-Repair Technique Determination: Decision via Soft Computing Versus Taguchi Methods," Proc. 4th World Automation Congress, (IFMIP 2000), June, 2000, Vol. 11, Maui, HI, USA, pp. 124-130.
216. S. Berman, Y. Edan, and M. JAMSHIDI, "Multi-Agent Strategy for Automated Guided Vehicle Systems in Material Handling," Proc. 4th WAC Congress (ISSCI 2000), June, 2000, Vol. 11, Maui, HI, USA, pp. 322-327.
217. C. Guo, P. Sarkar, and M. JAMSHIDI, "Soft Computing-Based Intelligent Controller And Its Application To A Nonlinear System," Proc. 4th World Automation Congress, (ISSCI 2000), June, 2000, Vol. 11 Maui, HI, USA, pp. 340-345.
218. W. J. Parkinson, P. J. Wantuck, M. JAMSHIDI, K. S. Patrick, and S. S. Jaffer, "Fuzzy Control of a Nonlinear Three-Tank- System," Proc. 4th World Automation Congress, (ISSCI 2000), June, 2000, Vol. 11 Maui, HI, USA, pp. 452-460.
219. A. El-Osery, R. Lecointe, M. JAMSHIDI, H. Nehrir, and Jie Lu, "A Takagi-Sugeno Type Fuzzy Model for Power System Stability Analysis," Proc. World Automation Congress, (ISSCI 2000), June, 2000, Vol. 11 Maui, HI, USA, pp. 534-539.
220. W. J. Parkinson, S. P. Abein, K. L. Creek, P. J. Wantuck, T. Ross, and M. JAMSHIDI, "Application of Fuzzy Set Theory for Exposure Control in Beryllium Part Manufacturing," Proc. 4th World Automation Congress, (ISSCI 2000), June, 2000, Vol. 11 Maui, HI, USA, pp. 588-597.
221. M. JAMSHIDI, "V-IP An Innovative Approach to Engineering Education," Proc. Northeastern University Conference on Practice-Oriented Education, Boston, MA, April, 2001.
222. M. JAMSHIDI, "V-IP An Innovative Approach to Minority Students Engineering Education," Proc. ASEE Conference, Albuquerque, NM, June, 2001.
223. M. JAMSHIDI," AUTONOMOUS CONTROL SYSTEMS –Applications to Space Systems," Proc. Int. Conf. on Soft Computing and Computing with Words, (ICSCCW 2001), June 2001, Antalya, Turkey, pp. 1-12. (Keynote Address)
224. M. JAMSHIDI, "Autonomous Control Systems –Applications to Optical Systems and Image Processing," Proc. SPIE Annual Conference, San Diego, CA, USA, August 2001. (Keynote Address)

225. M. JAMSHIDI, "Autonomous Control of Complex Systems -Space and Earth-Bound Applications," Proc. Int. Conference on Systems Science, Worclaw, Poland, September 2001. (Keynote Address)
226. M. JAMSHIDI, "Autonomous Control of Cooperative Robotic and Satellite Systems," -Space and Earth-Bound Applications," Proc. Int. on Cooperative Robotics, Padeborne, Germany, September, 2001. (Keynote Address)
227. J. Burge, A. El-Osery, M. JAMSHIDI, and M. Fathi, "V-LAB: A Virtual Laboratory for Distributed Robotic Modeling and Simulation," Proc. IEEE Int. Conference on Systems, Man and Cybernetics, Tucson, AZ, October, 2001.
228. A. El-Osery, J. Burge, M. JAMSHIDI, and M. Fathi "Stochastic Learning Automaton for Learning Control of Robotic Agents, "Proc. IEEE Int. Conference on Systems, Man and Cybernetics, Tucson, AZ, October, 2001.
229. S. Berman, Y. Edan, and M. JAMSHIDI," Decentralized Autonomous AGVS for Material Handling," Proc. Int. Conference on Production Research, Prague, Czech Republic, August 2001.
230. M. JAMSHIDI, "Autonomous Control Systems – With Applications to Optical Systems," (Invited Keynote speech), Proc. SPIE Conference, Sand Diego, CA, July, 2001.
231. A. El-Osery, C. Abdallah, and M. JAMSHIDI, "Time Delay and Power Control in Spread Spectrum Wireless Networks," Proc. IFAC Conference on Time-Delay Systems, Dec. 9-11, 2001, Santa Fe, NM, USA.
232. M. JAMSHIDI, "A Vertically Integrated Approach to Graduate Engineering Education, "Proc. World Automation Congress, Orlando, FL, June, 2002.
233. W. J. Parkinson, R.E. Smith, F. N. Mortensen, P. J. Wantuck, M. JAMSHIDI, and J. J. Ross, "Fuzzy SPC Filter for a Feed-Forward Control System for a Three-Phase Oil Field Centrifuge" Proc. World Automation Congress, Orlando, FL, June, 2002.
234. S. A. Taheri, F. Macedo, and M. JAMSHIDI," ANN-Based Sliding Mode Control for Robots", Proceedings of Space and Robotics 2000, Albuquerque, NM, March. 2000.
235. S. Akhavan Taheri, H. Sarjoughian, M. Jamshidi, and B. Zeigler, "Imitating the Human Immune System Capabilities for Multi-agent Federation Formation," Proc. IEEE Biomedical Conf., Mexico City, Mexico, 2001.
236. A. El-Osery and M. JAMSHIDI, "A Stochastic Learning Automaton Learning Autonomous Control of Robotic Agents," Proc. IEEE Int. Conference on Systems, Man and Cybernetics, Tunisia, October, 2002.
237. P. Sridhar and M. JAMSHIDI, "V-Lab® - a virtual laboratory for distributed simulation," Proc. IEEE Int. Conference on Systems, Man and Cybernetics, Washington, DC, October, 2003.
238. S. Beaty, U. Dole, and M. JAMSHIDI, "Hardware design of an all-train rover," Proc. IEEE Int. Conference on Systems, Man and Cybernetics, Washington, DC, October, 2003.
239. Y. Wang and M. Jamshidi, "Multispectral Landsat Image Classification Using Fuzzy Expert Systems," Proc. World Automation Congress – Multimedia track, June 28-July 1, 2004, Seville, Spain.
240. J. Liu, M. Jamshidi, and S. Pourbabak, "A Novel Auto Regression and Fuzzy-Neural Hybrid Method to Identify Cardiovascular Dynamics," Proc. World Automation Congress – Multimedia Track, June 28-July 1, 2004, Seville, Spain.
241. S. Beatty and M. Jamshidi, "Simulation of a Spacecraft Electrical Power Distribution System Using the Simulink Power System Blockset and Soft Computing Techniques," Proc. World Automation Congress – Automation and Control Track, June 28-July 1, 2004, Seville, Spain.

242. P. Sridha and M. Jamshidi, "A Framework for Multi-agent Discrete Event Simulation: V-Lab®," *Proc. World Automation Congress* – Automation and Control Track, June 28-July 1, 2004, Seville, Spain.
243. S. Sheikh-Bahaei and M. Jamshidi, "Discrete Event Fuzzy Logic Control with Application to Sensor-Based Intelligent," *Proc. World Automation Congress* – Robotics Track, June 28-July 1, 2004, Seville, Spain.
244. T. Song, X. M. Huang, R. R. Lee, C. Gasparovi, and M. Jamshidi, "A Hierarchical Tissue Segmentation Approach in Brain MRI Images," *Proc. World Automation Congress* – Biomedicine Track, June 28-July 1, 2004, Seville, Spain.
245. H. Berenji, D. Herrick, and M. Jamshidi," Data Mining Based Algorithms for Prognostic Studies of the ABL System," Paper presented at the *6th Annual Directed Energy Symposium*, October 22-24, 2003, Albuquerque, NM
246. M. Jamshidi, D. Herrick, D. Washburn, and H. Berenji, **"**Design, Control and Prognostic of a Laser Pointing System," *Proc. Laser Systems Technology Conference,* 12-16 April 2004, Orlando, Florida
247. Y. Wang, F. Benito, G. A. Vera, and M. JAMSHIDI,"Control Design for Diagnostic and Prognostic of Hardware Systems," *Proc. FUZZ-IEEE Conference*, Budapest, Hungary, July 25-28, 2004.
248. V. Stone, K. Meadows, and M. JAMSHIDI, "An Investigation of Health Monitoring, Prognostic, and Diagnostic Techniques Using Pattern Recognition and Data Mining," to be presented at *AMOS Conference*, September 13-17, 2004, Maui, HI.
249. H. Berenji, Y. Wang, D. Vebgerov, R. Langari, and M. JAMSHIDI, " Using Gated Experts in Fault Diagnosis and Prognosis," *Proc. FUZZ-IEEE Conference*, Budapest, Hungary, July 25-28, 2004
250. Y. Wang and M. JAMSHIDI, "Fuzzy Logic Applied in Remote Sensing Image Classification", *Proc. IEEE Conference on Systems, Man and Cybernetics,* Hague, Netherlands, October 2004.
251. Y. Wang, M. JAMSHIDI, S. Morain, P. Neville, and C. Bales, "Multispectral Landsat Images Classification Using a Data Clustering Algorithm", *Proc. International Society for Photogrammetry and Remote Sensing*, Istanbul, Turkey, July 2004.
252. Y. Wang and M. JAMSHIDI, "A fuzzy-neural approach to remote sensing data clustering," *Proc. Int. Conf. On Machine Learning and Cybernetics* – August 26-29, 2004, Shanghai, China.
253. J. Liu and M. JAMSHIDI, "Cardiac system identification via fuzzy-neural systems," *Proc. Int. Conf. On Machine Learning and Cybernetics* – August 26-29, 2004, Shanghai, China.
254. P. Sridhar and M. JAMSHIDI, "Distributed multi-agent modeling and simulation, "*Proc. IEEE Int. Conference on Systems, Man and Cybernetics*, The Hague, the Netherlands, October 2004.
255. Y. Wang and M. JAMSHIDI, "Remote sensing using fuzzy neural clustering techniques," *Proc. IEEE Int. Conference on Systems, Man and Cybernetics*, The Hague, the Netherlands, October 2004.
256. P. Sridhar, A. M. Madni, and M. JAMSHIDI, "Intelligent Monitoring of Sensor Networks Using Fuzzy Logic Based Control," *Proc. 2006 IEEE SMC Conference*, Taipei, Taiwan, October 2006, pp. 3394-3399.
257. H. Azarnoosh, B. Horan, P. Sridhar, A. Madni, and M. JAMSHIDI, "Towards optimization of a real-world Robotic-Sensor System of Systems," *Proc, 2006 World Automation Congress* (Robotics Track ISORA), Budapest, Hungary, July, 2006 (See Volume 19, TSI Press, ISBN 1-889335-33-9, San Antonio, TX, 2006), pp. 223-230.

258. P. Sridharl, A. M. Madni, and M. JAMSHIDI, "Hierarchical Data Aggregation in Spatially Correlated Distributed Sensor Networks," *Proc, 2006 World Automation Congress* (Robotics Track ISORA), Budapest, Hungary, July, 2006 (See Volume 19, TSI Press, ISBN 1-889335-33-9, San Antonio, TX, 2006), pp. 39-44.
259. P. Sridhar, A. M. Madni, and M. JAMSHIDI, "Intelligent Monitoring of Sensor Networks using Fuzzy Logic Based Control," *Proc. IEEE Conf. on Systems, Man and Cybernetics*, Taipei, Taiwan, 2006,, 6pp.
260. M. JAMSHIDI, "A Vertically Integrated Approach to Graduate Control Engineering Education of Ethnic Minorities," *Proc, 2006 World Automation Congress* (Robotics Track ISORA), Budapest, Hungary, July, 2006 (See Volume 19, TSI Press, ISBN 1-889335-33-9, San Antonio, TX, 2006), pp. 395-402.
261. P. Sridhar, A. M. Madni, and M. JAMSHIDI, "Intelligent Object-Tracking using Sensor Networks", *Proc. IEEE Sensor Applications Symposium (SAS)*, San Diego, 2007, pp.1-5.
262. P. Sridhar, A. M. Madni, and M. JAMSHIDI, "Multi-Criteria Decision Making and Behavior Assignment in Sensor Networks," *Proceedings IEEE Systems Conference*, Waikiki, HI, April 10-12, 2007, 6pp.
263. A. M. Madni, P. Sridhar, and M. JAMSHIDI, "Fault-Tolerant Data Acquisition in Sensor Networks", *Proc. IEEE System of Systems Engineering Conference*, San Antonio, April 16-18, 2007, pp. 389-394.
264. F. Sahin, M. JAMSHIDI and P. Sridhar, "A Discrete Event XML based Simulation Framework for System of Systems Architectures," *Proc. IEEE System of Systems Engineering Conference*, San Antonio, April 16-18, 2007, pp.382-388.
265. V. Raghavan and M. JAMSHIDI, "A GPS-Based Autonomous Rover Navigation," *Proc. IEEE System of Systems Engineering Conference*, San Antonio, April 16-18, 2007, pp.486-491.
266. B. Horan, D. Creighton, S. Nahavandi, and M. JAMSHIDI, "A Bilateral Haptic Teleoperation of an articulated track mobile robot," *Proc. IEEE SoSE Conference*, San Antonio, TX, April 16-18, 2007., pp.Z395-Z402.
267. H. Azarnoush, F. Mir, S. Agaian, and M. JAMSHIDI, " Alaryngeal Speech Enhancement Using Minimum Statistics Approach to Spectral Subtraction," *Proc. IEEE System of Systems Engineering Conference*, San Antonio, April 16-18, 2007, pp. 576-580.
268. T. Nanayakkara, C. Piyathilaka, P. Siriwardana, A. Subasingha, and M. JAMSHIDI, "Development of Advanced Motor Skills in a Group of Humans Through an Elitist Visual Feedback Mechanism," *Proc. IEEE System of Systems Engineering Conference, San Antonio*, April 16-18, 2007, pp.408-413.
269. A. Jaimes, J. Gomez, S. Kota, and M. JAMSHIDI, "An approach to surveillance an area using swarm of fixed wing and quad-rotor unmanned aerial vehicles UAV(s)", *Proc. IEEE SoSE Conference*, San Antonio, TX, April 16-18, *Proc. IEEE System of Systems Engineering Conference*, Monterey Bay, CA, June 2-4, 2008, paper # 1569111991.
270. M. Joordens, J. Serna, S. Songer, C. Friday, J. Hoy, R. Seiger, W. Madalinski, and M. JAMSHIDI, "Low Cost Underwater Robot Sensor Suite," *Proc. IEEE SoSE Conference*, San Antonio, TX, April 16-18, *Proc. IEEE System of Systems Engineering Conference*, Monterey Bay, CA, June 2-4, 2008, paper # 1569095252
271. A. Kumar Ray, M. Gupta, L. Behera and M. JAMSHIDI, "Sonar Based Autonomous Automatic Guided Vehicle (AGV) Navigation," *Proceedings 2008 IEEE SoSE Conference*, Monterey Bay, CA, USA, June 2-4, 2008.

272. P. Benavidez, K. Nagothu, A. Kumar Ray, T. Shaneyfelt, S. Kota, L. Behera, and M. JAMSHIDI, "Multi- Domain Robotic Swarm Communication System," Proceedings 2008 IEEE SoSE Conference, Monterey Bay, CA, USA, June 2-4, 2008.
273. T. Shaneyfelt and M. JAMSHIDI, "Towards Net-Centric System of Systems Robotics in Air, Sea and Land," Proc. IEEE SoSE Conference, San Antonio, TX, April 16-18, Proc. IEEE System of Systems Engineering Conference, Monterey Bay, CA, June 2-4, 2008, paper # 1569101444
274. J. Prevost, M. A. Joordens, and M. JAMSHIDI," Simulation of Underwater Robots using Microsoft's Robot Studio©," Proc. IEEE System of Systems Engineering Conference, Monterey Bay, CA, June 2-4, 2008
275. S. S. Mirsaeid Ghazi and M. JAMSHIDI, "Adaptive Modeling: a Statistical Approach in Designing a Mathematical-model-based Controller," Proc. IEEE SoSE Conference, San Antonio, TX, April 16-18, Proc. IEEE System of Systems Engineering Conference, Monterey Bay, CA, June 2-4, 2008, paper # 1569124597
276. K. Nagothu, M. Joordens, and M. JAMSHIDI, "Distributed Protocol for Communications Among Underwater Vehicles," Proc. IEEE SoSE Conference, San Antonio, TX, April 16-18, Proc. IEEE System of Systems Engineering Conference, Monterey Bay, CA, June 2-4, 2008, paper # 1569108743, pp. 374-379.
277. C. Parisi, F. Sahin, and M. JAMSHIDI, "A Discrete Event XML based System of Systems Simulation for Robust Threat Detection and Integration," Proc. IEEE System of Systems Engineering Conference, Monterey Bay, CA, June 2-4, 2008.
278. T. Shaneyfelt, M. A. Joordens, K. Manoj Nagothu, and M. JAMSHIDI, "RF Communication between Surface and Underwater Robotic Swarms," Proceedings World Automation Congress, Waikoloa, HI, September 28 – October 2, 2008.
279. M. Joordens and M. JAMSHIDI, "Low Cost Underwater Robot Sensor Suite," Proc. IEEE SoSE Conference, San Antonio, TX, April 16-18, Proc. IEEE System of Systems Engineering Conference, Monterey Bay, CA, June 2-4, 2008, paper # 1569095252
280. A. Kumar Ray, M. Gupta, L. Behera, and M. JAMSHIDI, "Sonar Based Autonomous Automatic Guided Vehicle (AGV) Navigation," Proceedings 2008 IEEE SoSE Conference, Monterey Bay, CA, USA, June 2-4, 2008.
281. P. Benavidez, K. Nagothu, A. Kumar Ray, T. Shaneyfelt, S. Kota, L. Behera, and M. JAMSHIDI, "Multi- Domain Robotic Swarm Communication System," Proceedings 2008 IEEE SoSE Conference, Monterey Bay, CA, USA, June 2-4, 2008.
282. T. Shaneyfelt and M. JAMSHIDI, "Towards Net-Centric System of Systems Robotics in Air, Sea and Land," Proc. IEEE System of Systems Engineering Conference, Monterey Bay, CA, June 2-4, 2008, paper # 1569101444
283. K. Manoj Nagothu, M. Joordens, and M. JAMSHIDI, "Distributed Protocol For Communications Among Underwater Vehicles," Proc. IEEE System of Systems Engineering Conference, Monterey Bay, CA, June 2-4, 2008, paper # 1569108743
284. A. Moghadas, and M. JAMSHIDI, "Telemedicine in Health Care Systems," Proceedings of WAC 2008, September 28-October 2, 2008, Waikoloa, HI, USA, Paper # 4561299642.
285. M. JAMSHIDI, "System of Systems Engineering – Land, Sea and Air Applications," Proceedings 8th International Conference on Applications of Fuzzy Systems and Soft Computing, Helsinki, Finland, pp. 46-64.2008
286. T. Shaneyfelt, M. Joordens, K. Manoj, and M. JAMSHIDI, "RF Communication Between Surface and Underwater Robotic Swarms," Proceedings of WAC 2008, September 28-October 2, 2008, Waikoloa, HI, USA

287. T. Shaneyfelt, M. Joordens, K. Nagothu, J. J. Prevost, A. Kumar, S. S. Mirsaeid Ghazi, M. JAMSHIDI, "Applications and Prototype for System of Systems Swarm Robotics," Proc. IEEE SMC Conference, Singapore, October 13-15, 2008, pp.1314-1319.
288. M. A. Joordens, T. Shaneyfelt, S. Eega, A. Jaimes, and M. JAMSHIDI, "Applications and Prototype for System of Systems Swarm Robotics," Proc. IEEE SMC Conference, Singapore, October 13-15, 2008, pp. 2049-2055.
289. K. Nagothu, M. Joordens, and M. Jamshidi, "Communications for Underwater Robotics Research Platforms," Proc. IEEE Systems Conference, Montreal, Canada, April 7-10, 2008.
290. K. Nagothu, T. Shaneyfelt and M. JAMSHIDI, "System of System Communication for Heterogeneous Independent Operable Systems," Proc. IEEE SoSE Conference, Albuquerque, NM, June1-3, 2009.
291. S. Eega, M. Joordens and M. JAMSHIDI, "Design of Low Cost Thruster for an Autonomous Underwater Vehicle," Proc. IEEE SoSE Conference, Albuquerque, NM, June1-3, 2009.
292. A. Kumar Ray, P. Benavidez, L. Behera, and M. JAMSHIDI, "Motion Coordination of Multi-agent Formation of Rovers," Proc. IEEE SoSE Conference, Albuquerque, NM, June1-3, 2009.
293. A. K. Ray, L. Behera, and M. JAMSHIDI, "GPS and Sonar Based Area Mapping and Navigation by Mobile Robots," Proc. IEEE INDIN Conference, Cardiff, UK, June 23-26, 2009.
294. B. Kelley, K. Manooj, and M. JAMSHIDI, "Broadband RF Communications in Underwater Environments Using Multi-carrier Modulation," Proc. IEEE SMC Conference, San Antonio, October 12-14, 2009, pp.2303-2308.
295. M. Joordens and M. JAMSHIDI, "Consensus-Based Control of Underwater Robots," Proc. IEEE SMC Conference, San Antonio, TX, Oct. 12-14, 2009, pp.3163-3168.
296. M. JAMSHIDI, "Control Paradigms for System of Manufacturing Systems," Proc. IPROMS Conference, Cardiff, Wales, UK, July 16-17, 2009.
297. A. Jevtić, D. Andinal, A. Jaimes, J. Gomez, and M. JAMSHIDI, "Unmanned Aerial Vehicle Route Optimization Using Ant System Algorithm," Proc. IEEE SoSE Conference, Loughbrough, UK, June 2010.
298. A. Jaimes and Mo JAMSHIDI, "Consensus-Based and Network Control of UAVs," Proc. IEEE SoSE Conference, Loughbrough, UK, June 2010.
299. D. Shahgoshtasbi, S. Agaian, and M. JAMSHIDI, "A Biological Speech Recognition System by Using Associative Neural Networks," Proc. World Automation Congress, Kobe, Japan, September 2010.
300. A. Jevtic, D. Andina, A. Jaimes, J. Gomez, and M. JAMSHIDI, "Unmanned Aerial Vehicle Route Optimization Using Ant System Algorithm," Proc. World Automation Congress, Kobe, Japan, September 2010. **(Won First Place – Best Paper Award)**
301. J. Gomez and M. JAMSHIDI, "Fuzzy Logic Control of a Fixed-Wing Unmanned Aerial Vehicle," Proc. World Automation Congress, Kobe, Japan, September 2010. **(Won Second Place – Best Paper Award),** paper# WED-PM2.
302. A. Panchul, D. Akopian, and M. JAMSHIDI, "Time-Space Optimizing Fuzzy Logic Algorithm for Dynamic Load Balancing in Distributed Three-Dimensional Graphics Engine," Proc. World Automation Congress, Kobe, Japan, September 2010.
303. H. Berenji and M. JAMSHIDI, "Fuzzy Reinforcement Learning for System of Systems (SOS)," Proc. Fuzz-IEEE, Barcelona, July 2010, paper # 2396456420.

304. M. JAMSHIDI, "From Large-Scale Systems to SYSETM OF SYSTEMS – Control Challenges for the 21st Century," Proc. IFAC LSS, July, 2010, Lille, France, paper # Key-Mon-1.
305. P. Gazi, M. JAMSHIDI, A. Jevitc, and D. Andina, "A Mechatronic System Design Case Study: Control of a Robotic Swarm Using Networked Control Algorithms," Proc. IEEE Systems Conference, San Diego, CA, USA, April 2010 **(Won First Place – Best Paper Award),** paper # MON-PM3.
306. H. Berenji and M. JAMSHIDI, "Fuzzy Reinforcement Learning for System of Systems (SOS)," Proc. Fuzz IEEE, Taipei, Taiwan, June 27-30, 2011, paper # 4459851723.
307. S. Bakhtiari, S. Agaian, M. JAMSHIDI, "A Novel Empirical Mode Decomposition Based Systems for Medical Image Enhancement," Prof. IEEE Systems Conf., Montreal, Canada, April 4-7, 2011, paper # 7754328641, pp. 4-9.
308. K. Nagothu, B. Kelley, J. J. Prevost, and M. JAMSHIDI, "On Prediction to Dynamically Assign Heterogeneous Microprocessors to the Minimum Joint Power State to Achieve Ultra Low Power Cloud Computing", Asilomar Conference on Signals, Systems and Computers, CA, Nov-2011, paper # 6542221049.
309. P. Benavidez and M. JAMSHIDI, "'Mobile Robot Navigation and Target Tracking Systems," Proc. IEEE SoSE 2011, paper #1569433191, Albuquerque, NM, June27-31, 2011
310. D. Shahgoshtasbi and M. JAMSHIDI, "Energy efficiency in a smart house with an intelligent Neuro-Fuzzy lookup table," Proc. IEEE SoSE 2011, paper #1569453099, Albuquerque, NM, June27-31, 2011, 6pp.
311. A. Kumar Ray, G. McGinney, L. Behera, and M. JAMSHIDI, "Coordinated Traffic Scheduling for Communicating Mobile Robots," Proc. IEEE SoSE 2011, paper #1569453769, Albuquerque, NM, June27-31, 2011, 6pp.
312. J. Prevost, K. Manoj, B. Kelley, and M. JAMSHIDI, "Prediction of Cloud Data Center Networks Loads Using Stochastic and Neural Models," Proc. IEEE SoSE 2011, paper #1569453089, Albuquerque, NM, June27-31, 2011.
313. M. Dziuk and M. JAMSHIDI, "Fuzzy Logic Controlled UAV Autopilot Using C-Mean Clustering," Proc. IEEE SoSE 2011, paper #1569453279, Albuquerque, NM, June27-31, 2011
314. S. Bakhtiari, S. Agaian, and M. JAMSHIDI, "The Integration of Alpha-Rooting and Ensemble Empirical Mode Decomposition for Image Contrast Enhancement," Proc. SPIE Conference, Orlando, FL, 2011, paper # 7653119840.
315. S. Bakhtiari, S. Agaian, and M. JAMSHIDI, "An Enhanced Empirical Mode Decomposition Based Method for Image Enhancement," Proc. IEEE SMC Conference, Anchorage, AK, October, 2011, paper # 7764329401.
316. T. Shaneyfelt, S. Agaian, M. JAMSHIDI, and S. Erdogan, "Quanterion number based vanilla system for automating pollination," Proc. IEEE Int. Conf. on System Science and Engineering, Macao, China, June 8-10, 2011, paper # Thu-IV-R1.
317. T. Shaneyfelt, S. Agaian, M. JAMSHIDI, and S. Erdogan, "Vanilla Hypercomplex Recognition Machine Vision for Pollination," Proc. IEEE SMC Conference, Anchorage, AK, October, 2011, paper # 6539982734.
318. E. Bonab, E. Morales, H. Krishnaswami, and M. JAMSHIDI, "Bi-Directional Multi-Mode Grid Tied Converter for Solar Energy Conversion Systems," Proc. WAC 2012, Puerto Vallarta, Mexico Paper EDS # 1569538081.
319. D. Shahgoshtasbi and M. JAMSHIDI, "Modified Intelligent Energy Management System in a Smart House," Proc. WAC 2012, Puerto Vallarta, Mexico, Paper EDS # 1569524133.

320. J. J. Prevost, K. M. Nagothu, B. Kelley, and M. JAMSHIDI, "Load Prediction Algorithm for Multi-Tenant Virtual Machine Environments," Proc. WAC 2012, Puerto Vallarta, Mexico, Paper EDS # 1569572457, Best Paper Awardee.
321. Y. S. Manjili, A. Rajaee, B. Kelley, and M. JAMSHIDI, "Optimization Using Fuzzy-Logic and Markov Chain Model for Micro-Grid Systems," Proc. WAC 2012, EDAS # 1569523895, Puerto Vallarta, Mexico.
322. D. Shahgoshtasbi and M. JAMSHIDI, "Energy efficiency in a smart house with an intelligent Neuro-Fuzzy lookup table," Proc. IEEE SoSE 2011, paper #1569453099, Albuquerque, NM, June27-31, 2011
323. Y S. Manjili, A. Rajaee, B. Kelley, and M. JAMSHIDI, "Fuzzy-Logic Based Control for Battery Management in Microgrids," Proceedings of UTSA COS Conference, Paper presented at UTSA COS Conference **(Best Paper in Energy and Environment Section),** 2011
324. A. Rajaee, K. M. Nagothu, B. Kelley, M. Jamshidi, "Throughput Analysis for AMI Meters on Cognitive Radio Networks," Proc. IEEE Systems Conference, Vancouver, BC, Canada, March 19-23, 2012.
325. E. E. Bonab, O. Ghasemi, M. JAMSHIDI, and Y. Jin, "Adaptive Estimation over Distributed Sensor Networks with a Hybrid Algorithm," Proc. International Conference on Machine Learning and Cybernetics, Xi'an, China, June, 2012
326. R. Crosier, S. Wang, and M. JAMSHIDI, "A 4800-V grid-connected electric vehicle charging station that provides STACOM-APF functions with a bi-directional, multi-level, cascaded converter," Proc. IEEE Applied Power Electronics Conference and Exposition (APEC), Digital Object Identifier: 0.1109/APEC.2012.6166020, 2012, Pp. 1508 – 1515
327. Y. S. Manjili, A. Rajaee, B. Kelley, and M. JAMSHIDI, "Intelligent Decision Making for Energy Management in Microgrids1 with Air Pollution Reduction Policy," Proc. 6th IEEE SoSE, Genoa, Italy, July 16-18, 2012, EDAS # 1569592101
328. M. JAMSHIDI, "Sustainable Energy Systems: Cyber-Physical Based Mangement of Micro-Grids by a Fuzzy Systems Approach," Proc. 2012 ICAFS, (Invited Presentation), Lisbon, Portugal, August 29-31, 2012 (INVITED).
329. M. JAMSHIDI, "Sustainable Energy Systems: Cyber-Physical Based Intelligent Mangement of Micro-Grids," Proc. 2012 Buda University Workshpop on Sytems Engineering, Budapest, Hungary, September 10-12, 2012. (INVITED)
330. M. JAMSHIDI, "Sustainable Energy Systems Management with Environmental Constraints," Proc. LINDI Conference, Slovakia, September 6-7, 2012. (INVITED)
331. M. Niknamfar, Y. Manjili Sahraei, M. JAMSHIDI, and M. Shadaram, "Cost Effective ROF Communication System for CATV Channels over WDM Network and Fuzzy Modeling of the System," Proc. IEEE Workshop on Computing, Networking and Communication, San Diego, CA, January 28-31, 2013.
332. K. M. Nagothu, A. Rajaee, B. Kelley, and M. JAMSHIDI, "Interference Aware Scheduling for Peak Channel Reuse and Max-Capacity In Smart Meter Networks," Proc. IEEE Workshop on Computing, Networking and Communication, San Diego, CA, January 28-31, 2013.
333. K. M. Nagothu, B. Kelley, and M. JAMSHIDI, "MIMO-Interference Aware Scheduling Enabling the Allocation of Unbounded Co-channels in Unplanned Networks," Proc. IEEE Systems Conference, Orlando, FL, April 15-17, 2013.

334. S. A. Henson, M. J. D. Henshaw, V. Barot, C.E. Siemieniuch, M.A. Sinclair, M. JAMSHIDI, H. Dogan, S. L. Lim, C. Ncube, and D. DeLaurentis, "Towards a Systems of Systems Engineering EU Strategic Research Agenda," *Proc. IEEE Systems Conference*, Orlando, FL, April 15-17, 2013.
335. M. Niknamfar, Y.S. Manjili, M. Shadaram, and M. JAMSHIDI, "Fuzzy modeling of an ROF communication system for CATV channels over DWDM network," *Proc. ABES 2012*, UTSA, TX. (Best Paper Award winner)
336. B. K. Tannahill, C. E. Maute, Y. Yetis, M. N. Ezell, A. Jaimes, R. Rosas, A. Motaghi, H. Kaplan, and M. JAMSHIDI, "Modeling of System of Systems via Data Analytics – Case for 'Big Data' in SoS," *Proc. 8thIEEE SoSE*, Maui, HI, June 2-4, 2013, EDAS # 1569744777
337. T. S. Shaynefelt, M. JAMSHIDI, and S. Agaian," **Hierarchical Robotic Crane System for Post Grid Array Environments,**" *Proc. 8thIEEE SoSE*, Maui, HI, June 2-4, 2013, EDAS# 1569736353
338. E. Bonab, O. Sadeghi, M. JAMSHIDI, and Y. Jin, "**Time Delay Estimation in Gene Regulatory Networks,**" *Proc. 8thIEEE SoSE*, Maui, HI, June 2-4, 2013, EDAS# 1569747703
339. A. Daali, Y. Huang, and M. JAMSHIDI, "**A System based Approach to Construct a Kaposi Sarcoma-Associated Herpesvirus (KSHV) Specific Pathway Crosstalk Network**," *Proc. 8thIEEE SoSE*, Maui, HI, June 2-4, 2013, EDAS# 1569747605.
340. J. Prevost, K. M. Nagothu, and M. JAMSHIDI, "Optimal Update Frequency Model for Physical Machines State Change and Virtual Machine Placement in the Cloud," *Proc. 8th IEEE SoSE*, Maui, HI, June 2-4, 2013, " EDAS# 1569744689.
341. P. Benavidez, J. Lambert, A. Jaimes, and M. JAMSHIDI, "Landing of a Quadcopter on a Mobile Base using Fuzzy Logic," *Proc. 3rd World Conference on Soft Computing*, San Antonio, TX, December 16-18, 2013, EDAS # 1569816599.
342. M. Ezell, A. Motaghi, and M. JAMSHIDI, "Alpha-Rooting Image Enhancement Using a Traditional Algorithm and Genetic Algorithm," *Proc. 3rd World Conference on Soft Computing*, San Antonio, TX, December 16-18, 2013, EDAS # 156981805613.
343. Y. Yetis, H. Kaplan, and M. JAMSHIDI, "Stock Market Prediction Using With Artificial Neural Network," *Proc. 3rd World Conference on Soft Computing*, San Antonio, TX, December 16-18, 2013, EDAS # 1569815967
344. S.A. Henson, M.J.D. Henshaw, V. Barot, C.E. Siemieniuch, M.A. Sinclair, H. Dogan, S.L. Lim, C. Ncube, M. JAMSHIDI, and D. De Larentis, "Towards a Systems of Systems Engineering EU Strategic Research Agenda," *Proc. 8th IEEE SoSE*, Maui, Hawaii, USA, June 2-6, 2013.
345. B. K. Tannahill and M. JAMSHIDI, "Modeling of System of Systems via Data Analytics – Case of 'Big Data' in SoS," *Proc. 7th International Conference on Soft Computing, Computing with Words and Perceptions in System Analysis, Decision and Control*, Izmir, Turkey, September 2-3, 2013, pp. 1-11.
346. B. Tannahill and M. JAMSHIDI, "Big Data Analytic Paradigms - From PCA to Deep Learning," *Proc. AAAI Workshop*, Stanford, CA, March 2014.
347. M. K. Muppidi, S. Vaishnav, M. JAMSHIDI, and M. Joordens, "**AUV location detection in an enclosed environment," EDAS Number** 1569917517, *Proc. WAC*, Waikoloa, HI, August 3-7, 2014.
348. A. Moussavi-Khalkhali and M. JAMSHIDI, "Leveraging Machine Learning Algorithms to Perform Online and Offline Highway Traffic Flow Prediction," *Proc. Workshop of Machine Learning for Predictive Models* (MLPM 2014), Detroit, MI, 2014.

7 Reviewed Journal Papers

1. M. JAMSHIDI, G. D'Ans, and P. V. Kokotovic, "Applications of a Parameter-Imbedded Riccati Equation," IEEE Trans. Auto. Control, Vol. AC-15, No. 6, December, 1970.
2. M. JAMSHIDI, "A Parameter Adjustable Regulator for a Winding Process," Int. Journal of Control, Vol. 8, No. 5, September, 1972.
3. M. JAMSHIDI, "Three-Stage Near-Optimum Design of Nonlinear Control Processes," Proc. IEE, Vol. 121:886-892, 1974.
4. M. Malek-Zavarei, M. E. Mokari-Bolhassan, and M. JAMSHIDI, "Optimal Scheduling for Industrial Processes," *Pazhoohandeh Research Journal*, Ministry of Science and Higher Education, Tehran, Iran, No. 6, September, 1974
5. M. JAMSHIDI, "A Three-Stage Design of Non-Linear Control Systems with Time Delay," *Int. Journal of Control*, Vol. 21, No. 5, 1975.
6. M. Razzaghi and M. JAMSHIDI, "On the Jacobi Condition for Control Problems with Input Delay," *Proc. IEE*, Vol. 122, No. 11, November, 1975.
7. M. JAMSHIDI and M. Razzaghi, "Optimization of Linear Systems with Input Time-Delay," *Kybernetika*, Vol. 11, No. 5, 1975.
8. M. JAMSHIDI, "Optimal Control of Non-Linear Power Systems by an Imbedding Method," *IFAC J. Automatica*, Vol. 11, No. 6, November 1975.
9. M. JAMSHIDI, "Analog Computer-Aided Design of a Constant Pressure Water Reservoir System," *IEEE Trans. Ind. Elect. and Control Instrumentation*., Vol. IECI-22, No. 3, August, 1975.
10. M. JAMSHIDI and F. Boettiger, "A Parameter Imbedding Solution of Algebraic Matrix Riccati Equation," *Int. Journal of Control*, 1976.
11. M. JAMSHIDI and M. Razzaghi, "Near-Optimum Design of a Class of Non-Linear Control Systems with Time-Delay," *Iranian Journal of Science and Technology*, Vol. 4, No. 4, 1975.
12. M. JAMSHIDI, "A Feedback Near-Optimum Control for Non-Linear Systems," *Information and Control*, Vol. 32, No. 1, September, 1976.
13. M. JAMSHIDI, "Application of Three Time-Scale Near-Optimum Design to Control Systems," *Automatic Control Theory and Applications*, 1976.
14. K. C. Chu, M. JAMSHIDI, and R. E. Levitan, "An Approach to On-Line Power Dispatch with Ambient Air Pollution Constraints," *IEEE Trans. Auto. Cont.*, Vol. AC-22, June, 1977, pp. 385-396.
15. M. JAMSHIDI, "Optimization of Some Dynamic Industrial Control Processes by Analog Computer," *Transactions of IACS*, 1976.
16. M. JAMSHIDI, "Optimization of Water Resources Systems with Statistical Inflows," *Proc. IEE*, Vol. 124, January, 1977.
17. M. JAMSHIDI and M. Heidari, "Application of Dynamic Programming to Control Khuzestan Water Resources System," *IFAC J. Automatica*, Vol. 13, No. 3, May, 1977.
18. M. JAMSHIDI, "A Large-Scale Optimal Stochastic program for Operation of Khuzestan Water Resources System," *Iranian J. of Science and Technology*, 1978.
19. K. C. Chu, M. JAMSHIDI, and R. E. Levitan, "Real-Time Urban Power dispatch with Ambient Air Quality Constraints," *IFAC J. Automatica*, Vol. 14, January, 1978.
20. M. JAMSHIDI, "An Imbedded Initialization of Newton's Algorithm for Matrix Riccati Equation," *IFAC J. Automatica*, Vol. 14, May, 1978, pp. 167-170.
21. R. J. Peters, K. C. Chu, and M. JAMSHIDI, "Optimal Operation of a Water Resources System by Stochastic Programming," *Journal Mathematical Programming*, Study g, 1978.

22. M. JAMSHIDI, M. Malek-Zavarei, and I. Vakilzadeh, "Problems of Automatic Control Education in Developing Countries," *IFAC Journal Automatica*, vol. 16, January, 1979, pp. 105-112.
23. M. JAMSHIDI, "An Overview on the Solutions of the Algebraic Matrix Riccati Equation and Related problems," *J. Large Scale Systems*, 1980, vol. 1, pp. 167-192.
24. M. JAMSHIDI, Book Review, "Structure of Interconnected Systems," *IEEE Circuits and Systems Magazine*, 1980.
25. M. JAMSHIDI and M. Malek-Zavarei, Book Review, ORACLS, A Design System for Linear Multivariable Control, *IEEE Control Systems Magazine*, vol. 1, no. 1, Mar. 1981.
26. P. Dorato and M. JAMSHIDI, "Some comments on optimal collection of solar energy," *Solar Energy,* 1982.
27. M. JAMSHIDI and C. J. Herget, "Computer-Aided Control Systems Engineering," *IEEE Control Systems Magazine*, Vol. 2, No. 4, November, 1982, pp.-2.
28. M. JAMSHIDI and M. Malek-Zavarei, "A Hierarchical control of large-scale time-delay systems," *J. Large-Scale Systems*, vol. 4, 1983, pp. 149-163.
29. M. JAMSHIDI and A. T. Bahill, Book Review, "Control of time-delay systems," *IEEE Control Systems Magazine*, vol. 2, No.2, Jun. 1982, pp. 21-22.
30. M. JAMSHIDI and C. M. Wang, "Optimization of large-scale nonlinear time-delay systems," *IEEE Transactions on System, Man and Cybernetics*, vol. SMC-14, no. 1, pp. 209, Jan., 1984.
31. J. M. Brideau and M. JAMSHIDI, "Optimal control of large scale nonlinear systems with time delay," *Int. Journal of Control,* 1984, vol. 7, pp. 33-46.
32. C. M. Wang and M. JAMSHIDI, "Optimal control of large-scale nonlinear systems with time delay," *Int. Journal of Control*, vol. 39, No. 4, pp. 683-699, 1984.
33. M. JAMSHIDI, G. S. Axelby, O. I. Franksen, and N.B. Nichols, "A Century of IEEE," *IEEE Control Systems Magazine*, Vol. 4, No. 4, November, 1984.
34. R. E. Salters and M. JAMSHIDI, "On the aggregation of large-scale stochastic systems with multiplicative noise," *J. of Large-Scale Systems,* vol. 11, pp. 31-42, 1986.
35. M. Shahinpoor, M. JAMSHIDI, and Y. T. Kim, "Exact solution to the inverse kinematics problem of a standard 6-axis robot manipulator," *J. Robotic Systems*, vol. 3, No. 3, pp. 259-280, 1986.
36. M. JAMSHIDI, R. Morel, T. C. Yenn, and J. Schotik, "Computer-aided design of systems and networks packages and languages," *Advances in Computer-Aided Control Systems Engineering*, M. Jamshidi and C. J. Herget (eds.), North-Holland, Amsterdam, 1985.
37. H. Seraji, M. JAMSHIDI, Y. T. Kim, and M. Shahinpoor, "Linear control of two-link nonlinear robot manipulators," *J. Robotic Systems*, vol. 3, no. 4, pp. 349-365, 1986.
38. J. M. Santiago and M. JAMSHIDI, "On the extensions of balanced method of model reduction," *Int. J. Control Theory and Advanced Technology*, Vol. 2, No. 2, pp. 207, 226, 1986.
39. M. Tarokh and M. JAMSHIDI, "Elimination of decentralized fixed modes with minimum number of interconnection gains," *J. Large-Scale Systems*, vol. 11, 1987, pp. 207-215.
40. F. Asamoah and M. JAMSHIDI, "Lyapunov stability of bilinear large-scale systems," *J. Large-Scale Systems*, vol. 11, pp. 69-79, 1986.
41. Y. T. Kim, M. JAMSHIDI, and M. Shahinpoor, "On near-optimum control of robot manipulators," *Int. J. Robotics and Automation*, 1987.
42. F. Asamoah and M. JAMSHIDI, "Stabilization of a class of singularly perturbed bilinear large-scale systems," *Int. J. Control*, 1987, pp. 1587-1594.
43. M. JAMSHIDI, H. Seraji, and Y. T. Kim, "Decentralized control of nonlinear robot manipulators," *J. Robotics and Autonomous Systems*, vol. 3, pp. 261-370, 1987.

44. Z. Huang and M. JAMSHIDI, "On linear aggregation of input output analysis," J. of Systems Engineering, Vol. 24, no. 1, 1987 (in Chinese).
45. M. JAMSHIDI, "Robots in Manufacturing," Robotics and Computer Integrated Manufacturing, vol. 6, No. 4, 1989.
46. S. O'Neill and M. JAMSHIDI, "Robot_S: An interactive robot simulation language," Robotics, Vol. 14, No. 3, pp. 245-256, 1989.
47. M. JAMSHIDI, "Multi-Arm Robotics," Robotics and Autonomous Systems, Vol. 5, No. 4, 1989.
48. M. JAMSHIDI, "Fixed and flexible robots," editorial, J. Robotic Systems, Vol. 6, No. 4, 1989.
49. M. Aldeen and M. JAMSHIDI, "On a decentralized control of large-scale systems with modal reduction," Int. J. Comp. and Elect. Engr., Vol. 16, No. 2, 1990, pp. 99-103.
50. H. Bolandi, R. L. Carroll, and M. JAMSHIDI, "Discrete-time model reference adaptive control of robot manipulators," Iranian J. Science and Technology, 1992.
51. J. N. Liou, M. JAMSHIDI, and G. P. Starr, "Adaptive edge-tracking control of an industrial robot," J. of Robotics and Computer Integrated Manufacturing, Vol. 6, No. 4, pp. 331-337, 1989.
52. F. Asamoah, and M. JAMSHIDI, "Response to comments on Lyapunov stability of bilinear large-scale systems," Infor. & Decision Technologies, 1990.
53. M. JAMSHIDI, "Robots in Unstructured Environments," J. of Robotics and Automation, 1990.
54. M. JAMSHIDI, "Intelligence in Robotics and Manufacturing," editorial, J. of Intelligent and Robotic Systems, Vol. 3, No. 1, 1990.
55. Z. Geng and M. JAMSHIDI, "On the expert system control of robot manipulators," J. Robotic and Intelligent Systems, 1991.
56. M. JAMSHIDI and C. S. Tseng, "On robust decentralized control of a 5-axis robot," J. Robotics and Computer-Integrated Manufacturing, Vol. 5, No. 4, pp. 309-316, 1990.
57. B. J. Oh, M. JAMSHIDI, and H. Seraji, "Two adaptive control structures of robot manipulators," IFAC J. Automatica, 1991.
58. B. J. Oh and M. JAMSHIDI, "Decentralized adaptive control of robot manipulators," J. Robotic Systems, Vol. 6, No. 4, 1989, pp. 461-483.
59. W. Horne, M. JAMSHIDI, and N. Vadiee, "Neural Networks in Robotics - a Survey," J. Intelligent and Robotic Systems, Vol. 3, No. 1, pp. 51-66, 1989.
60. Z. Geng and M. JAMSHIDI, "Learning control systems analysis and design based on a 2-D systems theory," J. Intelligent and Robotic Systems, Vol. 3 No. 1, pp. 17-26, 1990.
61. M. JAMSHIDI, "Sensors and Interfacing in Robotic Manufacturing," Robotics and Computer Integrated Manufacturing, Vol. 7, 1991.
62. M. JAMSHIDI, "Large-Scale Systems - Modeling and Control," in Circuits, Systems, and Information, M. Jamshidi, M. Mansour, B. O. D. Anderson and N. Bose (eds.), TSI Press, Albuquerque, NM, 1991.
63. C. Abdallah, D. Dawson, P. Dorato, and M. JAMSHIDI, "Survey of robust control of robots," IEEE Control Systems Magazine, Vol. 10, 1991.
64. M. JAMSHIDI and V. Salminen, "Mechatronics," editorial, Int. J. Computers & Electrical Engineering, Vol. 18, No. 3., 1992.
65. R. Lumia and M. JAMSHIDI, "Robotics and Manufacturing Research at National Institute for Standards and Technology," J. Intelligent and Robotics Systems, Vol. 4, 1992.

66. M. JAMSHIDI and C. Gosselin, "Trends in Robot Kinematics, Dynamics, Control, Sensing, Programming, and Simulation " J. Robotics and Autonomous Systems, 1992, pp. 199-200.
67. M. JAMSHIDI and D. Cherchas, "Flexible and Redundant Robots," J. Robotics and Computer-Integrated Manufacturing, Vol. 9. No. 2, 1992, pp. 85-86.
68. M. JAMSHIDI, "Research and Education in Robotics and Manufacturing Systems, ", J. Robotics and Computer-Integrated Manufacturing, Vol. 1, No. 1, 1992, p.1.
69. M. JAMSHIDI and S. Hayati, "Robots in Unstructured Environments," editorial, J. Robotic Systems, 1992.
70. M. JAMSHIDI, "Artificial Intelligent Robotics and Manufacturing," editorial, J. Expert Systems and Applications, Vol. 4, No. 3, 1992, pp. 271-272.
71. M. JAMSHIDI, "Automation in Manufacturing Systems," editorial, Int. J. Systems Automation Research and Applications, 1992.
72. M. JAMSHIDI and P. Dauchez, "Parallel and Multiple Manipulators," editorial, Int. J. Robotics and Automation, 1992.
73. J. White, R. Brittain, R. Kisner, and M. JAMSHIDI, "Advance Control Architectures for Nuclear Reactor," Control Theory and Advanced Technologies, Vol. 8, September, 1992.
74. B. J. Oh, M. JAMSHIDI, and H. Seraji, "Adaptive decentralized control of robot manipulators," J. of Intelligent and Robotic Systems, 1992.
75. J. Mullins, M. Shahinpoor, M. JAMSHIDI, G. P. Starr, and S. Kassicieh, "Robotics and Manufacturing Education and Research - The New Mexico Initiative," J. Robotics and Computer-Integrated Manufacturing, Vol. 9, 1992, pp. 15-25
76. M. JAMSHIDI, "Editorial on Adaptive Signal processing," Int. J. Computers & Electrical Engineering, Vol. 18, No. 3/4., 1992.
77. J. N. Liou and M. JAMSHIDI, "Adaptive robust force robot control of contour following robotic system," J. Robotics and Autonomous Systems, Vol. 5, 1992, pp. 283-297.
78. N. Vadiee and M. JAMSHIDI, "A new paradigm for intelligent control", Int. J. Computers & Electrical Engineering, Vol. 18, No. 2. 1992.
79. M. JAMSHIDI and Z. Geng, "A learning 2D control approach for robot manipulators," J. Expert Systems and Applications, 1992, pp. 297-304.
80. R. D. Colbaugh and M. JAMSHIDI, "An adaptive impedance control scheme for robotic solid waste handling," J. Robotic Systems, 1992.
81. S. R. Lee and M. JAMSHIDI, "On the stability of large-scale time-delay systems," Kybernetika, 1992.
82. S. R. Lee J. Davis, and M. JAMSHIDI, "On the Lyapunov stability of large-scale systems with time delays," J. Control Theory and Advanced Technology, 1991.
83. W. Honey and M. JAMSHIDI "ROBO_SIM - A simulation environment for robots, Robotics and Autonomous Systems, Vol. 9, No. 4, 1992, pp. 305-317.
84. J. Benitez-Read, M. JAMSHIDI, and R. Kisner, "Advanced control designs for nuclear reactors," J. Control Theory and Advanced Technology, Vol. 8, No. 3, 1992, pp. 447-464.
85. J. Benitez-Read and M. JAMSHIDI, "Adaptive input-output linearizing control of nuclear reactors," J. Control Theory and Advanced Technology, Vol. 8, No. 3, 1992, pp. 535-545.
86. Z. Geng, R. L. Carroll, and M. JAMSHIDI, "A learning 2D control approach for the EBRII nuclear reactor," J. Control Theory and Advanced Technology, 1992.
87. Z. Geng, R. L. Carroll, M. JAMSHIDI, and R. A. Kisner, "An adaptive learning control approach with application to water tank level control," J. Control Theory and Advanced Technology, Vol. 8, No. 3, 1992, pp. 577-592.
88. Z. Geng, R. L. Carroll, and M. JAMSHIDI, "An adaptive learning estimator and controller for nuclear reactors," J. Control Theory and Advanced Technology, 1992.

89. M. JAMSHIDI and M. Shahinpoor, "Environmentally Conscious Manufacturing," Int. J. Conscious Manufacturing, Vol. 1, No. 1, 1992, p. 1.
90. R. Marchbank and M. JAMSHIDI, "A Fuzzy Control Approach to a Two-Axis Mirror Laser Beam Alignment System," J. on Engineering Applications of AI, 1993.
91. W. Parkinson, P. D. Shalek, K. H. Duerre, G. F. Luger, and M. JAMSHIDI, "Two Intelligent Control Systems for Silicon Carbide Whisker Production " Journal of Intelligent and Fuzzy Systems, Vol. 1, NO. 3, 1993.
92. M. JAMSHIDI and M. Bayoumi, "Robot Control - Position, Force, and Impedance," J. Intelligent and Robotic Systems, Vol. 5, 1993.
93. M. JAMSHIDI, R. V. Patel, and B. Bavarian, "Neural Networks and Fuzzy Logic in Robotics and Manufacturing," editorial, Int. J. Computers & Electrical Engineering, Vol. 18, No. 2., 1993.
94. M. JAMSHIDI and C. C. Nguyen, (eds.), "Special Issue on Parallel Robots and Mechanisms," Journal of Robotic Systems, 1993.
95. M. JAMSHIDI and T. J. Ross, "Intelligent and Fuzzy Systems, an Editorial," Journal of Intelligent and Fuzzy Systems, Vol. 1, 1993.
96. J. Benitez-Read, M. JAMSHIDI, and C. Abdallah "Robust adaptive control of nuclear reactors," J. Iranica Scientia, Vol. 1, 1993.
97. D. Barak and M. JAMSHIDI "Fuzzy Control of an Environmentally Conscious Refrigeration System," Journal of Intelligent Manufacturing, 1994.
98. C. Hwang, M. Saif, and M. JAMSHIDI, "Fault Detection and Diagnosis of a Nuclear Power Plant Using Artificial Neural Networks," Journal of Intelligent and Fuzzy Systems, Vol. 2, 1995.
99. N. Vadiee and M. JAMSHIDI, "A Tutorial on Fuzzy Rule-Based Expert System (FRBES) Models 1 : Mathematical Foundations," Journal of Intelligent and Fuzzy Systems, Vol. 1, No. 2, 1993, pp. 171-188
100. N. Vadiee and M. JAMSHIDI, " Fuzzy Rule-Based Expert Systems 1: Inference Schemes," Journal of Intelligent and Fuzzy Systems, Vol. 1, No. 3, 1993.
101. A. Martinez, E. Tunstel and M. JAMSHIDI, "Fuzzy Logic Based Collision Avoidance of a Mobile Robot", Robotica, vol. 12, Part 6, 1994, pp. 521-527.
102. M. JAMSHIDI and L. A. Zadeh, "Welcome Editorial for Inaugural Issue," Iranica Scientia, Vol. 1, No. 1, 1994.
103. E. Tunstel and M. JAMSHIDI, "On Genetic Programming of Fuzzy Rule-based Systems for Intelligent Control," Int'l. Journal of Intelligent Automation and Soft Computing, Vol. 2, No. 2, 1996.
104. M. JAMSHIDI, A. Titli, and J. P. Martin, "On the Stability of Fuzzy Control Systems: Time and Frequency-Domain Methods," Intelligent Automation and Soft Computing, Vol. 4, pp. 109-125, 1998.
105. M. JAMSHIDI and L. A. Zadeh, "Fuzzy Control Systems - An Introduction, Iranica Scientia, Vol. 2, No. 4, 1996, p. vii.
106. K. Kumbla and M. JAMSHIDI, "Hierarchical Control of Robotic Manipulators," in Journal of Intelligent and Fuzzy Systems, Vol. 4, 1996.
107. S. Heger, A. S., N. K. Alang-Rashid, and M. JAMSHIDI, "Application of Fuzzy Logic in Nuclear Reactor Control, Part I: An Assessment of State-of-the-Art," Nuclear Safety Journal, Vol. 36, No. 1, 1996.
108. S. Heger, A. S., N. K. Alang-Rashid, and M. JAMSHIDI, "Simulation Studies of Nuclear Reactor Control Using Fuzzy Logic," Journal of Intelligent and Fuzzy Systems, 1995.

109. E. T. Tunstel, S. Hockemeier, and M. JAMSHIDI, "Fuzzy Control of a Hovercraft Platform," *Engineering Applications of Artificial Intelligence*, Special Issue: "Control Applications of Fuzzy Logic", Vol. 7, No. 5, October 1994, pp. 513-519.
110. M. JAMSHIDI, Book Review, Control Systems by P. Belanger, *IEEE Control Systems Magazine*, Vol. 16, 1996.
111. E. Tunstel, M. JAMSHIDI, "On Genetic Programming of Fuzzy Rule-Based Systems for Intelligent Control," *AutoSoft Journal*, Vol. 2, No. 3, 1996, pp. 215-222.
112. M. JAMSHIDI, "Fuzzy Control of Complex Systems," *International Journal of Soft Computing*, Inaugural Issue (Invited), Springer-Verlag, Germany, 1996.
113. H. Xue, M. JAMSHIDI, and E. Tunstel, "Genetic Algorithms for Optimization of Fuzzy Systems in Prediction and Control, "*Int. Journal of Knowledge-Based Intelligent Engineering Systems,* Vol. 1, No.1, 1997, pp. 13-21.
114. M. R. Akbarzadeh-T, and M. JAMSHIDI, "Evolutionary Fuzzy Control of A Flexible Link," *Journal of Intelligent Automation and Soft Computing*, Vol. 3, No. 1, pp. 77-88, 1997.
115. E. Tunstel, T. Lippincott and M. JAMSHIDI, "Mobile Robots: Fuzzy - Behavior Modulation and Evolution," *Int'l. Journal of Intelligent Automation and Soft Computing*, Vol. 3, No. 1, 1997, pp. 37-49.
116. M. JAMSHIDI, "Fuzzy Logic Systems," *Encyclopedia of Electrical & Electronic Engineers*, J. Wiley & Sons, J. G. Weber (ed.), New York, 1998.
117. M. JAMSHIDI, "Large Complex Systems," (McGraw Hill), *Encyclopedia of Science and Technology*, New York, 7th, Edition, 1999.
118. C. Guo, P. Sarkar, and M. JAMSHIDI, "Soft-Computing Based Intelligent Controller and its Application to a Nonlinear System," *AutoSoft Journal*, 2002.
119. C. Guo, Z. Ye, Z. Sun, P. Sarkar, and M. JAMSHIDI, "A Hybrid Fuzzy Cerebellar Model Articulation Controller Based Autonomous Controller," *Int. Journal on Computers and Electrical Engineering*, Vol. 28, No. 1, pp. 1-16, 2002.
120. M. JAMSHIDI, D. Padilla, and M. de Oliveira, "Intelligent Evolving Soccer Robots: Part 1 Fuzzy Logic Fundamentals," *Robot Science and Technology Magazine*, Vol. 1, Nov. 1998, pp. 8-13
121. M. JAMSHIDI, D. Padilla and M. de Oliveira, "Intelligent Evolving Soccer Robots: Part 2 Genetic Algorithms," *Robot Science and Technology Magazine*, Vol. 1, March, 1999, pp. 22-26
122. M. Akbarzadeh, K. Kumbla, E. Tunstel, Jr., and M. JAMSHIDI, "Soft Computing for autonomous robotic systems," *Int. Journal of Computers and Electrical Engineering*, Vol. 26, No. 1, 2000, pp. 5-32.
123. M. JAMSHIDI and J. Johnson, III, Editorial, "Science and Technology Research at NASA University Research Centers," *Int. Journal of Computers and Electrical Engineering*, Vol. 26, No. 1, 2000, pp. 1-4.
124. M. Kamel and M. JAMSHIDI, Editorial, "Autonomous Agents," *Intelligent Automation and Soft Computing*, Vol. 6, No. 3, 2000, pp. 171-172.
125. M. Kamel and M. JAMSHIDI, Editorial, "Distributed Manufacturing Systems," *Journal of Intelligent Manufacturing*, Vol. 11, No. 3, 2000, p. 235.
126. M. Johnson, K. Firoozbakhsh, M. Moneim, and M. JAMSHIDI, "Flexor Tendon Repair Techniques: Multi-Objective Decisions by Soft Computing," *IEEE Transactions on Engineering in Medicine and Biology*, Vol. 20, No. 1, March, 2002, pp. 345-360.
127. S. Berman, Y. Edan, and M. JAMSHIDI, "Decentralized Autonomous Automatic Guided Vehicles in Material Handling," *IEEE Transactions on Robotics and Automation*, 2003.

128. M. JAMSHIDI, "Autonomous Control of Complex Systems: Robotic Applications," Applied Mathematics and Computation, Vol. 120, Issue 1-3, pp. 15-29, 2001.
129. A. El-Osery and M. JAMSHIDI, "Image Enhancement Using a Fuzzy Expert System," Soft Computing, Vol. 7, pp. 97-106, 2002
130. S. Berman, Y. Edan, and M. JAMSHIDI, "Simulation in the Development of a Control Strategy for an Automated Guided Vehicle System," System Analysis Modeling and Simulation Journal, Vol. 42, No.2, pp. 23-39, 2002.
131. M. JAMSHIDI, "Autonomous Control of Complex Systems – Space and earth bound applications," System Science Journal, pp. 451-468, 2002.
132. M. JAMSHIDI, "Controls, Large- Scale Systems," (Academic Press), Encyclopedia of Science and Technology, San Diego, CA, 3rd, Edition, 2002, pp. 675-686.
133. A. El-Osery, J. Burge, M. JAMSHIDI, A. Saha, M. Fathi, and M. Akbarzadeh-T., "V-Lab – A Distributed Simulation and Modeling Environment for Robotic Agents – Control Through Stochastic Learning Automata," IEEE Transactions on Systems, Man and Cybernetics, Vol.32, No. 6, pp. 791-803, 2002.
134. P. De Rego and M. JAMSHIDI, "Evolutionary Algorithm Design method to Optimize Spacecraft Formation for a Single Pass Interferometric Radar Observation," submitted to International Journal on Computers in Industry, 2002.
135. M. de Oliveira and M. JAMSHIDI, "Modeling and Control of Robot Manipulators," Book Review IFAC Journal Automatica, Vol. 37, pp. 1681-1682.
136. A. Martinez, A. W. Doerry, M. JAMSHIDI, and D. Bickel, "Coherent Data Alignment and Baseline Calibration for Improved Two-Pass Interferometric SAR," accepted, SPIE Optical Engineering Journal, 2004
137. M. JAMSHIDI, S. Sheikh-Bahaei, J. Kitzinger, P. Sridhar, S. Beatty, S. Xia, Y. Wang, U. Dole, J. Liu, T. Song, E. Tunstel , Jr, M. Akbarzadeh, A. El-Osery, M. Fathi, X. Hu, and B. P. Zeigler, "V-Lab® - Distributed Intelligent Discrete-Event Environment for Autonomous Agents Simulation," Intelligent Automation and Soft Computing – AutoSoft Journal, Vol. 9, No. 3, pp.181-214, 2003.
138. M. JAMSHIDI, S. Sheikh-Bahaei, J. Kitzinger, P. Sridhar, S. Beatty, S. Xia, Y. Wang, U. Dole, J. Liu, T. Song, E. Tunstel, Jr., M. Akbarzadeh, A. El-Osery, and M. Fathi, "V-Lab® - Distributed Intelligent Discrete-Event Environment for Autonomous Agents Simulation," Simulation International Magazine, 2003.
139. S. Sheikh-Bahaei, J. Liu, M. JAMSHIDI, and P. Lino, "An Intelligent Discrete Event Approach to Modeling, Simulation and Control of Autonomous Agents," Intelligent Automation and Soft Computing – AutoSoft Journal, Vol. 10, No. 4, pp. 337-348, 2004.
140. T. Song, C. Gasparovi, X. M. Huang, R. R. Lee, M. Pattichis, and M. Jamshidi, "A Hybrid Tissue Segmentation Approach in Brain MRI Images," IEEE Transactions on Medical Imaging, 2003.
141. M. JAMSHIDI, S. Sheikh-Bahaei, J. Kitzinger, P. Sridhar, S. Xia, Y. Wang, J. Liu, E. Tunstel, Jr., M. Akbarzadeh, A. El-Osery, M. Fathi, X. Hu, B. P. Zeigler, "A distributed intelligent discrete-event environment for autonomous agents simulation," Applied system simulation, Kluwer Academic Publishers, January 2003.
142. J. Liu, M. Jamshidi, and S. Pourbabak, "An Intelligent Modeling Approach to cardiac Output," Medical Journal of Cardiology; 2004.
143. T. Song, M. JAMSHIDI, R. R. Lee, M. Huang, "A Modified Probabilistic Neural Network for Partial Volume Segmentation in Brain MR Image," IEEE Transactions on Neural Networks 18(5): 1424-1432 (2007)

144. Y. Wang and M. JAMSHIDI,_"A Hierarchical Fuzzy Classification Scheme for Remote Sensing Data," Intelligent Automation and Soft Computing, Vol. 13, No. 4, pp. 431-451, 2007.
145. J. K. Liu, M. JAMSHIDI, and S. Pourbabak, "A hybrid approach on identification of cardiovascular dynamics," IC-MED Journal, Issue 1, No. 1, 2007, pp. 49-60.
146. T. Song, M. Huang, R. R. Lee, and M. JAMSHIDI, "A Data-Adaptive Fuzzy Rule Base System for Putamen Segmentation in Brain MR Images," Intelligent Automation and Soft Computing, Vol. 12, No. 4, pp. 431-451, 2006.
147. T. Song, J.Y. Liu, and M. JAMSHIDI, "Fuzzy Logic and Application in Biomedical Image and Signal Processing," Chapter in Intelligent Techniques in Medicine, M. Pattichis, ed., Kluwer Publishing Co., Amsterdam, the Netherlands, 2007.
148. K. Hipel, M. JAMSHIDI, J. M. Tien, and C. C. White, "The Future of Systems, Man and Cybernetics: Application Domains and Research Methods," IEEE Transactions on SMC, Part C, Volume 37, No. 5, September 2007, pp. 726-743.
149. P. Sridhar, A. M. Madni, M. JAMSHIDI, "Hierarchical Aggregation and Intelligent Monitoring and Control in Fault-Tolerant Wireless Sensor Networks," IEEE Systems Journal, Volume1, No. 1, September, 2007, pp. 38-54.
150. D. De Laurentis, C. Dickerson, M. Di Mario, P. Gartz, M. Jamshidi, S. Nahavandi, A. P. Sage, E. Sloane, D. Walker, "A Case for an International Consortium on System of Systems Engineering," IEEE Systems Journal, Volume 1, No. 1, September 2007, pp. 68-73.
151. P. Sridhar, A. M. Madni, M. Jamshidi, "Multi-Criteria Decision Making and Behavior Assignment in Sensor Networks," IEEE Instrumentation and Measurement Magazine, Volume 11, No.1, February 2008, pp. 24-29.
152. A. K. Ray, L. Behera, and M. Jamshidi, "Sonar-based Rover Navigation for Single or Multiple Platforms: Forward Safe Path and Target Switching Approach," IEEE Systems Journal., Vol. 2, No. 2, 2008 (cover article), pp. 258-272.
153. M. JAMSHIDI, "SYSTEM OF SYSTEMS ENGINEERING – New Challenges for 21st Century," IEEE Systems Magazine, Vol. 23, No. 5, May 2008, pp.2-19.
154. M. JAMSHIDI, "Introduction to System of systems engineering," Chapter 1, in System of Systems Engineering – Innovations for the 21st Century, (M. Jamshidi, ed.), Wiley & Sons, New York, pp. 1-18, 2009.
155. S. Mittal, B. P. Zeigler, J. L. R. Martin, F. Sahin, and M. JAMSHIDI, "Modeling and Simulation of System of Systems Engineering," Chapter 5, in System of Systems Engineering – Innovations for the 21st Century, (M. Jamshidi, ed.), Wiley & Sons, New York, pp. 101-149, 2009.
156. P. Sridhar, A. M. Madni, and M. JAMSHIDI, "Advances in Wireless Sensor Networks: A Case Study in System of Systems Perspective," Chapter 12, in System of Systems Engineering – Innovations for the 21st Century, (M. Jamshidi, ed.), Wiley & Sons, New York, pp. 275-292, 2009.
157. M. JAMSHIDI, "Introduction to System of systems engineering," Chapter 1, in System of Systems Engineering – Principles and Applications, (M. Jamshidi, ed.), CRC – Taylor & Francis Publishers, Boca Raton, FL, USA, pp. 1-36, 2008.
158. F. Sahin, M. JAMSHIDI, and P. Sridhar, "A System of Systems simulation framework and its applications," Chapter 4, in System of Systems Engineering – Principles and Applications, (M. Jamshidi, ed.), CRC – Taylor & Francis Publishers, Boca Raton, FL, USA, pp. 107-132, 2008.

159. F. Sahin, B. Horan, S. Nahavandi, V. Raghavan, and M. JAMSHIDI, "System of autonomous rovers and their applications," Chapter 14, in System of Systems Engineering – Principles and Applications, (M. Jamshidi, ed.), CRC – Taylor & Francis Publishers, Boca Raton, FL, USA, pp. 366-384, 2008.
160. M. A. Joordens and M. Jamshidi, "Consensus Control for a System of Underwater Swarm Robots," IEEE Systems Journal, Vol. 3, No.4, pp. 65-73, 2010.
161. A. K. Ray, P. Benavidez, L. Behera, and M. JAMSHIDI, "Motion Coordination for a Formation of Rovers," IEEE Systems Journal, Vol. 3, No.3, pp. 381-393, 2009.
162. M. JAMSHIDI, "Twenty years of editorship of Elsevier CEE Journal," Editorial, International Journal of Computers in Electrical Engineering – CEE, Vol. 34, 2009, pp. iii-iv.
163. M. JAMSHIDI, "Control Paradigms for System of Manufacturing Systems," Proc. IEEE ININ, Cardiff, UK, 2010., pp. 1123-1133.
164. M. A. Joordens and M. JAMSHIDI, "Design of a Prototype Underwater Research Platform for Swarm Robotics," AutoSoft Journal, Vol. 17, No.2, 2010, pp. 111-132.
165. J. Gomez and M. JAMSHIDI, "Fuzzy Adaptive Control for a UAV," Journal of Intelligent and Robotic Systems, Vol. 62, No. 2, pp. 271-293., 2011.
166. A. Panchul, D. Akopian, and M. JAMSHIDI," Time-Space Optimizing Fuzzy Logic Algorithm for Dynamic Load Balancing in Distributed Three-Dimensional Graphics Engine, International Journal of System of Systems, Vol. 2, No. 4, pp. 309-320, 2011.
167. M. JAMSHIDI, J. Gomez, and A. Jaimes, "Intelligent Control of UAVs for Consensus-based and Networked Control Applications," in Special Issue on "Fuzzy Set Theory and its Applications" of the International Journal of Applied and Computational Mathematics of Applied and Computational Mathematics, Vol. 10, No.1, 2011, pp. 35-64.
168. M. JAMSHIDI, A. Jaimes, and J. Gomez, "Cyber-physical control of unmanned aerial vehicles," Iranica Scientia, in Special Issue on 6-Decases of Contributions of Professor Lotfi A. Zadeh to science, engineering, Technology and Academic Leadership," Vol. 45, 2011.
169. M. JAMSHIDI, Guest Editor," Iranica Scientia, Vol. 18, No. 3-D1, pp. 520-521, 2011 SPECIAL ISSUE on 6-Decases of Contributions of Professor Lotfi A. Zadeh to Science, Engineering, Technology and Academic Leadership", Elsevier, Amsterdam, the Netherland, 2011.
170. "Guest Editorial," Iranica Scientia, Vol. 45, 2011 (in Special Issue on 6-Decases of Contributions of Professor Lotfi A. Zadeh to science, engineering, Technology and Academic Leadership", Vol. 18, no. 3-D1, pp. 663-668, 2011.
171. M. JAMSHIDI, "From Large-Scale Systems to Cyber-Physical Systems," Journal of Internet Technology, Vol. 12, No.3, pp. 367-374, 2011.
172. R. Jia, C.-J. Qian, and M. JAMSHIDI, "Semi-global finite-time stabilization via output feedback of planar non-linear systems with application to MPPT in photovoltaic systems," Int. J. Automation and Control, vol. 6, No. 2, pp. 140-156, 2012.
173. K. Nagothu, B. Kelley, M. JAMSHIDI, and A. Rajaee, "Persistent Net-AMI for Microgrid Infrastructure Using Cognitive Radio on Cloud Data Centers," IEEE Systems Journal, Vol. 6, No. 1, 2012, pp. 4-15.
174. A, Jevitc, A´ lvaro Gutie´rrez, D. Andina and M. JAMSHIDI, ""Distributed Bees Algorithm for Task Allocation in Swarm of Robots", IEEE Systems Journal, Vo. 6, No.2, 2012, pp. 296-304.
175. T. Shaynefelt, M. JAMSHIDI, and S. Agaian, "A vision feedback robotic docking crane system with application to vanilla pollination, "Int. J. Automation and Control, vol. 7, No. 1/2, pp. 62-82, 2013.

176. Y. Manjili, R. Vega, and M. JAMSHIDI, "Cost-Efficient Environmentally-Friendly Control of Micro-grids using Intelligent Decision-Making for Storage Energy Management," Intelligent Automation and Soft Computing, Vol. 19, No. 4, pp.649-670, 2013.
177. M. Gupta, L. Behera, K. S. Venkatesh, and M. JAMSHIDI, "A Robust Visual Human Detection Approach with UKF Based Motion Tracking for a Mobile Robot," Submitted to *IEEE Systems Journal*, 2013
178. R. R. Nair, L. Behera, V. Kumar, and M. JAMSHIDI, "Multi-satellite formation control for remote sensing applications using artificial potential field and adaptive fuzzy sliding mode control," Submitted to *IEEE Systems Journal,* 2013
179. J.J. Prevost, K. Manoj, M. JAMSHIDI, and B. Kelley, " Energy Aware Load Prediction for Cloud Data Centers," Accepted in *IEEE Systems Journal*, September, 2013
180. A. K. Ray, L. Behera, and M. JAMSHIDI," Controller for a Coordinated Platoon of a Group of Mobile Robots," Submitted to *IEEE Systems Journal*, 2013
181. M. JAMSHIDI and Y Manjili, "Environmentally-Friendly Control of Micro-Grids Using Fuzzy Control for Storage Management," Chapter 4, Springer-Verlag, (R. Yager and S. Shahbazova, eds.), 2013
182. D. Shahgoshtasbi and M. JAMSHIDI, "A new intelligent Neuro-Fuzzy paradigm for Energy Efficient Homes," *IEEE Systems Journal*, Vol. 8, No.2, 2014, pp.
183. B. Tannahill and M. JAMSHIDI, "System of Systems and Big Data Analytics – Bridging the Gap," *International Journal on Computers for Electrical Engineering*, Elsevier, Vol. 40, pp. 2-15, 2014.
184. S. Chang, K. Nagothu, B. Kelley, and M. JAMSHIDI, "A Beam Forming Approach to Smart Grid Systems Based on Cloud Cognitive Radio," *IEEE Systems Journal*, Vol. 8, No.2, 2014, pp. 664-673
185. S. Bakhtiari, S. Agaian, and M. JAMSHIDI, "A Novel Method for Orientation Estimation in Highly Corrupted Fingerprint Images," *Int. Journal of Intelligent Computing for Medical Sciences and Image Processing,* Vol 6, No.1-2, Accepted to appear.
186. E. M. Bonab, M. L. Lindsey, M. JAMSHIDI, and Y.-F. Jin, "Application of graph Theory in Network Analysis for Biological Processes," *Int. Journal of Intelligent Computing for Medical Sciences and Image Processing,* Accepted to appear in 2014.
187. P. Benavidez, J. Lambert, A. Jaimes, and M. JAMSHIDI, "Landing of an ArDrone 2.0 Quadcopter on a Mobile Base using Fuzzy Logic," *Int. Journal of Complex Systems,* Vol. 1, No. 1, 2013, pp. 1-25.
188. J. Arumugam, S. Jeevanandham, and M. JAMSHIDI, "Damping of inter-area oscillations in multimachine systems," *Int. J. Automation and Control*, Vol. x, No. y, 2014.
189. A. M. Mayers, P. J. Benavidez, G. V. S. Raju, David Akopian, and M. JAMSHIDI, "Closed Loop Power Control Using Third Order Quadratic Approximator," International Journal of Wireless Information Networks, On-line Springer-Verlag (DOI 10.1007/s10776-014-0236-x), 2014
190. A. M. Mayers, P. J. Benavidez, G. V. S. Raju, D. Akopian, and M. JAMSHIDI, "A Closed Loop Transmission Power Control System using a Nonlinear Approximation of Power-Time Curve Approximator", IEEE Systems Journal, Vol 10, No. xx, 2014 In press

Name Index

FSC
www.fsc.org
MIX
Papier aus verantwortungsvollen Quellen
Paper from responsible sources
FSC® C105338

If you have any concerns about our products,
you can contact us on
ProductSafety@springernature.com

In case Publisher is established outside the EU,
the EU authorized representative is:
Springer Nature Customer Service Center GmbH
Europaplatz 3, 69115 Heidelberg, Germany

Printed by Libri Plureos GmbH
in Hamburg, Germany